T0093792

SPACELAB PAYLOADS
Prepping Experiments and Hardware for Flight

Michael E. Haddad and David J. Shayler

SPACELAB PAYLOADS

Prepping Experiments and Hardware for Flight

Springer

Published in association with
Praxis Publishing
Chichester, UK

PRAXIS

Michael E. Haddad
NASA Systems Engineer, Retired
Cocoa Beach, FL, USA

David J. Shayler
Astronautical Historian
Astro Info Service Ltd
Halesowen, UK

SPRINGER-PRAXIS BOOKS IN SPACE EXPLORATION

Springer Praxis Books
ISSN 2731-5401   ISSN 2731-541X   (electronic)
Space Exploration
ISBN 978-3-030-86774-4   ISBN 978-3-030-86775-1   (eBook)
https://doi.org/10.1007/978-3-030-86775-1

Front cover (top): Neurolab Rack Train, in the Operations and Checkout Building, Kennedy Space Center, being moved from Level-IV area to Level-III/II area, c. November 1997.

Front cover (bottom): Spacelab-1 in Cargo Integration Test Equipment (CITE) stand, Operations and Checkout Building , Kennedy Space Center, c August 1983. Being prepared to transfer to Orbiter Processing Facility for installation into the Space Shuttle *Columbia*. Spacelab Pallet with experiments in foreground (red "Remove Before Flight" tags), Spacelab Module, gold handrails in the middle.

Back Cover: The Neurolab payload for STS-90 undergoing further processing in the O&C Building at KSC, c. December 1997. Neurolab was designed to study the effects of microgravity on the nervous system. The crew, of Commander Richard Searfoss, Pilot Scott Altman, Mission Specialists Richard Linnehan, Dafydd (Dave) Williams, M.D. (Canadian Space Agency) and Katherine (Kay) Hire, and Payload Specialists Jay Buckey, M.D., and James Pawelczyk, Ph.D., flew STS-90 and Neurolab between April 17 and May 3, 1998, on Space Shuttle *Columbia*. Photo: NASA/KSC

Cover design: Jim Wilkie
Project Editor: Michael D. Shayler

This Springer imprint is published by the registered company Springer Nature Switzerland AG
The registered company address is: Gewerbestrasse 11, 6330 Cham, Switzerland

Contents

Foreword

Sam Durrance, May 16, 2019, with the Astro Log Book he used during the 1980s.

This book describes what "Level-IV" was; its purpose, and the people that did their job to make my mission, as well as all the other Spacelab missions, a success.

Mike Haddad was working the Level-IV mechanical engineering aspects of the Astro payload when I met him though his good friend Scott Vangen. Scott was a Level-IV Electrical Engineer who was responsible for preparing the flight hardware, as well as myself, for the Astro-1 mission. During that time, we developed a good working and personal relationship.

Riding a rocket into space, the exhilaration of zero-g, the satisfaction of operating the Astro Observatory, or the emotional impact of seeing Earth from orbit – how can I possibly describe space flight?

So many talented people had worked so hard for so long to make this moment possible. I felt extremely proud to be part of it. I was well prepared for all the technical aspects of our mission. But nothing could have prepared me for the sight below.

It was broad daylight somewhere over the Atlantic Ocean far below. The Cape Verde islands were speeding lazily by. An intricate pattern of brilliant white clouds laced the deep rich blue of the Ocean. The African continent was rapidly approaching from the horizon. The thin-blue line that defines the extent of the atmosphere was visible just over the bright limb of the Earth. We were a lone, magnificent, flying-machine, careening through the black void of space at five miles per second; and yet, somehow, at the same time, floating effortlessly above the Earth.

Soon our orbit carried us past the terminator. The setting Sun created a brilliant light show unmatched by anything I have seen. The full Moon cast a beautiful, serene glow over the Earth below. The airglow layer, out on the horizon, above the limb, seemed to blanket and protect the cities below, whose lights drifted silently past. It was time to go back to work; we came to observe the stars, but I couldn't keep my eyes off the Earth.

The amazing light show of sunrise or sunset flashed by every 45 minutes. I saw continents and oceans scroll by at a startlingly fast pace. I could see the imprint of geology on a global scale that seemed to make it come to life. And I saw the *impact of humanity* on this delicate, isolated, island of life, set in a black-black void of endless space. I was left with a sense of the beauty and fragility of planet Earth, and a feeling that it is imperative to protect and preserve it.

I left Earth with great anticipation, great excitement, and a little apprehension. I returned with a new feeling of wonder, a new perspective – a space traveler's perspective. It is something I will always have.

Also, to this day, Mike, Scott, and a number of Level-IV people have remained very good friends of mine.

Thank you Level-IV.

Samuel T. Durrance
Space Shuttle Payload Specialist
STS-35 (Astro-1) and STS-67 (Astro-2)

To all those men and women who worked so hard and sacrificed
so much to make the Space ab Program a success.

*"Every time I saw a Spacelab mission lift off, I got a very special feeling inside
that lasted about two weeks. The performance of the experiments and the Spacelab
was an unqualified success on every mission and it would be hard to be more proud
of the work we did. The fact that we got to work with a bunch of great people
from KSC, other NASA centers, and from around the world, as well as working
on a new complement of experiments every few months, made it a fun, exciting,
and interesting job that could easily be called a passion rather than work."*

Tracy Gill, Level-IV Electrical Engineer

Authors' Preface

It is widely said that each of us has the potential to write a book. That may be true, but although the prospect of such an idea may indeed be within us, getting that idea or desire down, traditionally on paper or more recently on a computer file, is another matter.

The gestation from idea to book is a long and involved one, with many hurdles, pitfalls, delays and disappointments before a final product is found on a shelf or available as a commercial download. To cooperate on a book project adds to the logistics and molding of the story into a final product with which to approach a publisher. The next challenge, to get a publisher to support the book's production, brings a new dimension to the conversion of a good idea into a physical book. Above all, each title is an incredibly personal journey from idea to shelf, one that takes a significant amount of time to be completed successfully, as recounted by the authors below.

Michael E. Haddad

There have been many books, journals, articles, papers, documentaries and films about the Space Shuttle Program. And why not? It was an amazing engineering achievement, as shown in the 1981 film "*Space Shuttle: A Remarkable Flying Machine*". The "payloads", which used to be called "cargo" before someone pointed out that the name did not quite suit the function since cargo is what is transported in the bowels of a ship, have also been documented a great deal. This covered not only what the payload was but also what they accomplished during each Space Shuttle mission.

The missing piece – and the reason why I, Michael E. Haddad, wanted to write this book – is because very little has been documented about *how* those elements, primarily focusing here on the Spacelab-related payloads, were prepared at Kennedy Space Center (KSC) for their mission. The timeframe is from the mid-1970s through to the early 2000s. The assembly, integration, testing, servicing and final closeouts – essentially making sure the payloads were 100 percent operational before launch – was called the "Level-IV" function at KSC. I know, because

I was hired into Level-IV as a Mechanical Engineer in 1982, near the beginning of the Spacelab program, and worked there for seven years. Level-IV referred to the starting point of the physical work begun at KSC, but our job was not just focused at that single location, or just at KSC in general. We would go to the design centers before the hardware and software arrived at KSC, support the mission at whatever control center was in charge of the on-orbit experiment, and then perform post-flight operations at landing locations. Because our responsibility included all these other locations and functions, the term was changed from "Level-IV" to "Experiment Integration".

The idea for this book began back in 2000 following the STS-99 Shuttle Radar Topography Mission (SRTM), the last Spacelab-related mission. I have written many professional papers and performed hundreds of presentations, but had no clue about how to write a book. So, I began to reach out to those who had such experience to see if they would take on the task of writing it, with my help to provide the information to be included. Most of the people I contacted liked the idea and provided me with information regarding their contributions and work performed during Level-IV, but were too busy with other commitments to take on the task.

Over the course of the next 17 years, a number of events took place where discussions of the book came up.

In October 2002, a group of NASA and contractor employees attended a Level-IV Reunion. In October 2010, Dean Hunter, my Lead Engineer during Level-IV, a super engineer and great friend, sadly passed away. Dean was the best mentor I had. I learned so much from him that I was able to use to carry me through the rest of my career.

Six months later, in May of 2011, we had a Level-IV Party at *Fish Lips* restaurant in Port Canaveral, Florida. With the end of the Shuttle program approaching, we wanted to get together for one last time before many of us either retired or moved on to other projects.

When Janice Voss, a very good friend and astronaut who flew five times on the Shuttle, also passed away, a Celebration of Life occurred at Johnson Space Center (JSC) in March 2012 which included a Tree Planting Ceremony in her name. In attendance were a number of us from Level-IV and many of her fellow astronauts. To honor her and the work she did, the book once again was talked about and a decision made to try other paths to get the project moving again.

Over the next two years, I reached out to a number of persons I thought could help get the ball rolling. I contacted Kenneth Lipartito, co-author for the book "*A History of Kennedy Space Center*", who had interviewed me for the part of his book that dealt with payload integration, including the Level-IV tasks. "I thought the sections of the book that dealt with payload integration on the Orbiter were among the most original and interesting," he told me. But like many times before, everyone thought it was a good topic but could not help.

Jay Barbree, a seasoned NBC reporter who covered the space program for decades, stayed in a home in Cocoa Beach when covering KSC activities. This was right next door to what would become known as the "Payload Beach House", not to be confused with the "Astronaut Beach House" located on-site at KSC. Located directly on the Atlantic Ocean in south Cocoa Beach, The Payload Beach House (its name was shorted to just the Beach House) was occupied at first by three Level-IV Engineers, Scott Vangen, Craig Jacobson and Jim Dumoulin. Eventually, 12 different people called it home, including myself, and it became the focal point for many social activities related to the Level-IV work and personnel. The most famous was as the location for the annual Payload Halloween Party, a tradition which began in 1986. This will be described later in the book. Talking the idea of the book with Jay, he stated "Mike, you will never get rich writing it, but it is a good story to tell." I was very sad to hear the news of Jay's passing on May 14, 2021.

I had just finished reading Mike Mullane's book, "*Riding Rockets*" and thought maybe he could help us. Mike was very cordial and responded: "Mike, I doubt that Simon & Schuster [his *Riding Rockets* publisher] would have an interest in this type of 'niche' book… in this case, one that's aimed at us space geeks." He suggested a few contacts and wished us "Good luck!" Very good feedback and helpful, but still no promising leads.

To celebrate the 30th Anniversary of Spacelab-1, we threw a huge celebration at the Kennedy Athletic, Recreational and Social I (KARS-I) park in November 2013. It was attended by more than a hundred people from all aspects of the Spacelab Program, not just Level-IV. Invitations even went out to the crew members from STS-9, many of whom responded but unfortunately could not attend.

For the next number of years, many events took place that would always bring up a discussion about our work in Level-IV, but finally in July of 2017, Mike Lienbach, a good colleague of mine, had written the book "*Bringing Columbia Home*" and so I contacted him and his co-author Johnathan H. Ward about the possibility of Johnathan writing this book. As with many others, Johnathan really liked the idea but could not take on the task. But he suggested I contact David J. Shayler, an accomplished author in the space field who has written many books about the American and Russian space programs and someone who was really interested in the Spacelab Program. First contact with David was in July 2017. He was very enthusiastic about the idea and as we talked it became apparent that he was in 100 percent but really wanted me to co-write it as well. Actually, instead of him taking the lead and me helping out, he stated that I needed to lead the effort. The idea was that his extensive background in writing space-related books and my background on really living Level-IV would make the perfect team to accomplish the task the correct way and help sell the idea to a publisher. More dialogue occurred and finally he convinced me to do this. Lots of discussions and formal processes needed to take place over the next few months, but by May 2018 we had a contract with Springer publishing. David and I were finally going to be writing this book.

David J. Shayler

A flight into space comprises many elements, the foremost being the mission itself from launch through − for a crewed mission − to the recovery. There is so much more to each mission that is often overlooked or simply not reported outside of the program itself. This includes: the preparation for the mission, in terms of developing the mission itself, preparing the hardware, creating the experiments and formulating crew training; the support during the flight, by teams on the ground in Mission Control, at communications centers and in various support rooms, and also during recovery with the vast network of search and rescue teams; and of course, postflight in ensuring the crew readjusts to life back on Earth, the results from the experiments are analyzed and the physical hardware is examined to understand how it stood up to the rigors of spaceflight. All of these elements then go into creating a better understanding of the mission flown against what was planned, and how these results can help improve subsequent missions and future investigations, or improve new hardware and refine procedures to obtain even greater return from the huge investments in spaceflight, both human and robotic.

As a long-term enthusiast of human space exploration, over the years I have become more interested in these ancillary aspects of each mission, which expand the understanding of what the flight itself is trying to achieve and what is involved not only in creating the mission, but also in supporting it and safely recovering the crew and vehicle at its conclusion. It is one thing to have an interest in such things but quite another to obtain the information and, more importantly, to understand it. To ensure accurate and in-depth material is at hand to create the core of what becomes a book, a significant network of contacts and sources is essential. In addition to the referenced material, first-hand contact with those who were "up close and personal" with a mission or program is just as important. Access not only to the flight crew but also the flight controllers, the launch and recovery teams, and the experiments all adds little pieces of information to fill in the gaps in the wealth of written data, and provides a priceless insight into how all this worked (or in some cases did not work) on the real mission. As a writer, I have also benefited from access to some wonderful archives, resources, contacts and personalities which have added special insights into the topic being written about.

As a youngster growing up with Apollo and its follow-on Skylab, I was disappointed when those programs ended but fascinated by the concept of the Space Shuttle and the anticipation of what it had to offer. Even though the reality of the program fell far short of what the original estimates were projected to be, following each Shuttle mission as they unfolded offered an intriguing insight into as near to routine space operations as the Shuttle program could offer. As the Space Shuttle Program developed, so did my writing career, and the opportunities to record aspects of the program beyond that of just logging each mission and those who flew them. I became fascinated with what happened to the Shuttle Orbiters on the ground in between the missions and in trying to track the process of placing payloads on the mission and assembling the hardware to fly it.

This type of research was useful when I was asked by fellow author Harry Siepmann to write what turned out to be my first book on the Space Shuttle,[1] having already authored articles, delivered presentations and compiled and published a monthly magazine called *Orbiter* through my company Astro Info Service. That book was followed the same year by a large coffee-table book on the Space Shuttle *Challenger*.[2] Both of these titles included detail of hardware and payloads, some of which in the second title was specially sourced from the contractors for the hardware elements. This type of detail and research helped compile the monthly *Orbiter* magazines published between 1984 and 1991.

Since those early days in my writing career in the 1980s, I have continued to collect and record data on the Space Shuttle ground operations, its missions, hardware, experiments and payloads, and of course the crews. This has created an archive and resource which has continued to be of assistance in compiling more recent titles I have produced, detailing aspects of the Shuttle Hubble Servicing Missions and the development of the Shuttle/Space Station docking and assembly missions. There are plans to continue this work in future titles and in publications by Astro Info Service, and so when I was approached by Michael Haddad with the view to cooperating on a book on Shuttle payload processing, I jumped at the chance.

Living in the UK made regular access to the Shuttle Launch Facility (SLF) and archives at KSC in Florida difficult, so having my co-author Michael write his story of being involved in the day-to-day activities down at the Cape seemed logical. When he explained that the account focused upon the Spacelab missions I was even more pleased, being both a Brit and a European. Spacelab was part of Europe's contribution to the Shuttle program and the Pallets were fabricated right here in the UK. I have often felt that Spacelab has been unjustly overlooked in the accounts of the payload system within the Shuttle program. Here was a change to redress that omission.

This is not a "users guide" to Shuttle payload processing, but a layperson's insight into how that processing came about and was operated, by those who were involved in making it happen. Without this team, the Shuttle Spacelab payload could not and would not have flown, and being mostly "under the radar", it is clear

---

[1] *From the Flight Deck 2: NASA Space Shuttle,* published in 1987 by Ian Allen.

[2] *Challenger, Aviation Fact File*, Salamander Books, 1987.

that the work they did was both effective and done well to achieve the results it did with so few failures in the whole system.

The Space Shuttle has had many critics, which is fine, but it also has many things to be proud of, and the Level-IV team at the Cape who processed the majority of the science payloads, up to the early 2000s, can justly be proud of what they accomplished. It is their personal involvement, dedication and sacrifice which helped make flying the Spacelab hardware look so easy, whereas in fact the truth was far from that impression.

This, then, is the story of Level-IV at NASA KSC, Florida, through the eyes and experiences of those who were directly involved, as told to Mike Haddad and molded by both of us into the account you read here. This is where you can learn just how challenging and involved getting an experiment or items of hardware off the ground and into space actually was. We hope you enjoy the journey.

Acknowledgements

As with any book project of this nature, there are a far greater number of people who are involved than those names who appear on the front cover and who are credited with the account herein. Each author therefore offers his thanks to those who made this book possible, and whose names number far too many to be listed.

Michael Haddad

This book is not an official NASA-endorsed account but a private recollection and explanation. Some of the information stated within the many interviews may differ from what people may have heard or opinions written in other documentation, but we have done our best to ensure all the data is correct and reflects the memories of the people who lived it.

The majority of the images used in this book originate from NASA, various military service organizations, this author's own collection and those of other personnel – those credited in the individual captions – unless specifically stated. However, despite extensive searches, we have been unable to determine the exact origin of some of the images and would therefore welcome any input to enable us to credit the appropriate source. We can provide copies of photos upon request.

Throughout this process, there have been a few people always willing to help, with Maynette Smith, a Level-IV engineer, being the main support. She had maintained the email distribution list we had created to stay in contact with as many ex-Level-IV personnel as possible. She also created a Facebook site to try and document our work and reach out to as many people as possible that worked Level-IV, to get correspondence started that would help populate a possible book with facts from those who lived it.

Dean Hunter, a super engineer, one of the best bosses I ever had and a good friend, taught me so much that is documented in this book. He was also greatly admired by others around him. Dean expected excellence and at times seemed to get really mad with me but, as stated to me by one of my co-workers, you always knew when Dean was really mad at you, something I'll get into in more detail in the Spacelab-1 section of this book. Dean passed away unexpectedly on October

8, 2010. The inscription on the most cherished of all his awards, the Spacelab panel with parts from many missions mounted upon it that was presented to him by the technicians and engineers that worked with him, as mentioned in the Preface, read: "Presented this day, March 23, 1995, to Dean Hunter. Inspired by your innovative leadership, technicians and engineers have continually pushed the edge of the envelope, repeatedly finding elegant solutions to seemingly insurmountable technical challenges. Represented here are the fruits of but a few of these unheralded innovations. Undoubtedly, these are among our best efforts during our finest hours."

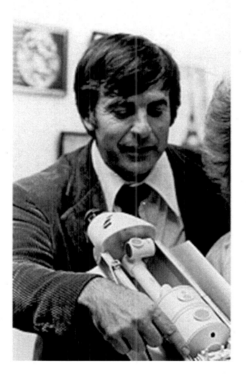

Dean Hunter, pointing out details of the Spacelab to a co-worker. [NASA/KSC.]

Janice E. Voss (b. 1956), known to her friends as "JV", passed away on February 6, 2012, aged only 55. As an astronaut, she flew five times on the Shuttle. I met JV through some friends at KSC who worked with her on a number of missions. She was a very kind person, a very smart person and a wonderful friend who would have made a huge contribution to this book. She loved her chocolate and she became very good friends with my cousin, Lisa Thomas, sharing lots of chocolate samples that JV acquired over the years through her travels around the globe. But one of my most vivid memories of JV was the day I received my Silver Snoopy

Award, October 22, 2001. The Silver Snoopy is an award for outstanding performance, contributing to flight safety and mission success. An astronaut always presents the Silver Snoopy because it is the astronauts' own award and less than one percent of the aerospace program workforce receive it annually, making it a special honor to receive this award. Mine was from the crew expressing their appreciation for the outstanding support that I had given to the Space Shuttle Program as a Flight Crew Extra-Vehicular Activity (EVA) and Intra-Vehicular Activity (IVA) Interface.

Janice Voss ("JV") at the CITE stand in the Operations and Checkout (O&C) Building, participating in testing for the STS-99 Shuttle Radar Topography Mission (SRTM) payload. Insert: Silver Snoopy award (l to r): Mike Haddad, Janice Voss, Sam Haddad. [NASA/KSC.]

The way this worked was that your family was informed about the award but you were not. At the time, they would have you attend what appeared to be a normal meeting with all your co-workers, but really it was a gathering for presenting the award, and your family was positioned outside the meeting room. When the astronaut arrived, they and your family would enter the room for the surprise and presentation of the award. My brother, Samuel Haddad, also worked at KSC and he was part of my family that entered the room with JV, and right away it hit me that I was going to receive a Silver Snoopy. What my brother didn't know was that he was also going to receive a Silver Snoopy. Normally, one crew member would present one award to one employee, but JV found out we both were going to receive the awards so she set it up to present the awards to both of us at the same time. She started by asking me to come up to read the reason for the award, which for my brother seemed weird. Why would Mike read for his own award? But as I started reading, it was details of why my brother was receiving the award, not me. So "Surprise," Sammy, you're receiving a Silver Snoopy. After the presentation to him, he was asked to stay at the front of the room and then read the reason for my Silver Snoopy. This was the first and only time in the history of NASA that two brothers received Silver Snoopy awards at the same time. All thanks to JV, a super person who is truly missed.

To all of you who contributed information to me to write this book, especially those that I interviewed, you provided incredible stories of how each of you performed your job and the innovative solutions to many problems encountered. A number of them made me laugh so hard my stomach hurt, but there were also the touching moments and heartaches that each of you endured during some very sad days at NASA. Thank you all.

I also appreciate Astro-1 and -2 Payload Specialist Sam Durrance for his excellent Foreword and Riley Duren for his Afterword.

Thanks also to my co-Author David J. Shayler, who has helped me so much on this journey of writing a book. His experience with the many books he has written and awesome suggestions for this process were huge in helping me to streamline tasks when I was heading in the wrong direction. A master of words, which I am NOT, our Project Editor Michael D. Shayler, Dave's brother, added so much of the personality to this effort, something I could have never done by myself.

Finally and especially, thanks to my wife Kathy, my soul mate, the love of my life, who put up with all my craziness during the writing and production of this book. I bounced so much off her during this multi-year process and she stuck it out the whole time, giving me great feedback and inspiration to keep going many times when I wanted to say the hell with it and walk away.

David Shayler

The majority of the content in this book was supplied by my co-author Mike Haddad and his colleagues who worked in the KSC Level-IV team and associated facilities. I also would like to thank them for their patience and assistance in detailing their part in the story.

For my own research and participation, I thank those at the KSC PAO (Public Affairs Office) who have assisted me over many years in detailing operations at the Cape. These include Kay Grinter, Margaret Persinger, Ken Nail and Elaine Liston. I also take this opportunity to thank PAO Specialist Manny Virata for his personal guidance around the facilities at KSC during a memorable first visit to the facility in November 1990 for the launch of STS-38, and a second in November 1993. Those tours brought the information I had gathered over the preceding two decades to life. The cooperation with my co-author in this current project has offered an even greater understanding of everything that I had read and later saw in person. In addition, my contacts at the NASA History and PAO departments at JSC have provided additional information on Shuttle operations over a period of four decades.

In detailing the various Spacelab missions for this and other projects, I am indebted to astronauts Vance Brand (Astro-1), Michael Foale (ATLAS-1 and -2), Robert Gibson (Spacelab-J), Tony England (Spacelab-2), the late Karl Heinz (Spacelab-2), Jeff Hoffman (Astro-1), Don Lind (Spacelab-3), Story Musgrave (Spacelab-2), Rhea Seddon (Spacelab Life Sciences-1 and -2), and the late Bill Thornton (Spacelab-3).

Thanks as always goes to my brother Mike Shayler, our Project Editor, for his skills and talent in turning our original draft into the polished, finished format you read here. Thanks also to Jim Wilkie for interpreting our cover design into the finished cover.

Thanks go to Hannah Kaufman from Springer for guidance through the publication process, and her predecessor Maury Solomon who initially saw potential for the project through the Springer portfolio and suggested the authors get together for a joint partnership. Thanks also to Clive Horwood of Praxis books for mentoring the project and to the referees who supported the early proposal.

Thanks to all those who worked at KSC over the years in preparing not only the Spacelab-related hardware in Level-IV but also in processing all the payloads and hardware at the Cape to enable each Shuttle to fly. This is a story focused on the Level-IV Spacelab processing, but it is also part of the history of operations at the Cape and the lives of those who worked there, and to their families who understood their dedication to their work to make each element as safe and as reliable as they possibly could. This is their story too.

Finally, thanks to my mother Jean, who for once was spared reading the proofs of one of my books because she was reading those from a different project, and to my wife Bel, who once again saw me embark on yet another book project, albeit less writing and more proof-reading and pre-editing this time, while at the same time working on other titles, taking on a new editorial role, and still having time to work in our kitchen and enjoy time with our beloved white German Shepherd Shado. Thank you once again to all for your support of my passion for recording the developing story of human spaceflight.

List of Abbreviations and Acronyms

Glossary

Airlock
An intermediate location to permit passage between two dissimilar environments.

Flight Hardware
Hardware that is qualified to fly in space.

Long Module
Two short, pressurized Spacelab Modules, comprising the Core and Experiment segments, attached together as one.

Mission Peculiar Experiment Support Structure
A structure that spanned the Orbiter Payload Bay and could carry smaller payloads not needing the larger Spacelab Pallets.

Orbiter
The crewed element of the Space Transportation System commonly referred to as the Space Shuttle. This winged vehicle provided accommodation for the crew during their mission in Earth orbit, facilities to support scientific experiments, and capabilities to handle large payloads.

Pallets
'U'-shaped structures that carried unpressurized payloads and were launched in the Payload Bay of the Orbiter.

Payloads
Scientific experiments, satellites, planetary spacecraft, and large sections of the International Space Station. Hardware that would be launched in the crew compartment and/or Payload Bay of the Orbiter.

Payload Bay
A 60-foot (18-m) long by 15-foot (4.6-m) diameter location for carrying cargo and payloads, such as Spacelab Modules and Pallets, in the Space Shuttle Orbiter.

Payload Specialist
A flight crew member who was not a professional career astronaut but a specialist to operate scientific equipment in orbit, usually for a single mission. Nominated by members of the Principal Investigators of the given mission.

"Ship-and-Shoot", or "Ship-n-Shoot"
A payload arriving at KSC that is launched with little or no pre-launch testing.

Short Module
A single Spacelab pressurized Module, namely the Core segment. This configuration was never flown.

Spacelab
European-designed and built pressurized Module and unpressurized Pallet system for conducting extensive scientific research for up to 18 days in Earth orbit from the Payload Bay of a Space Shuttle Orbiter.

Test Stands
Physical structure locations where payload assembly, integration and testing occurs.

Turnaround
Time between missions

User Rooms
Rooms assigned to payload customers and used as a control room for their payload.

AFD	Aft Flight Deck
AFRC	[Neil A.] Armstrong Flight Research Center (formerly DFRC)
ARC	[Joseph Sweetman] Ames Research Center, California
Astro	abbreviation for "Astronomy"
ATE	Automated Test Equipment (Level-III/II)
ATLAS	Atmospheric Laboratory for Applications and Science (STS-45, 56 and 63)
ATM	Apollo Telescope Mount (Skylab)
C/D	Design and development phases

CCAFS	Cape Canaveral Air Force Station, Florida
CCTV	Closed Circuit Television
CDMS	Command and Data Management System
CDR	Critical Design Review
CFES	Continuous Flow Electrophoresis System
CHROMEX	Chromosome and Plant Cell Division in Space
CITE	Cargo Integration Test Equipment
CSA	Canadian Space Agency
CRF	Canister Rotation Facility
CWA	Clean Work Area
DEVs	Deviations
DFRC	[Hugh L.] Dryden Flight Research Center, California
DOD	Department Of Defense
DR-xxx	Double Rack
Exxx or EM	Engineering Model
EAFB	[Glen] Edwards Air Force Base, California
EBA	European Bridge Assembly
ECLSS	Environmental Control and Life Support System
ED	Experiment Developer
EDO	Extended Duration Orbiter
EDL	Engineering Development Lab
EGSE	Experiment Ground System Equipment
EI	Experiment Integration
ESA	European Space Agency/Engineering Support Area
ESPS	Experiment Segment and Pallet Simulator
ELDO	European Launcher Development Organization
EMP	Enhanced Multiplexer/demultiplexer
EOM	Earth Observation Mission (became ATLAS)/End of Mission
ESR	Engineering Support Room
ESRO	European Space Research Organization
ESTEC	European Research and Technology Center
ET	External Tank (Space Shuttle)
EVA	Extra-Vehicular Activity (spacewalk)
Fxxx	Flight (serial number)
FMPT	First Materials Processing Test (Spacelab-J)
FOP	Follow On Production
FOST	Flight Operations Support Team
FU	Flight Unit
GAS	Get Away Special (experiment)
GIRD	Ground Integration Requirements Document
GIRL	German Infrared Radiation Laboratory (Spacelab D-4 unflown)

GRC	[John H.] Glenn Research Center, Cleveland, Ohio
GSE	Ground Support Equipment/Government Supplied Equipment
GSFC	[Robert H.] Goddard Space Flight Center, Greenbelt, Maryland
GSOC	German Space Operations Center, Oberpfaffenhofen, Germany
GOWG	Ground Operations Working Group
HMF	Hypergolic Maintenance Facility
HOSC	Huntsville Operation Support Center
HST SM	Hubble Space Telescope Servicing Mission (STS-61, 82, 103, 109 and 125)
IML	International Microgravity Laboratory (STS-42 and 65)
IPS	Instrument Pointing System
ISS	International Space Station
IVA	Intra-Vehicular Activity
JAXA	Japanese Aerospace Exploration Agency
JEM	Japanese Experiment Module
JSC	[Lyndon B.] Johnson Space Center, Houston, Texas
KARS	Kennedy Athletic, Recreation and Social
KSC	[John F.] Kennedy Space Center, Florida
LC	Launch Complex
LCC	Launch Control Center
LM	Long Module (Spacelab)
LMS	Life and Microgravity Spacelab (STS-78)
LPS	Launch Processing System
LSSF	Life Sciences Support Facility
MBB	Messerschmitt-Bölkow-Blohm, Germany
MCC	Mission Control Center, JSC, Houston
MDAC	McDonnell Douglas Aircraft Corporation
MDE	Mission Dependent Equipment
MDM	Multiplexer/demultiplexer
MFD	Manipulator Flight Demonstration
MHI	Mitsubishi Heavy Industries
MPE	Mission Peculiar Equipment
MPESS	Mission Peculiar Equipment Support Structure
MPLM	Multipurpose Logistics Modules
MPS	Main Propulsion System (Space Shuttle)
MSFC	[George C.] Marshall Space Flight Center, Huntsville, Alabama
MSL	Material Science Laboratory (became USML)/Material Science Laboratory (STS-83 and 94)
MMSE	Multi-Mission Support Equipment
MOU	Memorandum of Understanding
MVAK	Module Vertical Access Kit

NASA	National Aeronautics and Space Administration
NASM	National Air and Space Museum, Washington, D.C.
O&C	Operations and Checkout
OFT	Orbital Flight Test
OMI	Operation and Maintenance Instructions
OMRF	Orbiter Maintenance & Refurbishment Facility (OPF-3)
OMRSD	Operations and Maintenance Requirements Specifications Document
OPF	Orbiter Processing Facility, KSC
OSF	Office of Space Flight
OSS	Office of Space Science
OSTA	Office of Space and Terrestrial Applications
OV	Orbital Vehicle (Space Shuttle)
PCR	Payload Changeout Room
PCTC	Payload Crew Training Complex
PCU	Payload Checkout Unit
PD	Payload Developer
PDR	Preliminary Design Review
PETS	Payload Environmental Transportation System
PGHM	Payload Ground Handling Mechanism
PHSF	Payload Hazardous Servicing Facility
PI	Principal Investigator
PMA	Pressurized Mating Adapter
POCC	Payload Operations Control Center
PON	Payload Operations Network
PPCU	Partial Payload Checkout Unit
PPLF	Partial Payload Lifting Fixture
PR	Problem Report
PRCU	Payload Rack Checkout Unit
PSR	Payload Support Room
PSSIT	Pallet Segment Support Integration Trolley
PTE	Payload Test Engineer
R/R&D	Research/Research and Development
RAHF	Research Animal Holding Facility
RAM	Research and Applications Module
RFP	Request for Proposal
RMS	Remote Manipulator System (Canadarm)
RSS	Rotating Servicing Structure
SAEF	Spacecraft Assembly & Encapsulation Facility
SBIR	Small Business Innovation Research
SEB	Source Evaluation Board

SED	Spacelab Experiments Division, KSC
SEIS	Spacelab Experiment Integration Support
SLDPF	Spacelab Data Processing Facility, GSFC
SLF	Shuttle Landing Facility, KSC
SL-M	Spacelab-Mir (STS-71)
SLS	Spacelab Life Sciences (STS-40 and 58)/Spacelab Simulator
SM	Single Module (Spacehab)/Short Module (Spacelab) – not flown
SOM	Spacelab Operational Mission
SPCDS	Spacelab Payload Command and Data System
SR-xxx	Single Rack
SRB	Solid Rocket Booster
SRL	Space Radar Laboratory, (STS-59 and 68)
SRM	Satellite Retrieval Mission (STS-51A)
SRR	Systems Requirement Review
SRTM	Shuttle Radar Topography Mission (STS-99)
SSHIO	Space Station Hardware Integration Office
SSME	Space Shuttle Main Engine
SSPF	Space Station Processing Facility
STA	Structural Test Article (STA-099)
STS	Space Transportation System
TAP	Test & Assembly Procedures
TDRSS	Tracking and Data Relay Satellite (System)
TIM	Technical Interchange Meeting
TPS	Test Preparation Sheets
TSS	Tethered Satellite System (STS-46 and 75)
UA	Unexplained Anomalies
USA	United Space Alliance
USML	US Microgravity Laboratory (STS-50 and 73)
USMP	US Microgravity Payload (STS-52, 62, 75 and 87)
VAB	Vehicle Assembly Building, KSC
VAFB	Vandenberg Air Force Base, California
VFT	Verification Flight Test (Spacelab-1, Spacelab-2)
VPF	Vertical Processing Facility
WAD	Work Authorization Documents

Prologue
A new generation of 'Rocket Scientists'

In the early 1980s, NASA was preparing to launch the first Space Shuttle to begin a new era of human spaceflight by the United States, "after Apollo". The Space Shuttle was a combination of several elements: two reusable Solid Rocket Boosters (SRB) and three overhaulable Main Engines fueled from a single-use huge External Tank (ET), which combined was called the Main Propulsion System (MPS). This unique combination would launch a multi-flight Orbital Vehicle, one of a fleet of five, containing a human crew and its manifested cargo into low Earth orbit. From here, missions lasting from a few days up to nearly three weeks would be conducted before the vehicle was returned to Earth, landing on a runway for processing and repeating the operations, with a different payload, time and time again. The whole combination of Orbiter, SRBs, and ET was known as "The Stack".

This "manifested cargo" carried on the Orbiter consisted of new hardware being developed at the same time as the Shuttle which carried it. Called "payloads", these would be taken into low Earth orbit either in the Orbiter's large Payload Bay or on the pressurized "middeck" area. These payloads contained a multitude of science experiments from all over the world, a significant amount of which would be contained in a pressurized laboratory or on unpressurised Pallets under a program called "Spacelab", which was being developed by NASA in cooperation with the European Space Agency (ESA).

The question was "How would these payloads be prepared for launch?" The answer was to reintroduce the concept of allowing NASA personnel to perform the job that a contractor would normally perform. Instead of overseeing a contractor, NASA personnel would instead perform the engineering function themselves and get their own hands dirty. Thus was created the "Level-IV – Experiment Integration" organization at Kennedy Space Center (KSC). Many young NASA personnel, most of them right out of college, would be responsible for preparing domestic and foreign multi-million-dollar experiments for spaceflight. This book tells the story, from the engineers' point of view, of how that unique group accomplished this task, details the technical decisions made that contributed to the success of the Spacelab and science programs, describes where the experience gained was distributed throughout other areas of the space program, and reveals how knowledge of that work is being used for current and future spaceflight activities. This, therefore, is the inside story of how major scientific payloads were prepared and flown successfully and safely many times by the Space Shuttle system, creating a useful prelude for the successful development and operation of much more complex and long-term science programs on the International Space Station (ISS).

1

A Laboratory for the Space Shuttle

*A Spacelab flight unit is designed to have
a lifetime of 50 missions or five years.*
STS Press Information,
Rockwell International
March 1982

The statement above outlines the original intention for Spacelab in the early 1980s. Back then, it was predicted that nominal mission duration for the facility would be seven days, as it turned out to be. However, the hardware was designed to support missions of up to 30 days with suitable consumables and a power extension package, though that never occurred; the longest Shuttle Spacelab mission was 17 days. If 30-day missions had indeed flown, they would probably have come at the cost of a reduced scientific payload to accommodate the added supplies and facilities necessary to support such a mission. The option was never put to the test.

Whatever its capabilities, the success of Spacelab and its associated facilities is one of the achievements of the Shuttle program. That being so, it then begs the question why did the Spacelab missions stop flying at the peak of their success? There is no simple answer to that. Limitations in the hardware was one reason, despite the fact that they were totally reusable and therefore crucial in lowering the costs of re-flying the same elements several times, a core reason behind the Space Transportation System (STS) in the first place. In truth, the demise of Spacelab had less to do with limited scientific achievements or operational setbacks and more to do with politics in hallowed halls of power far away from the processing facility at the Kennedy Space Center (KSC) in Florida, where this story is focused.

© Springer Nature Switzerland AG 2022
M. E. Haddad, D. J. Shayler, *Spacelab Payloads*, Springer Praxis Books,
https://doi.org/10.1007/978-3-030-86775-1_1

This, then, is *not* a story of the development of the Spacelab hardware. For in-depth background to that evolution, the reader is guided to the titles listed in the bibliography at the back of this book. Neither is this an account of the mission-by-mission operations of Spacelab hardware on orbit, or the research fields the missions addressed, though these areas are touched upon throughout. Instead, this is a story of human endeavor, perseverance and dedication over decades; of long hours and separation from family members to ensure that the hardware destined to fly on the Space Shuttle did so effectively, usefully and, above all, safely. There are countless records of each of the 135 Shuttle missions and stories of the crews who flew them. Far fewer are accounts of the people on the ground who painstakingly prepared everything necessary to make sure the missions flew, and flew well. This is their story, a very different account of the Space Shuttle program, of how a generally youthful team of male and female engineers and managers created the Level-IV Spacelab processing system at the National Aeronautics and Space Administration's (NASA) KSC facility. With this in place, they then devoted themselves to becoming an effective force across what for some turned out to be their entire careers, before finally moving on when the Shuttle was retired in 2011.

But before we begin their story, let us take a brief look at the leading player in the tale – the European-built Spacelab module, its carrier pallets and associated equipment.

WHAT IS SPACELAB?

Spacelab was the Space Shuttle's scientific research laboratory, essentially a pressurized module and a series of unpressurized pallets carried in the Payload Bay of the Orbiter. Throughout the lifetime of the Shuttle program, the modules and pallets "provided the scientific community with easy, economical access to space and an opportunity for scientists worldwide to conduct experiments in space concerning astronomy, solar physics, space plasma physics, atmosphere physics, Earth observations, life sciences and material sciences."

In addition to the crewed module, unpressurized pallets were utilized as platforms for a range of experiments and equipment. Over a period of 22 years, at least 32 Shuttle missions carried or operated Spacelab components. Depending on how the missions are categorized, there were 24 dedicated Spacelab missions using either the large pressurized laboratory (termed the Long Module), a number of unpressurized pallets, or a combination of both. Though a smaller pressurized laboratory was available (termed the Short Module), it never flew. However, in addition to the 'designated Spacelab' missions, the second (1981) and third (1982) Shuttle missions under the Orbital Flight Test Program (OFT), as well as five later missions flown between 1998 and 2000, all had Spacelab components as part of the payload. The last designated Spacelab mission, with a Long Module, was flown in 1999, but pallet hardware was still flying on missions into 2008. From the beginning of Shuttle flight operations to the creation of the International Space Station (ISS), Spacelab was an integral element in the Space Shuttle program that, according to NASA, "paved [a] critical path to [the International] Space Station."

The headlines from Spacelab tell of the missions flown, the experiments performed and results obtained, but behind those headlines lies the real story of how this remarkable set of components were prepared for flight, the painstaking work involved in preparing the experiments and hardware to leave the launch pad, and what happened to them after the mission ended.

This has been a long-overlooked element of the Shuttle program – until now. This is the story of how a myriad elements were brought together to fly on a Shuttle mission and the human story behind that tale, and it is one which has its origins in the early years of not only the Shuttle program, but also space exploration itself.

The Shuttle Era 1981–2011

For 30 years, between April 1981 and July 2011, Space Shuttle missions left the launch pads at Launch Complex 39 within KSC in Florida for their missions in Earth orbit, ranging from a couple of days to nearly three weeks. Each carried a unique payload or re-flown set of experiments. For the first decade of operations, the primary goals of the Shuttle fleet were to establish a commercial market, provide the capability to deploy huge scientific payloads or operate extensive arrays of scientific experiments, and to be of service the nation's security.

Designed to be a reusable launch system, and marketed as an all-encompassing access to orbit, the strained launch and preparation process came to a grinding halt in January 1986 with the loss of *Challenger* and her crew of seven, just seconds after leaving the launch pad. After heartfelt internal examination of the whole system over a two-year hiatus to September 1988, the fleet returned to flight, and the program caught up with delayed payloads. The second decade of Shuttle operations looked very different to the first. Now the emphasis was on science and a closer look back at our planet Earth with a series of scientific missions. But in the early 1990s there were more hurdles to overcome and new challenges to address, most notably the spiraling cost of the proposed Space Station program which threatened not only the Shuttle and Spacelab programs, but the future of American human spaceflight itself.

The answer came in the beneficial cooperation with a former rival in the race to space and to the Moon. At the same time as the United States space program was in turmoil, on the other side of the world the break-up of the Soviet Union suddenly afforded the opportunity from within the new-look Russia to seek partnerships in many areas, including space exploration. After several years of doubt and debate on both sides, in 1993 Russia became the 16th global partner in the program to create a truly *International* Space Station. In its third decade of operations, the Shuttle would now become the logistics 'truck' for the assembly and supply of ISS.

The prize for this commitment and cooperation was a state-of-the-art expansive (and expensive) research facility in low Earth orbit, designed to prepare for extended human spaceflight and ventures far beyond Earth orbit, such as the long-awaited return to the Moon, the imaginative exploration of Mars and epic expeditions in deep space to visit the asteroids. To achieve the first step in that dream – the creation of a large permanent platform in Earth orbit – the cost was, in part, the loss of a cooperative Western asset to human spaceflight: the Spacelab system.

Spacelab was a joint US/European Space Agency (ESA) initiative to provide a pressurized scientific laboratory and unpressurized support pallets that could be flown in the cavernous Payload Bay of the Shuttle Orbiter on short-duration missions. Across the scientific, engineering and technical research fields, the multi-element Spacelab system was like a space-age train set, with the capacity to mix and match missions and payloads with hardware and objectives, most of which was totally reusable and interchangeable. From the benefit of hindsight, the program clearly had its problems and downfalls, but all in all Spacelab was a great system and when flown it worked like a dream, with few exceptions and failures. It was a perfect test bed for developing hardware, techniques, procedures and theories in advance of more in-depth research on a station. Indeed, Spacelab could have supplemented the work on station had there not been a limit on the number of Shuttle vehicles capable of assembling and supporting ISS and a corresponding drain on funds to support independent missions.

The Orbiter Fleet

Originally, NASA envisaged a whole fleet of Orbiters supporting launches every two weeks throughout the year. Initially, two were built for the test program; Orbital Vehicle 101 (OV-101) *Enterprise* was used in the series of atmospheric Approach and Landing Tests (ALT) in 1977 and various ground tests which followed. It was then scheduled to be upgraded to become the second orbital flight vehicle, but this proved too expensive and *Enterprise* was destined for a life of ground tests and simulations, never to reach orbit. *Columbia* (OV-102) was the vehicle which completed the first orbital tests from 1981 and instigated operational flights the following year. Over the next couple of years, *Columbia* was joined by the uprated Structural Test Article STA-099, the replacement for *Enterprise*. This was renamed OV-099 *Challenger* and entered service in 1983.

They were followed by OV-103 *Discovery* in 1984, OV-104 *Atlantis* in 1985 and plans for OV-105, but that is where the production line stalled, with just a set of structural spares provided but *not* a completed Orbiter. This all changed after the tragic loss of *Challenger* and her crew in January 1986, when authorization was given to assemble OV-105 from the structural spares. This became *Endeavour* in 1992. The problem was that there had been no directive to create another new set of structural spares to build OV-106 or beyond. As a result, only six Orbiters were ever built, and it left a fleet of just four after the grounding of *Enterprise* and the loss of *Challenger*.

For the next decade, the four Orbiters coped remarkably well with the transition from the commercially-orientated objective, through a science-based program, to the prelude to station assembly, including diverting to fly nine docking missions to the Russian space station Mir to provide America with hands-on experience in docking the huge Shuttle to a space station. However, tragedy struck once again in early 2003, when *Columbia* and her crew of seven were lost during a re-entry catastrophe, later determined to have been the result of damage caused during launch. They were just minutes from completing a highly successful and productive two-week science mission, one that had utilized the pressurized Spacehab middeck augmentation module instead of the larger Spacelab laboratory module.

The loss of *Columbia* and her crew hastened what had been postulated for some time, in the decision to terminate the Shuttle program, but only after the ISS had been completed. After the Shuttle was retired in 2011, it was not only the surviving Orbiters that were grounded and placed in museums, but also the whole Spacelab system. Irreversibly tied to the STS program almost since its inception, flown throughout its formative period, and then almost abandoned in its latter years in favor of the ISS, there was no other suitable carrier to continue flying the Spacelab hardware, nor any desire to incorporate elements of it on the ISS.

This was a bitter pill to swallow, not only for the hundreds of workers at the Cape who processed and prepared each Spacelab payload – who were now seeing perfectly good hardware languish in back lots seemingly forgotten and abandoned, with some sent to museums and some melted down for scrap – but also for the scientific community who had painstakingly prepared hundreds of experiments and research programs. Though these investigations flew on what by today's standards were relatively short missions, each gathered useful baseline data that would allow science investigations to fly on ISS more quickly and efficiently. But it was not just the Americans who lost Spacelab; Europe did too, as their early access to space. The investment Europe had put into the program from the late 1960s had been huge, but the access and return provided was not as great as expected or envisaged, even as late as the early 1990s. Lessons learned from Spacelab helped in the development of the European *Columbus* laboratory for ISS and the Italian Multi-Purpose Logistic Modules, but the loss of Spacelab was felt around the world as former users in Germany, Japan and France saw their opportunities to access human spaceflight slip away. It resulted in a cooperative program on Soviet and later Russian space stations, and a greater interest in ISS.

DEVELOPING A CONCEPT

The Spacelab era encompassed almost the entire Space Shuttle program, operating from STS-2 in 1981 through to STS-90 in 2000, with elements of Spacelab hardware continuing to fly until 2008 (see Table 1.1). Before exploring how the various elements of the Spacelab inventory were prepared to fly in space, it is worth recalling the origins and early development of the program as part of the US National Space Transportation System (NSTS).

Table 1.1: FLOWN SPACELAB MISSIONS

Quick look data				
Mission name	Orbiter	Launch Date	Spacelab Mission	Module/ Pallet
STS-9	*Columbia*	Nov 28, 1983	Spacelab 1	Module
STS-51B	*Challenger*	Apr 29, 1985	Spacelab 3	Module
STS-51F	*Challenger*	Jul 29, 1985	Spacelab 2	Pallet
STS-61A	*Challenger*	Oct 30, 1985	Spacelab D1	Module
STS-35	*Columbia*	Dec 2, 1990	Astro-1	Pallet
STS-40	*Columbia*	Jun 5, 1991	Spacelab Life Sciences 1 (SLS-1)	Module
STS-42	*Discovery*	Jan 22, 1992	International Microgravity Mission 1 (IML-1)	Module
STS-45	*Atlantis*	Mar 24, 1992	ATLAS-1	Pallet
STS-50	*Columbia*	Jun 25, 1992	US Microgravity Laboratory-1 (USML-1)	Module
STS-46	*Atlantis*	Jul 31, 1992	Tethered Satellite System-1 (TSS-1)	Pallet
STS-47	*Endeavour*	Sep 12, 1992	Spacelab J	Module
STS-52	*Columbia*	Oct 22, 1992	US Microgravity Payload-1 (USMP-1)	Pallet
STS-56	*Discovery*	Apr 8, 1993	ATLAS-2	Pallet
STS-55	*Columbia*	Apr 26, 1993	Spacelab D2	Module
STS-58	*Columbia*	Oct 18, 1993	Spacelab Life Sciences-2 (SLS-2)	Module
STS-62	*Columbia*	Mar 4, 1994	US Microgravity Payload-2 (USMP-2)	Pallet
STS-59	*Endeavour*	Apr 9, 1994	Space Radar Laboratory-1 (SRL-1)	Pallet
STS-65	*Columbia*	Jul 8, 1994	International Microgravity Laboratory-2 (IML-2)	Module
STS-64	*Discovery*	Sep 9, 1994	Lidar-In-Space Technology Experiment (LITE)	Pallet
STS-68	*Endeavour*	Sep 30, 1994	Space Radar Labaoratory-2 (SRL-2)	Pallet
STS-66	*Atlantis*	Nov 3, 1994	ATLAS-3	Pallet
STS-67	*Endeavour*	Mar 2, 1995	Astro-2	Pallet
STS-71	*Atlantis*	Jun 27, 1995	Spacelab-Mir (SL-M)	Module
STS-73	*Columbia*	Oct 20, 1995	US Microgravity Laboratory-2 (USML-2)	Module

(continued)

Table 1.1: (continued)

| Quick look data | | | | |
Mission name	Orbiter	Launch Date	Spacelab Mission	Module/ Pallet
STS-75	*Columbia*	Feb 22,1996	Tethered Satellite System-1 Re-flight (TSS-1R)/ US Microgravity Payload-3 (USMP-3)	Pallet
STS-78	*Columbia*	Jun 20, 1996	Life and Microgravity Spacelab (LMS)	Module
STS-83	*Columbia*	Apr 4, 1997	Microgravity Science Laboratory-1 (MSL-1)	Module
STS-94	*Columbia*	Jul 1, 1997	Microgravity Science Laboratory-1 Re-flight (MSL-1R)	Module
STS-87	*Columbia*	Nov 19, 1997	US Microgravity Payload-4 (USMP-4)	Pallet
STS-90	*Columbia*	Apr 17, 1998	Neurolab	Module
STS-99	*Endeavour*	Feb 11, 2000	Shuttle Radar Topography Mission (SRTM)	Pallet

There were 31 'dedicated' missions under the Spacelab series between 1983 and 2000.
16 were Long Module and 15 were Pallet/MPESS missions
Columbia (OV-102) flew 16 Spacelab missions (10 Module and 6 Pallet/MPESS)
Challenger (OV-099) flew 3 Spacelab missions (2 Long Module and 1 Pallet/MPESS)
Discovery (OV-103) flew 3 Spacelab missions (1 Long Module and 2 Pallet/MPESS)
Atlantis (OV-104) flew 4 Spacelab missions (1 Long Module and 3 Pallet/MPESS)
Endeavour (OV-105) flew 5 Spacelab missions (1 Long Module and 4 Pallet/MPESS)

Since the dawn of the space age in October 1957, access beyond the atmosphere had been by means of expendable launch vehicles or sounding rockets, with human spaceflight initially achieved using one-shot vehicles to sustain and support the small crew of one, two or three persons. These early human missions beyond our atmosphere were focused upon proving the concept that a flight into space (vacuum) was not only possible but also survivable. Naturally, engineering objectives were at the forefront while science research took more of a back seat for some time, due to the operational requirements of the missions, the limitations of the vehicles and the restrictions on crew time.

A Change of Direction

After the first few Apollo missions had proven the concept, the Moon landing program provided an opportunity to expand the scientific return from each mission. But inevitable budget restrictions curtailed such expansion plans almost as soon as they began, thereby losing the opportunity to gain a significant scientific yield from operations both on and around the Moon.

It would take the introduction of the first research platforms in space – space stations – before the opportunity to extend scientific research was afforded through prolonged exposure to spaceflight by successive crews. There were still challenges in these early space station missions, most notability in the reliability of the hardware, the efficiency of the crews, the regular resupply of consumables, and the amount of experiment and sample data that could be returned to Earth. All of this created huge logistical problems, a poor return on the investment in each vehicle and extremely high launch costs to place such hardware in orbit.

Ideally, what was needed was a low-cost, reusable and highly reliable system to ferry hardware, experiments, crews and supplies from Earth to orbit, to enable the resupply of a space station or lunar base and then return experiment samples, crews and unwanted material to Earth. This was the concept behind what became known as the American National Space Transportation System, later shortened to the Space Transportation System program, or more commonly the Space Shuttle program.

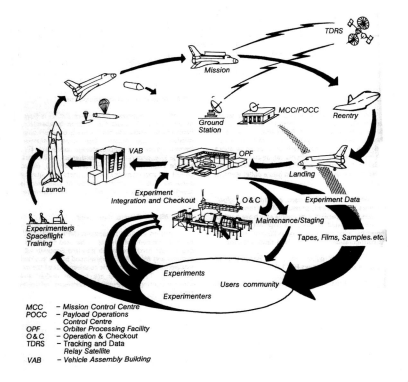

Typical Shuttle-Spacelab mission flow. [ESA.]

From RAM to Spacelab

In 1969, Europe was encouraged to participate in NASA's post-Apollo program. At that time, there was a proposed scientific orbiting laboratory concept known as the Research and Applications Module (RAM), which was to have been used in conjunction with a ten-man space station. Two years later, the idea of a separate space station had been abandoned (for now) and a small modular laboratory, called a "Sortie Lab" or "Sortie Can", was developed to fit into the Payload Bay of the Orbiter.

It was recognized early in the program that with a cavernous Payload Bay available on each mission, the Shuttle could benefit from a small pressurized laboratory module, within which short-term experiments and research could be conducted in a variety of scientific fields on some of the planned missions. On September 10, 1971, NASA's Space Station Task Force Director, Douglas R. Lord, requested that the Marshall Space Flight Center (MSFC), in Huntsville, Alabama, should begin in-house studies for a small pressurized crew module (laboratory), to be carried on short-duration missions in Earth orbit in the Payload Bay of the Space Shuttle. Over the next nine months, in-house studies continued to evaluate the concept and its possible applications within the Shuttle program.

Then, over June 14–16, 1972, a delegation from the European Space Conference visited Washington for high level discussions with US officials regarding the prospect of European participation in the Sortie Can project. As a result of these talks, the European Research and Technology Center (ESTEC) was assigned to establish the required resources needed for Europe to develop what was now being identified as the Sortie Module as a fully functional research laboratory. Over the next six months, officials on both sides of the Atlantic developed a plan to create an agency-to-agency agreement. In a move to coordinate all European space activities into a single agency, European space ministers agreed, over November 8–9, 1972, to merge the European Space Research Organization (ESRO) and the European Launcher Development Organization (ELDO) into what became ESA in 1975. By December 1972, European space ministers had made a formal commitment to the Sortie Lab, and the following month both NASA and ESRO complete the first drafts of a Memorandum of Understanding (MOU).

On January 18, 1973, the Council of ESRO voted to authorize a "Special Project" to develop the Sortie Lab, which the Europeans started calling "Spacelab" after "a laboratory in space". Throughout 1973, the refinement of the agreements continued between NASA and what at the time was the eleven European nations of ESRO. By September 24, the MOU was signed to implement the project and in October, the NASA Sortie Lab Task Force at Headquarters was renamed the Spacelab Program Office. A new era in US-European space cooperation had begun. Parallel studies had been conducted at NASA's MSFC in Alabama and at ESRO and the German aerospace manufacturer Messerschmitt-Bölkow-Blohm (MBB), but with the MOU, all US work ceased and the responsibility for the work was placed with nine European countries.

Over the next decade, the laboratory, its subsystems, the pallets and associated hardware were developed in parallel. Experiments were devised and procedures ironed out to enable the system to meet the specifications outlined in payload documents, leading to the opportunity to fly the module.

In March 1974, ESRO requested proposals from two German contenders, and on June 4, 1974, it awarded a six-year contract to an industrial consortium of companies from eight European countries to design, develop and manufacture the Spacelab hardware. This team would be led by VFW/Fokker ERNO Raumfahrttechnik in Bremen, Germany.

The team assembled to achieve this included subcontractors:

- *Aeritalia, Italy*: Module structure
- *Engins Matra, France*: Command and data management
- *AEG Telefunken, Germany*: Power distribution
- *Dornier Systems, Germany*: ECS and instrument pointing
- *Hawker Siddeley Dynamics (later British Aerospace), UK*: Pallet structure
- *Bell Telephone Manufacturing, Belgium*: Electrical ground support
- *Inta, Spain (subsequently taken over by SENEA)*: Mechanical ground support
- *Fokker VFW, Netherlands*: Airlocks
- *SABCA, Belgium*: Utility bridge, and
- *Kampsax, Denmark*: Computer software.

Delivery and Qualification

There were a number of development, test and flight units produced under the Spacelab program:

- An engineering model was used for maintenance and refurbishment procedures, experiment integration checks, crew training and as a mission simulator.
- A second engineering model was delivered to ESA.
- A single Spacelab mockup was also delivered to ESA for configurational studies.
- Two flight models were delivered to NASA.
- Delivery of the first flight unit was originally scheduled for 1979, with the first flight planned for 1980 on the 7th Shuttle flight (STS-7). It was designated as Spacelab-1, the first verification flight of the Long Module (actually flown as STS-9 in 1983)
- Spacelab-2 was to have been the second verification of the pallet-only configuration (no pressurized module) and Instrument Pointing System (IPS). During 1983, delays with the NASA Tracking and Data Relay Satellite

System (TDRSS) and a redesign of the IPS required scheduling the flight of Spacelab-3 (a NASA microgravity-dedicated mission) to fly in April/May before Spacelab-2. With the eventual success of Spacelab-2 in July/August 1985, ESA's development program was completed and the Spacelab system became operational.

- Eleven pallet units were prepared by Hawker Siddeley and delivered as five flight units, five test models and a single spare unit.

Spacelab delivery at KSC, December 1981. [ESA.]

THE COMPONENTS

The concept of the Space Shuttle was that of an orbital vehicle that could make regular trips to space and back again, and could support a crew for a week or more in Earth orbit to complete wide variety of missions, ranging from research to satellite deployment and retrieval, as well as some with classified Department of Defense (DOD) objectives. Key to this was the large Payload Bay of the Orbiter, located behind the twin-deck crew compartment and measuring 60 feet (18 meters) in length and 15 feet (4.6 meters) in diameter. Two large Payload Bay doors were firmly closed during ascent and decent but had to be opened during orbital operations, exposing the contents to the rigors of spaceflight. Thermal radiators were mounted on the inside of the doors and these needed to be exposed to space in order to control Orbiter temperatures. The main "flight crew" consisted of a mission Commander and a Pilot, who looked after the Orbiter and its systems, performed orbital maneuvers of the vehicle and rendezvous exercises, and ensured mission success and crew safety. They were joined by two or more Mission Specialists, who were primarily responsible for the payloads and research studies on each mission, operating the robotic arm to deploy and retrieve payloads, and conducting spacewalks. These were all career NASA astronauts, but there was a third crew category, called the Payload Specialist. These crewmembers were assigned from academia, the military, or major payload customers as required, mostly for a single flight before returning to their previous career shortly after the mission. Several crewmembers of Spacelab missions were from the Payload Specialist category, supporting the mission objectives.

The module of Spacelab was designed as a laboratory in which crewmembers could perform experiments in an internal atmosphere kept at comfortable temperature and humidity levels, allowing astronauts to work in a "shirt-sleeved environment".

A variety of experiments would be made available, some of which could be installed on unpressurized pallets in the Payload Bay. This meant that the components had to be flexible in design and compatibility to offer a large range of configurations and options to the customers to meet their needs, while remaining within the strict guidelines of flying hardware on a human spacecraft. The modular approach chosen offered the most flexibility to meet the needs of both NASA and its potential customers. To assist in preparing any equipment and experimentation to be flown on Spacelab, a Payload Accommodation Handbook was made available to providers of hardware and experiments.

Combinations of modules and pallets could be flown allowing for a variety of configurations, though not all were actually implemented during the Space Shuttle program. The range of subsystems offered to users included power supply, thermal control, and command and data management. It is from this inventory that the requirements of each mission were sourced, using standard Mission Dependent Equipment, or Mission Peculiar Equipment for that specific flight.

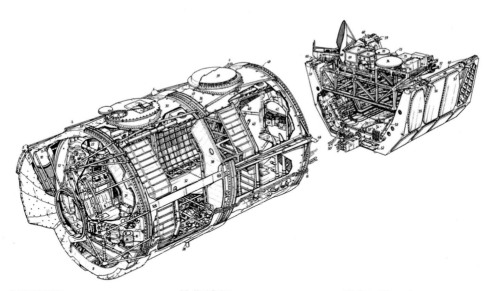

MODULE

1 Insulation blanket
2 Close-outs (skin/racks)
3 Cabin air ducting (from sub-floor)
4 RAAB
5 High Data-rate recorder
6 Handrails
7 Water/Freon heat exchanger
8 Utility tray
9 Gaseous nitrogen supply
10 Gaseous nitrogen tank
11 Temperature transducer
12 Forward end-cone
13 Module/Orbiter lower feed-through
   plates (two)
14 Insulation blanket supports
15 Freon pump
16 Water pump
17 Lithium Hydroxide cartridge
   stowage
18 Freon lines
19 Control-centre rack
20 Debris traps
21 Workbench rack
22 Stowage container (lowers for access)
23 Upper module/Orbiter feed through
   plate
24 Gaseous nitrogen fill-valve bracket
25 Gaseous nitrogen reducing valves
   (two-stage)
26 Position for double rack
27 Position for single rack

28 Keel fitting
29 Sub-floor
30 Aluminum alloy module shell
31 Electrical connectors for rack
32 Floor of aluminum-skinned honeycomb sandwich (centre panel fixed,
   outer panels hinge up for access)
33 Overhead duct channels
34 Viewport
35 NASA high-quality window
36 Fasteners for insulation blanket
37 Rack fire-suppression system
38 Double rack
39 Experiment
40 Airlock controls
41 Overhead lights
42 Avionics cooling-air ducts
43 Aft end-cone
44 Radial support structure
45 Fire extinguisher (Halon)
46 Portable oxygen equipment
47 Foot restraint
48 Module/Orbiter pickups (four)
49 Module-segments joints,
   incorporating seals

PALLET

50 Freon lines from Module
51 Pallet interface
52 Cable ducts
53 Cold plates
54 Inner skin-panels

55 Outer skin-panels
56 Pallet/Orbiter primary pickup
57 Pallet/Orbiter stabiliser pickup
58 Connector supports
59 Pallet hard-points
60 Handrails
61 Support systems Remote Acquisition
   Unit (RAU)
62 Experiment RAU (several)
63 Experiment power distribution box
64 Pallet/bridge supports
65 Experiment-supporting bridge
66 Electrical junction box
67 Integrally-machined aluminum-alloy
   ribs

EXPERIMENTS*

68 Synthetic aperture radar
69 Solar spectrum
70 X-ray astronomy
71 Solar constant
72 Charged-particle beam
73 Advanced biostack
74 Isotopic stack
75 Micro-organisms
76 Lyman Alpha
77 Waves
78 Low energy electron-flux

*Representative Experiments

Spacelab and pallet cutaway diagram and key. [Graphic from *Spacelab, An International Success Story* (NASA SP-487).]

In total, eight basic configurations of Spacelab were available to users, four of which were termed baseline design configurations that remained under configuration control within the Spacelab Program. The other four options were available as alternative configurations specific to the intended objective and investigations, should the need arise.

The four base line configurations were:

- Long Module configuration, which afforded the largest pressurized volume.
- Short Module plus three pallet segments, 29.5 ft. (9 meters), that offered the largest pallet mounting with a module. The three pallets segments were rigidly attached, forming a pallet "train" configuration. (Configuration 5).
- The 29.5-ft. (9-meter) Pallet configuration. This consisted of three independently suspended pallet configurations (Configuration 6).
- The 49.2-ft. (15-meter) Pallet configuration was the only possible configuration for an experiment platform in the Spacelab range that required exposure to the space environment (Configuration 8).

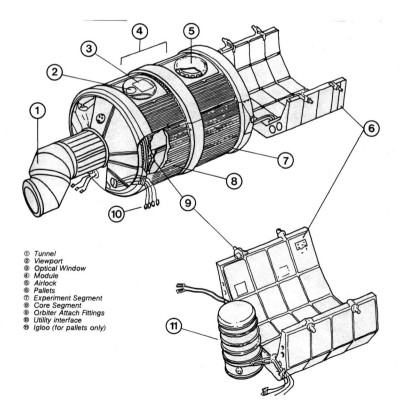

① Tunnel
② Viewport
③ Optical Window
④ Module
⑤ Airlock
⑥ Pallets
⑦ Experiment Segment
⑧ Core Segment
⑨ Orbiter Attach Fittings
⑩ Utility interface
⑪ Igloo (for pallets only)

Spacelab's main components

Supplementary equipment include the Igloo for pallet-only missions, which provided system support. For module missions, the short or long Spacelab tunnel was available with elbow connecting elements. The Shuttle Orbiter Extra-Vehicular Activity (EVA) airlock configuration was included on each mission (for planned and/or contingency EVA access) either inside the middeck or externally as part of the tunnel system. Mission rules stated that sufficient provision should be made available in the Payload Bay for EVA access from the airlock, so that pallet-only configurations would not impede the exit or entry of pressure-suited astronauts. Just one mission (STS-71 in 1995) required the inclusion of the Orbiter Docking System with the Spacelab tunnel, to allow unimpeded access between the Orbiter crew compartment, the Russian Mir station and the Spacelab Long Module, while still affording (contingency) EVA capability.

The Spacelab system was completely reusable, designed to feature multi-use applications, and could be stacked or fitted in a variety of configurations, including completely enclosed, completely exposed to the space environment, or a combination of both.

Pressurized Module

Access was possible from the crew compartment by means of the pressurized tunnel attached at the lower aft middeck (airlock) bulkhead. The internal features allowed crewmembers to work in laboratory conditions as if they were on the ground. The "module" could be made up from either short or long sections capped by two end cones. The forward cone connected to the Spacelab tunnel section, while the aft cone included a small viewport to allow observation of pallet-mounted experiments.

The forward, "core" segment could be used in either the Short or Long Module designs. This contained Spacelab subsystems, electrical power distribution, environmental control, command and data management etc. The volume available for user equipment in this core section was approximately 60 percent.

Both modules were 13.12 ft. (4 m) in diameter, with the Short Module being 13.97 ft. (4.26 m) in length with a center aisle of 8.82 ft. (2.69 m) and 268.3 ft.^{3} (7.6 m^3) volume available for payloads. In contrast, the Long Module measured 21.94 ft. (6.96 m) with a center aisle of 17.65 ft. (5.38 m) and a volume of 77.69 ft.^{3} (22.2 m^3). When assembled, the complete Spacelab Long Module version, with end cones, measured 23 ft. (7 m).

The Long Module was designed accommodate a maximum of 8,000 ft.^{3} (226.5 m^3) of experiment equipment, or a maximum weight of almost 13,000 lbs. (5,896.7 kg). The dry weight of Spacelab was about 17,750 lbs. (8,051 kg), with a (limited) integrated launch weight of 31,000 lbs. (14,061 kg) in order to remain within the 32,000 lbs. (14,515 kg) maximum cargo weight the Orbiter was capable of returning to Earth safely.

The second segment in the Long Module configuration was termed the "Experiment Section", with the facility to install an airlock on the upper part of this section, if required, to allow the exposure of small items to the space environment and their retrieval without the need for an EVA. There was capability to fit a high quality window/view port to the upper part of the segment for observation or high-optical measurements. All modules were affixed to the Orbiter Payload Bay at the keel and at each side.

Unfortunately, the Short Module was never flown on a mission. The later, commercially-developed Spacehab middeck augmentation module was not part of the Spacelab system and provided much needed additional locker space and logistics stowage on Shuttle missions. It also supported a number of smaller science research programs and experiments, but was still limited in its capacity, even over the unflown Short Module at 8.9 feet (2.71 m) long (Spacehab was 10 feet (3.05 m) long). Another element frequently flown on Spacelab missions was the Extended Duration Orbiter (EDO) cryogenic storage "pallet" or kit, to support extended Shuttle missions of between 14 and 18 days.

Racks and Stowage

The racks installed in Spacelab modules were of two designs and different dimensions. *Single racks* were a standard 19-inch (483 mm) width, while *double racks* featured two single racks mounted side by side as one unit. All racks were fitted with power, data and cooling interfaces. With a maximum volume available of 1.57 ft.^{3} (3.92 m^3), the racks offered the most useful way to mount the user's equipment and experiments, with 639 lbs. (290 kg) available in single racks and 1,278 lbs. (580 kg) in double racks.

Unpressurized Pallets

Due to their design flexibility, the British-built unpressurized pallets flown on Shuttle missions were not confined to Spacelab module or designated missions but, like the Canadian-built Remote Manipulator System (RMS), were an integral element in the STS system. Each single pallet measured 9.43 ft. (2.875m) in length and 14.27 ft. (4.35m) in width. Due to their 'U'-shaped design, the pallets matched the inner contours of the Orbiter Payload Bay and the maximum cylindrical payload measurement had a radius of 6.15 ft. (1.875 meters). The pallets were also laterally restrained when fixed to the Orbiter keel.

The versatility of the pallets included the capability to attach experiments on the inner skin panels using the structure's honeycomb design and insert arrangements, in a 5.5 x 5.5 inch (140 x 140 mm) square grid pattern. In total, if uniformly distributed, a single pallet could support a payload mass of about 1,362 lbs. (3,000 kg). Heavier experiments could be bolted directly to the pallet structure using some of the 24 latching points available across the pallet. This ensured far less vibration than for those mounted on the panels. Pallets mounted together formed a 'pallet train' but were positioned to enable up to 0.78 inch (20 mm) maximum deflection value from inertia loading, torsion load imparted by the Orbiter and variable temperature effects.

Each pallet segment included the basic features required for the operation of a user's experiments or equipment. This included power distribution boxes, housekeeping facilities, experiment data buses, remote data acquisition units, cold plates and thermal capacitors. When pallets were flown in conjunction with a pressurized module, these facilities were connected to the Spacelab subsystem equipment located in the core segment of the module. When there was no pressurized module flown, these services were located in the Igloo

Igloo

The Igloo was a cylindrical canister pressurized to 1 atm that was located in the Payload Bay. It contained a power control box and an emergency power box, a subsystem interconnection station, a data multiplexer, a subsystem power distribution box, a remote amplification and advisory box, two remote subsystem data acquisition units, three computers with two input/output units (subsystem and experiment) and a mass memory unit, and a high-rate multiplexer. Measuring 44 inches (1,120 mm) in diameter and 93.85 inches (2,384 mm) in height, it had a volume of 77.7 ft.^{3} (2.2 m^3). It featured primary structures and a secondary structure, a removable cover and Igloo mounting structures. Two units were manufactured by Belgian company SABCA and both were flown.

Utility Bridge

This was installed to carry electrical lines between the module and equipment installed on the pallets.

Instrument Pointing System

The IPS was a gimbaled three-axis pointing device that flew on Spacelab-2 and both Astro missions. It was used to point telescopes, cameras and other instruments precisely. Its pointing accuracy was to within 1 arc second (unit of degree) and its three pointing modes were Earth, the Sun, and stellar focused. Mounted in the Payload Bay, it measured 10 ft. (3.05 m) tall, with a diameter of 5 ft. (1.52 m) and a mass of 2,500 lbs. (1,135 kg). Two units were manufactured by Dornier, primarily from aluminum, steel and multi-layer insulation.

Tunnel

The Spacelab tunnel measured 3.28 ft. (1 m) in diameter, with its internal dimensions enabling the transfer of equipment measuring up to 1.83 x 1.83 x 4.16 feet (0.56 x 0.56 x 1.27 m) between the Spacelab module and the Orbiter middeck. Segmented sections enabled the tunnel to be lengthened or shortened depending upon the mission requirements for a Short or Long Module, positioning the Shuttle airlock internally or externally to the middeck, or the inclusion of the Orbiter Docking System. The airlock could be integrated into the tunnel to provide EVA access to open space, and while no EVAs were scheduled or completed on any of the Spacelab missions flown, contingency EVA planning and training was a requirement for all missions.[1] The tunnel system featured an adapter section which fitted directly to the exterior aft hatch on the middeck of the Orbiter and provided the attachment of the Shuttle airlock module on top to facilitate an EVA capability. From the adapter section, the tunnel ran longitudinally the required length to the final section, a 'jog' or 'dog-leg' to join with the higher centerline entrance to the Spacelab module.

Airlock

The Shuttle airlock alleviated the need to decompress the whole crew compartment to enable a team to perform EVA. For most missions the airlock was located in the aft middeck area, but for Spacelab missions carrying the pressurized module the airlock could be located externally as part of the transfer tunnel system, to provide additional volume in the middeck area and a convenient exit to open space when a tunnel was attached. A tunnel adapter was used and the airlock mated at the forward position in the Payload Bay. For the EVA, the airlock was positioned on top of the tunnel adapter, supported and stabilized to the aft bulkhead by structural adapters. There were two access hatches, one on the top of the adapter for access to the airlock and the other on the aft end towards the Spacelab laboratory.

---

[1] Every Shuttle mission included at least two NASA crew members trained to accomplish a range of contingency or emergency EVAs if required.

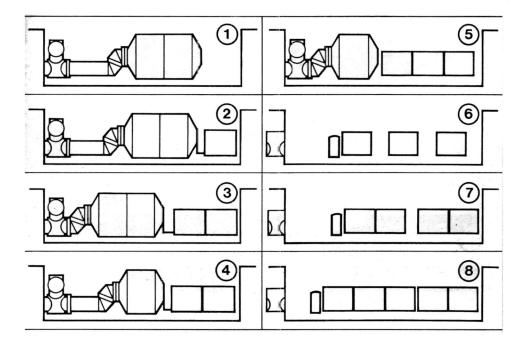

Spacelab basic configurations.

Sub Systems

Each module was provided with a number of subsystems, including, briefly:

- Electrical power: Derived from the Orbiter fuel cells and could vary depending upon the configuration flown and mission duration. DC primary power was nominally 28V but could vary between 24V and 32V. AC power was available at 115/220V.
- Thermal control: Heat rejection for user equipment in Spacelab was provided by the Orbiter. The cabin air loops provided conditioned air for the crew members and extracted the heat generated by the astronauts and equipment. The temperature was adjustable from 64.4 to 80.6 degrees Fahrenheit (18 to 27 degrees C). Experiments and equipment requiring venting and cooling loops formed part of this subsystem.
- Command and data management: This was a comprehensive system onboard the Spacelab and provided services to support the acquisition of data; processing and formatting; the transmission of data; recording, monitoring and display of experiment data; a command and control system; closed-circuit TV (CCTV); audio recording; and a caution and warning system.

- Software: Dedicated Experiment Processors (DEP) which ran within the experiment; Experiment Flight Application Software that ran in the onboard experiment computer; and the software run by the Level-IV test facility.

Additional Equipment

The Spacelab module, pallet and Igloo were be supplemented with additional equipment, termed *Mission Dependent Equipment* or *Mission Peculiar Equipment.*

- Mission Dependent Equipment (MDE): This equipment was flown in addition to the basic equipment (termed mission independent). The MDE was categorized as experiment racks, storage containers, film storage kits, experiment heat exchangers, cold plates, remote acquisition units, intercom stations, CCTV monitors, airlock, view ports, and high quality windows. The requirements for any science equipment were selected on an individual basis, appropriate to the mission being flown.
- Mission Peculiar Equipment (MPE): This type of equipment was identified as additional items necessary to connect the experiment hardware to the required Spacelab subsystems and MDE. MPE was developed in conjunction with the payload integrator to meet the requirements for the experiments. This type of equipment could include pallet bridges, other mounting substructures, harnesses and refrigerator lines.

Spacelab Stowage

As with every crewed mission into space, the restricted dimensions of the space vehicle means that volume is at a premium. In addition to the payload, fuel and consumables (oxygen and water) available to the crew, provision has to be made for food, personal clothing and equipment, hygiene facilities and waste stowage. All of this impacts the volume available for experiment hardware and their support equipment. On missions that did not carry the Spacelab, the limited volume of the Shuttle middeck could result in cramped conditions for a crew of up to seven for a week. STS-61A, also known as Spacelab D1, holds the current record for the largest crew − eight people − aboard any single spacecraft for the entire period from launch to landing.

Middeck lockers helped organize stowage, but for Spacelab missions (and later Spacehab flights) the additional stowage volume was both a useful and welcome addition to the mission, and a bonus to the crew. Stowage in the Spacelab module was available on the upper part of the equipment racks and in the ceiling of the module. Rack containers had a volume of 2 ft.^{3} (0.056 m^3) with a mass capability of 55.12 lbs. (25 kg). The ceiling containers had a volume of 28.25 ft.^{3} (0.8 m^3) and a load capacity of 72.76 lbs. (33 kg). The Long Module configuration included 14 ceiling containers but the Short Module configuration made only eight available for user equipment. The installation of an optical window or airlock reduced this availability to five or six ceiling containers, respectively. Both the rack and ceiling containers were utilized for film stowage during a mission.

Middeck Stowage

Equipment could be located in the middeck storage lockers in the main Shuttle crew compartment. In the middeck, container modules could be inserted in the forward avionics bay. On a nominal mission there were up to 42 containers provided in this area, though not all were available for experiments as the majority were filled with crew equipment, clothing and food, and camera equipment. In addition, there was an area to the right side of the airlock module where a further nine containers could be attached. This could vary significantly depending on the type of mission flown. With internal dimensions of 18.4 x 10.6 x 21.1 in (46.73 x 26.92 x 50.8 cm), these lockers provided a useful if limited stowage capability, but were just a short tunnel journey away from the Spacelab module.

A more in-depth account of the Space Shuttle Orbiter and Spacehab locker system was given by co-author David Shayler in 2017 [1]. Briefly, the middeck locker trays were available in two sizes, large and small, which had the same internal width (16.9 in or 43 cm) and depth (20 in or 50.8 cm). The small version was only 4.5 in (11.43 cm) tall, while the large trays measured 9.59 in (24.3 cm) tall. Each locker receptacle measured 18.4 in (46.8 cm) wide, 21.1 in (53.4 cm) deep and 10.6 in (26.9 cm) tall, allowing for one large tray, or two small trays on top of each other, to fit into one locker space in the rack. A hinged door flap and twist catches secured each locker until needed. There was a specific identification system assigned to pinpoint a particular locker location and it was marked on the door flap. Using the example of locker MF47C, this indicated it was M= Middeck; F= Forward; (there was also MR [Right] ML [Left], MO [Overhead] and MA [Aft]); 47 meant 47 percent across the forward bulkhead from left to right facing; with the number representing the percentage of whatever surface the locker was located in; while C indicated the distance, in 6-in. (15.24-cm) increments from the tip of the locker frame, with 'A' at the top and progressing down alphabetically. Even if a locker space was empty, it would still be taken into consideration in numbering the available lockers [2].

Fields of Research

Experiments conducted on Spacelab missions included research in the fields of astronomy, solar physics, space plasma physics, atmospheric physics, Earth observations, life sciences and material sciences.

FROM CONCEPT TO CREATION

The construction of the Spacelab hardware was begun in 1974 by ERNO, at that time a subsidiary of VFW-Fokker GmBH, Germany.

The first Long Module (LM-1) was "donated" to NASA in exchange for flight opportunities for European astronauts. NASA purchased the second Long Module (LM-2) from ERNO for its own use. The module was assembled from welded aluminum alloy, with an integral stiffening waffle panel machined on its inner face. An internal secondary structure was attached to the outer cylinder to carry subsystems and experiment equipment loads.

Hardware delivery was originally planned for 1979, with the first Spacelab mission being flown the following year. However, issues in both preparing the Shuttle system for flight and preparing the Spacelab hardware for shipment meant a three-year delay to complete delivery of all components. Finally, on December 4, 1978, the first pallet (serial number E0002) to be used in the OFT program was delivered to KSC. This Engineering Model (EM) was very basic in design and lightly equipped, with an Orbiter Freon® pump, a cold plate, a power control box and a command module. A second EM pallet (serial number E0003) arrived on dock at KSC on April 22, 1979. They would both fly as part of the OFT program to gather important data prior to the first operational Spacelab mission.

A year later, in December 1980, the EM of the Spacelab pressurized module was delivered to KSC. It was identical to the Flight Unit (FU) in all but the stringent flight testing required for certifying the hardware to fly in space. The EM was used by both agencies at KSC to certify systems and facilities in advance of delivering the first FU. A complete Spacelab 'FU' consisted of a Long Module and up to five pallets.

The delivery of the flight hardware was split into two shipments. The first arrived at KSC on December 11, 1981, and included the first Long Module (LM-1) and one pallet. Following a program of checks and tests, it was finally accepted by NASA on February 5, 1982. The 1973 NASA/ESRO MOU stated that NASA could procure a second Spacelab unit no later than two years before the first unit had been delivered. This option was taken up by NASA in a signed contract at the end of January 1980. On July 8, 1982, this second FU (LM-2) and the remaining nine pallets arrived on dock at KSC. Each unit was accompanied by its associated software, a set of spares and a range of Ground Support Equipment (GSE), such as specialized jigs and testing equipment. An option for further procurement of Spacelab hardware was not pursued.

The Spacelab tunnel system was fabricated for NASA by McDonnell Douglas and was delivered to KSC in December 1982. The first of the two Igloo units built by the Belgian Company SABCA was delivered to KSC in August 1982, while the first of the two IPS units fabricated by Dornier arrived at KSC in February 1984.

NASA finally certified Spacelab as a component of STS on January 13, 1983. Spacelab was finally about to fly and it was this suite of hardware which was expected to support at least 50 missions over the following years. Fifteen years later, only 22 *major* Spacelab missions had flown between 1983 and 1998, though a number of other missions up to the end of the program in 2011 used Spacelab pallets in support of other objectives. Despite its success, there were clearly lost opportunities – for a variety of reasons – in not flying further Spacelab missions. These lost opportunities are explored further in Chapter 12.

Integration Activities

A sequence of integration activities were created to ensure mission success for both the hardware and experiments.

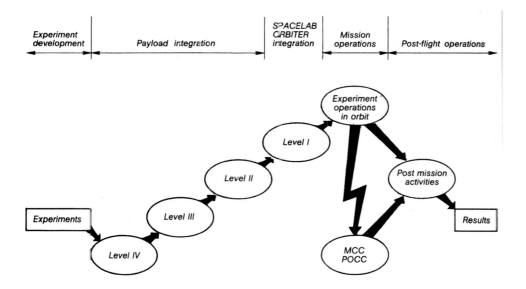

Integration activities and payload-related program phases. [ESA.]

To summarize the whole sequence:

- **Level-IV Instruments to racks and pallets (the main focus of this title)**
- Level-III Spacelab subsystem racks and module assembly
- Level-II Payload assembly to Spacelab
- Level-I Spacelab to the Orbiter for launch processing

Verification of the System

The Spacelab Verification Flight Test (VFT) program was created ensure the hardware complied with its design requirements. Spacelab-1 would be the VFT-1 mission, flying the Long Module configuration and one pallet. This was to be followed by Spacelab-2 as VFT-2, which would be flown to verify the Igloo multi-pallet configuration and the IPS. Spacelab-3 would therefore become the first Spacelab Operational Mission (SOM), allowing the program to progress as planned.

The VFT program was designed to determine the operational capability of the environmental control, structures, electrical power distribution and command and data subsystems; the habitability of the Long Module configuration; an evaluation of the module and pallet environments; the verification of the Igloo and IPS facilities; the compatibility of Spacelab exterior materials with the space environment; and to establish that contamination limits were not exceeded on sensitive optical surfaces and other experiments.

However, due to development issues with the IPS and other programmatic considerations, it was decided to fly Spacelab-3 before Spacelab-2. In addition, the pallet system was available in three configurations: the basic Multiplexer/Demultiplexer (MDM) Pallet where systems were reliant on the Orbiter; the Igloo Pallet where the Igloo supplied the subsystems; and the MPESS structure. As part of the OFT program, two EM Spacelab pallets were utilized, for STS-2 (carrying the OSTA-1 payload) in November 1981, and STS-3 (carrying the OSS-1 payload) in March 1982.

Spacelab Manifest

By 1983, the year the first Spacelab mission flew, there were expected projections for between six and eight Spacelab flights *per year* during the Financial Years (FY) 1984 to 1986, with similar projections for the following years. The manifest could, and indeed would, constantly be amended over the ensuing years as payloads and missions were delayed, inserted, cancelled, or of course flown. Missions were expected to include a combination of dedicated module and pallet configurations, and with Spacelab hardware flown on mixed-payload flights utilizing the pallet systems and special structures from the Spacelab inventory.

- Dedicated missions – were the missions where Spacelab hardware occupied the whole Payload Bay, either the pressurized module and/or pallets and support structures.

- Mixed payload – where elements of Spacelab hardware shared the Payload Bay and Orbiter resources with one or more other payloads.

The Spacelab Flight Division developed an early concept for Dedicated Discipline Laboratory missions. The idea behind this was to reduce the cost of creating individual missions each time, as well as offering the maximum opportunity to fly the same instruments, experiments and payload to maximize their scientific return and capital investment. The original idea was to fly and then re-fly the same group of compatible instruments several times, while retaining growth potential to upgrade and refine the assembled payload but with little integration or reassembly between missions.

The characteristics of these dedicated research missions were:

- To gather a group of compatible science instruments which had suitable interfaces with Spacelab hardware and software.
- These missions would be regularly flown at intervals of between six months to two years.
- Though the basic design of the 'mission' or the 'payload' would not require significant changes, there would be flexibility and growth built into this system for individual instruments to ease the deletion of outdated instruments or the addition of new and improved ones.
- Between missions of the same type, it was envisaged that the 'dedicated payload' would not be completely disassembled from the Spacelab hardware, thus saving processing time.
- The long term plan was to evolve these instruments, experiments and support hardware into the low Earth orbiting facility referred to at the time as the *Freedom* Space Station, which eventually evolved into the current ISS.

The ten 'dedicated discipline' Spacelab mission series in early planning documents were:

- Space Biomedical Laboratories (which became Spacelab-4 and then the Spacelab Life Sciences series).
- Space Plasma Laboratory (initially Spacelab-6 then cancelled).
- Space Telescopes for Astronomical Research Series (originally the Office of Space Sciences flights OSS-5, -6 and -7, which evolved into the Astro series).
- Shuttle High Energy Astrophysics Laboratory (SHEAL, subsequently cancelled).
- Shuttle Infrared Telescope Facility (SIRTF, which evolved into the Compton Telescope Facility).
- Solar Optical Telescope (SOT, later cancelled in favor of SunLab which was itself cancelled).
- Material Science Laboratory (in this format MPESS-only missions NOT laboratory missions).

- Environmental Observation Laboratory (later becoming the pallet/MPESS Earth Observation Missions that were renamed ATLAS)
- Shuttle Radar Laboratory (Originally OSTA-3/5/7 which became SRL 1 and 2), and
- International Microgravity Laboratory.

Summary

So that is Spacelab. With its concept and purpose defined, and the hardware developed and delivered, the next phase was to prepare to fly the components on a mission as early as possible to qualify the hardware for hopefully many operational missions. However, long before that date arrived, the concept of creating a system to handle and process the hardware and experiments had to be devised, and this is where our story of Level-IV at KSC unfolds across the following chapters, offering a different account of the highly successful Spacelab program.

References

In addition to the individual references listed below, the following publications were frequently referred to in the compilation of this section.

- *Spacelab User's Manual*, ESA, dp/st (79) 3, July 1979.
- *Pallet & Half Pallet, Shuttle Dedicated, Re-usable Carriers*, British Aerospace Dynamics Group, Space & Communications Division, Stevenage, England, 3rd Edition, 1981.
- *Spacelab Program Preparations for First Flight and Projected Utilization.* James C. Herrington, The Space Congress Proceedings, 1983.
- *Spacelab, Research in Earth Orbit.* David Shapland and Michael Rycroft, Cambridge University Press, 1984.
- *Spacelab: An International Success Story*, Douglas R. Lord, NASA 1987.
- *Space Shuttle Transportation System*, Press Information, Rockwell International, March 1982, pp. 43–48.
- *NASA Historical Data Book* Volume 5: NASA Launch Systems, Space Transportation, Human Spaceflight and Space Science 1979–1988, Judy A, Rumerman, NASA SP-4012, 1999, pp. 148–150 and 462–479.

1. *Linking the Space Shuttle and Space Stations. Early Docking Technologies from Concept to Implementation*, Springer-Praxis, 2017 pp. 169-191.
2. e-mail from Steve Hawley to David Shayler March 8, 2016.

2

"Ship-and-Shoot"

"Ensuring nothing fell through the cracks,
mitigating damage to the hardware
and harm to the crew."
Mike Haddad, former NASA Systems Engineer.

What does "Ship-and-Shoot" mean? It is a processing philosophy in which hardware and software are certified for flight by the vendor at the vendor's location and then shipped to the launch site, with no further testing required prior to integration into flight systems and/or the launch vehicle.

THE CASE AGAINST "SHIP-AND-SHOOT" PHILOSOPHY

For decades, "Ship-and-Shoot" (sometimes referred to as "Ship-n-Shoot") had been seen as a goal by program managers, from the Redstone Rocket Program in the mid-1950s, through the Mercury, Gemini, and Apollo Programs, to the Space Shuttle Program and on to the International Space Station (ISS). "Ship-and-Shoot" will save time, money, schedule and, some say, wear and tear on the hardware, with minimal risk to the mission. The "faster, better, cheaper" approach.

The certification of flight hardware and software by a vendor will uncover many problems that can be corrected by that vendor while still at their site. The equipment used by the vendor to test their payload or experiment could be flight-equivalent, a prototype of flight systems, or flight components of flight systems. This equipment is meant to simulate the actual flight vehicle, or the flight systems

© Springer Nature Switzerland AG 2022
M. E. Haddad, D. J. Shayler, *Spacelab Payloads*, Springer Praxis Books,
https://doi.org/10.1007/978-3-030-86775-1_2

to be integrated to the into the flight vehicle, thus giving the vendor high confidence that their payload will work with the flight vehicle or systems.

Jim Dumoulin observed that, "Ever since the first NASA payload was lofted into space there have been discussions about 'Ship-and-Shoot'. Payload developers have always wanted to ship their payloads to the launch pad as ready as possible, but schedule pressures and Murphy's Law [both] force compromises."

John-David Bartoe noted, "I think they talked about that ['Ship-and-Shoot'] at Marshall [Space Flight Center, Huntsville, Alabama] when we were working on Spacelab-2, and KSC [Kennedy Space Center, Florida] said, 'No, we're not going to do that'. And that was totally obvious when everybody came to KSC, how critical it was *not* to do 'Ship-n-Shoot'. But the interesting part was, when I moved on up to headquarters, they would talk all the time about [how] Space Station was going to be 'Ship-n-Shoot' and there was only one organization that consistently fought back and that was KSC. In these meetings, the KSC center director would say, 'Absolutely not, it will not work'.[1] So it still lingers out there, this idea. And who knows, with the next project, KSC is going to have to stand up against it yet again, in my opinion... I can take an example where it shows how it doesn't work to do 'Ship-n-Shoot' and that's the first launch of the Boeing crew Starliner. It had no End-to-End Test at the Cape. That's what did it! [led to the problems on the test launch]."[2]

Explaining "Ship-and-Shoot"

"*Ship*" means shipping to the launch site, in this case, KSC. It involves transportation of the payload from anywhere in the world to KSC, via ground, sea, air, rail, or any combination of these in multiple shipments.

Ground transportation was usually by an air-ride truck or tractor trailer, to protect the payload from any unwanted high loads or shock acceleration that may occur along the route. In some cases, specialized trailers would be built just for that payload. There are many road access points to KSC but some involve transportation over bridges with weight restrictions, which may require the payload to use one of the other access roads.

Sea transportation is only available through the access canal built for the Apollo program. The freight would have to enter the Banana River and head north to the Saturn Barge Channel, then southeast to the Turn Basin located near the Vehicle Assembly Building (VAB) and the press site at Launch Complex 39 (LC-39).

---

[1] KSC constantly pushed to do testing, which proved successful at catching numerous problems. Otherwise, if the hardware really had done "ship-n-shoot", there would have been many bad days on orbit.

[2] See the article at https://spaceflightnow.com/2020/04/06/after-problem-plagued-test-flight-boeing-will-refly-crew-capsule-without-astronauts/

From the Turn Basin, none of the roads leading to KSC facilities involve traveling over one of the bridges.

For air transportation, large cargo planes arrived at the three-mile-long Shuttle Landing Facility (SLF) located northwest of the VAB. There is a runway on the Cape Canaveral Air Force Station (CCAFS) called the Skid Strip that can accommodate payload arrival, but as it is only 10,000 feet (3,048 m) long, there is a limit to what can land there. As with those for sea transport, all roads leading to KSC from the air landing strips do not require traveling over one of the bridges.

According to Jim Dumoulin, "If you don't have the time to barge your payload to the launch site you are limited to flying in on a large transport aircraft. Once a payload (and its shipping container) gets wider or heavier than what can fly in on a Super Guppy [24 ft diameter, 54,000 lbs.], a C-17 [18 ft diameter, 170,000 lbs.] or a Russian Antonov AN-124 [17 ft diameter, 377,000 lbs.], you need to do launch site integration. If your payload is small enough and you can get it right every time, then 'Ship-and-Shoot' makes sense. The larger, heavier or more complex a payload becomes, the more economical it is to do final integration and testing at the launch site."

Rail access was via a single route from the Jay Jay Rail Yard to the north of KSC near Mims, Florida, then across the Indian River to the LC-39 area. This rail route was used almost exclusively for Solid Rocket Booster (SRB) segments, but was available for payloads if needed. The Jay Jay Rail Yard is the connecting link between KSC and the Florida East Coast (FEC) Railway, west of the Jay Jay Railroad Bridge and the Indian River.

Getting onsite at KSC means going through the security gates. The flight hardware almost always had a securing escort, for a number of reasons. First of all, most of the flight items were unique, one-of-kind pieces worth multi-millions of dollars and needed such protection. Another reason is that the containers used to transport them could be very large, which could cause traffic problems and similar issues. Some would have to be moved at slower speeds than the posted speed limit. Even though they could have arrived after having traveled for days over public roadways at much higher speeds, extra precautions were taken at KSC to ensure nothing would happen to the items once on site before they got to their final destination. The transportation of Ground Support Equipment (GSE) did not have such stringent requirements.

KSC consists of hundreds of buildings, so the hardware could go to any of those facilities based on a number of factors. Spacelab hardware almost always went into the Operations and Checkout (O&C) Building located on the southern part of KSC.

Having arrived at the O&C, an initial inspection would be performed to ensure no physical damage had occurred during transportation. This would mostly entail a NASA Quality person and a NASA engineer walking around the item or items to conduct a visual inspection. Examples of the kind of damage that was observed included holes in an item, which may have been caused by a forklift while loading or a piece of debris striking it during transport. Or a protective cover could have come loose and the item encountered rain on its journey, so it arrived wet. Co-Author Mike Haddad remembers, "One of the flight structures that I was going to use for mounting a set of experiments had just a plain blue tarp coving it. The tarp had come loose on one corner and the flapping of the tarp during transport had worn off all the protective paint on the structure, to the point [that] you could see the base metal. Very bad. [It was] Something we noted and had to fix before beginning work with the structure. This fact was relayed to the organization that shipped it, so as to provide better protection for the next similar item or other items that would be transported. This was also spread to all the other organizations that would be transporting hardware to KSC, to assure this would not happen again."

The O&C will be discussed in more detail later, but there were three main entrance points to get the large pieces of hardware from outside to inside. Smaller items could be brought into the facility via a number of the personnel access doors. Most hardware required offloading from its transportation device outside the building, but some transportation devices could be moved inside and then have the hardware offloaded. Once offloaded, it would be placed in a location that would allow it to be uncrated. Most of the time this required a crane operation, because the lid or structure of the transportation device was very heavy. Additionally, the crane was capable of very slow and precise movements, which could prevent any part of the container from damaging the flight hardware during uncrating or disassembly.

As the hardware was uncrated following shipment, it was inspected again for any physical damage. If any was found, photos were taken and it was then determined if any further actions should take place. Most of the time, no problems were discovered, so the hardware was placed in a protected location in preparation for the next phase of operations at KSC.

"*Shoot*" involves integration into the flight systems and/or flight vehicle, performing the necessary mechanical and electrical connections to the payload and verifying those interfaces.

For the hardware, this could be as simple as checking that it fitted. The electrical element merely involved making sure the electrical connections were mated properly and continuity existed across those interfaces. Software checks ensured that the vehicle could "see" the payload. None of this involved "End-to-End" testing of the payload against the flight vehicle to ensure all commands and operations could be performed successfully. After all, this had been done by the vendor at their location, so there was no need to do it again, right? WRONG.

Testing the hardware against simulators is fine, but it does not verify that the actual flight items to be integrated for the mission will function together in the same way and result in the same outcome. "Standing alone, components may function adequately, and failure modes may be anticipated. Yet when components are integrated into a total system and work in concert, unanticipated interactions can occur that can lead to catastrophic outcomes." [1] This is a fact many at KSC have known for a while, but convincing the rest of the world of this fact was a tough road to cross. Through all the programs previous to the Shuttle, final testing at the launch site against the launch vehicle and systems was proven to drive out unseen problems not discovered during all previous testing. A famous line the flight crews used all the time was "Test like you fly".

For the Spacelab program, it was decided early on that "Ship-and-Shoot" would not work due to the complexity of the Spacelab experiments, subsystems and systems that would all have to work together to ensure success. To do this testing properly, a series of "Levels" would be established.

"The Europeans wanted to do Spacelab integration at ERNO in Germany and built an entire control room of test equipment to do so," Jim Dumoulin told the authors. "However, a fully-loaded Spacelab module, protected in a shipping container, was too bulky and heavy to transport. It was also so complex, both internally and in its interfaces with the Shuttle, that if anything went wrong during testing, the round trip transportation time to send a payload back home for repair would wreak havoc on the launch schedule. 'Shipping-and-Shooting' Spacelab didn't make sense and the concept of Level-IV Integration was born. The control room ESA [the European Space Agency] built was shipped in its entirety and became the Level-III/II Automated Testing Equipment [ATE] control room."

The "Ship-and-Shoot" concept would rear its ugly head again during the planning for Space Station. The discussions and work that was undertaken to avoid that philosophy will be addressed later in the book, along with how testing performed at KSC not only saved small individual parts from failing on-orbit but also saved an entire Space Station mission.

CREATING A NEW PROCESSING CONCEPT

The terms "Level-IV", "Level-III/II" and "Level-I" refer to the different levels of operational activities that were required to process a Spacelab mission, a concept never performed before in the space industry. Each level of operational activities performed a unique function and provided a way of building on the previous level in preparing the Spacelab flight hardware and software for launch. It was similar to an assembly line in a factory, but very different in implementation. Not only was the function different for each level, so, too, was the location at which each level took place. This gave structure and definite transition points, so there was no confusion about what level and what function was being performed.

Each level had its own supporting GSE hardware, software and test equipment, and could involve different requirements documents, procedures format, and problem reporting format, while some involved completely different NASA and/or contractor organizations or a combination of both. Defining each level, and its role in processing organization responsibility and other operational processes for each of the flight elements, was critical to ensure a smooth processing flow while achieving all the goals required to bring the flight elements to a 100 percent operational status. The Spacelab missions could contain dozens of individual experiments and tens of thousands of parts for each mission.

There were questions that needed to be answered. For example: Would the same personnel follow the hardware they had worked with at one level on to the next level? Would the responsibility be transferred to the prime organization responsible for that next level? It was decided that the same organization and processes would follow the elements and be responsible throughout all the levels from wherever they began. So, as each level was being performed, multiple organizations now had to work together many times in parallel. This required a tremendous integration effort, so that all the requirements processes would be accomplished and nothing would "fall through the cracks" with the potential to cause damage to the hardware and harm to the flight crew. As will be shown later, not all payloads and experiments needed to pass through every level to achieve a flight-ready condition, due to the specific payload and experiment configuration for that particular mission.

Level-IV Location and Organization

A key issue to be resolved early was the location at which the assembly, integration, servicing and testing of the payloads containing the experiment hardware would take place, and who would be responsible for preparing them for a Shuttle mission.

It was decided to perform most of this work at KSC in Florida. As Scott Vangen explained, "Marshall was going to do Level-IV and then ship the integrated floor down to KSC and phase into Level-III/II, where the rest of the Spacelab hardware was being processed. Managers, particularly Bill Jewel and maybe a few other key individuals, advocated and won the argument, if you will. I'm sure Marshall wasn't happy at the time that that function [Level-IV] should best be at KSC. [But now] the Principal Investigators [PIs] and their team had a one-stop [shop]. They come, they bring their hardware to KSC, they see the flow all the way. It just made a lot of sense, and it's fun when things actually happen because they make sense. As a consolation, Marshall kept the sustaining engineering for the checkout unit and that [also] makes sense. They designed it, they were getting ready to use it, so they shipped it to the Cape."

Like all other organizations at KSC, a contractor needed to be selected and a NASA team formed to oversee the work of the contractor on the flight hardware – or so you would assume. There were other organizations at KSC that had NASA personnel performing the work of contractors, but they dealt with GSE. This time, though, the new group would perform processing of flight hardware with no contractors at all, except for the technicians. Engineers, Safety, Quality, Operations, Configuration Control and so on would only be performed by NASA personnel.

This reintroduced the unique concept of using NASA personnel to perform the "hands-on" work usually performed by contractors. "Hands-on" meant that NASA personnel would perform the engineering functions: turning the bolts, filling the fluid tanks, pushing the buttons and commanding the experiments. The concept would cut costs by having only one person performing the work instead of two (a contractor performing the work with NASA oversight). This would also attract young professionals and provide a training ground for future NASA engineers. NASA engineers would work side-by-side with contractor technicians, supported by NASA Quality and Safety personnel. Thus was created the "Level-IV" organization at KSC.

Level-IV group photo in the Operations and Checkout Building, Kennedy Space Center, December 1982. [NASA/KSC.]

"During Apollo and Shuttle, we had 'hands-on' capability at KSC. George Matthews ran an engineering lab and we took all the brightest electrical engineers and went there, and they built hardware and used to checkout Apollo and Shuttle. When we did the Launch Processing System [LPS], we built all the interface in-house between the large computers and small computers," explained NASA Payloads Director John Conway.

While the "hands-on" concept had been done before at KSC, it was all related to the GSE. What made the Level-IV "hands-on" effort different was that it dealt with flight hardware as well as the GSE.

Gary Power, NASA Level-IV Test Director, said, "It was just a dream, but the primary complaint of NASA engineers was that they come out of school and they're hired by NASA in an oversight role. Now they don't know what oversight is, but it's basically looking over the shoulders of a contractor like McDonnell-Douglas or Boeing at that time, those kinds of people. The new hires, what you would call fresh outs, were disappointed in their role of being an oversight. They were itching to do hands-on, and finally Level-IV came through for that purpose to get the NASA engineers. We did have help from Safety and Quality support, but primarily the hands-on role in Level-IV was with NASA engineers and their counterparts."

Level-IV Mechanical Engineer Tony Ornelas added, "They wanted to be able to put together a payload and have a training program for young NASA engineers, so they could get some hands-on experience to actually learn how to [do it], how procedures were written, why they were written a certain way and why you had to have certain document safety steps, the cautions and the whole nine yards. So first they wanted to do it at an economical basis. I guess they were trying to transfer trainers but also trying to save money, as opposed to giving the contractors big, big bucks to have them do the work, because they knew Engineering should be able to do some of the stuff anyway."

Herb Rice, NASA Level-IV Lead, Electrical Engineering, noted, "[For] those engineers, that would be the way, a good way, that the whole center could get good engineers and give them the experience necessary to be able to manage contractors. And also get *really* good ones instead of just average engineers."

It was certainly a risky idea to entrust unique, one-of-a-kind, multi-million-dollar flight hardware to young NASA personnel, many right out of college, and ask them to perform "hands-on" activities that had never been done before. But they rose to the challenge. Working with domestic and international universities, private companies, government organizations, and other space-related centers at locations all over the world, this was truly the first international effort in support of the new Space Shuttle program.

"Well we did [it]," said Herb Rice, "but there was a political thing that the whole Level-IV idea was started by Bill Jewel and it was accepted by Tom Walton, because Tom liked to do different things and that would make them be outstanding as a thing. And so he sold the whole idea to the Center Director as a way to attract capable engineers to the space center. That was about the time that the large interest in space within the United States was beginning to drop off, and instead of getting a thousand or so applicants for one job that was open, we were getting maybe 50 applicants for 35 jobs or something like that. So they felt like they were beginning to fall off."

In the West End

Level-IV was the starting point where most, or all, of the experiment hardware was assembled, integrated, serviced and tested, beginning with individual pieces. The physical location was at the west end of the O&C building at KSC. The O&C had a Low Bay that could house the main and large Spacelab components, and would be the prime location for all the Level-IV integration efforts. Also at the west end was an Apollo Telescope Mount Room (ATM). This was the location where the Skylab ATM was processed prior to launch in May 1973 and the room's name was never changed. There were also two rail systems (North Rails and South Rails), a Rack Integration Room and a Tunnel area under the main floor of the High Bay. Each of these locations will be discussed in greater detail in subsequent chapters. NASA personnel were responsible for almost all the work in this location. This included engineering, quality control, safety and logistics. The only contractor used for the Level-IV tasks was Boeing Spacelab Experiment Integration Support (SEIS).

Some of the SEIS team in the Level-IV area of Operations and Checkout Building, KSC, circa 1983. (Left to right): Gregory Haile, Ellis McCrorey III, David Anthony, Lori Vuyick, Merle Hammer, Andrew Petro, Kim Mackey, Jay Smith, Herbert Linhart, Walter Preston, Michael Grovier, James Nail, Dennis La Palme, Sharon Clanton, Danny Mills, Douglas Creech, Chris Madore, Anthony La Court, Bruce Burch, Edward Wagner, Richard Fritz, Michelle Falk, Ronald McGaha. [NASA/KSC.]

For some missions, the NASA Level-IV "hands-on" task required a different organization to perform work in the Level-IV location to share use of the specialized GSE, though that was rare.

Level-III was the stand where the basic Spacelab module shell and Igloo subsystems would be integrated and tested, with work performed by contractors. Level-II was the next stand, just east of Level-III, where the integration of the experiment systems with the basic Spacelab module shell, Igloo and subsystems would take place. Both were located in the center part of the O&C Low Bay, and these locations were the responsibility of the McDonnell-Douglas Aircraft Corporation (MDAC) as contractor-run operations with NASA oversight (NASA monitoring, with inputs as required). MDAC had a totally different set of requirements documents, procedure format and processing flows from those done by Level-IV. Level-III and Level-II were actually separate locations, but because both sites could perform the same role, they were known as Level-III/II.

Level-I meant operating the hardware with the real Shuttle Orbiter flight software at the Cargo Integration Test Equipment (CITE) area, at the east end of the O&C. Level-I also included all activities that occurred after the hardware left the O&C, for example at the Orbiter Processing Facility (OPF), the launch pads, or the landing sites. MDAC was responsible for CITE, while the OPF, launch pad and landing sites were the responsibility of the Shuttle processing contractor. Initially, that was Rockwell, but it was then transferred to United Space Alliance (USA) in 1995 through to the end of the Shuttle program. As far as Spacelab missions were concerned at Level-I, NASA Level-IV would perform experiment-specific tasks, MDAC would perform Spacelab systems and subsystems tasks, and Rockwell/USA would perform Shuttle Orbiter tasks. This meant that all three organizations would be working on different flight elements in the same location at the same time; an integration nightmare you would think, but one that worked wonderfully. Because the same personnel followed the experiment hardware through each level, and also because these same people worked other missions besides Spacelab, the term "Level-IV" was subsequently changed to "Experiment Integration" (EI).

EI at KSC started, quite literally in most cases, with the basic nuts and bolts of what would later become a large payload containing tens of thousands of parts, miles of wire and weighing over 10 US tons. As Mike Haddad recalled, "I know, because I was hired into Level-IV as a mechanical engineer in 1982, near the beginning of the Spacelab program, and worked there for seven years." The following chapters of this book will describe just how successful this risky concept was over the course of almost two decades.

Reference

1. Extract from the Columbia Accident Investigations Board Report (CAIB Report) in 2003, Volume 1, p. 187.

3

From the Ground Up

"Every facility, team and processing task
was setup as an assembly line
to reduce the number of shifts it took
to prepare a mission for space."
Jim Dumoulin
NASA Level-IV Electrical Engineer,
interview December 19, 2019

Most of the work performed by Level-IV/Experiment Integration (EI) personnel
was at the John F. Kennedy Space Center (KSC), but to get a complete and comprehensive understanding of the payload, some tasks would take EI personnel
beyond KSC to many locations both domestically and internationally. This chapter will describe the facilities (ground systems, test stands, control rooms, offline
labs and User Rooms), systems and subsystems that were at KSC, and mentions
some of the other locations around the world that were also visited as part of the
payload preflight, mission and postflight EI operations.

KENNEDY SPACE CENTER: AN OVERVIEW

Back in December 1959, KSC was originally called the Launch Operations
Directorate (LOD), and reported to NASA's Marshall Space Flight Center (MSFC)
in Huntsville, Alabama. In July 1962, the LOD was renamed the Launch Operations
Center (LOC), making it one of the NASA field centers. It received its current
name on November 29, 1963, a week after President John F. Kennedy was

© Springer Nature Switzerland AG 2022
M. E. Haddad, D. J. Shayler, *Spacelab Payloads*, Springer Praxis Books,
https://doi.org/10.1007/978-3-030-86775-1_3

assassinated in Dallas, Texas. Physically, KSC is located in Brevard County on the east coast of central Florida, adjacent to the Cape Canaveral Air Force Station (CCAFS). Its footprint measures 34 miles (55 km) long and roughly 6 miles (10 km) wide, covering 219 square miles (570 km²).

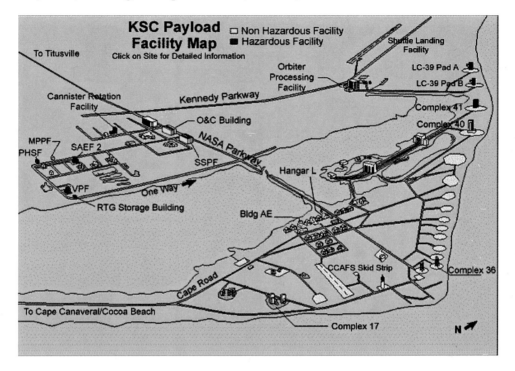

Graphic map of Kennedy Space Center (KSC). [NASA/KSC.]

As noted by Jim Dumoulin, "The design of the Space Shuttle and its large payload capacity was centered on the concept of routine access to space. Every facility, team and processing task was setup as an assembly line to reduce the number of shifts it took to prepare a mission for space. At KSC, entire buildings were built, such as the Canister Rotation Facility, just [to try to] shave a day or two off the turn-around time between landing and the next launch."

KSC is split into two different areas: the Launch Complex 39 (LC-39) area encompassing the northern part, and the Industrial Area to the south.

LC-39 contained the facilities (Vehicle Assembly Building (VAB), Launch Control Center (LCC), Launch Pads, Crawlerway, Barge Turn Basin) that prepared the Orbiter systems and the Orbiter itself for payload installation, launch, landing and payload removal postflight. The Industrial Area mostly contained the facilities that prepared the payloads for installation into the Orbiter, but also a few locations that worked on Orbiter components and Solid Rocket Booster (SRB) components. While these two areas contained many different facilities that supported the Orbiter, SRB, External Tank (ET) and payloads, only the facilities that interfaced with the payloads, and those the payloads traveled through during ground processing, will be addressed here.

Industrial Area, Southern KSC

The Industrial Area facilities included the Headquarters Building (HQ), the Operations and Checkout Building (O&C), the Vertical Processing Facility (VPF), the Space Station Processing Facility (SSPF), the Hypergol Maintenance Facility (HMF), the Spacecraft Assembly & Encapsulation Facility 2 (SAEF-2), the Canister Rotation Facility (CRF), the Payload Hazardous Servicing Facility (PHSF), the Launch Equipment Test Facility (LETF), the Parachute Refurbishment Facility (PRF) and the Merritt Island Spaceflight Tracking & Data Network Stations (MILA).

Operations and Checkout Building

A five-story structure containing 602,000 ft² (55,926 m²) of offices, laboratories, Astronaut Quarters, and payload High and Low Bay Areas, the O&C Building was located immediately east of the KSC HQ Building. The O&C was previously known as the Manned Spacecraft Operations Building, and was renamed the Neil Armstrong Operations and Checkout Building on July 21, 2014, during the celebrations of the 45th anniversary of the Apollo 11 landing. Dating back to the 1960s, the O&C was used to receive, process, and integrate payloads for the Gemini and Apollo programs, as well as for assembly and testing of the Apollo spacecraft during the Apollo and Skylab programs in the 1970s. It was modified for the Shuttle program and for initial segments of the International Space Station (ISS) through the 1990s.

One of the best known O&C operations to the outside world involved the Apollo and Space Shuttle astronauts' quarters and suit-up room that were on the third floor in the building. The crews were often shown eating and then going to the suit-up room. Once suited up, they would take an elevator down to the first floor, leave through a set of double doors and board the astronaut transfer van (Astro Van) located in the cul-de-sac between the north and south sections of the building, where the TV crews and well-wishers would be waiting to see them off. From there, the Astro Van would transport the crew to LC-39 and either Pad A or B.

The O&C Building's main function was horizontal payload processing for a variety of NASA and NASA customer payloads, and it could accommodate several payload elements being processed simultaneously. Payloads that were integrated and processed horizontally were received, assembled, and checked out in the O&C Building. They were then transported either to the Orbiter Processing Facility (OPF) for mating with the Orbiter, or to the VPF to be combined with vertically processed payloads.

The O&C Building was divided into three basic areas: Administration (north section); Laboratory, Control, and Monitor (middle sections); Bay Area, Service and Support (south section). For the Shuttle program, the O&C Bay Area was modified from the Apollo program to accommodate 'U'-shaped pallet-type payloads, including special structures, Spacelab module configurations, and certain other Space Shuttle payloads.

All three sections of the O&C Building, looking southwest. High Bay/Low Bay area to the left (large gray door, and high and medium roof); Laboratory, Control and Monitor in the center (two vertical gray pipes); and Administration on the right.

Actual hands-on mechanical and electrical experiment payload integration was performed primarily in the Low Bay and High Bay Areas. Laboratories and shops provided the offline payload support to the integration conducted in the Bay Area. Control and monitor functions provided support for the Bay integration. The Service area contained support systems for the Bay Area, such as shipping and receiving, as well as bonded storage areas.

The Bay Area in the O&C Building was 650 ft. (198.1 m) long and a uniform 85 ft. (25.9 m) wide, except in the High Bay at the east end where it was 38 ft. 5 in (11.7 m) wide. It was divided into a High Bay Area 175 ft. (53.3 m) long and 104 ft. (31.7 m) high, and a Low Bay Area 475 ft. (144.8 m) long and 70 ft. (21.3 m) high.

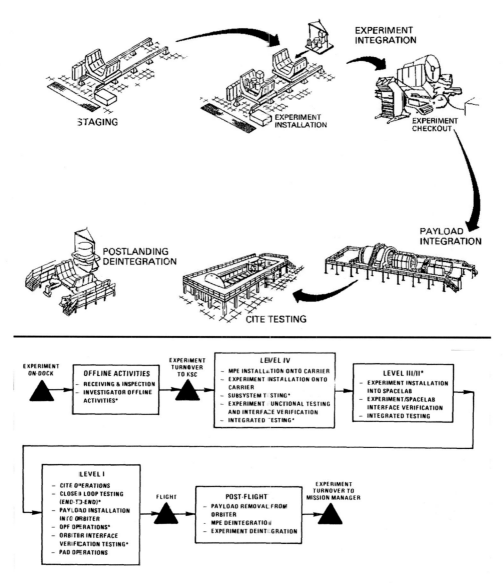

Top: O&C Bay Area payload processing. [Graphic from K-STSM-14.1.14-REVE-O&C, Revision E, November 1994 O&C Building Payload Processing and Support Capabilities.] Bottom: Horizontal Processing Flow showing tasks performed at each Level.

High Bay

The High Bay contained the storage and refurbishment area at the 26 ft. (7.9m) level for the horizontal sling kit and for the Multiuse Mission Support Equipment (MMSE) strongback, which was used for handling the horizontally integrated payload elements in the O&C Building and in the OPF. This required two cranes to use.

STS-90 Neurolab payload is moved into the Payload Canister using the strongback and two green overhead cranes; location is in the O&C High Bay. From here, it would be transported to the OPF for installation into the Orbiter. [NASA/KSC.]

A Partial Payload Lifting Fixture was also available for smaller payloads, and was designed to require the use of only one crane.

Partial Payload Lifting Sling moving the Spacelab tunnel section in the O&C Low Bay. [NASA/KSC.]

The High Bay also served as the parking area for the MMSE payload canister and transporter, during payload preparation and canister loading for transport. The High Bay door on the east end was a six-leaved vertical-lift metal door, 80 ft. (24.4 m) high and 40 ft. (12.2 m) wide, with pneumatically operated door seals. The Module Vertical Access Kit (MVAK) was also located in the High Bay inside the East Chamber, one of the two Apollo altitude chambers. The West Chamber was used for storage.

Low Bay

The Low Bay was the main area for horizontally processed payloads. The major facility elements in the Low Bay were: two Experiment Integration (Level-IV) work stands; two Integrated Assembly and Checkout (Level-III/II) work stands; the Cargo Integration Test Equipment (CITE) stand; the Rack, Floor, and Pallet stand; the End Cone stand area; the Tunnel Maintenance Area; and the Staging and Assembly Area.

Spacelab Payload Processing

Operations & Checkout Building
Kennedy Space Center, Florida

1. North Level IV Stands
2. South Level IV Stands
3. Simulate Aft Flight Deck
4. Spacelab Rack Train
5. MPESS
6. Spacelab Pallet
7. Spacelab Igloo
8. Rack Room
9. Level III/II Stands
10. CITE Stand
11. Three User Rooms & HITS
12. Control Rooms
13. Spacelab Module

O&C Low Bay overview looking east. [NASA/KSC.]

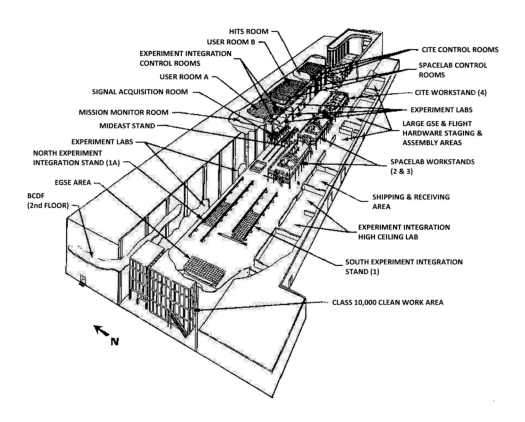

Graphic showing the layout of the O&C Building.

Work stands

Level-IV stands, one to the north and one to the south, served as support for pal-lets, pallet segments, special structures or flight carriers, and racks and floors, during experiment integration and in the early stages of assembly and testing.

The Level-III/II and CITE stands provided movable left- and right-hand longe-ron fittings and load monitoring capability, and they could accommodate the European Space Agency-supplied (ESA) Ground Support Equipment (GSE). The stands served as the primary structural interface between the trunnions and the simulated Orbiter bridge rail.

Level-IV (Experiment Integration) Stands

The Level-IV stands were used in the assembly, disassembly, and functional check-out of experiments on pallets or special structures and within racks. These stands were used with the integration trolleys that supported the payload pallets. Several single pallets for one mission could be integrated in one stand. Operations in these stands also included payload component mating, demating, staging, and refurbish-ing. Access to the payload was via the Bay Area floor using portable GSE.

Co-author Mike Haddad recalled that, "Level-IV had two massive sets of 60 ft. [18.3 m] long stainless steel rails [which Level-IV personnel referred to as the North and South Rails] that provided stable mounting for movable support fix-tures that matched exactly the attachment points in the Space Shuttle Payload Bay. At the west end of the South Rails there was a 25 ft. [7.62 m] tall structure over-looking the rails, called the Aft Flight Deck [AFD] simulator. The rails allowed nearly 360 degree access to payload fittings, and a raised floor area at the end of the rails housed racks of remote control switching gear and 500-Amp power sup-plies that simulated the 28V fuel cells on the Space Shuttle. For Spacelab-1, Level-IV only had one control room and one full set of rails [South]. There was a partial set of rails [North] that could be used for offline alignment and payload buildup. After Spacelab-1, a second complete set of Spacelab Hardware arrived [called the Follow On Procurement or FOP] and more room was needed to process Spacelab. New equipment was also added to the Integration Area in the O&C High Bay and the North Rails were extended so that Level-IV could support inte-grating two Spacelab missions at the same time."

The two South Rails were 79 ft. (24.3 m) long with a raised floor that enabled routing of utilities to the payload during experiment integration, and could support flight hardware. The two North Rails were 91 ft. 10 in (28.0 m) long and the same width as the South Rails, with the same raised floor and utilities.

The mechanical equipment, systems and services available at the rails included individual white- and yellow-colored access platforms (which were available for use during experiment integration), the subsystem floor and rack simulator, the trunnion support, fluids and gases, and a portable hoist that was shared among the

Bay Area rails. The AFD simulator racks in the Experiment Integration Area were used for the functional checkout of experiments on pallets and special structures, and in racks, or what was known as Partial Payloads. One AFD simulator rack was used for the South Rails, the other for the North Rails. The two AFD simulator racks were located in the dedicated Electrical GSE (EGSE) area. The blue-colored subsystem floor and tan-colored rack simulator were used in conjunction with the gray-colored flight Spacelab experiment rack train for all module configurations. It simulated the flight workbench rack, control center rack, and flooring — all of which stayed in the flight core segment of the Spacelab module located in the Level-III/II stands. The simulator was positioned on the rails in the proper Orbiter Xo, Yo, and Zo location, in reference to the flight racks and floors. The blue-colored trunnion support structure was used in the Experiment Integration stands to support one Spacelab (MPESS) or pallet-type structure. Handling fixtures and Special fixtures, or frames, were available for handling sets of racks — six double racks and four single racks. The fixtures were used to install racks on the floor. The North and South Rails had several fluids and gases available for the payload user.

Gray racks (single on the left and in the center of image and double rack on the right) with gold handrails in the blue handling fixture, located in the Rack Room south of the Level-IV area. [NASA/KSC.]

The Mechanical GSE (MGSE) Area was on the west side of the north stand, beside the class 50,000 Clean Work Area. The water and Freon® servicers were also located there and fed the services to the North and South Rails. As co-author Mike Haddad recalls, "These units were where I cut my teeth when I started Level-IV. Roberto Tous was in charge of them at the time and then brought me in to take over day-to-day operations, but Roberto was still the expert. These servicers contained large reservoirs that we would have to fill and then circulate through their filters, and do sampling to make sure it met specifications before being supplied to the flight hardware. At that time, JSC SE-S-0073 Space Shuttle Fluid Procurement And Use Control was used for all gases and fluids. These servicers provided the cooling support during testing of the flight hardware. Freon® was used for external loops on the pallets and MPESS; water was used for internal loops on the Spacelab modules."

At the far west end of the Mechanical Area there was a small west door leading to the outside and small cargo elevator that went up and down between the Tunnel and the Low Bay floor. A large cargo elevator on the east side of the Mechanical Area was the main elevator used to move heavy and large items between the Tunnel and the Low Bay.

Portable GN$_2$ (gaseous nitrogen) and GHe (gaseous helium) K-bottles could also be used at any locations required. As Mike Haddad explained, "We would order the K-bottles, stating the proper specification, and after we received them we would have to install our own pressure regulators to reduce the pressure from around 2200 psi (151 bars) for standard pressure bottles, or 3600 psi (248 bars) for high-pressure bottles, down to whatever pressure was required. Any additional hosing was placed at the end of the regulator, and we would then sample and test the gases coming out the end of the hoses that would eventually connect to the flight hardware. We wanted to make sure all the GSE was clean and the gas met the proper specification coming out the very last GSE item leading into the flight hardware."

For the electrical power requirements, the GSE at the rails provided 120/208 volt AC power up to 60 amps, 60 hertz and 400 hertz, and 28 volt DC power up to 125 amps was available for payload use.

The EGSE area was located west of the North and South Rails, next to the Apollo Telescope Mount (ATM) clean room, and provided approximately 431 ft^2 (40.1 m^2) of raised floor in support of dedicated experiment integration activities.

The EGSE area contained AC and DC power (including AC power supply and Hz power converters), distributors, timing signal conditioning, control panels, TV, Payload Checkout Unit (PCU), and associated GSE.

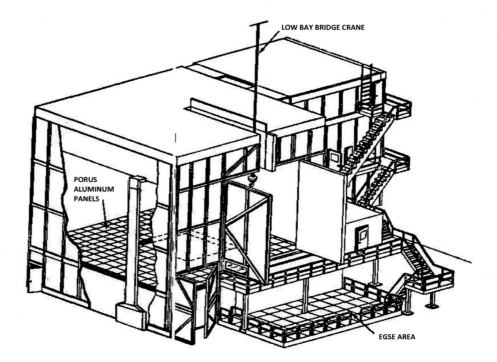

LOW BAY BRIDGE CRANE

PORUS
ALUMINUM
PANELS

EGSE AREA

ATM and EGSE area. [Graphic from K-STSM-14.1.14-REVE-O&C, Revision E, November 1994 O&C Building Payload Processing and Support Capabilities.]

Communications and data handling capabilities at the Experiment Integration stands consisted of the PCUs and the Partial Payload Checkout Unit (PPCU).

The PCUs were used to provide a simulation of the Spacelab Command and Data Management System (CDMS) to the experiments, and to provide monitor and control functions for the EGSE, ECE, or flight hardware. The PCUs could be used independently in support of dual flow processing of Spacelabs in the North and South Rails.

The PPCU performed checkout of partial payloads and their carriers located within the Level-IV test stand area. It was a modular, distributed-processing system consisting mainly of commercially available hardware and software. Custom hardware was required to support the unique interfaces to the payload/carrier, and custom software to support specialized processing requirements. The equipment was located in three main areas:

a) The Payload/Carrier and some facility interfaces were contained within the Payload Interface Rack, located at the Level-IV test stand.

b) The majority of the system was located in rooms 1289H/1291/1293 of the O&C. Commercial equipment was housed in office environment-type packaging, while custom equipment was housed in several standard equipment racks.
c) Workstations were located in the Payload User Rooms (A, B, and C) on the fourth floor of the O&C building.

Several interface panels and the system cabling were located between these areas in the O&C building. Various data cables also terminated in the High Rate Data Equipment (HRDE) room (room 1263) for routing to user-selected locations.

Level-III/II Stands

These two stands were designed to accommodate the buildup of the Spacelab elements into a payload. Because they contained much of the same equipment and services, they are presented together. Each stand was 82 ft. (24.9 m) long and 33 ft. (10.1 m) wide. Access to the periphery of the payload was provided by the white walkway around the stand on the north, west, and south. The basic work stand deck for these stands was identical to the CITE stand.

On the mechanical side, these work stands provided hoisting equipment, access platforms, trunnion support assemblies, fluids, and gases.

For the electrical side of things, the Level-III/II work stands provided AC power for payload use. DC power was provided by GSE. The AC power receptacles were both rail and pedestal mounted. DC power was provided by ESA's Spacelab Payload Command and Data System (SPCDS) for Spacelab payloads, and the equipment provided for experiment integration. The Raised Platform was dedicated for use with Spacelab system GSE. Grounding for equipment and instrumentation was provided via numerous ground plates throughout the area.

For communications and data handling, the Level-III/II work stands contained portions of the Automatic Test Equipment (ATE), which was used for Spacelab detailed verification and diagnostic testing, real-time telemetry processing, data display, test sequencing, control and command generation, timing, rapid fault-finding, and signal routing. It included the required operational and diagnostic software necessary to fulfil operational and self-testing requirements. The ground power unit and the Orbiter interface adapter provided power to Spacelab during the test phases before Orbiter-Spacelab mating, and simulated the electrical functions of the Orbiter-to-Spacelab interface, respectively. The remaining portions of the ATE were located in a control room remote from the work stands.

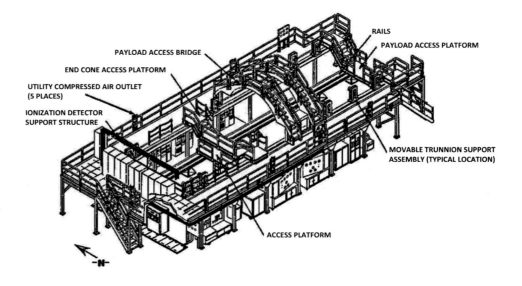

RAILS

PAYLOAD ACCESS PLATFORM

PAYLOAD ACCESS BRIDGE

END CONE ACCESS PLATFORM

UTILITY COMPRESSED AIR OUTLET
(5 PLACES)

IONIZATION DETECTOR
SUPPORT STRUCTURE

MOVABLE TRUNNION SUPPORT
ASSEMBLY (TYPICAL LOCATION)

ACCESS PLATFORM

-N-

Level-III/II work stands. Payload access bridge over the top of the Spacelab module; end cone access platform used to gain entry/exit from the module. [NASA/KSC.]

CITE Stand

The CITE stand simulated the Orbiter electrical accommodations for all payload interfaces. The CITE was used to verify compatibility between payloads and the simulated Orbiter electrical and electronic interfaces.

When required by the payload, closed-loop Payload Operations Control Center (POCC) interface tests were conducted with the payload in this stand. CITE testing would include an Interface Verification Test, a Mission Sequence Test (MST) if appropriate, required payload element tests, and operations to validate planned online pre- and postflight operations, including emergency, contingency, and scrub and turnaround procedures. A time-compressed MST would be included when it was judged to be an effective demonstration of multiple system compatibility, either payload-to-payload or payload-to-Orbiter. During CITE testing, those participating would include payload and carrier owners and flight crewmembers. A block of serial time was provided for payload-specific operations.

The mechanical equipment located on this stand included an AFD structure, a forward Orbiter bulkhead assembly, an aft bulkhead, Orbiter simulated cable trays, fluid lines, and a midbody assembly. Openings were provided to duplicate

Orbiter access hatches. The equipment included the AFD Simulator, which provided a working platform and support for a Payload Specialist Station (PSS), an On-Orbit Station (OOS), and a Mission Specialist Station (MSS). Space was available for mission-specific consoles and electrical interface panels, and cool air was supplied to the AFD modules. Rockwell International (RI) receptacles, or equivalent payload-provided receptacles, were used for installing the payload panels. There was space for two payload-dedicated modules. In the PSS, six panels were payload-dedicated, while three panels were available for the payload in the MSS. Other items available were the heat exchanger, and fluids and gases. The only fluid provided to the CITE stand was facility chilled water through the water servicer or with the heat exchanger. GN$_2$ and GHe gases were provided by pneumatics panels and compressed air stations. Compressed air was available at the CITE stand in five places.

The CITE stand electrical supply provided AC and DC power for payload use, while 28V DC power was provided by GSE. There was a rack for DC power distribution and AFD electrical distribution panels. The electrical distribution panels were available for payload use at the MSS, OOS, and PSS. Instrumentation and equipment grounding was provided at the CITE stand, as was lighting.

For communications and data handling, the payload-to-Orbiter interfaces verified in CITE were: scientific data, launch (same as the firing room) data bus, Pulse Code Modulator Master Unit (PCMMU) data bus, payload data bus, caution and warning, timing signals, and AFD AC and DC power.

The CITE functional equipment located on the floor under the CITE stand on the west end consisted of wideband terminal distribution, video and data assembly, Hardware Interface Modules (HIM) and patch panel, interface terminal distributor, DC power suppliers and distribution, AC power distribution, RF test set, GSE interface distribution, and the CITE test set. Additional CITE equipment was located in the CITE control room.

Rack, Floor, and Pallet Stand

The Spacelab Rack, Floor, and Pallet Stand (also called the Mideast Stand) was located along the north Low Bay wall. The stand could accommodate various configurations of single and double electrical rack buildups that housed the payload experiments. It could also accommodate pallets on the integration trolley.

End Cone Stands

An area measuring 10 ft. by 16 ft. (3.1 m by 4.9 m) was available north of the CITE stand for storage of the Spacelab module end cones.

Other Areas

A Tunnel under the High Bay and Low Bay floor was used mostly for storage of small flight hardware items, and a logistics area for kitting (organizing) flight hardware. It contained a secure area for small GSE instrumentation. Access to the Tunnel was via several sets of stairs throughout the Low Bay and High Bay. Mike Haddad recalls: "In the early days, all the flight hardware piece parts were located in the cage area of the Tunnel, where we would kit what we required as stated in the procedures we wrote. Then we would just come and go to get the hardware as needed. As the number of payloads we worked increased and we got busier and busier, it was not possible for us to keep up organizing the hardware into kits, so a logistics person was hired. We would hand that person a copy of our procedure, listing the hardware required, and they would kit it for us."

A women's restroom was added to the Low Bay because there was only a men's restroom that could be accessed from there. Before this was added, women would have to leave the Low Bay to use the restroom in the main hall, a leftover from the Apollo days.

The Horizontal Sling Kit for Spacelab was stored on top of the altitude chambers on the north side of the High Bay, while an area north of the CITE stand was for work on the Spacelab Igloo. Portable servicing equipment could be used in this area. In the High Bay east of the CITE stand was an area used for maintenance of the Spacelab Transfer Tunnel, while on the south side of the Bay Area, east of the Shipping, Receiving and Inspection (room 1469), there were three processing areas. These were used for staging and mechanical assembly of payloads but were not served by the Bay Area bridge cranes, so large flight hardware and GSE was moved by means of the air-bearing pallet system, and smaller items via wheeled carts and an electric forklift.

Apollo Telescope Mount Room Class 10,000 Clean Work Area

The Apollo Telescope Mount (ATM) Room was originally built for the Apollo Telescope that flew on Skylab. It was left over to use during the Shuttle program and never renamed. The Clean Work Area was located at the west end of the Bay Area and was used for offline experiment integration. The Clean Work Area was a two-level structure, 36 ft. 8 in by 36 ft. 10 in (11.1 m by 11.2 m) and 38 ft. (11.6 m) high, that could be maintained in accordance with Class 10,000 Clean Work Area conditions. The second-level floor was made up of a grid of porous 2 ft. by 2 ft. (0.6 m by 0.6 m) aluminum panels at an elevation of 10 ft. 6 in (3.2 m). Removing selected panels provided access between the first level ante-room and the second level. On the first level, there was an ante-room which served as an airlock entry area for equipment being brought into the Clean Work Area. There was a seismic pad on the first level, located approximately beneath the opening. This Clean Work Area had sound-absorbing wall panels.

A pneumatic door could be opened to allow movement of hardware. Movable panels in the front top wall and the roof opened to provide access for the Low Bay 27.5-ton (24.9 metric ton) bridge crane. A 5-ton (4.5-metric ton) bridge crane traversed the Clean Work Area. A personnel air shower, 3 ft. 1.5 in (7.8 cm) wide, located inside the access room, was provided for entrance into the Clean Work Area.

The air-conditioning system provided area temperature and Relative Humidity (RH) as a standard service in the Clean Work Area. If requested, the clean area air-conditioning could be activated as an optional service. This air-conditioning consisted of High Efficiency Particle Accumulator (HEPA) filter diffusers, flexible ducts, and a dehumidifier. Temperature could be maintained at a nominal 72 degrees F (22.2 degrees C) and RH at 50 percent, if the air-conditioning system was activated. Heat-actuated devices detected fire and activated the alarm system for the Class 10,000 Clean Work Area.

Personnel Access

Specific operational facilities and areas had strictly controlled access. Each area had an area permit unique number and obtaining a number to gain access to these areas required specific training and certifications. Access to the payload processing and support areas of the O&C Building was controlled and monitored, as Mike Haddad recalls: "When I started in EI in 1982, the access was controlled with security guards only. They would sit there all day and just check that everyone had the proper numbers. I felt for them; that had to be one of the most boring jobs but it had to be done. Even though we came and went dozens of times a day, they would check to make sure we still had the proper numbers." A short time later, the guards were replaced by either access control monitors or by electronic devices, and then eventually just electronic devices. A KSC badge and area permit with numbers 40 and 73 was required for access. For the electronic access, a Personnel Access Control Accountability System (PACAS) magnetic card was required, in addition to the KSC badge and area permit, to gain access on all shifts.

Personnel access to the Bay Area was obtained on the north side from the first floor corridor, monitor posts A1 and C7, and from the south side at monitor post D14. The PACAS was activated and monitored at all times for access control during operational hours only. Co-author Mike Haddad noted: "Most thought operational hours were 8am – 4:30pm, but it became apparent real soon that operational hours for EI usually meant at least two shifts and sometimes 24/7." Third floor access was limited because the Astronaut Quarters were near the west end of the building. However, labs and control rooms were accessible from the middle and east end of the third floor corridor. Several stairwells connected these areas and the other floors.

Personnel that were from offsite or needed temporary access to the O&C were issued "To Be Escorted" Temporary Area Authorization (TAA) badges, usually pink. Such personnel had to be escorted by a properly badged person at all times while in the O&C High Bay/Low Bay. The maximum number of persons that could be escorted by one person was three. As Mike Haddad remembered, "We could not always leave the Low Bay to escort a person, so we would have an office mate or someone else escort that person into the Low Bay and then we would take responsibility from there, then reverse the process if that person had to leave the Low Bay but we could not. The TAA personnel became a real pain logistically – not really for us but more for them waiting to be escorted – so we encouraged any person that was to be at KSC for a period of time to take the training necessary for un-escorted access, or they would have to stay with us for the long hours while we worked in the Low Bay. We could be working on several different payloads at one time, so they would have to stand by and wait for us to get [around] to their payload work, which was not fair to them."

Equipment Access

Large items of GSE and flight equipment entered the O&C Building through the east vertical lift door of the High Bay Area, or the Shipping, Receiving and Inspection Area (room 1469) in the Low Bay. Mike Haddad recalls: "Very early in the Level-IV days, equipment access through the personnel access door caused a real problem. The door opening was the width of two doors, but what was not shown on the drawing was a large metal post that was a middle support for both doors. So Experiment Developer (ED) personnel would see the drawing and think that [it] had a full width of two doors and planned the access through there instead of the larger, more complex east door or 1469. Of course when they arrived, the equipment would not fit through the opening. We learned real quick to inform the EDs of this error, until the doorway was modified and the support removed. Some would ask, why not just update the drawing? We felt having the larger opening would definitely be better operationally to get large items into the Low Bay. Also, to avoid all the paperwork [caused by] modifications to a facility drawing, we bought it off as 'bringing the hardware up to print'."

Emergency Egress

On the south side, first floor, there were seven emergency egress routes to the outside of the building. According to Mike Haddad, "Drills would be performed but were usually planned ahead of time. This was done to assure no hazardous or critical operation was taking place, the idea being that the drill did not endanger or increase the possible damage to flight hardware. For example, if a lifting operation was taking place, for the drill it was not desirable to lock the crane out and leave the flight hardware hanging from the hook. Every procedure we wrote and ran had an Appendix Z: Emergency Instructions. In there, it would state what to do in case of a fire and such. Using the example above, if we had only just lifted the flight hardware from its support GSE, we would immediately reverse the operation and set the flight hardware back into its support GSE, lower the crane a slight amount to release the load, then lock the crane out and evacuate. If we were close to installing the item, we would gently set it down into its flight position and perform any quick temporary securing [if possible], lower the crane a slight amount to release the load, then lock the crane out and evacuate. I don't remember ever having to evacuate the Low Bay for an emergency, but I do remember a time when one of the control rooms had to be evacuated. We were in the middle of a test and I was down in the Low Bay monitoring the channels on the Operational Intercommunication System [OIS]. All of a sudden we heard a voice start choking severely and then it was cut off. We had performed so much testing [that] we knew each other s voices over the OIS, so right away we knew who was having difficulties. Very quickly, we learned one of the control rooms had been evacuated and the person choking was ok. [It turns out] old style temperature and humidity paper charts located in the control room started to leak ink and the smell had a choking effect on people. I think this resulted in using different ink or going to electronic means of recording the information."

Floor Loading Profile

"There was a tunnel under the floors of the Low Bay and High Bay," Mike Haddad recalls, "but only in certain areas, so the floor loading requirements [for these areas] were put into place so nothing too heavy would fall through the floor. Black-and-yellow striped floor loading markings were painted on the floor relating to the underlying tunnel, so personnel would know not to overload the floor in those locations." Trucks and heavy equipment entering the O&C High Bay through the east vertical door could not exceed the stated wheel load markings painted on the floor. The maximum uniformly distributed load on the assembly and test floor could not exceed 250 pounds per square foot (1,220 kgs/m^2).

Restrictions

The O&C Building Bay, Laboratory, and Control Areas had controlled environments, and no manufacturing operations were permitted in these areas. Simple manufacturing tasks could be performed in the shops and Service and Support areas. Hazardous operations were kept to a minimum.

Operating Regulations

Access to the stands and controlled zones in the flight hardware processing areas were controlled via barriers, and sometimes had a person acting as a monitor to control access in and out of the area. There were general work area rules for personnel working on the stands that contained flight hardware. Entrance areas were provided with tacky mats which had to be used prior to ingress (walking over the tacky mats would remove any dirt from the bottom of the shoes). The smock exchange station was located at the High Bay east end Instrumentation Library, which made it easy to exchange clothing and not have to leave the Bay Areas. Tobacco products, food, beverages, chewing gum, and flame-producing devices were prohibited. All personnel working in the controlled area were required to log in with the access monitor. Control areas were established to keep strict control of who and what went in and out of the area around (and on) the flight hardware. Completion of the sign in/sign out log, listing the particular Work Authorization Document (WAD), tools, equipment, solvents, and chemicals to be utilized, was also required.

All tools had to be tethered when working over flight hardware. Badges had to be placed on the badge board in the proper slot, indicating where personnel would be working (i.e., work stand, pallet, module, upper level or ground level). Items containing mercury or glass could not be taken near flight hardware. The maximum number allowed in the Spacelab module at one time was ten people, while pallet loading was a maximum of six persons with all panels installed. Use of flammable liquids and hazardous substances required coordination with Operations. Approved safety harnesses and lanyards had to be worn whenever personnel were required to work close to an unprotected edge of an elevated platform, stand, or other structure where there was a danger of falling. Personnel were required to don clean room garments, coveralls or smock, cap, foot coverings, and gloves, and to tether "eyeglasses" and remove or tape rings and watches. All protective clothing had to be removed before leaving the controlled area. The hardware being accessed determined what clean room garments had to be worn.

Paging and Area Warning

All areas of the O&C Building were part of the KSC administrative Paging and Area Warning (P&AW) System. Speakers were located throughout the building: in the office or Administrative Area; the Laboratory, Control, and Monitor Area; the Bay Area; and the Service and Support Area. The P&AW System was used to inform personnel of emergency conditions such as adverse weather and fire alarms, as well as for public announcements.

Environmental Control

The High and Low Bay Areas provided a visually Clean Working Area (CWA). Air quality was rated as Class 100,000 clean. This meant that there could not be more than 100,000 particles greater than size 0.5 microns in one cubic foot of air. The ATM Room was rated as a Level 3 CWA, although it could be maintained as a Level 2 CWA if stringent controls were adhered to. The test stands and specific work areas were Level 4 CWAs. Both the assembly and test area, and the offline labs, were Level 4/5 CWAs (see Table 3.1).

Table 3.1: CLEANLINESS REQUIREMENTS

Clean Work Area Levels		Level #2	Level #3	Level #4	Level #5
			Non-		Non-
Parameter	Airflow Type	Laminar	Laminar	Non-Laminar	Laminar
Maximum Airborne Particulate Counts Per M³ (Per ft³)	Req. 0.5 ≥ 0.5 μm	10,000	1415.9 [50,000]	2831.7 [100,000]	8495.1 [300,000]
	Req. 5.0 ≥ 5.0 μm	65	8.5 [300]	19.8 [700]	28.3 [1,000]
	Monitoring	Continuous	Continuous	Continuous	Monthly
Temperature (degrees C) (degrees F)	Requirement	21.7 ± 3.3 71 ± 6	21.7 ± 3.3 71 ± 6	21.7 ± 3.3 71 ± 6	21.7 ± 3.3 71 ± 6
	Monitoring	Continuous	Continuous	Continuous	Monthly
Relative Humidity (RH) (%)	Requirement	50 Max	50 Max	50 Max	50 Max
	Monitoring	Continuous	Continuous	Continuous	Monthly
Maximum Particle Fallout	Goal*	Level 200	Level 500	Level 750	Level 1000
	Monitoring	Continuous	Continuous	Continuous	Every 6 Mo.
Maximum NVR 0.1m²/mg month	Requirement	1.0	1.0	1.0	2.0
	Monitoring	Continuous	Continuous	Continuous	Annually
Maximum Volatile Hydrocarbons (ppm) (v/v)	Requirement	15	15	15	N/A
	Monitoring	Every 2 weeks	Every 2 weeks	Every 2 weeks	N/A
Minimum Positive Pressure	Requirement	0.05 in H₂O	0.05 in H₂O	0.02 in H₂O	N/A
	Monitoring	Daily	Daily	Daily	N/A
Minimum Air Changes	Requirement	20/Hour	6/Hour	4/Hour	2/Hour

*Levels per KCI-HB-5340.1 continuous monitoring

In summary, the floor level of the High/Low Bay was CWA 5, so personnel were allowed to work in normal street clothes. For the work stands, personnel were required to wear smocks (which looked like lab coats). Inside a Spacelab module or on a Spacelab structure, a full "bunny suit" covered the person from head to toe. Wearing gloves was mandatory whenever flight hardware was touched. All these restrictions were put in place to keep the flight hardware very clean. As Mike Haddad observed, "We took this very seriously. Oil from your hands could destroy sensitive optics, or a hair getting into an electrical connector could cause havoc. At first, only the astronauts could go without the protective clothing, but after a few processing flows it was determined they, too, needed to 'suit up' around the flight hardware. I remember one time the NASA Administrator came into the Level-IV area and wanted to get a close-up look at some of the flight hardware. He, too, had to put on a smock to climb the access stands located next to the flight hardware."

Illumination

The High and Low Bays were lit by ceiling-mounted fixtures. Each fixture had a 1000W incandescent lamp and a 1000W metal halide lamp with an integral constant wattage mercury-vapor transformer. Mike Haddad commented on working under these lights: "While the lights were very good at illumination, they were hard on the eyes after a while. Many times, I would spend all day in the Low Bay, so I would have to close my eyes for minutes at a time just to let them rest."

GSE and Services

Mechanical and electrical GSE and services required to support the payload assembly and testing were located in and around the work stands. The GSE and services available included AC and DC power, compressed air, fluids, a gas vent system, central vacuum cleaning, and handling and access equipment.

Grounding

All structures, equipment, and instrumentation in the area were grounded. Copper ground plates were at various points in the processing rooms, and on or near the work stands. Each plate had four connector lugs. The equipment ground plates were mounted to structures in the area, while the instrumentation ground plates were insulated from the building steel by insulating bushings or washers. Payload-unique GSE and instrumentation had to be grounded when used in the area.

Fire Protection and Safety

Manual pull stations were located at all marked exits. The area was equipped with photoelectric (smoke) detectors in the supply and air ducts. When activated, these detectors shutdown the air handling units and sent a silent alarm to KSC's central monitoring station for a fire department response.

Crane and Hoist System

Three overhead electrical bridge cranes serviced the Bay Area (see Table 3.2). One crane serviced only the High Bay Area, while the other two cranes serviced both the High and the Low Bays.

Table 3.2: BRIDGE CRANE DATA

Crane Coverage (number)	Capacity m ton (ton)	Hook Height m (ft-in)	Trolley Travel m (ft)	Bridge Travel m (ft)	Hook Speed Full Load m/min (ft/min)
High Bay (1)	24.9 (27.5)	24.1 (82-3.5)	19.5 (64)	45.1 (148)	0.6 (0 to 2)
Low Bay* (2)	24.9 (27.5)	14.6 (47-9.5)	19.5 (64)	190.5 (625)	0.6 (0 to 2)
JIB Crane	0.9 (1.0)	6.7 (22)	N/A	N/A	N/A

*Also serves High Bay

"Everyone used to ask, why a 27.5-ton limit on the overhead cranes? Why not a nice even number like 27-ton or 28-ton?" recalled Mike Haddad. "The story we heard was [that] between the Apollo and Shuttle program, when the lids of the altitude chambers had to be removed using what at that time were 25-ton overhead cranes, they noticed the weight of the lids had exceed the 25-ton max load of the cranes. They determined the lids weighed 27.5 tons, so the facility people bumped up the max load of the O&C overhead cranes to 27.5 tons."

The minimum distance between the east and west Low Bay hooks was 22 ft. (6.7 m) and that between the Low Bay hooks and the High Bay hook varied from 12 ft. 9 in (3.9 m) to 13 ft. 10½ in (4.2 m). "This minimum distance was critical," continued Mike Haddad, "because some of the payload developers would be designing slings for the payload [that] required two hooks. But if the spacing was less than the 22 feet, it would not work in the O&C. If they had a design like this, we would tell them to try and design the sling for one hook operation or, if two hooks were still required, space out the lifting sling attachment point to greater than 22 feet."

Air-Bearing Pallet System

An air-bearing pallet system was available for internal transport within the Bay Area. The system consisted of a cart that could be connected to facility air, and 16 small pallets, each of which could support 4000 lb. (1,814.4 kg). The system could move the Pallet Segment Support Integration Trolley (PSSIT), the long and short rack and floor trolley, and the pallet transport cage with dedicated flight hardware on board, into and out of the service or processing areas on the south side of the High and Low Bays, where ceiling heights were too low to use the bridge crane for hoisting. In addition, there were four large pallets, each of which could support 10,000 lb. (4,536 kg).

Vent Systems

The "clean" Experiment Vent System for experiments was in place at each work stand (Level-IV, Level-III/II, and CITE stands) in the Bay Area. The Vacuum Vent System provided vacuum venting of vacuum pump exhausts, de-servicing of Freon® 114 cooling systems, and other GSE. Each of the work stands had a pair of 2-in (5.08-cm) galvanized ports.

Vacuum Cleaning

Vacuum Cleaning outlets were located at each work stand in the Bay Area, the class 10,000 ATM, the north wall, and for the shops and processing rooms (A, B, and C).

Communications and Data Handling.

The Operational Intercommunication System (OIS) was a multichannel voice communication network that interconnected operational areas required for payload processing at KSC and the Cape Canaveral Air Station (CCAS). The digital OIS in the O&C Building tied the Bay Area, the High Rate Multiplexer (HRM) Test Station and Closed-Circuit Television (CCTV) Room (1263), the Payload Checkout Unit (PCU) control rooms, the Spacelab and CITE control rooms, the Customer Management Center (CMC, room 4269), and the User Rooms.

Each OIS unit could access two selectable channels. If the OIS was in the active-monitor mode, all four headsets could talk and listen on one channel while simultaneously monitoring a second channel. In the dual mode, two headsets could talk and listen on the same or a different selected channel with no monitor capability for either pair of headsets. The OIS units were mounted on the railings, pedestals and columns of the stands in the area, as well as on the north and south area walls near work areas. The OIS also had the capability to link to LC-39 areas.

Operational Television (OTV) could provide video surveillance and recording of payload processing in all operational areas of the Industrial Area to the O&C Building. Payload processing in the OPF, the Rotating Service Structure (RSS), and at the launch pad could be observed through the OTV system. OTV distributors in these areas routed the video by way of the Payload Television Video System (PTVS) Video Routing Switcher System (VRSS) to monitors in the O&C Building control and monitor rooms.

The Bay Area also contained commercial telephones in several places, available from floor level at the three access control monitors' desks, at the control desk for each stand, and on the west side under each stand.

The Paging and Area Warning System of the High and Low Bays was part of the KSC P&AW System (this is different to the one mentioned earlier). Speakers were mounted on the walls to inform personnel in the Bay Area. In addition, each work stand and operations desk had its own paging system for use during specific work in the stand.

Timing, in Inter-Range Instrumentation Group (IRIG) A, B, and H formats for Greenwich Mean Time (GMT) and Mission Elapsed Time (MET), was provided in the Bay Area. Countdown clock displays were located on or near each work stand.

The Work Stand Bulkhead Interface Assembly for each work stand had provisions or interface with the wideband transmission lines for Experiment Checkout Equipment (ECE) command and data transmission, between User Room ECE and work stand ECE, or experiment and ECE lines.

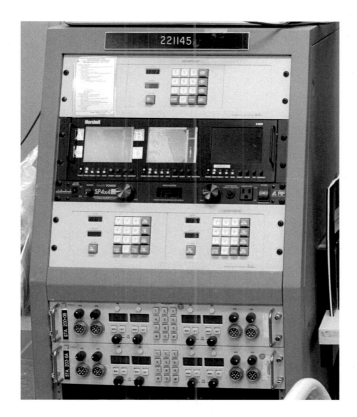

OIS-Rackmount-PCU (OTV). OIS STA 200-0B & STA 203-6A. OTV panel center of console, white and gray buttons. [Jim Dumoulin.]

Portable Access and Handling GSE

The Ground Support Equipment (GSE) available in the Bay Area consisted of both launch-site and ESA-supplied equipment for access to and handling of the payloads. For Spacelab, ESA constructed handling, access, and shipping equipment for the racks, modules, end cones, Igloo, and pallets. McDonnell Douglas in Huntington Beach furnished the Tunnel dolly (yellow). The ESA GSE was painted a medium blue. In addition, an overhead strongback was available for handling the Spacelab elements.

Module Vertical Access Kit /Vertical Access Simulator

The east chamber (Chamber L), located on the north side of the High Bay, housed the Vertical Access Simulator (VAS) used to train personnel in gaining access to the Spacelab module while it sat vertically in the Orbiter at the launch pad. The

VAS contained a mockup of the Orbiter middeck, the Spacelab transfer tunnel and the Spacelab module. The Module Vertical Access Kit (MVAK) was used in the VAS for training, as well as in the flight vehicle as GSE for installing or removing flight hardware and specimens. The MVAK consisted of pulleys, several harnesses, the joggle positioning ring, brace, and seat, and a ladder and safety net. This all allowed a person to be lowered into the module from the middeck, to make the transfer through the 90-degree bend in the tunnel, and to gain access to the module. Entry to the VAS could be obtained from the High Bay floor or through the top of altitude chamber "east".

Control and Monitoring Areas

Within the O&C Building were the Control and Monitoring Areas for horizontal payload processing. This section included the Experiment Integration control rooms, the Spacelab control rooms, the CITE control room, and the User Rooms. Because of the electronic equipment in these control and monitoring rooms, a water deluge fire suppression system was used. The system included ionization detectors, piping for water dispersion, alarm bells, and revolving beam lights. The air handling units were automatically shut down when the system was activated.

LEVEL-IV AND OTHER CONTROL ROOMS

Jim Dumoulin outlined the flow that Spacelab experiments followed to be installed on a mission. "Spacelab experiments were built on an assembly line in the O&C. They went from Offline Labs to Level-IV [Experiment Integration], to Level-III [Subsystem Integration], to Level-II [Experiment and Subsystem Joint Testing], and then to a high-fidelity Orbiter Payload Bay simulator called CITE [Cargo Integration Test Equipment]. After CITE, the payload was ready to be integrated into an actual Orbiter at the OPF or the pad." Each step in the process had a dedicated control room on the O&C third floor and a dedicated set of test stands in the O&C High Bay. Each processing level had two control rooms connected to two test stands. Two missions at a time could be supported. For CITE, there were two control rooms but three test stands [one in the O&C and two in the VPF]. Even when a mission went to the OPF or the pad, these control rooms were still used for all monitoring and control. The Firing Room only had one console [the C-1 console] shared by all payloads launched on the Shuttle.

"Just like in the Shuttle firing rooms, all of our consoles were built out of Apollo era-type consoles. They just flipped them upside down. The panels we used in our Level-IV control rooms were mostly designed for some Apollo function and then they were retrofitted at Marshall and shipped back to us in Level-IV. Back then, there weren't a lot of computers controlling things; everything was switch based, so you had these panels with switches where you could change out the functions, the lights etc., We used these switches and panels initially at Level-IV, at least for the first couple of missions, and they were Apollo era panels that were just redone for Spacelab. They had something called networks back then, but they weren't computer networks. They were 28-volt DC networks, so that when you flipped the switch, it commanded a relay that turned on something else that turned on something else and it activated the payload. So you could actually turn on a massive amount of stuff with just a few switches and no software on the activation side, on the flight side. Now, when you activate and test an experiment, there's a ton of software involved in that, but back then, [even with something like] powering up the Orbiter, like the onboard, an astronaut would go to the Spacelab panel. There were a few flip switches to turn on an entire payload on board and we could just do those same things in Level-IV."

Payload Processing Activities

The area processing activities included staging, experiment integration, payload integration and verification, and post-landing de-integration for horizontally processed payloads. The Bay Area was capable of supporting several payloads at the same time in different stages of integration and de-integration.

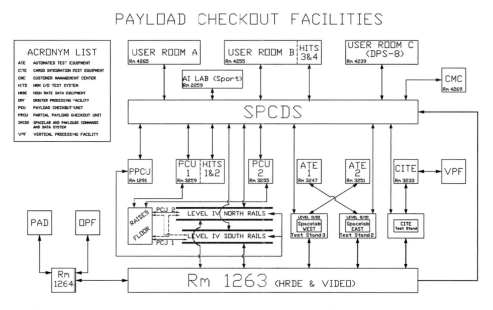

Payload Checkout Facilities. This shows the flow of information between the different locations in the O&C and other areas of KSC (VPF, pad and OPF) and how each of the components interacted with each other. [Jim Dumoulin.]

Level-IV Offline Labs

While subsystems were being validated preflight, experiments that came from all over the world were shipped to offline labs at KSC, where they were checked out by their developers. Today, NASA refers to this as "Pre DD-250 handover" of a Payload to the government, but back in the Spacelab days the interface was less formal. Before handover, Level-IV Experiment Engineers worked with Principal Investigators (PI) to learn everything they could about the experiment and to teach the PIs everything they needed to know about how their experiment would be operated while in space. During this phase, it was the Experiment Engineers' job to write the acceptance test procedures that would be used to safely power up and test the payloads after they were integrated. Every experiment function and every pin of every wire harness had to be analyzed. If the integration process required the removal or rerouting of a cable, a test procedure needed to be generated that retested every function that went through that interface.

Once a PI deemed that an experiment was ready, it was turned over to the Level-IV Experiment Integration team. An experiment's unique testing hardware was relocated to the O&C fourth floor in an area called the "User Rooms", while the flight hardware was moved to the Level-IV area in the O&C Low Bay.

Jim Dumoulin also explained how the control rooms were utilized. "Telemetry from the system was sent to a ground telemetry station on the O&C first floor [Room 1263] where it was recorded, de-mixed using a high rate demultiplexer and sent to the User Rooms on the fourth floor and to other centers. All processing activities in the High Bay were controlled from sets of control rooms on the O&C third floor that had floor-to-ceiling bay windows overlooking the test stands. These control rooms were previously used to test and checkout the Command and Lunar Modules during the Apollo program.

"The CITE control rooms were positioned overlooking the CITE stand [near the O&C's two Apollo era altitude chambers]. These control rooms also controlled test stands in the Vertical Processing Facility [VPF] that processed non-Spacelab scientific payloads and satellites. The Level-III/II control rooms overlooked the Level-III/II test stands in the middle of the High Bay where the non-experiment unique parts of Spacelab were serviced. Since ESA's initial plan for Spacelab module re-servicing was 'Ship-and-Shoot', all the test equipment in the Level-III/II control rooms and test stands was designed to operate in Europe. The racks and test equipment were bright yellow and looked totally different than the NASA-designed systems in CITE and Level-IV. They also operated on European 50Hz power, so nothing in the room was interchangeable with US equipment.

"Finally, the Level-IV control rooms were adjacent to the back doors of the Astronaut Quarters and overlooked the North and South Rails. Each Level-IV control room had a Test Conductor's Console, two Perkin Elmer/Concurrent Real Time computer systems [PCU, HITS], a Spacelab Flight Computer, a rack of hardware simulators and a Data Display Simulator [DDSS]. The Payload Checkout Unit [PCU] was used to automate the power up of the Spacelab avionics system and the High rate I/O Test Set [HITS] processed and recorded the downlink telemetry. The Spacelab Flight Computer [a Mitra 125s and, post-*Challenger*, an IBM AP-101SL] ran the actual flight load of the Spacelab Experiment Computer Operating System [ECOS]. The meeting room next to the PCU control room [O&C Room 3255] was dismantled and a second PCU control room, called the PCU-2 control room, was created. Dave Sollberger [now retired from the position of Chief Engineer for the Launch Services Program] did the layout for PCU-2 and placed every PCU-2 console in the exact mirror location of PCU-1, with a window between the control rooms. The layout was so precise that you could do a 'Disney World Haunted Mansion' effect in the control rooms; with proper adjustment of the lighting in one control room, you could make 'ghosts' seem to sit at the mirror location in the other control room."

Jim Dumoulin then went into greater detail. "The Level-IV control rooms were amazing works of engineering. Every aspect of Spacelab and the Shuttle had to be simulated so that the experiments under test would think they were talking to an actual Spacelab module and Orbiter. For Spacelab-1, every part of every system was being tested for the first time. When problems arose, it was always a detective challenge to determine if the problem was a hardware issue, a flight software issue, a ground software issue, a misconfigured cable, a telemetry/database issue, an experiment issue or incorrect documentation. The system was designed so that any of the simulated systems could be pulled out and replaced with the actual flight element. Most of these simulators were housed in a 4 ft³ (1.21 m³) cube called the Computer Interface Device [CID]. The CID was designed by IBM/Huntsville and housed in an IBM mainframe storage cabinet, with hinged circuit bays that opened like pages of a book. The CID had hardware models that simulated Spacelab's Pulse Code Modulated Master Unit [PCMMU], Multiplexer/Demultiplexer [MDM], Master Timing Unit [MTU], Mass Memory Unit [MMU], Remote Acquisition Unit [RAU] and High Rate Multiplexer [HRM].

"These hardware modules had tens of thousands of wire-wrap socketed TTL logic chips that were designed by IBM/Huntsville's proprietary automated computer logic generation software. As new flight software was developed and idiosyncrasies detected in the actual flight units, a Level-IV Electrical Engineer was assigned to implement wire-wrap changes to the hardware modules. The system schematics were page after page of individual logic flows [called Automated Logic Display or ALDs]. They were mathematically correct hardware simulators but were nearly impossible to decipher by humans. One wrongly placed wire and the CID could be out of commission for days. Wire wrapping was done with a wire-wrap tool or a battery powered 'Wire Wrap Gun'. You placed a colored laminated thin wire in the tool and twisted it on posts on the back side of the socket holding the logic chip. Usually, you would use a different color for the mods so it was easier to track changes.

"[As I said] The hub of the control room was this box called the CID [that] was designed by some IBM guys... but to understand where all the wires went [involved] stuff that you didn't learn in school. In electrical engineering, you had a way to look at schematics and a way to look electrical stuff, but the IBM folks that built that system were just experimenting with computer design logic, so they were on the forefront of some of that. They used a computer just to write out the simulation or how the hardware would be... and then they would have us make a modification by saying, 'well, we need to hook this wire to that wire'. But we wanted to understand why they were making all these changes, so we were trying to flip through all of these computer generated documents that nobody was intended to read to try to figure out what modification was being done to fix things. The reason you want to do that is because when something glitched and it didn't work, you needed [to be able to] flip over to the actual huge board, put a scope on it, look at a signal and say, 'that's why this isn't working'. So you had to understand that what was going on.

"One of the things we had was a red line book [in which] we had to know what the mods were and then verify that the mods that we did matched the flight configuration. [That's why] we used multicolor wires. The original layout was one wire and every time you made a mod, you would grab a different color wire and you put those in there. I was eventually reassigned to the PCU, which are connected to the CID, so I was involved in the testing, but very seldom did I actually do the wire rack.

"A typical mission test started by powering up the computers and simulators in the PCU Control Room with a mission-unique Spacelab Flight Software load from MSFC. Level-IV then ran a series of mission tests with simulated subsystems and simulated payloads. Once the flight software load worked in simulation, real Spacelab components and experiments were integrated into the system and replaced any of the simulated units. Once the flight build-up was complete, a multiday test called the Mission Sequence Test [MST] validated the system under actual flight conditions. The MST consisted of slices of actual mission profiles and was a good indication that all experiments and flight software were properly installed and working."

Experiment Integration Control Rooms

These rooms were used for control and monitoring of experiment integration testing and for processing and monitoring data transmitted via the Spacelab Experiment Computer Input/Output (ECIO) channel during all subsequent levels of integration. The PCU control equipment and High Rate Multiplexer Input/Output Test System (HITS) 1 and 2 were located a separate room.

PCU control room, c 1939. [NASA/KSC.]

A second PCU was located in one room and HITS 3 and 4 were in User Room B. There was also a conference room and library area.

The purpose of the PCU was to test Spacelab experiments using a simulation of the Spacelab Command and Data Management System (CDMS), and Orbiter avionics interfaces. It also allowed monitoring and control of EGSE and MGSE within the Experiment Integration area. The PCU also provided post-test data reduction, done on a non-interference basis with real-time support. The two PCUs provided the capability to process the experiments for two Spacelabs concurrently. Either PCU could support either a module or an Igloo mission, in either the North or South Rails, independently of the other PCU and test stand.

Spacelab Control Rooms

Spacelab Control Rooms 1 and 2 housed the ATE, which was computer-controlled equipment used for electrical power-up testing of the Spacelab in the Bay Area. Payload integration test and checkout activities were controlled from these rooms. Overall test activities and integrated tests were controlled by the test engineers who operated and monitored Spacelab subsystems from the ATE consoles.

CITE Control Room

The CITE control room contained hardware similar to that developed for the KSC Launch Processing System (LPS). This LPS-type hardware was integrated with the CITE electrical and electronic set test stand hardware to provide display, monitoring, control, data conversion, and data processing for the payload interface verification. The test stand hardware served as the simulated Orbiter interfaces to the payload. Payload interface verification for both vertically and horizontally processed payloads was conducted from the CITE control room. The equipment consisted of test consoles containing keyboards, color CRT displays, and mini-computers with extensive disc storage for flexible parallel testing and monitoring by the test engineers. The Record and Playback Assembly could provide a magnetic tape recording of downlink data collected during the testing, once the testing was complete.

Engineering Support Area

The purpose of the Engineering Support Area (ESA, not to be confused with the European Space Agency) was to allow the customer to monitor the payload and upper stage parameters independently of control room monitors during integrated testing of the payload in the Bay Area or at the VPF. The customer could then participate in problem troubleshooting as necessary. The ESA also provided technical support to the systems engineers in the CITE control room during CITE testing. The ESA contained television monitors of a multi-function CRT display system (referred to as the AFD displays), CRT monitors of control room LPS displays and customer-requested ESA displays, OIS, telephones, and the LPS master console.

High Rate Multiplexer Test Station and CCTV Room

The High Rate Multiplexer (HRM) Test Station and CCTV room was located on the first floor of the O&C Building and contained equipment that handled the distribution of Spacelab and experiment integration HRDM data and clock channels to the User Rooms and to the HITS in those rooms; reception and distribution of the Spacelab HRM output signals after Spacelab was installed in the Orbiter; distribution of CCTV to the payload video switch and the User Rooms, and routing of ECE between the User Rooms and the test stands; Spacelab onboard CCTV system checkout and troubleshooting; and processing and transfer of data to various locations during CITE and OPF testing of Spacelab.

User Rooms

The User Rooms A, B and C were located on the fourth floor of the O&C Building. Each room could be divided into several areas, with a test operations room occupying a central area and serving as a focal point for all test operations. The remaining user areas could be separated by sound-absorbing movable partitions that could be arranged to accommodate the space allocations of the experimenters assigned to the mission.

Documentation regarding the User Room Support Provisions was provided to familiarize the experimenters and payload personnel with the facilities available during their occupancy in the assigned areas. These provisions included: lighting, fire suppression, environmental control, electrical power distribution, grounding. OIS, experiment TV monitoring, Master Timing Signal distribution, and Instrument GSE (IGSE) lines

The experiment data processed and included in the output of the HRM consisted of the following: Traffic of low-rate data controlled by the Remote Acquisition Units (RAU) and the ECIO channel; 16 experiment-dedicated channels and the direct access channel (DACH) for the high-rate data; burst time (the GMT sample retrieved from the status word of HRM formats) and the HRM format sync pulse; three experiment voice channels and experiment TV and analog channels.

Payload Customer Management Center

The Payload Customer Management Center (CMC) provided space for meetings and CCTV viewing of Bay Area operations during payload processing and CITE testing. Use of this room was controlled by CS-PTS.

Laboratories and Shops

The center section of the O&C Building contained laboratory and shop areas to support the Space Shuttle and payload ground processing activities, as well as the routine operations and maintenance of the KSC facilities. Some of the areas were dedicated to payload support; other areas could support payload activities on an as-available basis only.

Experiment Laboratories

The experiment laboratories were a group of rooms on the first and second floors of the O&C Building that were assigned for particular experiment offline operations. They were used for both preflight and post-landing operations. The air handling equipment in these laboratories, as for most of the O&C Building, was set to maintain the temperature at 71 ±6 degrees F (21.7 ±3.3 degrees C). The cleanliness level of these laboratory areas was maintained as at least a Level 5 CWA. All areas were equipped with telephones.

Biomedical Support Areas

On the third floor of the O&C Building, on the southeast end, were the biomedical examination and laboratory areas, containing medical facilities and laboratories that could be used by the investigators on a pre-arranged and shared basis. The specific rooms and their purposes were: physical exam room (office); clinical physical examination; clinical labs; biomedical lab; bio-instrumentation, electronics, and mechanical fabrication; and physical examination or technical office space.

Each room had a vast amount of medical equipment to perform specific tasks. The availability of selected areas within the Medical Suite was subject to the approval of the Director, Biomedical Operations and Research Office.

Baseline Data Collection Facility

Several rooms on the west side of the second floor were designated as the Baseline Data Collection Facility (BDCF). This 4600-ft^2 (427.5-m^2) area was used to conduct ground-based tests equivalent to the in-flight protocols for non-invasive human life sciences flight experiments. The BDCF was located beneath the Crew Quarters for ease of access and proximity. Data was collected before, during, and after flight, if required. The noise level within the BDCF could be maintained to less than 65 dBA. On the southwest corner of the BDCF were folding walls, to provide flexibility in the use of the area.

Material Science Laboratories

The material science laboratories were located on the first and second floors of the O&C Building. These labs were used for non-routine material testing, microchemical analysis, and malfunction investigations in support of Shuttle launches and KSC operations. These labs were staffed by civil service personnel who were professionals in the disciplines of chemical analyses, physical and mechanical properties, evaluations, and malfunction investigations. These services were available to support payload activities. The laboratories could be divided into three types: Microchemical Analysis Laboratories (MCAL), material testing, and malfunction analysis.

Each lab contained a substantial set of equipment to support all operations. Other labs included: Electrical Power Laboratory; electronics laboratories; Environmental Testing Laboratory; fluid mechanical laboratories; Lubricants and Polymers Testing Laboratory; Materials Testing Laboratory; Metallurgical Laboratory; Metrology Laboratory; Metallographic Laboratory; Coatings Evaluation Laboratory; Liquid Oxygen Testing Laboratory; MCAL Equipment; and MCAL Utilities.

Special Purpose Areas

Several rooms were used for the electrical support shop, the Test and Inspection Record (TAIR) Center, and the Instrument Pointing System (IPS) Laboratory.

KSC Instrument Calibration Laboratory

This contained the KSC Instrument Calibration Lab and the Staging and Loan Pool for instruments. Payload instruments could be calibrated here, and instruments for payload support could be borrowed from the pool.

Experiment Integration Electrical Shop

The southwest end of the Bay Area was the electrical/electronics shop for experiment integration support. Cable fabrication in support of experiment integration activities was performed here and the shop contained AC power receptacles, compressed air, and facility water.

Astronaut Breathing Air System

The Astronaut Breathing Air System was supplied by 24 K-bottles located in the O&C cul-de-sac. A stainless steel tube was routed from the K-bottles up the north wall of the Laboratory, Control and Monitor area, providing 100 psi (6.89 bar) breathing air to the room. This was fed to a master panel which supplied eight interface panels. Each interface had a 3 psi (0.20 bar) and 100 psi (6.89 bar) outlet panel and was used to perform cooling and leak checks on the astronauts' partial pressure flight suits.

Hazardous and Controlled Waste

In advance of their arrival, customers would fill out a KSC "Process Waste Questionnaire", for any hazardous and controlled waste they expected to generate at KSC during processing. All such waste would be managed in accordance with the requirements of Hazardous Waste Management procedures. Once a customer had identified launch site waste generations, a Satellite Accumulation Area (SAA) would be set up in facilities denoted as points of generation of these wastes. These SAAs would be established in order to comply with the intent of the Resource and Recovery Act of 1976, which was created to institute a national program to control the generation, storage, transportation, treatment, and disposal of hazardous and controlled waste. Regulations for the use, control, and disposal of waste at the launch site were strictly enforced. A Material Safety Data Sheet (MSDS) was required for each hazardous substance brought to KSC for payload processing. Inventory and accountability of all hazardous substances would be managed in accordance with the requirements the KSC Environmental Control Handbook. This data was used to comply with the Superfund Amendments and Reauthorization Act, 1986 (SARA Title III).

Storage and Other Areas

A single room was used for receiving and inspection of the payload flight equipment and GSE at KSC. The room was equipped with two large sliding doors, one leading to the outside of the building on the south side and one into the Bay Area on the north. A personnel door was located to the left of the north sliding door. A 4-ton (3.6-metric ton) electric monorail hoist was installed in the room, which provided service through the center of the entire length of the room and to the outside door. The room had both AC power explosion-proof receptacles and compressed air available, located on the south, west and north walls.

Second Floor Areas

Five rooms were used for the instrument loan pool and instrument calibration lab. Instruments from the pool could be loaned for payload support.

Third Floor Areas

On this floor, one room contained the Lyndon B. Johnson Space Center (JSC) Flight Data File; another contained the astronaut suit room; two more comprised the astronaut pre-/postflight physical examination room; and one additional room contained the Flight Crew Equipment (FCE) labs. These were controlled access areas.

Fourth Floor Areas

The fourth floor contained the User Rooms, as described earlier, and rooms for tape certification. In addition, there was an area of locked storage in the lobby of the fourth floor. Two rooms were devoted to a headset dispensary and repair.

Transportation Systems/Multi Mission Support Equipment

The Payload Canister was a Payload-Bay-sized container that was used to transport fully integrated Shuttle payloads, in a controlled environment, in the vertical position from the VPF to the launch pad's Payload Changeout Room (PCR), or in the horizontal position from the O&C to the OPF. Two identical Canisters were available for use. Each was 65 ft. (19.8m) long, 18 ft. (5.5m) wide and 18 ft. 7 in (5.6m) high. They had the capacity to carry processed payloads up to 15 ft. (4.57m) in diameter and 60 ft. (18.3m) long, matching the capacity of the Orbiter Payload Bay, and could carry payloads weighing up to 65,000 lbs. (29,483 kg).

The Payload Canister Transporter was a 48-wheel, self-propelled truck designed to operate between and within Space Shuttle payload processing facilities. Two were used to transport the Payload Canisters and their associated hardware throughout KSC. Each was 65 ft. (19.8m) long and 23 ft. (7.0m) wide. Their flat-beds could be lowered to 5 ft. 3 in (1.6 m) or raised to 7 ft. (2.13m) plus or minus 3 in (0.08m). Each transporter's wheels were independently steerable, permitting it to move forward, backward, sideways or diagonally, or to turn on its own axis like a carousel. They had self-contained braking and stabilization jacking systems. A bare transporter weighed 230,000 lbs. (104,326 kg), but with a full load of die-sel fuel and with the environmental control system, fluids and gas service, electri-cal power system, and instrumentation and communication system modules mounted, the transporter had a gross weight of 258,320 lbs. (117,172 kg). The transporter was steerable from diagonally opposed operator cabs on each end. Its top speed unladen was 10 mph (16.09 kph) and the maximum speed of the fully loaded transporter was 5 mph (8.04 kph). Because payload handling would require precise movements, the transporter had a creep mode that allowed it to move as slowly as 0.25 in/s (0.0063 m/s) or 0.014 mph (0.022 kph). Traveling *between* facilities utilized a 400 hp (298.3 kw) diesel engine. Traveling *within* a facility employed a 109.9 hp (82kw) electric motor. The transporter could carry the Payload Canister in either the horizontal or vertical position.

Payload Canister in vertical orientation, on the road heading towards Launch Complex 39. [NASA/KSC.]

The Payload Environmental Transportation System (PETS) was a self-contained payload carrier capable of transporting payloads over the road for either short or long distances. Payload environmental parameters inside the PETS (a constant temperature and humidity level) were maintained during transport operations via the on-board environmental control system.

The remaining facilities of the Industrial Area will be touched on only briefly, since very few Spacelab/experiment operations were performed in those facilities.

Vertical Processing Facility

The Vertical Processing Facility (VPF), as the name implies, was the location where vertical processing of payloads occurred. These were usually hazardous payloads that were assembled, integrated and tested in the vertical orientation, as opposed to non-hazardous horizontal processed Spacelab payloads. For certain Space Shuttle missions, there could be a combination of horizontal and vertical payloads on the same flight. A combination processing flow usually occurred by having the canister go to O&C and load the payload horizontally first. The canister would then be rotated to vertical and transported to the VPF for vertical payload installation, before being transported to the pad for operations there.

Canister Rotation Facility

During the early years of the Space Shuttle Program, canister rotation was performed in the Vehicle Assembly Building (VAB). Two cranes and much hands-on maneuvering were required to rotate the canister. In addition, this required a 15-mile (24.14 km) round trip from payload processing facilities located in the Industrial Area to the VAB.

Rotating a Space Shuttle Payload Canister and the payloads it housed was no easy task, even for the highly skilled Multi-Mission Support Equipment (MMSE) team, better known as the "Can Crew." A special building – the CRF – was built in 1993 in the Industrial Area to help the team handle the challenges of canister rotation. The 142-foot (43.28-m) High Bay included a 100-ton (74,654 kg) bridge crane and other specialized equipment required for lifting the canister.

LC-39 AREA, NORTHERN KSC

The facilities referenced here will only contain brief descriptions of their capabilities, focusing on only payloads operations performed. There are many more facilities in the LC-39 area, but most of these involved no payload operations, or performed only a support role for payloads, so they will not be discussed here.

Turn Basin

The Turn Basin was originally constructed to allow barges to offload the huge S-IC first stage and S-II second stage of the Saturn V launch vehicle. Once offloaded, the stages were then rolled into the VAB for processing and stacking. During the Shuttle era, the Basin was used to do a similar thing for the Space Shuttle External Tank (ET) and could also be used to offload large payloads. Located to the east of the VAB, the Turn Basin provided a direct water connection between the Atlantic Ocean and space launch processing activities at KSC. The docking area was used to receive hardware at KSC that was too large to travel over open roadways, rail or by air transport.

Orbiter Processing Facility 1, 2 and 3

The Orbiter Processing Facility (OPF) was located west of the VAB. Shortly after landing and safing at the Shuttle Landing Facility (SLF), the Orbiter was towed to a vacant bay in the OPF to go through safing procedures, such as removing residual fuels and explosive ordnance items. Only then was the mission payload removed and the vehicle fully inspected, tested, and refurbished for its next mission. Originally expected to have a two-week turnaround, it transpired that these functions took up approximately two-thirds of the time between missions. The remainder was devoted to the installation and checkout of the payload for the next mission. Power-up testing of Orbiter vehicles in the OPF was controlled from consoles in the Launch Control Complex (LCC). The OPFs were also occasionally used to store an Orbiter between its processing flow, when room inside the VAB or out at the pads was either occupied or temporarily out of service.

OPF-1 and OPF-2 consisted of two 29,000 ft^2 (2,700 m^2) High Bays separated by a 23,600 ft^2 (2,130 m^2) Low Bay. Originally, OPF-3 was called the Orbiter Maintenance & Refurbishment Facility (OMRF) and was intended to support Shuttle operations at Vandenberg Air Force Base (VAFB), California. When Shuttle operations planned for VAFB were terminated, the OMRF was disassembled and transported to KSC, where it was reassembled as OPF-3. OPF-3 was a 50,000 ft^2 (4,645 m^2) facility located across the street from OPF-1 and OPF-2. It consisted of a 95 ft. (29 m) tall High Bay Area and a two-story Low Bay Area.

The installation of large payloads into the Orbiter while located in the OPF began with canister and payload strongback transfer into an OPF bay via a large door. With the canister doors open, preparations could continue for payload removal. Using one of two OPF overhead cranes, the strongback was lifted with drop links over the canister containing the payload. The system was then lowered until the drop links were attached to the appropriate payload trunnions. Next came a very slow movement upward, gradually adjusting the center of gravity weights on the strongback to ensure the payload was lifted up normal to gravity. Once clear of the canister, the payload was lifted over the OPF access structure and then carefully lowered into the Orbiter, constantly adjusting the counterweight. Access to the trunnion/Orbiter CIL interface was achieved via Level-13 platforms.

After being mechanically secured in the Orbiter, the Payload Canister and strongback were removed from the OPF. At this point, electrical connections were performed.

Access to the Payload Bay could be by several means: through the Shuttle airlock from the middeck; via steps from the Level-13 platforms; using door 44 between the left Payload Bay door and the left wing; and also through an access door in the main landing gear compartment, which was used during (STS 51-F) Spacelab-2 post-landing operations.

Servicing and testing was then accomplished and closeout operations were completed. Most Spacelab modules were closed out for flight in the OPF prior to transportation to the pad, though there were some missions which required late access to the module on the pad. When Spacelab modules were not carried, work would still be performed on hardware at the pad, followed by the final closeout procedure.

Normally, once the payload and Orbiter operations were complete, the payload doors would be closed and the Orbiter towed over to the VAB for stacking operations with the ET and SRBs. In the early days of the program the Orbiter would be moved on its own wheels, which were then stowed during crane operations in the VAB. Subsequent to the termination of Vandenberg Shuttle operations, the large yellow VAFB Orbiter transporter was brought to KSC and used for Orbiter rollover.

Vehicle Assemble Building

The Vehicle Assembly Building (VAB) remains the largest single story building in the world. During the Shuttle program, the four High Bay locations were utilized as follows: High Bays 1 and 3 were used for integration and stacking of the complete Space Shuttle vehicle. High Bay 2 was used for ET checkout and storage, and as a contingency storage area for Orbiters. High Bay 4 was also used for ET checkout and storage, as well as for Payload Canister operations and SRB contingency handling. In addition, there was the Low Bay Area, which contained maintenance and the overhaul shops for the Space Shuttle Main Engines (SSME), and was also utilized as a holding area for SRB forward assemblies and aft skirts.

During the build-up of the vehicle inside the VAB, integrated SRB segments were transferred from nearby assembly and checkout facilities, then hoisted onto a Mobile Launcher Platform (MLP) in either VAB High Bay 1 or 3 and mated together to form two complete SRBs. Each ET arrived at KSC by barge, and was unloaded and transferred to High Bay 2 or 4 for inspection and check-out. They were then transferred to High Bay 1 or 3 and attached to the SRBs already in place on the MLP. With this completed, the designated Orbiter was towed over from the OPF to the VAB transfer aisle, raised to a vertical position, lowered onto the MLP and then mated to the ET/SRB combination. This assembly of Orbiter, ET and SRB was known as "The Stack". Once the assembly and checkout had been completed, the crawler-transporter entered the High Bay, picked up the platform and assembled Shuttle vehicle, and carried them to the launch pad to continue the processing towards launch.

Very few payload operations were performed in the VAB. The only work would be any support required for the payload, such as maintaining a continuous purge on science instruments located in the Payload Bay or middeck. Details of how all this would be accomplished was planned well ahead of time to ensure all requirements were met.

Pads 39A & 39B

Launch Complex 39's (LC-39) Pad A and Pad B were originally designed to support the Apollo program and were modified during the mid-1970s for Space Shuttle launch operations. Major changes for the Shuttle program included building a new Fixed Service Structure (FSS), the addition of a Rotating Service Structure (RSS), and replacing the single Saturn flame deflector with three new flame deflectors. The original Saturn V Launch Umbilical Tower was too tall for Shuttle operations, so the upper portions were removed from two of the Apollo Mobile Launchers and installed at each pad to serve as the FSS. Fuel, oxidizer, high pressure gas, electrical, and pneumatic lines connected the Shuttle vehicle with Ground Support Equipment and were routed through the FSS, RSS and MLP.

In support of payload accommodations at the Pad, each RSS featured provisions for loading the payloads vertically. Mounted on a semi-circular track, each RSS could rotate through an arc of 120 degrees on a radius of 120 ft. (36.6 m). A hinge on the FSS provided the pivot for the RSS to swivel until the Spacecraft Changeout Room fitted flush with the Orbiter's Payload Bay. This room permitted payloads to be installed and/or serviced under contamination-free or "clean room" conditions.

Payload Changeout Room

The Payload Changeout Room (PCR) featured the enclosed, environmentally-controlled portion of the RSS. This provided the opportunity for the late delivery, in the launch processing sequence, of cargo and payloads while the vehicle was on the pad. These would be inserted vertically into the Orbiter Payload Bay. To prevent exposure to the external air and potential contaminants, a seal fitted around the mating surface of the PCR and against the Orbiter Payload Bay area, allowing the Payload Bay or canister doors to be opened and the payload removed or installed. A clean-air purge in the PCR maintained environmental control during PCR cargo operations.

Operations at the pad required a delicate touch. Firstly, the canister was hoisted to the proper elevation in the retracted RSS and locked into position. The environmental seals of the PCR were then inflated, sealing against the sides of the canister. As a precaution, the space between the closed doors of the PCR and the canister was purged with clean air, thus ensuring the required cleanliness before the doors of the PCR and canister were opened to the internal environment.

Once the integrity of the area was confirmed, the Payload Canister could then be transferred into the PCR via the Payload Ground Handling Mechanism (PGHM). The PGHM was then retracted into the PCR and the canister and PCR doors were closed, followed by deflation of the environmental seal. The canister was then lowered onto its transporter and taken to a storage facility. Next, the RSS was then moved into position to enclose the Orbiter's Payload Bay and establish environmental seals. The space between the closed doors of the PCR and the Payload Bay was again purged with clean air.

Once the payload was on the PGHM, the required operations could be performed. Telescoping platforms at fixed levels provided access to the payload. An elevator could access each of the fixed levels of the PGHM, while "J" IUS platforms and 'diving' boards could be installed at varying different levels of the PGHM. These were accessible via a special stairway or steps from the fixed levels of the PGHM. Occasionally, this may have required fitting after installation into the Orbiter to preclude any interference that could occur during transfer operations. At this point, various services such as facility ECS, gases, special gas line and vents, or grounding connections would be completed, as well as any unique GSE that needed to be connected depending on individual payload requirements.

With the PCR doors and Payload Bay doors open, work could be performed on the payload, especially for those locations that could not be accessed once it was in the Orbiter.

With all assigned work completed, the PGHM extended the payload into the Orbiter Payload Bay, where mechanical and electrical connections were made. Environmental control was maintained during this phase of pad operations, with the Payload Bay purged with clean temperature- and humidity-controlled air and any services maintained.

Next came the testing between the payload and Orbiter to confirm compatibility. Once this had been tested, checked and verified, the final operations would be performed on the payload. Depending on the specific payloads, this could include a final top-off of fluids and gases, battery charging, and completion of other close-out operations. With the PGHM fully retracted, the two Orbiter Payload Bay doors and the PCR door were closed. Normally, the RSS remained up against the Orbiter for added weather protection until during the T-11-hour hold, nominally one day before launch. Then it would be rotated back to the launch location in preparation for launch.

Launch Control Center

The launch of the Shuttle from KSC was the responsibility of the Launch Control Center (LCC) a four-story building that is the electronic "brain" of LC-39. Attached to the southeast corner of the VAB, it is 18,159 ft. (5,535 m) from Pad 39A. At the time it was constructed, advances in electronics had made it unnecessary to continue locating blockhouses adjacent to launch pads. Firing Rooms 1 and 3 were configured for full control of launch and Orbiter operations, while Firing Room 2 was usually used for software development and testing. Firing Room 4 was only a partial firing room and was primarily used as an engineering analysis and support area for launch and checkout operations. The fourth floor of the LCC contained conference rooms, offices and mechanical equipment. The LCC firing rooms contained various consoles that supported different components of the Shuttle (SRB, Main Engines, etc.). The main management consoles, positioned in a "stair-step" fashion, faced away from the outside windows towards the system consoles located on the LCC floor. The C-1 console located along the front row of the systems consoles was dedicated to payload operations and was used for all Spacelab and other payload operations.

Shuttle Landing Facility

Orbiter landings at KSC were made on one of the largest runways in the world. The runway is located 3.2 km (2 miles) northwest of the VAB and is 15,000 ft. (4,572 m) long and 300 ft. (91.4 m) wide. It has 1000 ft. (305 m) of paved overruns at each end and the paving thickness is 15 inches (40.6 cm) at the center.

Payload activities that took place at the SLF included removal of middeck experiments from the Orbiter post-landing (approximately one hour after landing) and retrieval of time-critical experiment items from Spacelab and Spacehab modules in the Payload Bay (approximately 4–6 hours after landing). These activities took place on the runway before the Orbiter was towed to the OPF.

At times, the SLF was used to deliver and pick up payload items via aircraft.

OFFSITE EXPERIMENT INTEGRATION LOCATIONS

There were locations other than on KSC property (termed "offsite") at which the flight hardware would be prepared before arrival at KSC, operated in space after launch from KSC, or worked on postflight once it left KSC.

PI Sites All Over The World

Principal Investigator (PI) sites across the world would be used for preflight, mission support and postflight operations. Preflight consisted of any experiment work or task to be performed before arrival at KSC. Mission support would be minimal, with most PIs and their teams at the Payload Operations Control Centers (POCC), but some data could be routed to a PI site if specially established before launch. Following the mission, experiments would be sent back to PI sites for all postflight operations once they left KSC.

Hangar L

Hangar L, also called the Life Sciences Support Facility (LSSF), supported Life Sciences flight experiments for the Space Shuttle Program, as well as ongoing research in Controlled Ecological Life Support System (CELSS) and plant space biology. Hangar L supported Orbiter middeck and Spacelab Life Sciences experiment processing, and also had accommodations and lab space for specimen holding and the handling of small mammals, aquatic animals, plants, cells, tissues, and micro-organisms. The LSSF also contained limited areas for surgery, X-ray, data management, storage, synchronous ground control, and flight experiment monitoring. Hanger L was located on Hangar Road at the Cape Canaveral Airforce Station (CCAFS) and was the first hangar south of Industry Road. The building was a concrete block and steel frame hangar, with two-story structures connected to the north and south sides.

PAYLOAD OPERATIONS CONTROL CENTERS

A Payload Operations Control Center (POCC) was the location from which the payload would be controlled during the mission. A great deal of planning needed to take place at the POCC so that it would be prepared for the mission and be ready to support any changes or problems that occurred while it was ongoing, as well as any postflight support needed for the payload.

MSFC POCC (Huntsville Operation Control Center, HSOC)

Spacelab science missions were not controlled by flight controllers and researchers at Mission Control, Houston, but at the POCC in Huntsville, Alabama, beginning with STS-35, the Astro-1 payload.[1] The cadre of engineers and scientists

---

[1] The German Spacelab D1 (1985) and D2 (1993) science operations were handled from the West German DFVLR center near Munich, via TDRS and Intelsat satellites.

directed all NASA science operations, sent commands to the Shuttle, received and analyzed data from the experiments on board the Orbiter, adjusted mission schedules to take advantage of unexpected science opportunities or unexpected results, and worked with the Shuttle crew members to resolve any problems that occurred with the Spacelab experiments.

A lot of planning went into each mission, especially the multi-discipline science missions. Before the mission, a schedule of activities – called a timeline – was devised for the crew to follow on orbit. All onboard resources, such as power, TV downlink, and crew time had to be budgeted carefully to maximize the scientific data gathered during the mission. The timeline was also used to perform engineering analyses, to predict how much propellant would be needed, operating temperatures for the Orbiter and payload, and other safety and performance-related issues. This timeline was used to run "simulations" of the mission, when everyone in the mission team developed their interfaces with one another and performed a realistic run-through for the mission. This timeline was then developed and documented in several pre-mission products (for example Crew Activity Plans) which were then used by the crew during the flight.

Real-Time Changes

Once the mission was in orbit, a team worked around the clock to support the crew as they followed the timeline, and to respond to any unexpected situations. These on-the-fly "real-time" changes were done in the Payload Control Room (PCR, not to be confused with the Payload Changeout Room at the pad).

12-Hour Replanning

For longer-term changes, planning occurred every 12 hours in the Payload Support Room (PSR). The Replan Cycle was a quick, miniature version of the pre-mission timeline planning. At the beginning of each 12-hour cycle, the scientists submitted their requests for changes in the schedule, then met in the Science Operations Planning Group (SOPG) to discuss an overall plan. While this was being conducted, the Orbit Analysis Engineers (OAE) received updated tracking data from the Mission Control Center at JSC, and ran computer predictions of the timing of sunrise/sunset and any required observation start/stop times. The rest of the cycle time was spent in adjusting the timeline to satisfy all these new requirements. The same products developed in the pre-mission timeframe were generated and used by the team in the PCR during the following 12 hours, until a new plan emerged.

Johnson Space Center POCC

The POCC was located in Building 30 at JSC and was the command post for the management of Spacelab-1, Spacelab-2 and Spacelab-3 scientific payload activities during the mission, after which this was transferred to the MSFC. The POCC was similar to the Mission Control Center (MCC), which had overall responsibility for the flight and operation of the Orbiter. POCC and MCC personnel coordinated their efforts to ensure a successful mission. Members of the Marshall mission management team, and PIs with their research teams, worked in the POCC in either three 8-hour shifts or two 12-hour shifts. Using POCC equipment, they monitored, controlled and directed experiment operations aboard Spacelab.

German Space Operations Center

The German Space Operations Center, operated by the German Aerospace Center (DLR) at Oberpfaffenhofen near Munich, had an impressive track record with many years of invaluable experience. GSOC, as it was known, had been responsible for operating spacecraft for more than 35 years, playing a key role in countless crewed and uncrewed missions. The GSOC employed some 300 people, making it DLR's most prominent space establishment at its Oberpfaffenhofen site.

During the first Spacelab mission, FSLP (First Space Lab Payload, 1983), the GSOC was linked to Mission Control in Houston as a user center. In 1985, when the first German Spacelab mission D1 was launched, it served as a POCC, successfully demonstrating the capabilities and functions of a control center, particularly for European science. GSOC was solely responsible for the scientific aspects of the Spacelab D1 mission, making it the first control center in the western hemisphere outside of the USA to be directly involved in manned spaceflight missions.

A new building, equipped with state-of-the-art facilities, was designed and built at Oberpfaffenhofen for the next planned German mission. The building was first used for the Spacelab D2 mission in April 1993, and was also used in support of other ESA crewed missions, such as the EuroMir94 and EuroMir95 long duration residencies on the Russian Mir space station. The GSOC took part in the STS-99 SRTM mission in 2000, with German ESA astronaut Gerhard Thiele on board.

Because of its proven track record in human spaceflight, and building on the experiences with Spacelab, ESA selected GSOC to operate the *Columbus* research module, Europe's most significant contribution to the International Space Station.

Over a long period, the stringent demands of human spaceflight have resulted in the development of a number of new and innovative concepts and measures at

GSOC that have had the effect of enhancing overall system availability and operational safety. This also stands GSOC in good stead for the key requirements of other projects.

OFFSITE END-OF-MISSION LOCATIONS

NASA's Dryden Flight Research Center (DFRC), located within Edwards Air Force Base (AFB) in California, continued to support NASA's human space flight program as an alternate landing site for the Space Shuttle Orbiters throughout the program.[2]

Dryden was selected as the site for the Approach and Landing Test (ALT) program and the initial orbital landings because of the safety margin presented by Rogers Dry Lake and its lakebed runways. After operational landings resumed at KSC, Dryden continued to be an alternate site when unfavorable weather in Florida or special circumstances prevented a landing there. It was also a landing site on missions when developmental tests were being carried out, or specific payloads in the Orbiters required a lakebed runway.

White Sands Space Harbor

White Sands Space Harbor (WSSH) was a backup Space Shuttle runway, and was the primary training area used by Space Shuttle pilots practicing approaches and landings in the Shuttle Training Aircraft (STA) and T-38 Talon aircraft. With its runways, navigational aids, runway lighting, and control facilities, it also served as a backup Shuttle landing site when the primary landing site was Dryden. Level-IV personnel had to be ready at WSSH many times for any potential landing. WSSH was a part of the White Sands Test Facility, which is located approximately 30 miles (50 kilometers) west of Alamogordo, New Mexico, within the boundaries of the White Sands Missile Range runway facilities. While the Space Harbor was activated as a backup landing site for STS-116 due to poor weather conditions at both Edwards AFB (high cross-winds) and KSC (clouds and rain), White Sands was only used for one landing of the Space Shuttle, by *Columbia* on March 30, 1982 for STS-3.

---

[2] In March 2014, the Dryden Flight Research Center was renamed in honor of former NASA test pilot and astronaut Neil A. Armstrong (1930–2012).

Ferry Flight Requirements

Ferry flights generally transported the Orbiters from Edwards AFB to the SLF at KSC where the Orbiter was processed. This was common in the early days of the Space Shuttle program, when weather conditions at the SLF prevented the Shuttle from landing there. Some flights started at the DFRC following delivery of the Orbiter from Rockwell International to NASA from the nearby facilities in Palmdale, California. NASA operated two modified Boeing 747 aircraft, originally manufactured for commercial use, as Space Shuttle Carrier Aircraft.

Ferry Flight Environment

When the Orbiter used any landing site other than KSC, the payload generally remained in the Payload Bay and returned to the launch site via a ferry flight. This was accomplished by hoisting the Orbiter up in an open-air Mate/Demate Device (MDD) outside the hangar and attaching it to a Boeing 747 Shuttle Carrier Aircraft (SCA) that was rolled in beneath the Orbiter. The SCA provided electrical power to the Orbiter during the ferry flight.

During ferry flight operations, payloads in the Payload Bay were exposed to ambient conditions that were not controlled or monitored. Payloads were normally not powered, heated, or cooled. Customers were required to specify any unique payload requirements with regard to thermal control, as payloads could be subjected to significant variations, such as temperatures within the Payload Bay likely to be encountered during the ferry flight, any Payload Bay purges, and postflight ground processing.

The maximum duration of any flight segment was approximately four hours, during which time the Payload Bay environment was not controlled. However, as a nonstandard service, the payload temperature could be biased at takeoff within a reasonable range, using conditioned air supplied to the Orbiter Payload Bay via the Orbiter purge system while the Orbiter and SCA were on the ground. Ground purges during ferry flights could also be provided. Time on the ground between ferry flights also had to be considered, and this could vary from a few hours to more than a day, with the Payload Bay temperature potentially varying from about +10 to +125 degrees F (-12 C to 51.6 degrees C) as the result of diurnal and seasonal variations.

During one ferry flight to KSC from Edwards AFB, the Passive Optical Sample Assembly (POSA) flown during the mission remained exposed. A passively-deployed array of contamination-sensitive samples was mounted in a different location in the Payload Bay at DFRC during postflight operations prior to the ferry flight of *Columbia* back to KSC. The arrays were flown to aid in the assessment of contamination hazards of the Shuttle Payload Bay. The difference between these and the PCSA in place throughout the whole flight yielded a particle distribution that could be attributed just to the ascent/on-orbit/descent part of the flight.

The measured optical changes in the ferry flight samples were so low as to be in the realm of measurement uncertainty, and could be attributed to the deposited particulate This indicated that film formation or nonvolatile residue was very low, much like what has been found at KSC in the OPF and PCR facilities.

POSTFLIGHT OPERATIONS

Postflight operations included activities accomplished from the time the Orbiter performed a nominal End-Of-Mission (ECM) or an intact abort landing, until it was returned to an OPF. These operations included runway support of the crew, Orbiter, and payloads; Orbiter safing, de-servicing, and preparation for ferry to KSC; payload support/removal when required; Orbiter ferry to KSC; and post-mission payload activities.

Landing Site Support

Landing site support would vary depending on landing site selected, i.e., primary or alternate EOM, etc. KSC would be the primary landing site unless a mission-specific exception to land at Edwards was approved by the Space Shuttle Program (SSP). Under nominal conditions, payload purge and ECLSS coolant would be applied to the Orbiter.

Payload Bay Early Access

The Orbiter vehicle included physical access to the Payload Bay through the crew compartment and airlock hatches, to satisfy crew habitable module and similar payloads early access requirements. Payload Bay physical access requirements for a given mission also had to be further defined and were unique to that mission. Any special provisions required (e.g., special access platforms) would be supplied by the payload customer.

References

Numerous sources were used in the compilation of this Chapter. The most frequently referred documents being:

- John F. Kennedy Space Center, Brevard County, Florida. September 2010 Volume 1
- Countdown! NASA Space Shuttle and Facilities, 2007, Information Summary (IS-2007-04-010-KSC)

- Orbiter Processing Facility Payload Processing and Support Capabilities, (K-STSM-14.1.13-revd-OPF), October 1993
- O&C Building, Payload Processing and Support Capabilities. K-STSM-14.1.14-REVE-O&C, Revision E, November 1994
- NASA FACTS, Kennedy Space Center, FL, IS-2004-09-014-KSC (Rev. 2006)
- NSTS 21492, Space Shuttle Program Payload Bay Payload User's Guide, December 2000

4

The Men and Women of Level-IV

> *"We had more diversity than in any organization on the Cape.*
> *We had that for two years in row, the occasional awards*
> *and cost saving awards; we had some really good people."*
> Bill Jewel, "Father of Level-IV"

This book is not just about the system, procedures, and hardware which created the opportunity to fly Spacelab missions on the Space Shuttle. It is also about the people who made that happen in Level-IV at the Kennedy Space Center (KSC). They came from a variety of backgrounds and career paths, but all blended together into a coherent workforce whose dedication and devotion to the job in hand created the framework upon which the missions were able to achieve such outstanding results. In this chapter, before we explore the story of Level-IV operations further, we introduce (mostly in their own words) the key personalities you will encounter throughout this story. This is the human element of Level-IV, the men and women who ensured Spacelab flew well and safely throughout the Shuttle program.

As an organization, Level-IV consisted of many different offices and functions. Some of those included and often discussed in this book are: Electrical Engineering and Software; Management; Mechanical Engineering; Operations; Quality; and Technicians (Spacelab Experiment Integration Support, SEIS). This chapter will introduce some of the people that worked most of those functions and are quoted throughout this book, with additional information contained in Appendix I. While reorganizations changed titles and the names of offices and functions, in this chapter they are listed under general categories, and in first-name alphabetical order.

© Springer Nature Switzerland AG 2022
M. E. Haddad, D. J. Shayler, *Spacelab Payloads*, Springer Praxis Books,
https://doi.org/10.1007/978-3-030-86775-1_4

PAYLOAD OPERATIONS DIRECTORATE

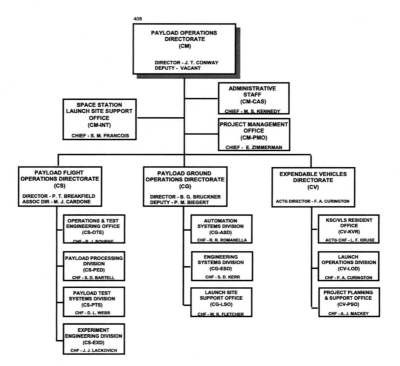

This organization chart for this time period of Level-IV would have fallen under the Payload Operations Directorate (CS). [NASA/KSC.]

Payloads had its own directorate and, like other KSC directorates, was one step below KSC's Center Director. There were other directorates, divisions and offices below the Payloads Director. Level-IV had many offices in the Payloads Directorate.

Bill Jewel: "I signed on to NASA in December 1959, my senior year at University of Tennessee. I started work in June 1960 at Langley Research Center, VA. I gained a BS in Engineering Physics and worked mainly in re-entry and meteor physics on solid rockets at Langley and Wallops Island, VA, until 1964. Then I transferred to KSC and was assigned to assembly, test and launch operations for the Saturn, Saturn 1B and Saturn V." The recognized "Father of Level-IV" explained that he was working a Source Evaluation Board (SEB) in late spring of 1980. He, the Center Director and Ike Rigel were called to a meeting where they asked Bill if he would put together a plan for an organization and operating procedures for performing Level-IV. "I was asked to develop an operation plan, origination and to help sell the concept to NASA management for KSC to assume the role of assembly, test, integration of experiments, and processing of Spacelab missions. That was projected at the time to involve about 60 percent of Space Shuttle missions. This was accomplished in the summer of 1980. At about the end of 1981, the Level-IV Division assumed the added responsibility of all Spacelab and experiment processing at KSC, becoming the Spacelab and Experiments Integration Division. Some just called it 'Hands-On'."

Herb Rice: Herb had previously led a group that wrote the ground software for handling the Shuttle, and saw an advertisement for Branch Chief over in the new Payload Directorate. He was a personal friend of the director at the time, so he decided he just had to do it. "It was kind of fun because that meant that the people who were doing the supervising [were] also actually involved directly in [the payload] hardware. So that was much more interesting than running it through contractors, which is what I had to do on Shuttle."

John Conway: "I started work for NASA right out of college in 1962, and I worked at Langley Research Center for about three and a half years before transferring down to the Cape in 1966. After Apollo, when NASA decided they were going to do the Shuttle program, they wanted to give KSC something to do because it was a big downtime. And so it was decided to give KSC responsibility for developing the checkout system for Shuttle, which turned out to be the launch processing. So I managed all of that for about 12 years. And then, when Tom Walton retired, they sent me to Payloads. I was called up by the Center Director and he said, 'I've decided to put you in that job'. So I went over to Payloads in 1985."

ELECTRICAL ENGINEERING AND SOFTWARE

Electrical Engineering dealt directly with the electrical and testing lead aspects of the payloads and experiments, whereas the Software aspect was programing the GSE and flight hardware to control the payloads and experiments.

Cheryl McPhillips: "I was graduating from a school with a master's in Electrical Engineering and a bachelor's at the same time. It was a five-year program. So I had all these jobs working defense and this one guy at Martin Marietta said, 'You're going to have to think about what you're going to tell your friends you do and if you believe in it or not'. So I came home and kept thinking about that. I never heard back from NASA, but I had an interview with Harris Corporation down in Melbourne [Florida], and I called up NASA and said, 'I'm coming over for an interview with Harris, but I really would like to interview with you guys. I never heard back from you'. They said, 'Okay, come on in', so I came in and I met with Herb Rice. He just took me to some back room; I can't remember where it was, it was like a small room with a big computer in it. Jim Dumoulin was in there and I think Scott Vangen and they were just so passionate about what they were doing. It was me and my roommate [both interviewing at that time]. She took the job at Harris, which was the highest offer we had, and I took the job at NASA, which was the lowest offer we had.

Cindy Martin-Brennan: Cindy started her career in journalism and did that for a couple of years before taking an introductory computer science course, taught by someone who actually worked at KSC. "I took my first field trip, from UCF [University of Central Florida] out to Kennedy Space Center. That's when I fell in love with computer science, fell in love with space, and within a couple of weeks I went back to the college and changed my degree to computer science. That set me on a path to eventually want to work at the space center, not realizing I'd be doing it right after I got out of college. So that was pretty cool."

Craig Jacobson: "Scott Vangen and I had been in college together and both of us wanted to work for NASA. He applied and so he was three or four months ahead of me in the pipeline. As usual, you ignore what they tell you because there's always people looking for people, and that's how it was with Level-IV. They were looking for people, but for some reason the personnel organization just was telling everybody they weren't hiring. And so I called them too, saying, 'I want to work for NASA, and I know Scott Vangen and he gave me your name, so I'm considering working for NASA'. Scott had done his interviews and he just really latched onto the Level-IV thing. And then he told me about it and I said, 'Well that really does sound like the thing to do', it was very exciting. So I bought an airplane ticket, went to KSC and talked to Bill Jewel about Level-IV. It was November 1, 1982 when I started, and then I just did the rest of the Level-IV and on with the rest of the career. Well, it was interesting because what I remember is that I came [to KSC] and we interviewed and everything, and then they put me on the PCU console to run the High Rate Data Multiplexer [HRDM] during a test. So they just started making me work right away. I was wearing my interview suit, working in PCU, doing that work. It was funny because Peter Dubock, who was the Spacelab-1 [Software and Avionics] engineer, was really impressed with my suit."

Dan Kovalchik: "After I obtained a BS in Computer Design from University of Houston [1983], I moved to KSC to process payloads with McDonnell Douglas. I saw a posting for a 'principal engineer' for a brand-new McDonnell Douglas section being formed to help NASA with its increasing Level-IV duties. Principal engineer postings were rare, but more to the point, their pay scale was on par with my management slot. I had made my decision to apply."

Diane Calaro: "I started January 2, 1991 as a Co-op. I came into checkout systems, which supported Level-IV. I started in the software group with Jim Dumoulin in his AI [Artificial Intelligence] lab, where I was an electrical [engineer] and specializing in computers. So I loved all of that that was going on. He was starting up the PON [Payload Operations Network] and I was learning LISP [LISt Processor programming language]. Right next door was the NI [Natural

Intelligence] lab and I started working my second Co-op term in there. I got the opportunity, as a Co-op, to design a Spacelab Bus Data Generator, basically a checkout tool that spit out the format for the data for the Spacelab Bus. So they were actually pieces of hardware. They were able to take out and test some piece of GSE before it was delivered to KSC. Also, we would fry a few computer chips here and they would end up in chip heaven – stuck in those Styrofoam tiles in the ceiling. I graduated in December of 1992 and I started right after that, I think it was January 23, 1993, as an engineer."

Donna [Dawson] Bartoe: "I graduated in August 1980 and I started working in October of that year, and that's about the same time the other original gang came together. I remember Catherine Clinton, she was HITS software; Ed Oscar, Maurice Lavoie, Jim (Hoppy) Cassidy. There was Ann Bolton who was PCU software and Herb Brown. I think Enoch Mosier was our fearless leader right from the get-go. Glenn Snyder was also, I believe, in the original class."[1]

Elias Victor: Elias felt fortunate to get hired by NASA as a Co-op student in 1988 after STS-26, the Return-to-Flight following *Challenger*, especially being hired into the Level-IV group under Scott Vangen and the PCU-1 group. "I met the greatest people in the world. Scott Vangen, Jim Dumoulin, Craig Jacobson, Riley Duren and Bill Sheredy. Those guys were the absolute best people that I had ever met in my life. I had spent 10 years going to school in Miami after coming from Cuba in 1980."

Gerardo Rivera: "Actually, I heard about Level-IV from Hector Borrero, one of the mechanical engineers. I started with NASA in 1979, doing facilities, electrical work at [Launch Complex] LC-39 from 1979 until about halfway through 1981, and Hector was in Payloads. He talked to me about Level-IV, about the cool work they were doing, all hands-on. And that definitely triggered my interest. So I looked into it and before I knew it I was transferred to Level-IV, working there from 1981 to about halfway through 1985."

Jim Dumoulin: "I started in Level-IV in December of 1982. After a number of interviews at several NASA centers, I was starting to get disillusioned. It seemed that actual NASA employees did not do any hands-on work and just managed contracts. I picked three areas to interview: the Shuttle Advanced Planning and Technology Office [Andy Pickett]; the Design Engineering Directorate [Frank Byrne]; and the Spacelab Experiments Division [Bill Jewel]." The first two interviews went fine, but the afternoon of his final interview in the Spacelab Experiments Division was a whirlwind. Bill Jewel ushered Jim into his office and told him he was not looking for NASA engineers to manage contracts. Bill said, "You can't manage a contract until you've done the work you are asking the contractor to perform." He said that Spacelab was new and the experiments on each mission

---

[1] Donna Dawson married STS-51F/Spacelab 2 Payload Specialist John-David Bartoe, after working with him during Level-IV operations.

were going to be unique. KSC was going to use contractors for the routine subsystem turn-around processing work (Level III/II) and young NASA engineers for all the unique experiment engineering tasks, called Level-IV. Bill had 99 openings for electrical engineers, mechanical engineers, cryogenic engineers and software engineers and at the time Jim applied, he had managed to fill only about half of them. Bill had to leave for a meeting, so he sent Jim over to Herb Rice in the Level-IV Spacelab Electrical Engineering Branch. Herb sent him to Enoch Mosier, the section chief of the Payload Checkout Unit section. Jim Dumoulin recalled that, "Everyone in the office was running around like crazy. Enoch snagged Donna Dawson [Bartoe] and asked her to show me around. She grabbed two OIS headsets, and I followed her over to a clean room where the first Tracking Data and Relay Satellite [TDRS] was undergoing final end-to-end compatibility tests, just before rollout to the pad for launch on *Challenger's* inaugural flight [STS-6]. By the time we got out of that test it was well into second shift. I didn't do much supporting, I was just observing. We then went to the Orbiter Processing Facility [OPF], got some bunny suits, and entered Space Shuttle *Columbia*, which had just returned from space on STS-5 and its tiles were streaked from the intense heat of reentry. Donna needed to inspect some pins on an electrical connector in the Payload Bay and the only time Level-IV could get Orbiter OPF time was on third shift. By the time we got back to the Operations and Checkout [O&C] Building it was well after midnight, my badge had expired and I had missed my return flight. Although I didn't get back to the personnel office to turn in my paperwork, I somehow got a job offer and started the next month."

Jim Sudermann: Jim was living in Huntsville, Alabama when he took a class on the Spacelab flight computer with a group of people from the Cape, including Scott Vangen, Donna Dawson, and a couple from McDonnell Douglas. By the end of the eight-week class they had gotten to know each other and talked about Level-IV. That was in the 1983 timeframe. "So, I got the phone call and was asked if I would like to come down, and I did. I owe a huge debt to Donna and to Scott because I think they really pushed to get me."

Joni Richards: Joni arrived in December 1989 when Astro-1 testing was in progress. Prior to that she had been working in the Shuttle program. She completed the Accelerated Training Program (ATP) but was going to leave NASA to find a place to do "real" electrical engineering when she heard about Level-IV. She talked to Dave Sollberger and was hired in. Dave introduced her to Cindy Martin-Brennan, who really needed a software programmer. "I didn't really want to do software, but Cindy was pretty insistent. So I was in charge of making sure that hardware ran appropriately in a Level-IV testing area."

Juan Calaro: "NASA came over to the University of Miami… and I think the one line that I remember from that interview that put me into Level-IV was that I wanted to work on something where I could actually see the results of my work.

And the person who interviewed me said, 'Oh, I think there's a perfect place at NASA where you would fit, where you could actually see that'. He was thinking Level-IV. Then I had a phone interview with John Batcher. He called me and asked if I was interested in [the job], and I remember telling him that I would think about it. Then I told my dad, 'NASA called me but I'm gonna see if I can negotiate the salary'. He said, 'You don't negotiate anything. You get back on the phone and you accept the job with NASA'."

Juan Rivera: Juan was developing software for the computer in the Firing Room when he got a call from his boss regarding a new group called Level-IV, and was advised to go report to Bill Jewel to attend a meeting of this new group.

Kevin Zari: "I came over from Expendable Vehicles, [where I started] my career in Hangar AE as a Co-op. The Hangar was a bit backwards with respect to technology – still using paper strip charts in 1991. My role was to rip off the paper from the strip charts after launch, hang them on a huge magnetic board, and examine the telemetry for separation of boosters and other events [reading timing codes, circling events on paper]. I wanted more. When my supervisor gave me a 'frequently annoys coworkers' review for continually asking why we didn't do things differently, I knew I needed to find another home. That is when Jim Sudermann [Zoot] said, 'Well, we'll take him', and offered me the position as a Co-op in the Level-IV Electrical branch."

Maurice (Mo) Lavoie: "In January 1979, I got a combination Electronics and Electrical Engineering degree from the Florida Technological University [now known as University of Central Florida], and went straight to work for Mr. Eugene Nelson learning the Spacelab's systems. Spacelab wasn't even at Kennedy Space Center yet. It was over in ERNO. Richard Smith was our Center Director and he gave a call over to Eugene Nelson in November of 1980 and he said, 'We're having a big meeting up here in my conference room and headquarters'. First time I'd ever been to the Center Director's conference room. He said, 'Bill Jewel was recommending that you attend, Enoch Mosier is recommending you attend. Eugene Nelson is recommending that you attend, will you attend?' And I said, 'Yes, not a problem. You need my help doing the French computers, the input/output units' displays systems, I can be available'."

Maynette Smith: "Herb Rice hired me at KSC, and I reported in on June 13, 1983, the week before STS-7 launched, when Sally Ride got her first flight. So, I started work on that day and I was put in the electrical branch. I worked the HITS/ECEP Ground Systems as well as later on the PITS system, the payload integration tests set, doing the partial payload testing."

Mike Wright: "I had always wanted to work for NASA, and particularly at KSC, but I was employed by the payload contractor. At that time, I learned from some friends about their positions in Level-IV, and thought that would be a good opportunity to both work for NASA and do more meaningful and interesting work.

There were some openings, so I applied. I was lucky enough to be carpooling with one of the NASA payload staff who served as a good reference, and I was selected."

Rey Diaz: "The NASA Deputy Administrator [Dr. Hans Mark] came to Puerto Rico and he wanted to highlight the need for engineers as part of the drive to recruit engineers from different parts of the United States. He'd talk about the need to have engineers for development of a Space Shuttle program. I was one of 60 people that he interviewed and the only one selected out of my group. I decided to join NASA because I wanted to be part of this space exploration."

Rich Jasnocha: Rich applied for the Co-op program in September of 1989 at the University of Illinois at Chicago, and received a phone call from Fred Head offering him a job at NASA. "[That] Just blew my mind. Didn't matter what I was going to be making, I was going to be working for NASA. I worked the first two quarters back to back, then I went back to school and then did another quarter, and then finished up school and graduated and started full time."

Riley Duren: "I started at Kennedy Space Center on September 12, 1988. I remember because that was about two weeks before the Return-to-Flight with *Discovery*. I hired in as a Co-op student. I was an undergraduate at Auburn University, studying electrical engineering. I was hired into Enoch Mosier's section and continued working as a Co-op until I graduated in June of 1991, and then I returned to Kennedy. But what was really funny is he [Enoch] said, 'I noticed that you're in the marching band in Auburn; the band director is my brother-in-law. We've got a community band here at Brevard County Community College, and we could use another trombone'. So, part of me thinks that the only reason I ever got into the space program is that Enoch needed a trombone player in the community band."

Roland Schlierf: "As a Co-op, I started off with the payload managers. I wanted to get plugged into something a lot closer to the hardware and all that, so I switched over to Level-IV as a Co-op with Craig Jacobson. He was my immediate lead at the time. So that was probably 1984, somewhere in there. I'd already kind of learned about flight hardware, because you can't really know how GSE works unless you know how the flight hardware works."

Scott Vangen: "The interesting thing is, when I came to KSC it was on the eve of the launch of STS-4, and I met a friend here and we saw them launch, which was fantastic. So it was my first human launch experience. I mean [that] really, really sealed the deal, if I wasn't sold already. I interviewed at KSC, I think it was the next day, with Frank Burn, who used to head up the EDL [Engineering Development Lab]. When we met, I said the EDL is doing interesting work, but it's still not spacecraft or payloads or science. And he said, 'Oh, well I need to introduce you to a friend of mine, a division chief name of Bill Jewel'. We went upstairs to his office and Frank handed me off to Bill. Bill took me up to the PCU control room. Enoch Mosier was there, who would become my first boss. Thank my lucky stars. Enoch was in the control room and Bill said, 'Here's an engineer, he's looking for the kind of stuff we're doing. Make sure he gets the full show and then bring him back to my office'. And so Enoch [Mosier] introduced me to people that we worked with in the early days; Donna [Dawson] and Eileen Lunceford and Mo Lavoie and Glen Snyder and the kind, of Level-IV electrical. They talked about the control room... 'This is the checkout system'. I said, 'Oh, that's very cool, what about the flight hardware?' That was the next step on the tour. Enoch took me down to the floor and Herb Brown was down there hooking up EPDBs [Electrical Power Distribution Boxes], getting the subsystem floor ready for Spacelab. [He said] 'Did you know we're going to build up the racks with our mechanical engineers and we're going to put them in these experiments? And as a test conductor, you'll get assigned experiments and you'll be hands-on writing the procedures'. It was a dream come true. I mean, this couldn't have been better. They took me back to Bill Jewel and he said, 'What do you think?' I said, 'When can I start?' and he said, 'When *can* you start?' I just needed to fly home and pack a bag and he said, 'Do it'. And so within two weeks I was back and started work at KSC. I could not have found a better job with better people at that time in my career."

Tracy Gill: John Conway used to vacation at the hotel that Tracy Gill's grandmother managed, which is how he knew about Shuttle Payloads. He was attending the University of Florida when one of his roommates said he had a call from Fred Head at NASA. Tracy Gill recalled, "[I said], There's no way there's anybody named Fred Head from NASA. You guys are jerks. So then, couple of days later, he called me again and he said, 'This here is Fred Head from NASA'. So I came in to the networks group, under Fred, and worked with Herb Brown and folks like that, [where I] started doing networks and electrical networks and power systems, power supplies."

MECHANICAL ENGINEERING

The Mechanical Engineering function dealt with the nuts and bolts of assembling and integrating payloads. Any structural work was performed by the mechanical group as well as any fluids and gases work, which included servicing payloads with cryogenic and hazardous propellants.

Aaron Allcorn: "I started my career at KSC in the Shuttle APU/Hydraulics group, then moved to the Level-IV area of the Payloads Directorate because I heard the NASA engineers there got to do 'hands-on' work."

Alex Bergoa: "I graduated in Mechanical Engineering from the University of Puerto Rico, and around December 1990 time frame I got a call. I think at the time there were a couple of positions open. One was for the mechanical group in Level-IV, another one for the fluids group in Level-IV. Actually, it was Luis Delgado who I believe called me for the fluids position. Based on my background, I decided to go with the fluids group. February 25, 1991, I started working Level-IV."

Angel Otero: "My first day at Level-IV was February 1, 1983. Historically, 20 years later, [that] ended up being the day of the *Columbia* accident, but it's kind of a hush reminder from my anniversary. I was hired right out of college. I got interviewed at the University of Puerto Rico by a generic NASA interviewer. I did every NASA center that was there and to be honest, my preference at that time – and ignorance is bliss – was to work at Huntsville because, being a fresh graduate in Mechanical Engineering, my idea was design. I wanted to design and the KSC interviewer was honest with me. He said, 'You know, we're not a design center, we're an operation center'. Later, I got a letter from KSC offering me a job and I thought, 'Well, at least I'm getting to NASA and then I can move from the NASA center in Florida to Huntsville'. Back then, they put you in a pool and rotated you in your first year at KSC to different jobs. I think you were working three months in [one] office and then went to the next office and so on. After I was a Level-IV for a month, I said, 'I'm not going anywhere else'. I couldn't imagine anywhere else any more fun than what I saw the people around me doing. So I asked to not be moved [and] I was never moved. I started Level-IV and stayed there my whole time at KSC."

Bob Ruiz: "I'd heard about it [Level-IV] from an engineer or project manager in the Shuttle side. I went and talked to Dean Hunter. Dean said, 'We're doing this new operation here, doing hands-on, and we're traveling to Germany and then Italy. I thought, 'Wow, that sounds great'. I was hired by NASA in December of 1980 and I went into Level-IV in March or April of 1981. When I showed up in there, Louis Moctezuma and Tony Ornelas and Barry Bowen were already there."

Bruce Morris: "I started down at Kennedy on July 11, 1983. NASA was the only place I wanted to work. I was lucky, [because] my grades weren't that good. They were okay. I was working in the aero engineering lab at Ohio State and I was lucky enough to get some verbal interviews. One of them was with Bill Mahoney, who was the branch chief for Level-IV at that time, and he told me about two different jobs. One was over in the Spacelab office and one was in Level-IV, and he was smart enough to tell me he really want to put me in at Level-IV because my hands-on experience would be more applicable in this kind of activity. So I was lucky there too. Bill was pretty straightforward about where he wanted me and it sounded like a lot of fun, and it proved out to be all of that."

Damon Nelson: "I had been a Co-op, pre-STS-1, at the Johnson Space Center. I did two stints at JSC. For my second Co-op tour, I worked APU [Auxiliary Power Unit] qualification for the Space Shuttle and then, when I hired on full time with NASA KSC in May of 1983, I was hired on as an APU systems engineer for a short period of time. And then I got wind of an opportunity in the Level-IV payloads group and that's when I transferred over, in the summer of 1983. When I moved into the Level-IV group, they had just announced this payload called the Orbital Refueling System that just happened to use excess APU parts. So it seemed like kind of a natural fit that I had that APU background."

Darren Beyer: "I met Joni Richards during a payload operation out at one of the launch pacs. She said, 'You ought to come work at Level-IV. It's a great group doing all the hands-on work.' [So] I moved over from what was the Shuttle-side payload operations group over to the experiments side for payloads. And that was the best move I ever made. I absolutely loved my time there. It would have been early 1991."

Debbie Bitner (Wilhoit): "I was hired by NASA, sight unseen. I sent an application to NASA right after I graduated from college. I got an offer letter in the mail; no interview, nothing. I wanted to see what I'd be doing, so I told them I couldn't accept the offer without coming down for an interview. I had them fly me down, went and interviewed on the Shuttle side ECS, and eventually ended up meeting Bill Mahoney and decided to come work Spacelab. They had me working on the Spacelab side overseeing the contractor for a while until OSTA-3. And that's when they gave me the ground support cooling unit to get up and running. So that was my first experience with Level-IV."[2]

Eli Naffah: Eli was attending the University of Central Florida in the early 1980s, and was working as a Co-op in design engineering at Kennedy Space Center when he heard about a new office that was being established under Bill Mahoney to give hands-on experience to young engineers. That sounded very exciting to him and actually it was his cousin, Sam Haddad (incidentally, the co-author's brother) who knew some people over there and arranged for Eli to talk to them. "They were just starting the office at that time. Bill Mahoney was the Branch Chief, Dean Hunter and Joe Lackovich were Section leads. They had Bob Ruiz and Antonio Ornelas as engineers and Barry Bowen who was also a Co-op student. They called him Coop-1 and I became Coop-2."

Frank Valdez: "I was recruited by Bill Mahoney. He was on detail to the Orbiter side for the first launches. I was working Orbiter back then, and he said that he was starting a group and they were going to be doing some hands-on work. So after I was able to be released from the Orbiter side, I went to work over there, starting in April 1982."

Gene Krug: "Bill Mahoney was watching how I interacted with the Rocketdyne techs, the NASA people and everybody. I worked a lot with the techs and he [Mahoney] had yet to get started down in Level-IV. I went to see Mahoney and he said, 'I'm starting a new branch down in the O&C Building and we're going to be working on payloads for the whole world, the European Space Agency and everybody, [so] how about coming down here with me?' He had Dean [Hunter] with him. Back then, I thought Dean was just a technician or something, I didn't know he was an engineer. We talked back and forth and I got to know Dean a little bit. They said, 'We sure would love for you to come on board', so I agreed. I got

---

[2] Debbie Wilhoit met and later married Gary Bitner while working Level-IV.

talking with Dean and they said, 'You got the job'. There was no announcement or nothing. That's how things worked then in the real world."

George Veaudry: "I started really in Spacelab in 1981, doing the checkout of the Spacelab Ground Support Equipment that was going to be utilized for the test, and the checkout of the Environmental Control Life Support System. I then turned to the testing and checkout of the same systems. I got my first opportunity to do a hands-on work for two payloads, the Astro-1 mission and MSL."

Jeannie Ruiz: Jeannie had been a Co-op at Auburn with NASA in the VAB Site Management before starting in Level-IV. She had only a phone interview with Bill Mahoney and was hired in the fall of 1982.

Johnny Mathis: Johnny started as a pre Co-op, which is a scholarship from high school, in malfunction analysis (at KSC) during the summer of 1984. It was a very interesting job. "I met Dean Hunter and he took me down to the Bay, showed me around, and I just fell in love with it. I heard a gentleman that now I understand was [from] Quality, asking Roberto Tous [Level-IV Mechanical Engineer], 'What says you can do that?', and Roberto held up his class ring and said, 'This does'. And I [thought], 'I want to work here'."

Josie Burnett: "I came back from The University of Florida in January 1987 after the Christmas vacation. I came back a little bit early, a week before everyone else did, because I wanted to prepare for these interviews. I didn't know that NASA was hiring because of what had happened with the *Challenger*. Well, lo and behold, NASA was there that week. They came a week early [and] I got an interview that week. It was almost like I wanted to say, 'Give me a job', because there weren't as many people there. So I interviewed with an HR person who said, 'Well, you'll hear back from us.' Later, I got a phone call. It was Dean Hunter who called me and who ended up hiring me. Dean has since passed, but he was just a wonderful man. I basically was telling him, 'No, I don't want the job. I want to see what it is before I hire in'. I didn't know what I was doing, but got introduced into NASA and introduced into Level-IV. I saw people like Mike Haddad, young people that looked to be my age and there was a diverse group of people there. There were women. They just looked like a very fun group. I got hired in May 1987. After the *Challenger* accident, there were two other female mechanical engineers before me that were hired year before me, Tracy Hill and Carrie Backus. So they were kind of my mentors I suppose at the time."

Luis Delgado: "I got hired on just before Christmas 1984. I got assigned to the Vertical Processing Division first and I'd watched a couple of operations there when Dean Hunter came back from being out of the office. I talked to him and said I wanted to do hands-on and then I ended up moving to there. He asked me, 'Do you want to be fluids, or mechanical, or structures?' And I said immediately, 'Fluids, they are a lot more dynamic'. I was lucky. I mean, I got handed all these jobs because the person that had all the jobs before was overwhelmed."

Luis Moctezuma: "I guess I was at the right place at the right time. I was part of the original mechanical group, with Dean Hunter, Bob Ruiz, Tony Ornelas, and our Co-op Barry Bowen. My experience and knowledge of surveying instrumentation may have been one of the things that got me into this group. I used to work as a surveyor in Puerto Rico. Although optical alignment is more precise than surveying, I guess managers at that time were looking for somebody who at least had knowledge of the instrumentation. Training and certification would do the rest."

Mark Ruether: "Prior to coming to Level-IV, I started with NASA at KSC in December of 1983. I had worked as a fluids and pneumatics systems engineer in the launch systems organization over in the Shuttle part of the Shuttle program, and then came over to Experiment Integration. Mike Kienlen, one of the engineers that I worked with when I started over in the launch systems group, came over to Level-IV integration and he loved it. He said that was a great place to be because you get the types of hands-on work that a lot of young engineers are looking for coming out of college. We really didn't have that over in the group that we started with, and he said, 'You've got to come over. It's great work, great people, and so forth'. And so finally he convinced me to come over and that was the person that got me in. That was October 1985."

Mike Hill: "Well, actually I was in college and we had a mandatory Co-op program, and there was this guy that had interviewed with Dean Hunter and turned down a Co-op with the Level-IV guys. I told my counselor that I was really wanting to work for NASA, so I actually got in as a second opportunity. I guess I didn't have the first chance at it. That's how I actually got in. Apparently, Dean Hunter hired me because I said I played softball, of all things. Softball was a big thing back then. So I started as a Co-op, I guess in the spring of 1984."

Michael Stelzer: Michael was hired in as a summer intern with the NASA KSC Spacelab & Experiments Division (SED) in May of 1983. As a level GS-2 summer intern, he was low man on the totem pole -- incredibly, even a notch down from the NASA Co-ops. But he would not have known it with how the management, leads, engineers and other team members treated each other. There was an abundance of respect and confidence shared across the organization. This, along with a 'can do -- pull ourselves up by the bootstraps' approach, created a phenomenal work team environment. As Michael Stelzer recalled, "The Spacelab program was ramping up quickly and they needed a bunch of help, so even with my complete lack of spaceflight experience, they put me to work right away. Working with David Steele, Michael Gisondi and later Bob Ruiz and Bruce Burch, I was assigned to the development of a set of ground-based equipment called the Spacelab Module Vertical Access Kit (MVAK)."

Sharolee Huet: After graduation from Ohio State in June 1988, and working odd jobs, Sharolee started in Level-IV in November 1988. Information from her in this book came from an interview performed in 1999. Sadly, Sharolee passed away on October 11, 2018.

Tia Ferguson: "When I was in college at Tulane in 1989, I submitted an application to both Marshall Space Flight Center and the Kennedy Space Center. It was Kennedy who called me back first. They interviewed me over the phone and offered me a job and I said, 'No way, not until I go see this place'. And so I paid my own money, got in the car, and drove down to Kennedy Space Center, where Carrie Backus took me around. There were two areas offered, fluids, and mechanical design, in Level-IV. And it was interesting when I was being interviewed, because I really didn't know what integration was but I knew that there were some cool racks down there [and] that I was going to get to get some hands-on experience. So I jumped at the opportunity. Interestingly, the day that I accepted the offer, Marshall Space Flight Center called me back and I had to tell them that I had already been hired. February 5, 1990 was my first day there in the mechanical group."

Tony Ornelas: "I guess they were looking for people that were ready to go to work, and Dean Hunter asked me if I wanted to join a group. I had been picked up already by the organization, NASA, and he asked me if I want to try this group, and I said, 'Yeah, heck yeah, let's just do it. It sounds like a lot of fun'. I don't think I had more than six months experience. And then Ernie Reyes said to Bill Mahoney, 'You come pick this guy up, or not, because if you do not, I'm going to pick him up'. So I guess he kind of pushed the button to get me signed on officially with the experimental section; that's how I got started at that time. It was just basically me and I think real soon after that we picked up Luis Moctezuma and then we picked up Bob Ruiz."

OPERATIONS

Operations dealt with support for the engineers to help accomplish their tasks, as well as being the interface between Level-IV personnel and other organizations. They also ran a number of tests, and created and maintained the schedules used by Level-IV.

Gary Powers: "We didn't have a test and operations at the time, so my job was to come in and implement a plan and pitch to not only the NASA team, but to the foreign national teams because they, especially the foreign nationals, did not understand how testing operations went."

Lori Wilson: "I started out working as a Co-op student for Mr. George English while I was still in school. I started late in life, after my husband died. I guess that's why some of the guys called me 'Mom'. I remember that there was talk that the EEO [Equal Employment Opportunity – anyone could get into any job without discrimination] Office was getting up to speed about the same time that Level-IV was being planned. The idea was to introduce and train new engineers to work for NASA. Coincidentally, the group was formed predominately of blacks, women and Hispanics. I look at it as having changed 'the face' of NASA at KSC."

Mike Kienlen: "Starting from a previous KSC job that was not satisfying, I talked to friends and I said, 'Where can I go work at Kennedy Space Center that does real engineering?' and everybody said, 'You've got to go to Level-IV'. I went down, got a meeting with [Bill] Mahoney and he said, 'Well, I don't have any openings in Mechanical Engineering, but if you really want to come here, I could put you with Gary Powers in Operations'. I said, 'Fine, I'll go anywhere'."

Todd Corey: Todd gained a BS in Aerospace Engineering and Mechanics from the University of Minnesota, and began work in Level-IV in November 1988.

QUALITY

Quality was the second set of eyes for all operations performed by Level-IV and was responsible for ensuring that work performed via the Level-IV procedures was documented correctly. They were also responsible for documentation of any problems and ensuring engineering closure of those problems.

Bob Raymond: "Actually, it goes back to the 1970s when I was in the air force. We were doing initial operational test and evaluation on a new airplane. We had a test pilot out of Edwards who came in with us, and we were working to open up the air refueling envelope of this new airplane. He told me that I really ought to consider going to NASA. He explained to me he was coming to NASA. 'I'm going into the astronaut program, so [you can] kind of drop my name around and see what's out there'. His name was Dick Scobee, he was the Commander when we lost *Challenger*. So I took that loss personally and always tried to do the best I could, especially in those circumstances, to make sure that these guys got to fly safely. So that was my in to NASA. I left the air force, I did what I agreed to do. I flew as a corporate pilot for a little while and did some other things. Then one day I got a call from NASA, Kennedy Space Center, who said, 'We heard you may want to come work for us'. I pretty much said, 'When do I start?' And I came on board in 1985, in Quality Assurance in the Spacelab and experiments division."

TECHNICIANS

Initially the only contractor working Level-IV, technicians provided the direct hands-on work for the hardware (mechanical and electrical). They would be responsible for tasks like the turning of the wrenches and fabrication of the cables, a job the engineers could perform if necessary. Spacelab Experiment Integration Support (SEIS) technicians also performed some clerical, monitoring and logistics functions, but only the technician comments will be provided here.

Bruce Burch: Around 1980, I was working for Boeing. I got a job with Boeing working in the plant that did the air conditioning for the VAB [Vehicle Assembly

Building]. I was working in the chiller plant there when Boeing went on strike. They launched the first Shuttle while we were on strike. When I came back from being on strike, because I was an air conditioning mechanic I got laid off. While I was laid off, I got contacted by Boeing management about another contract they were starting up in Level-IV, and they needed an air conditioning mechanic. They had some rack and ground conditioning units that needed somebody with a refrigeration background. So I went in and talked to them and got picked up to work in Level-IV and that's where I started. [I joined] Level-IV to work on air conditioning units."

Don Dobey: "I started in August of 1984. I'd been trying to get on there for a while and basically I was trying to get on through Jay [Smith], because he was the only one I had any connection to out there. I sent my resumé in and everything [but] it's hard to get out there unless you have some kind of connection to pull you in. I was able to get an interview and then finally got hired in August of 1984, and I was there till we got laid off in February 1987 after *Challenger*."

Gary Bitner: "I just applied at the [KSC] space center. I was living here and I got a call from Boeing Aerospace for an interview. I was thinking, 'This isn't going to be good' as I walked in the door and everybody was sitting there in suits and ties. Skip Montana called me in and we talked for a long time. He was impressed with my fluids experience, because I've done a lot of big commercial air conditioning work with fluids, and we talked about a few things, pressure, temperature relationship and things like that. He said, 'I'll let you know'. Let's say six months went by when I got the call to go back down [to KSC] for a quick interview and that's how [I got in]. Just Skip Montagna and Paul Sharpe, I think, were the two guys in charge and I met Bruce Burch. He was an air conditioning guy, so he was my buddy right off the bat."

Jay Smith: "I had previously built up a lot of skills with machinery, metal, and equipment and stuff, learning how to be a sheet metal mechanic. It was precision stuff. I learned a lot about measuring and precision manufacturing, in other words super-tight tolerances. I went and interviewed with Skip Montagna first time but did not get hired, but in another interview a couple of months later I was a little more prepared, and was able to articulate my experience and what I could do for him. So he ended up hiring me. I was a valuable part of the team."

With this team of people, Level-IV went on to perform a number of tasks that would lead to very successful Spacelab missions. Mike Haddad noted: "I don't think there could have been a better mix of people and personalities that worked so well together, overcoming some very big problems, to get the Spacelab missions off the ground."

Even after both the Spacelab and the Shuttle programs ended, this tight group of individuals remain good friends to this day and still stay in touch with each other. Many times, they get together for dinners or ski trips, and to talk about what is going on today in their lives and all the good times experienced during their Level-IV days.

5

Creating a System that Worked

*"We worked through numerous options
for juggling our assets to try and
accommodate the Shuttle program."*
Todd Corey, former Payload Operations Manager.

This chapter deals with the generic ways of doing business in Level-IV. It will give a general overview of processes that were created, tasks that were performed, tools used to accomplish those tasks, outside activities required to perform and/or prepare for Level-IV tasks, and other important information on the Level-IV job. Think of it as "a day in the life of Level-IV".

OPERATIONAL TASKS, INTEGRATION, AND CHECKOUT FLOWS

The operational tasks required to support flight experiments involved many aspects of planning, pre-flight operations, mission support and postflight operations. The Level-IV/Experiment Integration (EI) personnel were major players in all these tasks, which were essential to ensure no details were missed and everything that could be done was being done to make each mission a success.

Selecting Experiments for Flight

A very structured and rigorous process was required to select any item for approval to fly on the Space Shuttle. While details of that process will not be discussed in this book, a few topics are mentioned here to provide a history and timeline of how the process was influenced, and needed to be considered, by Level-IV personnel.

© Springer Nature Switzerland AG 2022
M. E. Haddad, D. J. Shayler, *Spacelab Payloads*, Springer Praxis Books,
https://doi.org/10.1007/978-3-030-86775-1_5

Several early Space Shuttle missions flown between 1981 and 1982 involved Spacelab components, but there were only a limited number of Spacelab modules, pallets, and Mission Peculiar Equipment Support Structures (MPESS) available. At that time, only two Orbiters were operational, *Columbia* (OV-102) and *Challenger* (OV-099). The third flight vehicle *Discovery* (OV-103) would not arrive until November 1983, followed by *Atlantis* (OV-104) in April 1985. Selecting the experiments and supporting hardware to fly on those Spacelab elements, and which Orbiter to use, had to be carefully considered when planning the manifest.

Once the experiments were approved to fly, the next step was selecting the mission to which they would be assigned. This selection process was called "*Manifesting*". Once an experiment was manifested, this would usually signal the start of the planning effort for Level-IV, but sometimes personnel would be assigned before this time to assist in the manifesting effort. An experiment could be a non-Spacelab experiment but manifested to fly using Spacelab components, so Level-IV personnel would be responsible for those experiments as well.

The lead Multi-Flow Planner for payload operations oversaw a small team of NASA and contractor personnel for the utilization of Spacelab assets, such as the Spacelab modules, racks, floors, test stands, and checkout systems. Based on Shuttle flight manifest options provided to the team by the Shuttle program staff at JSC, they would determine how best to utilize the assets to support the flight manifest, and any proposed changes to the flight sequence or calendar spacing. "We worked through numerous options for juggling our assets to try and accommodate the Shuttle program," recalled Todd Corey, a former Payload Operations Manager who became one of the Multi-Flow Planners. "Based on some of our hardware availability constraints, the Shuttle program made adjustments to their proposals."

Long before the hardware arrived at the Kennedy Space Center (KSC), EI personnel were assigned an experiment (sometimes multiple experiments) for each payload. Starting approximately 2–4 years ahead of time, they would travel to locations all over the world to participate in the many design reviews, meetings, schedule reviews, preliminary testing and other events required to prepare the hardware for arrival at KSC. This included not only the flight hardware, but also the Ground Support Equipment (GSE) that was required to support the flight hardware operations. In some cases, the experience of the EI engineers shaped how the flight and/or GSE hardware was designed and built [1].

As NASA Level-IV Electrical Engineer Roland Schlierf noted, "For me, in Level-IV you got a full dose of both flight and ground, even to a small extent as a Co-op, because you had to understand the flight in order to do the ground."

Level-IV engineers always worked multiple jobs. They were assigned to an Experiment Ground System Equipment (EGSE) subsystem based on their field of study, as well as to one or more primary and backup flight assignments. Jim Dumoulin recalled, "If your primary mission was next in line, you concentrated on the flight assignment and someone assigned to a later mission picked up your EGSE duties." In the early 1980s, NASA was working towards a goal of launching the Shuttle every two weeks, with an equal mix of scientific, commercial and military payloads. At that rate, there would be a scientific payload every six weeks, half of which were expected to use some variation of Spacelab hardware. In the event, launching a Spacelab every three to four months proved to be highly unrealistic; the build-up of payloads in the pipeline was always kept as full as possible.

Establishing a Network of Contacts

This was also the time to meet and get to know the people EI personnel would be working with over the next several years. This included managers, engineers, technicians and, most importantly, the Principal Investigator (PI), who was the main point of contact for each of the experiments. A PI was usually responsible for only one experiment; rarely would they have multiple experiments. The experiment could originate from academia at universities, from a company, or from some foreign nation or their domestic space agency.

For some of the larger Spacelab missions, a single EI person could interact with many different PIs at the same time in different locations. Establishing a good relationship with the PIs was critical to assuring success of the mission. The PI and EI personnel worked together to accomplish all that was required to achieve a 100 percent operational experiment before flight. Most often, the PI became friends with their EI counterpart, and that friendship lasted well beyond the end of the mission.

As Alternate Payload Specialist (APS) Scott Vangen explained, "Put yourself in the Principal Investigator's position. You've got your million-dollar instrument and this young guy or gal comes from the Kennedy Space Center, with just maybe a few years' experience under their belt, explaining they were going to work with you to make sure your experiment is physically integrated onto the Spacelab system at KSC. They were going to test it to both your and their satisfaction, help you and us troubleshoot any problems, and get you ready for launch. The experimenter was probably thinking, 'We'll get a real veteran from the Apollo era'. Well, that changed really quickly. I think, probably without exception, we were always embraced because they quickly learned — whatever they thought was going to happen at KSC — [that the reality] overwhelmed them when they arrived, and to have an advocate that was going to be there to pull them through that was huge. I understood their position and I think most of us did. It was their baby and they didn't want to literally hand their baby over to a stranger who says, 'We're going to take care of your baby'."

NASA Level-IV Mechanical Engineer Tony Ornelas added his own recollections. "I experienced that kind of feedback from some of the PIs, but not JAXA [Japan Aerospace Exploration Agency]. Maybe because it was only a physical integration onto the hardware and onto the platform Orthogrid, and maybe it was because they were out there [in Japan]. But we had a really good working relationship. I think once they saw that we could run those procedures and accommodate them — and being part of that effort is trying to get comfortable with it — I don't think I remember anybody ever being hesitant in turning over their hardware."

It was also essential that Level-IV personnel learnt the systems that would test the PI's experiments and established contacts with the people that designed and built those systems. For Spacelab, Bremen in Germany was the original systems integration site, so before they sent to anything over to KSC, it was checked out there, and many from Level-IV traveled there to get familiar with hardware and software.

The Tethered Satellite System (TSS) was another good example. Its multiple components were being integrated and worked on in Turin, Italy. Scott Vangen explained that a half dozen people were assigned to that mission. "We kind of went there as a force. We were backups to each other on that one as an example."

Not many jobs would give you this much exposure to such a unique group of highly specialized people from all corners of the globe. Mike Haddad recalls, "I would say to myself many times, 'Wow, this is so neat working Level-IV, because where else would I ever run into these people, doing this kind of work and being part of laying the groundwork that resulted in making missions to space successful?'."

It was an incredible experience to begin a relationship with these people from all around the world, but while exciting, it also posed some problems that needed to be overcome. Most would learn English, as the Level-IV personnel were less likely to learn multiple additional languages. That created barriers at times that had to be worked through, but there was also body language, as Scott Vangen explained. "I remember, to communicate with the Italians, [the issue] was the head nodding, in that a vertical head nod didn't mean yes, nor did a horizontal nod mean no. It was much more subtle than that. And it's almost opposite, but not quite. The vertical nod was acknowledging that they heard you, but not that they agree with you, and the horizontal nod was not so much a 'No', more 'I'm pondering what you're saying'. The Japanese were similar, where a nod was 'I acknowledge, I have heard what you've said, but that is not feedback that I concur with your statement'. Another thing [we] experienced with the Japanese team and JAXA was they needed to work it through their group to some senior person, who agreed and concurred on a course of action. There's a protocol in the Japanese that you're unlikely to just go quickly get an answer from someone and execute."

CONFORMING TO KSC PAYLOAD PROCESSING GUIDELINES

The capability of Spacelab, and the possibilities for performing virtually any kind of science research in space with a number of countries participating, made for an exciting time in the Shuttle program.[1] The fields of science to be performed on Spacelab missions included: astronomy/astrophysics, atmospheric sciences, life sciences (plant, animal, and human), materials sciences, low-temperature physics, Earth sciences, remote sensing, solar physics and space plasma physics, space technology demonstrations, fluid physics, and crystal growth.

EI personnel worked with domestic and international universities, private companies, government organizations, and other space-related centers, at locations across the globe. While the science aspects of Spacelab were eagerly anticipated, having EI personnel dealing with such a wide variety of disciplines and processes from so many different cultures was equally exciting, but also presented a really difficult challenge. It should be noted that different processes existed even between NASA's own field centers, let alone all the domestic organizations.

So how could EI personnel learn all their different processes? Simple − they did not. It would have been literally impossible to learn all those processes, so it was decided early on that just one process would be established at KSC, and that all customers would have to conform to that process. This concept often went down like a lead balloon with the experimenters. After years of working using their own processes, they now found they would have to learn a whole new process with NASA. Initially, it was difficult for them to accept this, but eventually they realized it was necessary.

The other very controversial process was that after the experiment was handed over to KSC, the EI personnel would become the responsible person for the experiment, not the PI as you (and they) might think. Naturally this was very difficult for the PIs to accept, as it was mandatory for this change of responsibility to take place. But why? Yes, the PI knew their experiment like the back of their hand, but the EI personnel knew the STS/KSC/NASA systems, and what was required to make sure that experiment interacted correctly with these systems to ensure everything possible was completed pre-flight at KSC, to aim for 100 percent success during the mission. This change did not mean the PI would not be involved. Even though EI personnel controlled what was to occur with the experiment, the PI would be working side-by-side with the EI personnel the entire time. One other fact to note. Unsurprisingly, each PI naturally thought their experiment was the most important project and should be given the highest priority in all aspects of KSC operations. The EI personnel had to be diplomatic in juggling the experiment operations based on what was occurring at the time, so the experiment requiring

---

[1] These countries included: Austria, Belgium, Canada, Denmark, France, Germany, Italy, Japan, the Netherlands, Spain, Switzerland and the United Kingdom.

the most immediate attention would frequently change, often with no love to the PIs of the other experiments. These concepts were not suddenly sprung on the experimenters at the last minute, but were relayed to them at the earliest possible time in the planning effort so the team could get over that hurdle and start working together. The result was that the EI personnel became as knowledgeable on the workings of each experiment as the PI.

Best Bang For The Buck

According to Scott Vangen, the mission managers advocated that "they wanted the best science bang for the buck. And they would, for instance, push against the flight team flying the Shuttle, saying, 'We want to do this attitude and this maneuver and do these things as well'. In the end [the important thing was] crew safety, bringing them home, [having] a successful mission and making the PIs happy. If you kind of kept those priorities, it worked." It was not easy but decisions sometimes had to be made, such as turning off a science experiment on orbit because it jeopardized others. "By the time you ran through the logic, through the system, that Principal Investigator team understood. [But] There were other times, too, where the Shuttle gave up things. [Maybe] we really wanted to do a roll maneuver for some thermal aspect, but if we did that, we'd lose the science target and we would not get it again. 'You're saying you can't live with 10 degrees more for 30 minutes?' But then someone would give in. The PIs would sincerely thank you, saying, 'This will allow us to write 20 new papers and keep ten people busy for five years writing the science'." According to Vangen, everything had to be integrated, and that was where having good flight directors and good mission managers made all the difference.

Ground Integration Requirements Document

As an experiment neared completion by the Payload Developer (PD), the PIs began to work with EI personnel to ensure experiment requirements were documented correctly in the Ground Integration Requirements Document (GIRD). The GIRD was the formal document in which PD requirements were communicated to KSC. One GIRD existed for each payload, so the requirements for all experiments on the same flight were included in the same GIRD. All payload activities and support services at KSC had to be requested in the GIRD. These included, but were not limited to: mechanical; thermal; electrical power; command and data handling; contamination control interface requirements; installation of the experiment onto the carrier; testing of the experiment; any required calibration; and any necessary maintenance. If any activity was not listed in the GIRD, it would not be performed at KSC unless it resulted from an unforeseen problem [1].

These requirements were then translated into support requirements and documented in KSC's formal response to the GIRD, called the Launch Site Support Plan (LSSP), which was Annex 8 to the NASA/Johnson Space Center Payload Integration Plan (JSC PIP). If KSC was unable to fulfill some requirements based on existing resources, this was thoroughly discussed with the PI and the Mission Manager to determine a solution before the experiment arrived at KSC.

The GIRD also contained a section on contingency landing requirements. The list of deliverable items and the contingency landing requirements were used by KSC to develop the Off-Site Plan, in the event the Shuttle could not land at the primary landing site for that specific mission. The Off-Site Plan would include: timelines; support hardware; support personnel; and a variety of other requirements needed to ensure the experiments were taken care of properly if a contingency landing took place.

Operations and Maintenance Requirements Specifications Document

The Operations and Maintenance Requirements Specification (OMRS) for the Space Transportation System (STS) were listed in the Operations and Maintenance Requirements Specifications Document (OMRSD). This included requirements for the External Tank (ET), Solid Rocket Boosters (SRB) and the Orbiters. This book will only discuss the OMRSD and how it related to the Orbiter component of the STS, because experiments and payload had no interfaces to the ET or SRBs. While direct experiment requirements were documented in the GIRD, the OMRSD was used for any Level-I Spacelab-to-Orbiter requirements or Orbiter support requirements. Types of OMRS would include: Spacelab-to-Orbiter electrical and mechanical interfaces; payload envelope checks; bonding requirements; contamination control; Orbiter-to-Spacelab Integrated Verification Test (IVT) requirements; access requirements; Launch Commit Criteria (LCC) requirements; landing requirements; and Shuttle Carrier Aircraft (SCA) ferry flight requirements (for those missions that did not land at KSC).

A Paper Mountain of Documents

In the days before digital documentation and online storage, paper was king... and there was a lot of it. To illustrate the sheer volume of documentation that EI and experimenter personnel had to understand and conform to for processing each piece of hardware at KSC, the *majority* but not *all* documents, if stacked on top of one another, would reach from the floor to the ceiling. A few examples of these documents included: Management of Facilities; Systems & Equipment Handbook; Unescorted Access and Personnel Reliability Program; KSC Security Handbook; Technical Operating Procedures Policy; KSC Radiation Protection Program; Space Transportation System Payload Ground Safety Handbook; and many others.

Some offices setup their own library of documents for all to reference as needed. As changes came in for those documents, Level-IV personnel would need to read and understand those changes and confirm the work required to satisfy those changes.

According to Rey Diaz, "We evolved it into what was called the Experiment Project Engineer. [I] took the role of looking at the set of experiments and then looking at it from the mission level. The main thing that I had to work here at the Kennedy Space Center was to ensure that we understood all the mission requirements for ground processing, all the test requirements it was supposed to go through, the Level-IV/Experiment Integration testing, the Spacelab subsystem testing through the Level III/II test stand, and then finally, all the mission requirements that interfaced with the Orbiter."

PAYLOAD/EXPERIMENT REVIEWS

There were many different reviews that occurred in preparing the experiments for the mission, though only the major ones are mentioned here. These reviews focused on the engineering design and operations of the hardware.

Systems Requirement Review

The System Requirement Review (SRR) examined functional requirements and performance requirements defined for the system, and the preliminary program or project plan, and ensured that the requirements and the selected concept would satisfy the mission.

Preliminary Design Review

The Preliminary Design Review (PDR) demonstrated that the preliminary design met all system requirements, with acceptable risk and within the cost and schedule constraints, and established the basis for proceeding with detailed design. It showed that the correct design options had been selected, interfaces identified, and verification methods described.

A number of documents were examined by EI personnel as part of the PDR process, but one of the most important EI responsibilities included reviewing and commenting on the experiment drawings and schematics. At that time, these were NOT computerized but were all on large sheets of paper, containing many sheets (paper pages).

To review these, EI personnel normally had them spread out across their desks, or taped individual sheets to the office wall to allow viewing of multiple sheets at the same time. Despite this now antiquated approach, it did help to identify problems and prevented the need to flip back and forth between sheets. From a mechanical engineer's standpoint, this required many weeks of studying the drawings, to become familiar with what the hardware should look like once assembled. EI personnel had to figure out the best way to assemble these pieces together into the Spacelab racks, onto the pallets and/or MPESS, which frequently revealed errors in the drawings and overall design. Some of the errors would be as simple as a misplaced reference number, quantity of screws, or the wrong size washer suggested for the type of bolts selected. More complex problems involved cable harnesses that changed locations depending on which sheet of the drawing was being viewed; total experiment assemblies that interfered with Spacelab support structures or hardware; hardware that extended beyond the payload envelope requirement, thus interfering with other hardware; wrong torque values for the hardware being torqued; pressure values beyond what was expected; wrong hose sizes or type of fittings; and many others.

The same was true for electrical schematics, wiring diagrams and electrical components. A great deal of time was required to understand the function of each electronics box, how it was supposed to "talk" to the other electronics in the Spacelab, the cabling in between, and the software used to command the systems. As with the drawings, this process frequently eliminated errors in the schematics and overall design.

Critical Design Review

The Critical Design Review (CDR) demonstrated that the maturity of the design was sufficient to support proceeding with full-scale fabrication, assembly, integration, and testing. CDR determined that the technical effort was on track to complete the flight, and that ground system development and mission operations continued to meet mission performance requirements within the identified cost and schedule constraints.

Ground Safety Reviews

Prior to the first Shuttle launch in April 1981, NASA and United States Air Force (USAF) safety officials met to determine what ground safety requirements should be established for payload processing, and the associated GSE used to prepare the payloads for launch. The decision was made to develop a single requirements document that could be used for payload processing on NASA or USAF property. The Kennedy Handbook (KHB) 1700.7/Space & Missile Test Operation (SAMTO) Handbook S-100 was published to establish the single requirements document [2].

During the Shuttle era, there was a three-phase (Phase I, Phase II and Phase III) safety review process, run by the Ground Safety Review Panel (GSRP). Central to this process was the payload's Ground Safety Data Package (GSDP). The complexity and depth of the GSDP was commensurate with the level of hazard presented by the payload during ground operations. GSDPs could be as small as a few pages – basically a letter on single sheet paper – or as large as five three-inch-thick (76.2 mm) binders crammed full of leaves of double-sided paper. The requirements of those reviews applied to any Payload Organization (PO) processing any flight hardware and GSE at any location under the jurisdiction of NASA's KSC. These included the Cape Canaveral Air Force Station (CCAFS) and at alternate or contingency landing sites. The requirements applied during each launch campaign, from arrival at KSC until final departure of the experiment after the mission.

The first of these was "offline", where the PO conducted hands-on operations with its payload prior to turnover to KSC for Shuttle integration, and after Payload/Shuttle de-integration until departure from KSC. The other was considered "online." where the payload was under the control of KSC, although this did not relieve the PO from the responsibility of ensuring their payload was safe. Some of these safety areas included pyrotechnics, high pressures, radiation, industrial hygiene, human factors and crane operations.

Timing the submission of the GSDP was imperative for the payload to receive its approval for ground processing. The critical date for determining submissions was the on-dock date, which was the date on which the first of the PO's hardware arrived at KSC. This did not have to be solely flight hardware. Per program requirements, the PO had to complete the Phase III Safety Review 30 days prior to arriving on-dock [2]. In reality, this did not happen very often, with final documentation often not arriving until just days before the hardware. This required NASA KSC safety personnel to literally take the documents home, sometimes spending all night reviewing the information to be able to complete the Phase III before hardware arrival as there were never enough hours in a working day. "We only got upset about the lateness of the packages [and] the amount of time you had to review them," recalled Roland Schlierf. "Right from day one, it was always just a week or two that you had to review these packages." If this could not be completed in time, smaller payload shipments would be stopped at the perimeter gate and not allowed to enter KSC. This could not be done for the very large payload shipments, so the hardware would arrive on site but would be impounded where nobody could touch it until the required review was completed.

MEETINGS

Several meetings were required during the planning effort in preparing for the mission, to work out problems but also to get members of the team together in a working relationship that would last many years. Scott Vangen recalled how challenging this sometimes became. "You were a member of their team for that mission. Many times, you put yourself in a position that you were their sole advocate for when tough decisions had to be argued on their behalf. I had many distinct meetings where, while I worked for NASA, I also worked for this team. And you had to be like a lawyer almost at times. It was not a question of loyalty to one or the other, but what was the right thing to do. So when a Level-IV Experiment Engineer stood up at a review board for a 'Use-as-is, or change flight' report, or getting ready for a Flight Readiness Review [FRR] prior to launch, they were speaking from two positions, not split loyalty. You were checking off both boxes. It was good to fly in both directions, so that the experiment was going to be successful and NASA was going to have a safe flight and successful mission."

Ground Operations Working Groups

The Ground Operations Working Groups (GOWG) were a forum for all those involved in a mission to meet in one room, often for several days to a week, to discuss all aspects of what was planned and due to be performed during ground operations at KSC. Mostly, these meetings would occur at KSC and were run by the Launch Site Support Manager (LSSM), but they could occur at other sites depending on specific topics to be discussed, with the LSSM leading the KSC team. These meetings would include all disciplines from local KSC and Experiment Developer (ED) personnel. The GOWGs drove out many problems and solutions to those problems. They focused on high level topics. For any very detailed technical topics that came up, a separate Technical Interchange Meeting (TIM) would be scheduled for discussions and solutions. EI personnel participated as a team member in the GOWGs and were usually the main players for any of the TIMs that were required.

Technical Interchange Meetings

As the title states, the TIMs were a place where technical personnel would meet to discuss very specific detailed technical issues and concerns, or to figure out solutions to very technical problems. They could involve some aspect of testing, software issues, servicing the experiment with special fluids or gases, handling of very sensitive equipment, cleanliness for specialized optics to be flown, and very critical alignments of telescopes or cameras. The EI person was at the heart of the discussions and solutions. As they were responsible for the experiment, they needed to be a main player in what and how tasks were to be performed to ensure a successful and safe mission, as well as ground operations both preflight and postflight.

Pre-Ship Review

Before hardware was transported to KSC, a pre-ship review would be held at the ED's location. These reviews would contain all the documentation, detail the configuration of the flight hardware and GSE at time of departure from ED's locations, and outline any work left over (called "traveled work") and incomplete at the ED locations and requiring completion at KSC that was not part of the plans discussed at the GOWGs and TIMs. All was usually fine, except the traveled work that was *not* planned. In this case, the questions to be answered were: What is the work to be done? How long would that work take (now impacting KSC schedule)? What are the resources necessary to complete the work, and does KSC even have those resources?

All this had be considered, and required the EI and other KSC personnel to scramble at the last minute to try and determine solutions to all the processing problems generated by the traveled work. It is easy to see how this could overwhelm the EI personnel working several experiments for one mission, as well as all the other missions they were responsible for processing. Generally, EI personnel accepted and accomplished this challenge.

SCHEDULES

To some organizations, schedules are used as placeholders. For EI personnel, the schedules were mandatory and needed to be accurate to ensure all the work could be completed in time to make the launch date. Tony Ornelas recalled that "the schedule" was the biggest concern: "What was [the] whole schedule going to be like? Always press for schedule, always had to get something done, done right, quick and so forth. Then, on top of that, something new would come up and so now we've got to do this. It was a lot of scheduling and a lot of trying to get things done to the schedule."

Working with the Spacelab side of activities provided considerable exposure to working with contractors, in addition to their NASA counterparts. As an operations engineer, Todd Corey explained that he spent considerable time coordinating with NASA and Lockheed personnel on the Shuttle side of operations, to integrate payload activities at the Orbiter Processing Facility (OPF), launch pad, and runway. "This would include activities such as the installation of the overall Spacelab into the Orbiter at the OPF, integrated Orbiter/payload testing, and installation of payloads into Spacelab at the launch pad via the Module Vertical Access Kit [MVAK]." The planning and scheduling of Spacelab module and Orbiter middeck late-stow activities at the launch pad was very challenging. Todd Corey continued: "We had to work very closely with the Shuttle NASA Test Director Office, and contractors, to incorporate our payload work into the Shuttle's already very busy countdown schedule. The negotiations were not always easy, but we all worked well together to make it doable."

Master Schedule

This was the overall timeline that had to be established for all activities that would need to be performed at KSC.

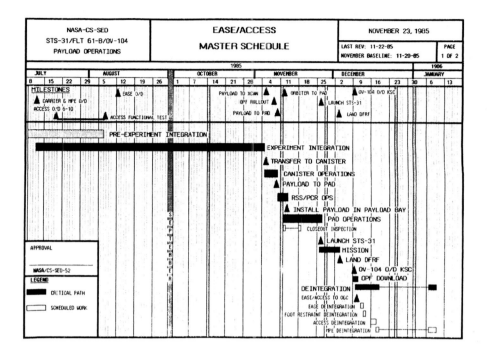

An example of the Payload Master Schedule, in this case for the STS-61B EASE/ACCESS payload.

Since the launch date was considered a "hard" or immovable one, it was logical to begin the schedule there and work backwards to determine the time required to complete all activities before launch. At the beginning of the Shuttle program, of course, none of these activities had ever been done before, so the only option was to guess how much time would be required to complete a test or assembly task and work from there. As more missions flew, payloads were carried and experience increased, the team became more accurate at estimating future timelines.

72-hour/11-day Schedule

The 72/11 Schedule was a split-column document which, as the name implies, showed all the detailed operations and support required for the next three days on the left side, and then 11 days out on the right side. This was used to see what near-term tasks had to be performed and then those a number of days in the future. As problems arose and things did not work out quite as planned, items would be shifted around to try and maintain a forward flow of work.

Daily Schedule

The Daily Schedule contained more detailed information for each of the tasks on the 72/11 Schedule, covering all the work and support that needed to occur on that specific day. Like the 72/11 Schedule, many item times would need to be shifted around based on what was happening real-time. It should be noted that the Operations personnel were responsible for keeping up with all these changes, so at the end there was an "As-Run Schedule" that showed exactly the worked performed on each day.

WAD, TAP, AND DEV

At KSC, nothing was done and nobody approached or attempted to work on the flight hardware and/or GSE unless there was an accompanying paper to perform the work and a check off on the steps accomplished.

Work Authorization Documents

One of the types of paper trail was called the Work Authorization Document (WAD). There were many different types of WADs at KSC, with the ones discussed here most frequently used. The "Task Leader" was the main person in control for implementing the steps of the WAD and/or sections of the WAD. A single WAD could have dozens of task leaders, depending on type of tasks stated in the WAD. Usually, the Task Leader was also the person that wrote the WAD.

Test and Assembly Procedures

The Test and Assembly Procedure (TAP) was the more formal procedure used by EI personnel. The TAP would contain steps to satisfy the requirements stated in the GIRD, but could also be used to satisfy OMRS requirements. It went through a formal review process, starting with a first draft on which comments were received. Then there was a second draft to ensure those comments were incorporated or discussed – and any changes or rejection of comments noted – and then a final version that was signed by all organizations that reviewed it. After obtaining all the signatures, it was released for implementation. The TAPs could be a dozen pages or so for simple minor operations, up to several hundred pages for testing of an integrated Spacelab module. TAPs were used to process flight hardware, as well as operate GSE.

   The different types of TAP included one-time use non-hazardous and hazardous, or multi (use) non-hazardous and hazardous. As the name implies, one-time TAPS were planned for a single usage and then "closed out". All TAPs were one-time unless specifically titled Multi-TAPs, so from this point on, one-time use TAPs will just be referred to as TAPs. An example would be powering up the Freon® servicer to start Freon® flowing through a piece of flight hardware or GSE. The Multi-TAPs could be imbedded in a TAP as needed; for example, if a certain flight's Freon® fluid system needed to be serviced, the main TAP would call out the Multi-TAP to power up the Freon® servicer, and then details of the flow's settings and temperature would be called up in the main TAP. If that same fluid system needed to be running to cool flight hardware during testing, it would usually be powered up and down several times while supporting testing operations over the course of many days or weeks. This saved lots of repetitive wording in the main TAP and allowed flexibility to conduct many servicing operations on experiments, payloads and missions. Non-hazardous TAPs were for most operations, such as torquing bolts, installing cable harness, checking cable continuity, etc. Hazardous TAPs were for hazardous procedures, such as crane operations, high pressure fluid operations and handing pyrotechnic devices. Being used for hazardous operations, these TAPs went through a higher, more complex level of review to ensure all the proper wording and controls were called out, so that no damage occurred to the flight hardware or GSE, and no harm would come to personnel. TAPs were formatted into main sections I–V and an appendix.

Test Preparation Sheets

Test Preparation Sheets (TPS) were a less formal procedure that did not go through the many review processes, but were still a means to work on the flight hardware and GSE.

Mike Haddad's TPS cover sheet, used to perform leak checks during Spacelab-1. [Mike Haddad.]

A TPS could also be used to satisfy requirements in the GIRD, and signatures were still required as with TAPs, so what was the difference between a TAP and TPS?

The TPS would be used for real-time tasks that need to be performed as soon as possible, tasks that were not covered in TAPs, or extra tasks that were not planned but needed to done to continue the processing flow. There were two types of TPS, Type A and Type B. Type A would be a TPS that performed a configuration change of what it was written against. Type B was a non-configuration change TPS, basically performing steps to bring the hardware to a planned and documented configuration. Most TPS were Type B. Like TAPs, a TPS could be as short as one sheet or as many as dozens and dozens of sheets, all depending on the task or tasks to be performed.

Operation and Maintenance Instructions

Formal procedures used by most of the rest of KSC, Operation and Maintenance Instruction (OMI) were rarely used by EI personnel. Like TAPs, OMIs would satisfy some requirements from the GIRD, but always from the OMRSD. OMIs would also call out TAPs to satisfy certain requirements. The OMI would go through a formal review process like TAPs, but would involve integration across many more organizations and functions, so the review process was a great deal more involved than for TAPs.

Deviations

Deviations (DEVs) to WADs could be temporary or permanent depending on the circumstances. An example would be that, while performing a test using the main WAD, there was a need to record a temperature every hour for the given heat load during the test. If the temperatures were consistent over many hours and nothing changed, a Temporary DEV could be written against the WAD to permit the temperature to be checked every two hours for that specific test, but would return to recording temperatures every hour for the next test. If, during that same test, it was noticed that the flow rate was not providing enough cooling for the heat load during the test, and future tests would have the same level or higher level of heat loads, a Permanent DEV would be written to increase the flow so the existing test and all future tests had enough cooling.

Along with DEVs came DEV runners, the people that would take the original DEV, make copies and then "run" them (well, walk them) to all the people that were tied into the operation of that procedure, so that everyone would have the most up-to-date documented version of what was being performed. This was mostly done during testing when many people were involved. It was not a fun job at all, but very necessary, especially for the procedures that ended up having possibly hundreds of DEVs, and personnel on different floors and in different areas of the building.

PRE-FLIGHT OPERATIONAL TASKS

These tasks began at hardware arrival to KSC, and were encompassed by Receiving and Inspection. Usually, "offline work" (not to be confused with an "offline laboratory"; offline work was a function, an offline laboratory was a location) was performed *before* the hardware was "officially" turned over to EI personnel. Online work was work performed *after* turnover of the hardware to EI personnel. So, there could be online work in an offline lab, a potential for confusion. Once online, the hardware proceeded through assembly, integration, servicing, testing, late stowage and finally closeout operations. EI personnel became the task leaders for all these operations.

Offline

Offline, as a responsibility, referred to the hardware being located at KSC but which was not yet KSC's responsibility. It also referred to any location that was not part of the main processing areas.

Receiving and Inspection: Once "on-dock", the specific item went through a Receiving and Inspection process to determine if any damage had been done to the flight hardware or GSE, and if all the paperwork (Ground Safety Data Package, drawings, procedures, travelled work, known problems, etc.) was in order. If not, a problem report, described later, was written documenting the problem and any resolution that had to be found. Mike Haddad recalls: "Receiving the experiment was like getting a gift on your birthday or at Christmas. I could not wait to open it, in this case removing it out of its shipping container, to see the experiment itself." An initial inspection of the container was done to look for damage. Un-crating it could be done manually or using a crane, depending on the size or weight. As coordinated early in the planning process, and finalized at the pre-ship review, an offline area, usually a lab or room, was used by the PI to complete any offline work before turnover to EI personnel. The lab contained the required resources to support the PI's activities. Sometimes, additional unplanned resources were required, which had an impact on EI personnel functions and caused another scramble at the last minute to complete. Another frequent impact that would occur during offline operations was that the time for the PI to complete their work became longer than planned. This compressed the EI online schedule of operations due to the fixed launch date, leaving less time to perform all the pre-flight tasks.

Turnover: This was the term for the official transfer of the flight hardware, GSE and paperwork from the Payload Developer to NASA/KSC. This was where online operations began, and would continue until after launch and postflight online operations. If a major problem occurred during online operations, the PI and EI personnel would do everything possible to avoid going back offline. Some tasks needed to be performed only by the PI, so the hardware could be transferred back offline until the next turnover, as NASA Level-IV Electrical Engineer Tracy Gill explained: "Sometimes, something went wrong, and we'd have to go back offline and let them [PIs] go back to their lab and tear it up, fix it and bring it back to us. Sometimes you'd help them there, even though that wasn't really your responsibility. But they were your buddy and most of us were engaged somehow to help them make sure they got what they needed."

EI personnel would monitor this offline work to understand the fix and ensure it would not impact or change any online operation(s). For example, if the weight was increased (or decreased) by the fix, thus probably changing the center of gravity (cg) as well, this new information needed to be factored into the overall payload weight and cg calculations. Another example could be that the time for an experiment to heat up properly took longer than before. Now the EI person had to factor that into any of the testing/operational activities preflight, and the change would have to be relayed to the flight crew and Mission Control so the mission times could be changed. Depending on how much more time was required, this could drastically impact the ground and flight integrated testing/operations.

Online Level-IV Experiment Integration

This was where the fun began for EI personnel. After the years of planning, reviews, meetings, conflict resolution and procedure development, it was time to take responsibility for all the assembly, integration and testing required to make the launch. This could be months, or over a year, depending on the complexity of online operations. Implementing the TAPs and TPS that EI personnel had developed, and with an aggressive success-oriented schedule, work began. Almost immediately, different types of problems would arise, causing additional work and a compression of the schedule to make launch.

Assembly and Integration

Level-IV operations began at locations in the west end of the Operations and Checkout (O&C) Building, where several payloads were *assembled*, *integrated*, *serviced* and *tested* at the start of this last push towards their journey into space.

- *Assembly* meant obtaining the parts necessary to put the experiment together. Depending on the mission, the parts could come from across the globe.

- *Integration* meant bringing together many different experiments and/or sub-systems that would support the experiment(s). Integration operations ranged from installing multiple small experiments into one rack, to integrating multiple telescopes on a pointing control system, as in the Astro-2 mission (STS-67, 1995).
- *Servicing* referred to supplying a certain commodity to an experiment/system, such as pressurizing a gaseous nitrogen bottle, loading Freon® into a cooling loop, or supplying liquid helium to a flight storage system used to keep sensitive flight detectors healthy.
- *Testing* could mean powering up a single experiment in a rack, running a full module of racks through a dry run of on-orbit operations, or passing fluid through a system looking for leaks.

As with offline, one of the first online tasks was receiving and inspection of the flight hardware that needed to be installed before the experiment. These included the bolts, screws, washers, nuts, cables, fluid lines, brackets, support structure, etc. Due to the very sensitive nature of the experiments, all this hardware had to be cleaned to high levels and double bagged to prevent any contamination during transport to KSC. Upon receipt, every piece of hardware was inspected to ensure it was double bagged, included the cleaning certification, and contained a flight tag certifying it was cleared to fly into space, with all the proper engineer and quality stamps to prove it. Each part was inspected for any damage or off-normal appearance, and any discrepancies noted. If necessary, new parts would be ordered, which of course could cause a delay or change the sequence of planned operations. Just this first step and already EI personnel had to scramble to continue the work – "Only the best of the best would fly". Mike Haddad recalled that, "Some of the attach hardware, like bolts, were made of stainless steel and had a dark lubrication on them to prevent any galling during installation. At first it would appear to be dirty, but we knew the difference between dirt and lubrication so we would make sure nobody would try to clean them, because if they did it would present serious problems down the road."

The assembly, integration servicing and testing activities required procedures to be written and performed, and Level-IV personnel did both. This had to be done for all levels of payload processing (Level-IV through Level-I), which included preflight integration and postflight de-integration, as well as support and direct input for on-orbit operations during the mission. NASA Level-IV Mechanical Engineer Tia Ferguson noted that when drawings were received from Marshall or JSC, and there was a difference between the two, Level-IV would have to interpret the drawings and write the procedures. "That [the procedures] had everything listed: every bolt, every washer, every wipe that you used, every alcohol that you cleaned, the type of alcohol you cleaned with. Every little step had to be documented. And then we worked with the contractor Quality and the NASA Quality and the technicians. And all of them had to stamp every single step so that we knew exactly what was going into every little piece. And that was the procedures."

Then each had to be officially rubber stamped, as Riley Duren explained: "You had an inspector along with you who provided a really important quality control assumption. They had a stamp, and they would stamp the different steps and the procedures you went along. Every inspector had a unique stamp. Usually after these tests, there was a fair amount of cleanup of paperwork. One day, one of these guys hunted me down and said, 'There's a couple of hundred pages that we have to go through and sign off'. [I had a 'non-official' stamp] I think it was a *Smurf* stamp, so I stamped it, and he said, 'What are you doing?' I said, 'It's much faster if we stamp this than if I have to initial it, so this is my stamp'. On the front page, I had written my name and signed it and I noted, 'This is my stamp'. The inspector said, 'You know, they're going to send this back' and I said, 'Let's just try it. What do we have to lose?' We went through to get the whole thing done. So somewhere there's a doc, a pile of paperwork, that has purple *Smurf* stamps all over it. And I got away with it. They never made me go back and fix it."

Each inspector could use one of a number of stamps, depending on the situation. A triangle stamp meant conformance, all was good. A hexagonal stamp mean non-conformance, something is wrong. A "D" stamp signified that the item was void.

Problem Solving and Documentation

Due to the one-of-a-kind nature of these payloads, problem solving was a major part of the everyday job. Each day was different from the previous day, because new problems occurred, or twists on old problems would challenge the ingenuity of the Level-IV team. Also, every payload complement was unique, so each payload (even for re-flights) was a whole new adventure. However, as more payloads were processed, the EI personnel experience base grew and grew. This allowed them to "tap" knowledge obtained from one payload and apply it to the next. The Problem Reporting And Corrective Action (PRACA) process was used for problems. Each problem would be documented on an individual Problem Report (PR) or Interim Problem Report (IPR), and it was the responsibility of the EI personnel to determine and carry out the solution to the problem. The difference between a PR and IPR was that a PR was a means to document a known problem: maybe the bolts were too short to attach an experiment to its support platform, or a pin bent on a connector while trying to mate it to an electronics box. PRs for a single payload could total into the thousands, with one double rack alone logging more than 100 PRs before it was ready for flight.

The IPR was for problems that occurred for no apparent reason, or with an indeterminate cause. For example, a command being sent to start up an experiment and then nothing happens. An IPR allowed for troubleshooting to determine cause, after which it could be "upgraded" to a PR for repair and retest, if necessary. If all kinds of fixes were attempted and a requirement could still not be met, a waiver would be requested. Taking a wavier was never desired. Depending on what the wavier involved, some were easy to justify and you went on your way. Others could be very complex and not easy to get approved. Some problems would be recorded as "Unexplained Anomalies (UA)" because they occurred only one time, or could not be reliably repeated, and so a solution was never determined. As an example, on Spacelab D-1 (STS-61A, 1985), an experiment would lock up for an unknown reason, and required a restart to "fix". EI personnel needed to weigh all the risks and make a tough engineering call on whether or not the scientists and the astronauts could live with this condition, which fortunately they did. "The Unexplained Anomalies were really hard. Of course, we did have some with data," Cheryl McPhillips recalled. "We used to always do the jiggle test – it's not working, so jiggle it to see if it'll work. A lot of times the old jiggle test would work and then you never could get it to fail again. You'd be thinking, 'Well, why did the data fail?' You were looking at the pins – is it bad? Is it a bent pin? You couldn't figure it out. You never knew. Sometimes, we never figured out why something didn't work one time, and then it would work again, and then it wouldn't work. So you'd have to go through the UA, [and it] was a hard process to get a UA approved. You had to have the most likely cause of what the ramifications would be if it happened in orbit, and what you would do if it did happen in orbit." Rey Diaz, put it more simply. "When you were out of options, and didn't want to break configuration, your first choice was to do the jiggle test. The whole idea here was that if you break configuration you're doomed, you'd have to retest the whole thing."

Other problems involved the Government Supplied Equipment (GSE, not to be confused with Ground Support Equipment), some of which was found to be older than some of the persons operating it. In one case, an old power supply built in 1964 caught fire in the O&C Clean Room Area. Tracy Gill recalled the incident: "I think it was right after holidays and I remember the fire detection [had gone] down in a building. That's one of the things that happens when you rely on an old power supply. It was two failures; the contact welded, and then some of the fans failed and basically overheated. The way they found out about it was because the Clean Room went out of spec because of the smoke, which was not picked up by the fire detection system that was down at that time. It was on a Sunday before we came back from the holidays. So on Monday, [we came back to find] fire trucks parked out front. Not necessarily funny. It's funny now. It wasn't then."

Left: Level-IV GSE power supply rack damaged by internal fire (thermal and smoke). Right: A good power supply, North Rail. [NASA/KSC.]

So, the EI personnel had to learn about the outdated power supplies to establish preventive maintenance for the rest of the equipment. EI engineers also had to keep track of each and every part, including those lost somewhere inside the Spacelab. One PR even described finding a part that was not lost. An electrical clip that was discovered in a pallet should not have been there, so it was recorded officially as "a found but not lost clip". EI work required making hundreds of decisions a day on how best to assemble, integrate, service and test the hardware [3].

Once the problems were solved, they usually involved some change in the design or operation of the flight hardware and/or GSE. These changes somehow had to be relayed back to the designers so the drawing or schematics could be updated and the paperwork reflected the most current configuration of the hardware. This was done via a Field Engineering Change (FEC). The PRs and IPRs would kick off an FEC or multiple FECs to document the current configuration. So, in the end, EI personnel would have a stack of paper consisting of PRs, IPRs, UAs and FECs.

Work Days

The days were long as well, most being 12 hours and many extending to 16+ hours. Sometimes, team members would sleep at work instead of going home. Due to the high launch rate and limited number of personnel at the time, multiple experiments, payloads and missions were being worked on at the same time. Mike Haddad remembers, "At the peak, I was working three different payloads at three different locations in the O&C Building, all at the same time on the same day." Electrical engineer Tracy Gill, who worked on the International Microgravity Laboratory (IML-1), Spacelab-J (Japanese Spacelab) and United States Microgravity Laboratory (USML-1) payloads simultaneously, remembers having to ask which module the team was talking about during a meeting. Some jobs would require multiple technicians, so the engineers quickly ran out of contractor technician support. As Mike Haddad recalled, "The work could not stop or else the hectic schedule would not be met, so we were trained to perform the technician role, which I did many times. The Quality Assurance person would read the steps from my procedure and I would perform the work. Another engineer was the only person with small enough hands to access very tight areas. One mechanical engineer spent the 1984 Christmas holiday installing electrical cables on the Spacelab-2 payload. It was a feverish pace, but most of us were young and could go flat out for months at a time because we lived and breathed the space program."

Scott Vangen recalled that he was living in New Smyrna Beach at the time. "That's over an hour's drive, so that's two hours added to 16 or 18 hours a day. And it was seven days a week. I got a paycheck one time and there was hardly any money in it. It was just ridiculously low. I handed it to my boss Enoch [Mosier] and he said, 'I'll look into it'. So he called the payroll office and they said I had rolled over 100 hours of overtime in the two-week period, and the computer had truncated the third digit. I had 102 hours, so it dropped the '1' and it busted the program. The payroll computer had not seen a number that high since the Apollo days, so it truncated my check, basically just zeroed it out because of the overtime hours. Some kind of bug in the thing. That was 50 hours of overtime in a week."

Assembly took a great deal of time because KSC was usually the first place the parts came together. Tasks that appeared simple on paper sometimes became very difficult once the actual assembly started. The challenge was staying on that estimated timeline, even when problems occurred and the hardware was not performing as designed on paper. EI personnel worked closely with personnel from all over the world as needed, to talk with the designers about a proposed fix to their hardware. Sometimes this was in person, or via phone, fax, or later e-mail to the host country, such as Italy for the Tethered Satellite System (TSS), or Germany for the Spacelab D1 mission.

Individual Rack Assembly and Integration

Spacelab racks arrived at KSC as basic, empty structures. For each mission, the racks would be populated with hardware assigned to fly on that specific mission. Assembly and integration of the rack components took place in the Rack Room of the O&C Building.

Rack Room. Two double racks being worked on using white access stairs (Engineer) and white step-up platform (Quality), with gray step-up stool (Technician) hard to see behind white stairs. To the right, a Technician and Engineer review kitted parts from a gray tote bin, to be used for rack integration. [NASA/KSC.]

Initially, this location was used just to assemble racks, but as more Spacelab missions flew, the room held over a dozen racks in some form of integrated or de-integrated configuration. Each Level-IV engineer could be assigned multiple racks per mission, so one engineer could be assembling racks for a future mission and disassembling racks from a previous mission, all at the same time. The more complicated racks could take several months to assemble and fit all the hardware as designed, including solving any problems that cropped up all along the way. There were structures that had to be added to the racks to support the hardware that would be installed later. The racks required fluid and air lines to be installed that were used to cool the equipment in the rack. Cabling, clamps, and brackets all had to be integrated into the racks as well. All this work had to be done and any problems solved before the experiments could be installed. One general problem was that cables were often larger than shown on the drawings, which impacted other hardware in the rack and made routing the cables very difficult. This also impacted the installation of other hardware. In these cases, EI personnel needed to be sure the fix for one piece of hardware did not cause other problems somewhere else in the rack.

Tia Ferguson observed, "When putting the ducts together, you had to put the copper seal in between. You did a leak check and some of them leaked, that's just how it was. You had to write steps to take it out and write steps to dispose of the one that didn't work. You had to write steps to fit the new one and steps to install the new kitting, and all that had to be documented."

After all the build-up of the rack came the experiment installation and, like before, some would fit while some would not, so changes would have to be made to complete the integration. Maybe an air-cooling hose was too short to reach the rack air cooling ports, or a bracket was just a little too wide causing a hardware interference. Many of the experiments were light enough so they could be manually installed, which gave the team lots of flexibility to move them around as needed to get them to fit. Some were too heavy for manual installation so GSE was used (most of the time, PI designed), but some would interfere with the GSE supporting the rack and modifications had to be made so no damage to the experiment or rack flight hardware occurred while installing the experiment. Again, EI personnel would have to solve the problems before the next step (rack flow balance) could occur.

Tia Ferguson remembered, "One time we had a problem and we'd pretty much de-integrated the whole rack on a PR. These were handwritten and included carbon copies. I hand-wrote probably 500 pages on one PR, basically for removing a whole rack and re-installing the same thing. Writing by hand was a lot of work and effort. This was when computers were just starting to come around, but they wouldn't let us use them [even though] everybody had one on our desks. I asked, 'Why can't I just print this out and staple it?' Nope, Quality said it has to be on the carbon copy form, you cannot print it out and staple it to the PR. We got hand cramps from writing. I was writing as fast as the technician could install stuff."

Tool control was also very important. A main tool box containing only approved tools would be used for integration of the flight hardware. For example, a crescent wrench was never to be used on flight hardware, so that tool would not be part of the tool box. The drawers of the tool box contained Styrofoam cut-outs for each tool. Each tool had a number, and each tool had to be logged when entering a control area and verified as removed when leaving the controlled area, to make sure you exited with the same tools you entered with and then replaced them back in the tool box. At the end of the day, the tool box was checked to verify that all the foam cutouts had a tool. This made it easy to see if one was missing. Anything that could come apart was tethered to something else. There was one incident early on where a Level-IV person was using an untethered flat-head screwdriver over flight hardware, and Murphy's Law always seems to apply at the most inappropriate time. Sure enough, the screwdriver slipped out of the person's hand and fell, striking some flight tape on a flight cable. Luckily, the cable itself was not damaged. Most of the time, these tethers had to be added to tools that were not really built for a tether, so Level-IV technicians had to come up with innovative ways for attaching them.

Once integration was complete, the next step was rack flow balance. This was the process of adjusting the small valves in the rack that controlled the amount of air going to each section of the rack. Each section required a certain amount of air flow to keep the equipment, experiments, and their subsystems at the proper temperature. This was an art of its own, because when one valve was adjusted the flow would change for the whole system, so you had to wait until it stabilized, take a reading, then adjust again until the proper flow was obtained for the whole system – a very tedious and time-consuming task. As Mike Haddad recalls: "The one engineer in charge of rack flow balance, Bob Ruiz, worked for hours and hours attempting to get the flows correct. He would reach the point of total frustration and would want to give up, but instead he had to just walk away for a period of time and return later to finish the task, achieving the correct flows. A job well done."

Following rack flow balance, some racks were tested individually to ensure all systems in each rack would talk and work together as planned. Once completed (and sometimes while still uncompleted), each individual rack would be added to the Spacelab module rack flight floor, called the "Rack Train", and a flow balance of the entire Rack Train would occur in preparation for integrated testing. Due to schedule conflicts, some rack integration would be completed after addition to the Rack Train.

The sequence of which racks would be installed first was planned out early, but due to problems during rack integration the sequence order would change, depending on which racks were completed and which were not. This re-sequencing also had to factor in which location of the flight floor was complete and ready for rack installation, a real scheduling nightmare. The Rack Train contained up to 10 experiment racks mounted to the Spacelab flight floor that was being integrated by EI personnel in the South Rails area of Level-IV, at the same time as rack work was performed in the Rack Room. The flight floor contained all the Spacelab experiment subsystems that supported the rack experiments. Two simulated Spacelab module control racks, Racks 1 and 2, competed the 12-rack Train. The two flight control racks were part of the Spacelab module being integrated by McDonnell Douglas Astronautics Company (MDAC) in the Level-III/II area of the O&C. The Rack Room was separate from the O&C High Bay and the South Rails where Rack

Train integration took place. So, the racks were rolled out of the Rack Room, on the same blue GSE stands that were used for rack assembly and integration, and placed next to the South Rails. Then a crane operation was required to get the rack from the floor level up to the Rack Train located on the South Rails.

Left: A Spacelab rack on the crane prior to being integrated with the rest of the Rack Train. Right: Spacelab rack installation onto the flight floor. [NASA/KSC.]

It should be noted that the racks could not support their own weight, so GSE brackets supported the racks until just after floor installation into the flight module in the Level-III/II location. During this crane operation, the GSE support brackets separated at the bottom of the GSE, with the weight of the rack supported from the top. The crane operation was not an easy task to complete, because all the interfaces on the bottom of the rack had to match the interfaces on the top of the flight floor. A push here and a tweak there persuaded the racks into place. Support would then be transferred to other support brackets, and the hardware used to lift the rack would be removed, placed back on the base, and rolled back into the Rack Room for later use. This process was repeated for each rack, until the full complement of racks was installed. Many times, racks were being installed on one location of the flight floor while work was continuing on the flight floor itself in another location. However, crane operations required all personnel to clear an area, usually 25 feet (7.62 m) around the items being lifted, so floor work would have to pause until the crane was disconnected from the rack. It really turned into an orchestrated dance of doing multiple tasks in the same locations.

Individual experiment tests (functional and interface) and then integrated testing (discussed in more detail later) would take place once all the hardware was together. The testing would continue until all the experiments were operational. Most of the time, the racks were closed out at this point and were ready for the next level of integration and testing, but sometimes the back panels had been removed to allow access to the back of the rack. Reinstalling those panels was not easy; in fact, most had to be match-drilled because the racks had been distorted or twisted just enough to the prevent the original holes from aligning. But as time went on, Level-IV people came up with better solutions to problems like this. Alex Bengoa recalled an improvement that Joe Delai came up with, which was "mapping for all the plate closeout covers for the back of the rack. They had to be match-drilled, so he came out with a piece of Lexan, and he would use that as a template to mark the bolt hole pattern and then use that to match-drill the actual flight panel."

Following testing, the Rack Train had to be prepared for the move to Level-III/II. Because the top of the racks supported by the GSE needed to connect to the inside of the flight module, the support for the racks went internal, using GSE called the Internal Braces Kit (IBK). Then the Rack Train would be moved, via overhead cranes, to the Level-III/II area in preparation for installation into the flight module.

Experiment Rack train (Racks 3–12) lift to Level-III. GSE version of Racks 1 & 2 are tan dogleg racks in foreground. Flight Racks 1 & 2 are located in the white module in the background. A gray upside down "U" internal braces kit supports the rack from the middle, and a large blue upside down "U" supports the top of racks during integration and lift. [NASA/KSC.]

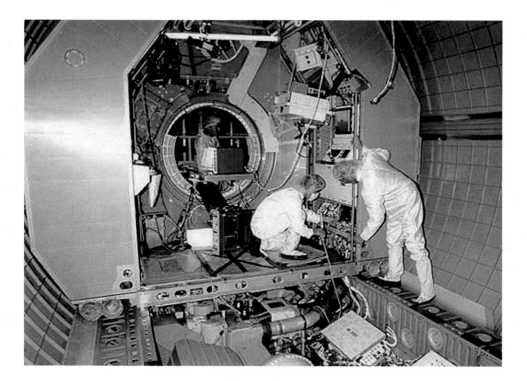

Spacelab Flight System racks 1 & 2 located inside the Spacelab module. Rack 1 to the left, Rack 2 to the right. The open volume is where Racks 3–12 were placed once rolled into the Spacelab module. The view is looking forward, and the round opening is where the Spacelab tunnel would be mated. The equipment below is the Spacelab Module flight subsystems. [NASA/KSC.]

Pallets

The Spacelab pallets, built at British Aerospace in the UK and supplied via ESA, were used to support either Spacelab or non-Spacelab missions. The Spacelab missions could consist of a single pallet, such as the Office of Space and Terrestrial Application-3 (OSTA-3) flown on STS-41G (1984), a double pallet like the Astro-1/STS-35 (1990) mission, or a triple pallet configuration as flown on Spacelab-2/STS-51F (1985). A non-Spacelab mission that used a single pallet was the LIDAR In-Space Technology Experiment (LITE-1) payload flown on STS-64 in 1994. The basic 'U'-shaped structure remained the same, but what was mounted to the pallet changed from mission to mission. While there was more volume available on the pallets than the racks, it did not take long for assembly and integration tasks to fill up this volume. This made it difficult to access mounting locations or connect cable to boxes as integration neared completion. Sequencing of hardware integration was critical to avoid "painting yourself into a corner", and having to remove previously installed flight hardware to get another piece of flight hardware fitted. Hardpoints were the main load-bearing locations, where EI assembly would install mounting brackets or struts that would secure honeycomb-shaped support structures, called Orthogrids, that the experiments, cables, fluid lines and electronic boxes would be mounted to.

STS-45 ATLAS-1 pallets at Experiment Integration, October 1990. [NASA/KSC.]

The first set of hardware to be mounted was usually the cooling system. This consisted of cold plates, fluid lines and fittings. They would be placed end-to-end to make up the cooling "loop" which Freon® would flow through as the cooling medium. Installation was normally straightforward. The cold plate was mounted first, and then the fluid lines would be routed, via fitting, from cold plate to cold plate and torqued to a specific range, usually starting at the lower end of the torque range. This would permit high torques if the fitting leaked, which was the main problem encountered during fluid loop assembly. Instead of making a few connections then leak testing, then adding a few more connections and leak testing again, it was better operationally to assemble the entire loop first and then leak test each connection. To perform leak testing, the loop was filled with helium at flight pressure of between 145 and 203 psi (10 and 14 Bar). Then a mass spectrometer that detected helium using a probe would test each connection for any leaks. A maximum leak rate was 1×10^{-4} cubic centimeters/second.

The type of fitting usually determined whether a connection would leak. As Mike Haddad recalled, "The cooling loops I assembled and tested used a cheaper type 'AN' fitting that used a nickel washer called a flare saver. This rarely sealed the first time, [so] I could write a waiver to use a stainless steel flare saver that would usually work. We did not want to torque the fittings while pressure was on the system, so we depressurized the system first, torqued all the leaking fittings to the intermediate torque range, repressurized the system and performed the next round of leak checks, but only on those connections we worked on. We did not want to mess with the connections that already passed and cause them to start leaking. Like the first round, some would seal, most would not, so we would repeat the process one more time, torquing the fittings to the maximum value, and a few more would seal. For those that still leaked, we needed to replace the metal seal, so the system was depressurized, the fluid line was removed from the fittings, a new seal put in, and the whole leak check process started again. We would continue to do this until all connections passed the leak test.

"The sealing surfaces on the AN fitting usually had dull look to them, while the sealing surfaces on the more expensive MS fitting were nice and shiny and we found rarely leaked. Based on this, I would always suggest on any drawing we reviewed that contained AN fittings to change them to MS fittings. This change in the design phase saved us a great deal of time and work during the assembly and integration phase, and the small increase in cost was negligible to the overall cost of other components."

Once the cooling system was completed, next up would be installation of the electronic boxes that mounted on top of the cold plates. Several of the electronics boxes were found to be over the maximum weight allowed for lifting by hand, so another crane operation was required. Once in place, bonding checks needed to be performed to keep the equipment free from such hazards as electrical shock and static discharge, as well as allowing for reliable fault clearing paths and the suppression of electromagnetic interference. Many items contained a bonding strap because the physical mounting locations could be on isolators or other locations that would not allow direct bonding of the box to the structure. This may seem like an easy task, but at times the strap would be too short to reach its attachment point, so either another attachment point had to be selected or a longer strap obtained; usually the former. The next problem was getting a good reading below the maximum allowed resistance of 2.5 milli-ohms, a military specification, MIL-B-5087B class 'R' bond, used for basically every box installed. Many would pass but some would not, so something needed to be done to get a good bond. Cleaning the end of the strap and its attachment location would normally solve the problem, but in a few cases the bonding location was painted, thus not allowing a good connection to the structure, so the paint had to be removed. A simple operation? Well, no. Only the paint right at the area of bonding could be removed, and all material removed had to be collected.

The next step could be the cable harness installation. These cables could be anything from a few feet in length up to 30+ feet (10+ meters) and weighing up to 220 lbs. (100 kg) or so. The process would start by laying out the cables along the surface as shown on the drawings, but as with many other tasks before, practice did not always match what was shown on paper. So, adjustment to the routing had to be performed, taking into consideration the limited mounting locations for the clamps that would secure the cables to the structure. The clamps were either too large or too small. Too large was an easy fix, by just wrapping flight-approved tape around the cable harness to make up the difference. Too small required a new clamp.

Mike Haddad recalls: "Early in the Level-IV days, drawings were very specific with very little leeway, so we had to make corrections and document them on PRs. After running into this problem several times, we suggested to the designers that while reviewing their drawings during the PDRs and CDRs for future missions, [they should] add special notes allowing us the tape trick and the option of adjusting clamps up and down a few sizes from those stated on the drawing for each location. These changes helped operationally in a number of ways. It avoided the need for more paperwork, saved more time and alleviated more signatures, and created a buildup of spare flight hardware. Because the same style clamps, bolts, washers, nuts, or whatever were used on many different missions, we would retain the extra hardware to possibly be used on a future flight, thus creating our own flight hardware logistics area. So when a different size clamp, bolt, washer, or nut was required, instead of waiting for the Payload Organization to send more hardware, causing a delay in operations, [we could] walk down to the flight hardware logistics area to select the hardware needed [and] be back up and running on the next task [in minutes]."

One other special note that was added dealt with being able to position the clamps in different orientations to that shown on the drawings. This was another huge operational help to the EI personnel. This allowed cable locations to be moved, for example flipping the flight clamp 180 degrees to avoid interference, without having write more paperwork.

Once the cable harnesses were in place, except for near the connections to the experiment location, the experiments could then be installed. Mike Haddad again recalls: "I would always leave the last clamp off the end of the cable harnesses near the experiments. This way, once the experiment was installed, I could position the cable harness as needed to ensure it would mate to the experiment, then install and secure the last cable clamp. The clamps were held down with a small bolt and washer, then torqued into place. At this point, the access to locations was getting tight due to all the hardware that was being integrated onto the pallet. Getting the clamp in place, the bolt through the clamp with the washer between the bolt head and clamp, getting the bolt started into the insert – we did not want to strip out the insert, that made for a very bad day – and then positioning a torque wrench on the bolt head to do the torque, I recall a few times where we had to use a mirror to see the actual bolt head."

Experiment installation most always required a crane operation. Before Level-IV was started in the early 1980s, it was decided that personnel involved in crane operations were required to use hardhats. After a short period of time performing crane operations and adding in safety measures, hardhats were no longer required. For one thing, it would have been very embarrassing if a cheap plastic hardhat had fallen off and damaged a multi-million-dollar, one-of-a-kind flight hardware experiment. Once again Mike Haddad explains: "Crane operations required a hazardous approved procedure to perform, which I wrote and performed many times. As Task Leader, I led the team assembled to perform the crane operations. This included the crane personnel of one up in the crane cab above our heads and at least two on the floor, one of whom would relay my movement commands to the crane cab operator, with the other holding an emergency stop button if required for any reason. Multiple technicians depended on the lift: NASA Quality, NASA Safety, the PI(s) and maybe their technicians, a still photographer (video recording would only be performed for major lifts, or rotation of large payloads between horizontal and vertical) and then myself. Some crane ops needed a dozen or more people. Many times, managers wanted to observe the crane ops, but it was my responsibility to determine that only the people required would be involved and in the control area. All others had to watch." As Mike Hill recalled: "'Hummers!' You always have these people that come and watch, and think they know how to do it better than you do. And they would always be there going 'Hmm'. That's why they were called 'Hummers'."

Using the crane was an involved operation by itself. The control area established around the item to be lifted was 25 feet (7.62 meters) in diameter. This also included the route the experiment would take, starting from the floor level of the O&C, rising up to over 30+ feet (10+ meters), sometimes traversing around flight hardware and GSE, but *never* over a person. A person under a suspended load was an Occupational Safety and Health Administration (OSHA) violation. Very rarely would a waiver to this rule be required, but sometimes there was simply no way of doing part of the lift without a person under the load. The Task Leader would try to avoid going over other flight hardware at all costs, but the Level-IV location became a very busy place with lots of flight hardware, so finding a path that did not contain flight hardware became almost impossible. The sequence of events went like this: A pre-task briefing was led by the Task Leader, who would get everyone together near the experiment to discuss the operation that was about to take place, who was doing what, and pointing out any special tasks, concerns, safety items, etc. Then the Safety and Quality people would talk if needed, and everyone would be asked if there were any questions or concerns. Finally, after an OK from everyone, the lift could begin. At this point, the Safety person would set up the control area using chains, blinking lights, etc., so that only those involved were inside the control area. If the control area included another payload being worked on, all personnel would be cleared until the lift operation passed and the payload work area was clear of the control area. The managers and anyone else there to watch remained outside the control area. If the preflight servicing (if required) and testing were successful, this would be the last crane operation on this experiment until it was ready to be pulled off during postflight operations.

MPESS and Experiments to MPESS

Some experiments and their supporting hardware did not require the larger pallets and all that volume, so the smaller, box-shaped MPESS would be used. Almost everything described above for pallets could be repeated when integrating flight hardware on an MPESS, though a few major differences existed. The MPESS structure could be difficult to access when installing hardware, and the bottom of the MPESS would be used many times, thus requiring installation from the bottom up (kind of upside down). All surfaces of the MPESS (except the bottom) could be accessed easily from scaffolding or special access platforms, and with the smaller size as compared to the pallets the entire width could be reached from either side, thus rarely requiring anyone to actually get on to the MPESS for work. Some work would require personnel trying to wedge themselves in between the support structure, to gain access to the underside of the top surface.

Special Structures

One of the special structures was the Unique Support Structure (USS) for the Spacelab D1 mission, developed and built by MBB/ERNO. While it looked similar to an MPESS, with a box-like appearance, it was very different in many ways. It was a carbon fiber carrier structure instead of aluminum as with the MPESS, and had only three mechanical attachments to the Shuttle (two forward on top, one at the keel), unlike the five for an MPESS (two forward on top, two aft on top, one at the keel). It did have three attachment fittings for ground handling (two forward and one aft on top – called the GSE trunnion) but the aft attachment was only used so the whole structure would not rotate about the forward fittings when being lifted.

Igloo and Supporting Subsystems

On flights where the habitable Spacelab module was not flown, but pallets were, a pressurized cylinder known as the "Igloo" carried the subsystems needed to operate the Spacelab equipment.

Alignment

Some of the experiments containing telescopes or observation hardware required critical alignment to a reference point on the Shuttle or to each other. The accuracy had to be done to the arc-second (1/3600 of one degree of a circle) in most cases, and this was another task for EI personnel. The expert EI person on this was Luis Moctezuma. He was responsible for setting up the GSE instruments, taking the readings, performing the math calculation, and directing personnel to perform adjustments to achieve alignment requirements, to ensure all flight hardware was within specification. Because the alignment was so critical, any movement around the flight hardware or GSE instruments would throw off the reading, so most of the alignment work was performed on either second shift, third shift or at weekends when nobody but alignment people were working. To do all this, Luis had to establish a very stable reference point in the Level-IV area, which was not afforded by the floor itself. The O&C had huge support columns that ran the length of the building and were used to support the large overhead cranes. As Mike Haddad recalls, "I remember how Luis used to come in on weekends to take measurements during the day. At first, for some reason, the measurements that should have been the same were different between the morning and evening, and it really frustrated him trying to figure out why. He finally figured out that the huge solid, stable support columns that his reference target was placed on weren't so solid and stable. The temperature difference between morning and evening on the outside of the building caused the columns to expand and contract the slightest amount, enough to throw his alignment numbers off. That was how critical and accurate the alignment had to be. To fix the problem, alignment operations were only done on the second and third shift at night, when the outside temperature was more stable"

Cleanliness

Cleaning, cleaning, cleaning. It was very important to keep the flight hardware as clean as possible and minimize contamination. EI personnel went to great lengths to keep the hardware clean. Like most operations in the space program, nothing was easy. Contamination comes in all forms and EI personnel had to deal with all of them: airborne particulate matter: surface particulate matter; non-volatile residue; volatile hydrocarbons; and microbial contamination. There was also cleanliness for the flight hardware itself, including fluids, gases and such that were contained inside the flight hardware, as specified in document JSC SE-S-0073, Specification, Fluid Procurement and Use Control.

For airborne cleanliness, it was required to keep the hardware at the 100K level or less (Clean Work Area Level 4). That meant the particle count per 1 ft^3 (0.03 m^3) of air could not exceed 100,000 particles of 0.5 micron and larger; basically, it was *very* clean. Control areas were set up around the flight hardware using plastic chains (they used white picket fences in the Apollo days) or other forms of barriers. The control area could have a Control Monitor person that would be responsible for ensuring all personnel signed in and out when entering or leaving the control area.

Bunny-Suited Astronauts

"In the early days of Level-IV there were only so many people around, so the control areas were self-controlled," recalled Mike Haddad. "We still had to sign-in, log our times and such, but there was no control monitor person. As the workload increased, and more control areas were being established, the SEIS [Spacelab Experiment Integration Support] contract included control area monitors to keep everything in order." The control areas went from the floor level all the way up to the ceiling. Smocks, with hair nets or caps, were required when near flight hardware, and latex or plastic gloves were used if personnel were touching flight hardware. Bare hands *never* touched the flight hardware. A person would enter the control area in normal clothes, then don a smock (a kind of clean lab coat) or bunny suit (full body covering) as needed. Most of the rack work, activities at the floor level, and work on access platforms adjacent to − but not over − flight hardware was performed in smocks. Tacky mats were used at the bottom of stairs and entrances to access platforms above the floor, as a means of cleaning off the bottom of shoes to prevent any floor contamination being transferred to the platforms. Bunny suits were required when on flight hardware, like the pallets or inside the Spacelab modules. Tia Ferguson recalled an awkward moment: "A crew member on their first time down to KSC was taken into the module. We had to watch and make sure they didn't touch anything. We said, 'Put on these bunny suits and we'll go into the lab', and he proceeded to take his pants off in front of me. 'No, no, no, no, no, no! Don't do that! You don't have to take your pants off. You can put the bunny suit on over your clothes'." Mike Haddad recalled that "A normal day's attire for me was a t-shirt, jeans and tennis shoes. When you're in a smock and/or bunny suit for eight or more hours a day, climbing on scaffolding, and in or on flight hardware, good clothes and shoes get trashed pretty quick."

The Visibly Clean specification was used for surface contamination, even though higher levels could be used. The definition of '*Visibly Clean*' was: "A clean surface as seen without optical aids (except corrected vision) when viewed from a distance of 6 to 18 inches (15.2 to 45.7 cm) using white light." Special rags were required to keep the flight hardware surfaces clean using only approved cleaning solutions, most commonly denatured ethyl alcohol and Freon® 113.

Servicing

Another major activity, and sometimes a very hazardous operation, was servicing experiments and payloads with fluids and/or gases. Freon® and water were some of the room temperature fluids employed, but the team also dealt with cryogenic fluids like extremely cold superfluid helium (–458°F/-272°C) on United States Microgravity Payload-1 (USMP-1), Lambda Point Experiment (STS-50, 1992), and Spacelab-2 Infrared Telescope (IRT) and Properties of Superfluid Helium Zero-g (SFHe). Very hazardous hydrazine propellant was used for the Orbital Refueling System (ORS) experiment that flew on STS-41G (1984). Gases ranged from standard nitrogen and helium to the complicated gas mixture required for the Cosmic Ray Nuclei experiment on Spacelab-2 (STS-51F, 1985). Because of the sensitive nature of the experiments and other flight systems, following the documentation detailing specifications for the fluids and gases was critical. That document was JSC SE-S-0073, Specification, Fluid Procurement and Use Control. Servicing could be required periodically for a payload during its entire time on Earth, before launch and after landing. EI personnel required special training to be able to handle some of these commodities.

Three main pieces of GSE were used to service flight systems, as well as to simulate flight systems during testing: the Rack Conditioning Unit (RCU), the Freon® Servicer, and the Water Servicer. The RCU, which was a fancy name for a huge special air conditioning unit, was used to provide cooling air to the racks during testing. It simulated the flight air supply system that was part of the Spacelab module being assembled in the Level-III/II area. The air provided to the racks had strict temperature, humidity and contamination requirements, and it was the EI person's responsibility to ensure requirements were met and maintained over the many days of testing. Mike Haddad recalls: "Because cooling to the racks was required before power could be applied and the rack experiment testing started, the RCU was one of the first pieces of GSE started. We had to be one of the first support teams onsite and working, and the last to leave at the end of testing. This made for some very long days."

The Freon® servicer performed a multitude of support tasks. It simulated the Spacelab module external Freon® cooling system being assembled in the Level-III/II area for module missions that included a pallet, and simulated the Igloo Freon® supply system for pallet-only missions. Not only did it contain Freon®, it also provided gaseous nitrogen (GN2) and gaseous helium (GHe) to flight systems for leak checks. The servicer was located away from the flight hardware, so long lines ran from the servicer to the interface of the flight systems through the Tunnel Area of the O&C. Mike Haddad recalls: "We wanted to make sure the GSE systems did not contaminate the flight systems, so before connecting the GSE, we placed a jumper at the ends of the supply and return lines where the GSE connected to the flight hardware. We would flow gases and fluids through the system for a period of time to remove any contaminants, and then sample. Once the sample passed, the jumper would be removed and the GSE connected to the flight systems. I remember after servicing one flight system, as we began flowing Freon®, we saw a large pressure rise in the Freon® servicer. After some quick analysis, we figured the only thing that could have caused it was a clogged filter in the servicer. After removing the filter for inspection, we noticed lots of contamination. How did that happen? It wasn't our GSE, so it had to have come from somewhere on the flight system. But all the flight pieces I assembled were precision cleaned. We later found that a cooling plate had not been cleaned properly. The GN2 purge and GHe leak check of the system did not dislodge the contamination, so it wasn't until we flowed liquid Freon® that the contamination was dislodged."

The Water Servicer performed a similar functions to the Freon® servicer, simulating the Spacelab module internal water cooling system for module missions. Pallet-only missions used Freon®, so the Water Servicer fluid was not used.

PHASED TESTING

Testing of the flight hardware was done in phases, starting at the assembly level and going up to major integration testing with the Orbiter at the OPF or launch pad. It involved hundreds of personnel. As an example, for pressurized module missions, the Rack Trains were tested in the Level-IV area to highlight any problems between racks and simulated Spacelab subsystems. The Rack Train would then be placed into the flight Spacelab module (Level III/II) and tested against the flight Spacelab subsystems, to ensure an entire healthy payload before proceeding to Level-I, Cargo Integration Test Equipment (CITE) and/or the Shuttle Orbiter. Once the payload was installed into the Orbiter, it would be tested for the last time before launch. This scenario would change depending on whether the mission was a Spacelab module mission, pallet-only and/or MPESS mission, or other non-Spacelab missions using associated hardware.

Lori Wilson noted, "When the physical installation was complete, we took on the role of Experiment Test Director, coordinating the functional and integrated testing of the payload. Our office then followed the payload through Level-III/II work and installation into the Orbiter, coordinating Level-IV personnel to work on the experiments as required."

Interface Verification Test and Mission Sequence Test

Nobody wanted the astronauts to take up valuable science-gathering time fixing problems that could have been detected and fixed on the ground. The tests could range from powering up small individual experiments, like a single middeck experiment, to operating large payloads containing 70+ international experiments, like Spacelab-1 (STS-9, 1983). The EI engineer often knew more about the operation of their individual experiment than the PI themselves, primarily because the EI engineer understood how it worked within the Spacelab environment. The PI certainly knew how to run it on the bench back in the design lab, but that did not always involve an adequate simulation of the Spacelab Command and Data Management System (CDMS) avionics, flight software, power systems, and other systems in which the EI engineer was an expert.

Another benefit provided by the EI engineer was their knowledge of KSC operations amid the multitude of organizations and paperwork systems. The EI engineers would guide the PI teams through the often confusing processing flow, so in the end the PI teams would have learned how to interact in the Spacelab/Shuttle world (something that would prove invaluable during mission operations). In a way, the team helped train the PIs.

"The Software Development Facility [SDF] at Marshall Space Flight Center [MSFC] was where all the software was written for the commands and the control and the systems," Scott Vangen observed. "They wrote the flight code for the computers as well and they'd ship the code down to our team of Jim Dumoulin, Cindy Martin-Brennan and those folks, who would load it onto the computer. Then us Electricals would make sure all the system had checked out and interfaced to the Spacelab, the rack floor or the pallet systems. One cool thing about Level-IV, and there's many, is although we may have had specific tasks, we were multiplexing, covering each other, and so the distinction of 'This is your rack' or 'This is your responsibility' or 'This is your black box' quickly disappeared when you were testing and executing."

These phases of testing were also very useful for the crew. Most of the time, this was the first opportunity for the astronauts to see the hardware close to flight conditions.

Jim Dumoulin recalled that "The Payload Checkout Unit [PCU] was a hybrid Perkin Elmer 8/32 [later a Concurrent], running a Real-Time Operating System and connected to custom racks of simulators and ground versions of flight computers running the actual Spacelab Flight Software Experiment Computer Operating System [ECOS]. The PCU simulators were developed by Apollo-seasoned IBM engineers under contract from MSFC, using their knowledge of everything that was needed to checkout Apollo spacecraft. Not only did the system emulate and/or control all Spacelab systems, it was better instrumented and had better visibility into how the hardware and flight software interacted than the real thing. At the time, the Spacelab avionics were the most advanced ever flown, and had constant internal hardware and software monitoring systems known as Built-In Test Equipment [BITE].

"When something went wrong, a BITE Error would be sent down the telemetry stream, but the PCU had access to all the internal workings that led up to the error. All telemetry needed to be viewed via one of four separate High-rate Input/output Test System [HITS] systems [another Perkin Elmer Computer]. While this forced mission operators to troubleshoot with only the tools they would have during an actual mission, it was an incredible waste because the PCU had access to much more information than what was possible to see in the downlink. During Level-IV testing, failures were not always the experiment's fault. At this stage of the testing, it was just as likely that a problem was with untested flight software, or that the ground database was mapping PIDs to the wrong place in the telemetry. During Level-IV testing, pressing a screen print from a PCU ECMON display, or looking at the ECMON log files, provided detailed data to close out paperwork or report flight software errors. During flight, to do the same thing would require ground controllers in the Payload Operations Control Center [POCC] in Houston [for Spacelab-1] or MSFC [post-Spacelab-2] to be alerted to a problem and then quickly send uplink commands to dump onboard memory locations before they were overwritten."

Level-IV personnel were always looking for ways to improve and streamline the processes used during Experiment Integration.

"So, even though the Payload Checkout Unit was a wonderful system, it did require a team of people to run," Jim Dumoulin continued. "When I first got to Level-IV, the PCU was flying blind. It had no displays. You could power things up, but it would be expected that you were going to use and see the flight Data Display Units [DDUs] for actual visual observation. You didn't see the downlink coming from the spacecraft. You had to put in another system which was called the HITS system [High rate Input/output Test System], and that was a whole other team. It was still part of our Level-IV electrical group, but it required folks like Polly Gardner and Ed Oscar and three or four other people, and they had to power up things. But before those guys could see the instruments, you had to power up a comm system, and that was called the High Rate Data Multiplexer [HRDM]. That required turning on a box. And we had these things called LOLI logs – limited operation life logs – where you could only power those things up at certain times. When I first came on board, that was the way we did it. You scheduled with all those people. I decided there may be some better ways to view things using just the Payload Checkout Unit, so I started by building these assembly language programs called ECMONs that let us look at what the PCU was doing. The PCU would send a command and then we'd get talkback from the vehicle. I had to decide whether it was a valid thing. So you got a lot of visibility into actually how flight stuff worked, but with no display, it was really difficult and kind of boring to run.

"NASA was really worried at some point about the Japanese getting ahead of them, ahead of the government, in artificial intelligence. So Congress threw millions of dollars at NASA, and said 'Each NASA center, can you assign somebody to do that?' So they gave me a couple of million dollars. I bought a bunch of computer systems, some Symbolics computers, and I ended up writing this program called Smart Processing of Real Time Telemetry [SPORT]. It was written in artificial intelligence-type of languages and it analyzed the telemetry streams. It was basically the ECMONs, but instead of being an assembly language, it was using the first really high-res graphics, and this was back in 1984. It was HDTV. Basically, it was a 2K by 2K color display. These were boxes the size of two refrigerators and they cost a lot of money, and there were no strings attached. You could just do whatever you wanted. So I used that to help analyze telemetry from Spacelab, and it drove TV screens in the control rooms up on the top that had orbital tracks. It could do the graphical display of when you were flying a mission in Spacelab during the Level-IV simulations. We had some very archaic technology that was built in the 1970s. It ran an orbital model, but the orbital model was run on card punch tapes."

Jim explained further that sometimes the mission would fly over parts of the Earth that had weird electrodynamic properties, such as the South Atlantic Anomaly (SAA), and it was possible to see the dynamics of that by analyzing this telemetry. SPORT was an external system that added an Artificial Intelligence / Expert System capability to the telemetry collected by the ECMON assembly language programs. Eventually, the refrigerator-sized computers were replaced by a dozen or so networked IBM personal computers.

While SPORT was very successful and very useful for Level-IV, Jim did not stop there, finding other ways to improve testing. BITE code errors were generated by a device which would diagnose itself when there was a problem electrically and would set these codes. It was basic developmental flight instrumentation that gave an indication that something on board was wrong, or was not fast enough, or could not move. It was constantly doing a health and status check of the Spacelab computer, internal to the equipment, and storing these codes.

But nowhere in the PCU system, or even the HITS system, was actually reading these codes. They were just developmental, as Jim Dumoulin explained: "By building this smart processing system, we could look at these codes to try and understand them, and say, 'That's what that means', and then be able to graphically display it. So we had visibility into the systems much better than the folks that were sitting on the ground doing that during the mission. We had the luxury in Level-IV in that everything was done through a GMT time tab, and in Missions Sequence Tests, we would do the same mission sequence over and over and over until we got it right. So we got very comfortable with certain mission timelines. We got rid of all the boring side, like when the crews sleep or things are shut down, and we just did all the exciting things. When a payload was powered up, we would sequence each one. You could lay out a complete mission and say, 'Okay, I'm going to pick the certain sequences and times where there are multiple things happening'. And that's what became the Mission Sequence Test."

Jim Dumoulin had a way of just making things happen, but he could also see the big picture of Level-IV activity and created things nobody had thought of. "In the process of adding this telemetry processing and networking capability to Level-IV, we ended up building a high-speed computer network with hundreds of connected computers packed with high resolution graphic displays and large memory boards. Between missions, when these computers were not needed to view real-time telemetry, they were used for word processing to create mission procedures, producing mission schedules, and for sifting through massive amounts of data looking for trends. As computer prices came down in the 1980s, the KSC Payload Organization was well ahead in the field of computer networks and office automation.'

Many other disciplines made up the Level-IV team that tested the flight hardware, and all of them worked very hard together to ensure success on every mission. While many in the trenches made things happen, it still needed overall integration and control of everyone during these testing phases.

"I was running the procedure, it was the engineer who put it together," explained Mike Kienlen of NASA Level-IV Operations. "It was their design team, with the know-how to design everything. I built the schedule for it, but the technicians, the Quality guy, the engineering [built it]. There were probably 15 engineers out of Mechanical and Electrical to design and build that hardware, working with the science team. All the Ops people did was to coordinate the activities so the engineers could be as efficient as possible."

Scott Vangen described how they covered each other, because that led into the second responsibility which most, if not nearly all of the electrical engineers, and eventually many of the mechanical engineers, got assigned to; that of Experiment Engineers. The Experiment Engineers, as the name suggests, were assigned one per experiment (with a backup) that they would follow 'womb-to-tomb' with that hardware arriving at KSC. They prepared for that by traveling to the Principal Investigator's site, whether that was domestic or overseas, as Scott Vangen recalled: "First time I ever left the country was to support the experiments, and I've had probably a dozen and a half trips since associated with experiments. So it's pretty amazing when you think about the responsibility. Starting as a fresh out of college GS-7, within the first year of my career, if not the first months, I not only had the responsibility for the ground checkout system for the maintenance, the operations and execution thereof, but was also assigned one or more flight experiments, flight hardware, experiments, for writing the checkout procedures and troubleshooting any problems."

Tia Ferguson tells how it all fell into place for her, too. While still in the mechanical group, she was pulled into support testing, which early on was the only electrical assignment. "They started letting us have those experiences. I remember asking them specifically because I want to do it. Several of us got to do that."

"Due to pileups in the processing flow, it was not unheard of to be required to be in three places at once," explained Jim Dumoulin. "The vehicle teams at JSC and MSFC, and the contractor-heavy Shuttle organization, were assigned to Orbiter flows. They had enough personnel to rotate three separate shifts while in space or on the launch pad, and two shifts at the OPF. Many could even afford to take a few days off between missions. However, the Level-IV Payload Organization was staffed only one shift deep. A Level-IV engineer was typically supporting an experiment that was in space [or being de-integrated after returning from space]. At the same time, they had testing responsibilities for missions in the final pre-launch integration phase, at one of the launch pads, in one of the three OPFs, or undergoing integration testing in the O&C. Even while all that was going on, experiment engineers almost always had an additional experiment undergoing offline lab build-up, or in early development at a university."

Testing also encountered problems that were the result of an innocent accident, as happened with Cheryl McPhillips and "The Notebook". There was a notebook on top of the power console in the Payload Control Unit (PCU), and by accident she knocked that notebook off the console and it hit the switch that powered down the whole Spacelab. "And of course then you heard people coming onto the OIS [Operational Intercommunications System, the communication system used throughout KSC]. Basically if a person had a headset on, they were using the OIS. Well, they all started talking: 'We just lost data, we just lost telemetry'," Cheryl explained. "It was 'hurry up and wait' when you were testing. A lot of times it took hours to get going, get everything powered up and get into the test. So when it was powered down like that unexpectedly, you were not sure if everything was going to come back up right. And you knew you just cost the testing team a couple hours. After that, they put the switch guards on. They named them the *McP Switch Guard.*"

"While Level-IV was staffed by young engineers, our management made sure we had a few experienced folks that had been involved in NASA since the early days," explained Jim Dumoulin, "We were not always sure what these folks did, but we didn't question the process. One such person was Ray Norman. Every time we designed some hardware or got a new piece of equipment delivered, we had to send the schematics to Ray. One day, when we were about to power up Spacelab for a major integration test, the team was worried we would have to shut down for quite some time due to the long lead time nature of this equipment. It turns out Ray was the 'Spares Analysis' guy assigned to Level-IV, and his job was to purchase spares for everything that came across his desk. He looked up the assembly number and said, 'No problem, how many of these units do you want? I have three in the warehouse'. We were up and running the next day."

As Scott Vangen explained, "You would go in and say, 'I need X hours of testing', to either complete your integration test, or troubleshoot some problem report or whatever, and you'd personally be jumbling yourself and your multiple experiments in that case. And others were doing the same. We kind of all worked it out, so if I was busy I could ask my backup 'While I'm preparing that, can you work with the other experiment to prepare them for this afternoon when we power them up, get that procedure ready?' And that really did work. It was not the time to be off sick though."

As Tracy Gill stated, the team did take the workplace rules seriously. "We did try to adhere to the spirit of the rules all the time. Meaning sometimes if you needed it to be 16 hours or 18 hours you'd do it. One time I was writing a Mission Sequence Test. That's the big test and we had to get it in by the deadline, [even if it meant being there all night]. So I was in my office and worked the procedure all night, because I had to get it in the very next morning to get it into publishing. Then I went home at eight in the morning and took that day off, because I had to get it in to get the book released in time."

A Mission Sequence Test (MST), as the name implies, meant running a slice of the mission timeline. This was late in the flow, and the mission manager's team, along with the JSC people from Mission Control, would have put together some timeline of X number of hours; a day in the life − or a slice of a day in life − of what this would look like on orbit. Scott Vangen added, "So we would power up all the experiments and run them through their paces to see that they all played well together, and we would find different set of problems, such as integrated problems at a different levels over multiple experiments."

Tracy Gill recalled that, "The significant findings and the problem corrections we made ended up being reflected in the crew procedures or flight notes, such as 'If this happens, then do this'. So that stuff happened all the time."

Astronauts' Participation in Testing

"When we were doing testing during Level-IV," said Scott Vangen, "the Payload Specialists by far spent the most time with us, because they were payload people and they were going to be there if their instrument was powered up. The Mission Specialists would participate with some of the testing and that depended largely on two things, namely their time availability and how much their Commander prioritized that. Some Commanders saw Level-IV as better than any simulation that would happen, [thinking] 'I want my crew there for any time we're doing value testing and certainly during the Mission Sequence Test', which was the 'dress rehearsal' of the mission, if you will. Now [with] the Commander and Pilot, the chances are that if they came they would observe, but they had limited interaction with the actual commanding and the execution of the timeline, because that wasn't their responsibility."

OTHER LEVELS

All the Online Level-IV work, described above, had to be completed before proceeding to Level-III. While Level-IV dealt with the Spacelab experiments, Level-III/II dealt with how the Spacelab subsystems integrated with those experiments.

Online Level-III (Subsystem Integration)

Worked by MDAC, this Level prepared the Spacelab subsystems for integration with the Spacelab experiments, which would occur in Level-II. Because the Level-III and Level-II stands could perform both the Level-III and Level-II function, these were frequently grouped together and stated as Level-III/II operations.

Online Level-II (Subsystem to Experiment Integration)

This was where the flight Spacelab racks met the Spacelab module subsystems, a total flight-hardware-to-flight-hardware integration. Now with a full flight compliment, tests were performed to ensure all worked as desired. Once complete, it all passed to Level-I, to determine whether all the Spacelab flight hardware would work with the Orbiter it was flying on.

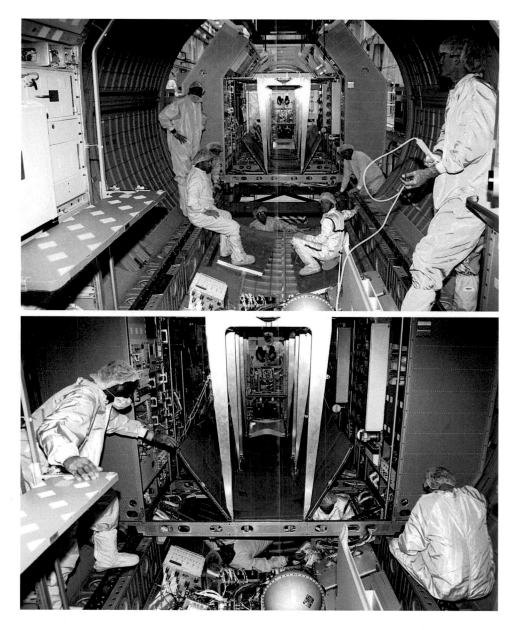

Spacelab Experiment rack train installation into Spacelab module. Top: Experiment Racks being prepped for installation. Flight Rack 2 is to the left in foreground. Bottom: Rack train partially installed. Flight floor panels raised to allow access to underside of flight floor (rectangle metal piece with oval holes, bottom center of photo). Clear view of internal braces kit supporting racks during installation. Once racks were fully secured to inside of module, the braces kit was removed. [NASA/KSC.]

Online Level-I (Simulated Orbiter Interface Testing)

The Level-I side was basically a high-fidelity orbital simulator called the Cargo Integration Test Equipment (CITE), controlled from the O&C test stands. This was used to test the interface between the full Spacelab flight hardware and a simulator of the Orbiter, to work out any problems before being transported to be integrated with the actual flight Orbiter.

Weight and Center of Gravity

Weight and Center of Gravity (WT&CG) was performed on every payload near the end of the O&C flow, so that the information could be sent to the Orbiter people to confirm the overall WT&CG of the payload and vehicle was within predicted values. Most of the time this was done by MDAC personnel, with Level-IV personnel in support for all the large payloads. Smaller payload WT&CG would be done by Level-IV personnel.

Payload Envelope Check

On the Orbiters, there were static (87-in, 221-cm radius) and dynamic/thermal (90-in, 228.6-cm radius) envelope requirements around the payload which ensured that whatever payload was installed would not impact any of the Orbiter components. Static was always inside the dynamic/thermal envelope because, as the names state, 'static' applied to when the Orbiter was in a static environment; 'dynamic' in a dynamic environment such as launch/landing; and 'thermal' applied to launch/landing as well as on-orbit. There were times where a payload would violate the static envelope, but was "stiff" enough not to exceed the dynamic/thermal envelope.

Sharp Edge Inspection

Sharp Edge Inspection was one of the final operations performed on the flight hardware. This would include both inside the Spacelab module and all external components that made up the Spacelab payload. The inspection determined whether any sharp edges existed that could harm the crew and, like everything else in the space program, a document existed that stated what constituted a sharp edge. Mike Haddad, who dealt with sharp edges, recalled, "Visually, you could see some sharp edges, but the main process we used was to take the latex gloves we were wearing and rub them across all potential sharp edge locations. If the glove snagged, we noted the location and fix it, documenting all the work we did. If the glove ripped, we would put on new gloves and continue the operation. The fix was usually as simple as placing a piece of flight tape over the sharp edge. Rarely would we need to file off the sharp edge or remove material."

IN THE ORBITER PROCESSING FACILITY

When all C&C operations were complete, the payload was placed into the Payload Canister and transported to the Orbiter Processing Facility (OPF). Tia Ferguson recalled the first time she saw the Shuttle in the OPF: "All you see is a really low black ceiling, and then you realize that [you're looking at] the tiles of the bottom of the Shuttle."

The Payload Canister entered the OPF and then effectively a reverse of the O&C operations occurred, removing the payload from the canister and installing it into the Orbiter. Once installed, and with most of the Shuttle operations in the OPF done during the first shift, the payload operations and testing was usually scheduled for second or third shifts, supported by Level-IV Engineering, Safety, Quality, Operations and SEIS technicians.

The Operations (Ops) person was a coordinator, who was not specifically responsible for anything. It was always an engineer who actually had the responsibility. The Ops person was there to coordinate and make the job as efficient as possible, so that when the engineer showed up the work could actually start, the technicians were ready, the equipment was there, and the tools they needed were on hand. Mike Kienlen said, "If you help the engineer, then they'll tell you what's going on, you'll have a good working relationship, and they'll want you around. The Ops people were the ones who would work with all the engineers, collect the data, and then sit with management and report everything that happened. [If] You act like an idiot, you act like a boss, like you're the king and you're the program manager, then the engineers won't tell you what's happening and it'll be your fault. All of a sudden, you'll be reassigned to the hall closet monitor in Headquarters on the third floor."

Todd Corey recalled that, "From my years of experience working at the OPF, the constantly changing Orbiter schedule made it difficult to keep our payload operations occurring on schedule. Sometimes we got out-prioritized in real-time by Orbiter work. It was always a balancing act to maintain an assertive posture with the Orbiter folks without being an irritant; to maintain our agreed upon times and yet recognize you had to show flexibility and be realistic about reacting to unforeseen problems. I tried to counter the 'hurry-up-and-wait' posture the OPF Ops folks would try to place us in. Once on site, as an Ops person, we would do whatever we could to support the team doing the work, such as running to logistics to get a tool, or grabbing a bunny suit if someone didn't bring one. Sometimes, just staying out of their way was good too."

During 1997, EI engineers performed a complete mission turnaround in the OPF for the two Microgravity Science Lab-1 (MSL-1) missions STS-83 and -94. Normally after a Spacelab mission the hardware returned to the O&C for de-integration, but because the STS-83 mission was shortened due to an Orbiter problem, the payload remained in the Orbiter for all re-flight preparations, a task never performed before or since in the 30-year history of the Shuttle program. EI contributed to the successful re-flight as STS-94 only three months later.

End-To-End Testing

End-to-End (E-T-E) testing was a test that would check to see if every link in the system would work as designed, as it would be used during the mission. Information from the experiments would transfer through the Spacelab subsystems, into the Orbiter systems, from the Orbiter to ground communications systems, on to the in-space (TDRS) communications, back down to the ground communications systems, and finally end up at JSC and POCC at MSFC. The command and data transfer system would be tested both ways, up and down. Like the MST, it would make the hardware think it was in space. This would be the last major test of the hardware before launch.

Middeck Experiments

The middeck experiments were small, about the size of a microwave oven, and were placed inside lockers located in the middeck area of the Orbiter. These experiments could also fly in place of the lockers. They could be processed at many locations around KSC and would then be transported to the Orbiter while it was in the OPF or at the pad.

Most of the middeck experiments required power, which would be supplied from a piece of GSE in the Spacelab or directly from the lab, depending on what capabilities that lab contained. The trick was trying to keep them powered during transport from the lab to the Orbiter. To solve this problem, a mobile GSE power box was created.

Rey Diaz designed the Middeck Power Control Box. "I created that before *Challenger*, (STS-51L, 1986). At the time, we had all these experiments requiring continuous power. How do you do that? We started using already-flown 28-volt batteries from Solid Rocket Boosters to power the middeck lockers from the point they were ready to leave the offline area to be installed into the Orbiter, and these batteries provided power for several hours. Initially, we didn't have a good control box. All we had was a very crude set of cables, so in trying to come up with a better way to manage this, we wanted it to have monitoring of power and to cause you to have to check the voltage to provide a level of protection to the payload. We had to do some basic connections in the early stage when we created this box. The whole idea here was that you now had fuses, you had a volt meter as part of the power control box. You had a power-on switch where you could do that. And also you could do a check without powering the payload. You could do a check on the battery, and they would tell you [whether] they had enough juice, enough power, to go through the entire set of operations you had to do before installation. I designed that back in 1985."

These power requirements could be very strict, so OMRS were setup to ensure the experiment was only without power for a short period of time during transition from GSE to mobile, and then from mobile to Orbiter power. If power was lost for more than a certain amount of time, so was the experiment, so it was critical for it to have that power. The experimenter could have supplied their own portable power setup, but it may have not been able to pass all the requirements necessary to use it at KSC, especially once it was around and/or in the Orbiter. Rey Diaz continued: "This was a way for us to provide some Ground Support Equipment that could help them set up their own continuous power and ground systems. It served many purposes, because it meant there was a ground system available to developers, so that they didn't have to try and come up with their own system of batteries, control box and all that."

When the middeck experiment was ready for installation into the Orbiter, it was switched from ground power at its integration area to the battery and control box, then placed into a transportation vehicle and transported to the OPF. Once there, it was unloaded and taken inside. Once the middeck payload arrived in the OPF on battery power, it would be located in the White Room adjacent to the Orbiter side hatch (the hatch shown on TV when the crew entered the Shuttle at the launch pad.) Orbiter personnel would have the final location prepared for the payload (tools, Orbiter power cable ready, etc.) and be inside waiting for it. The payload would be hand-carried through the hatches – with a long power cable still attached – and handed over to Orbiter personnel who did the installation. Once installed, and the Orbiter systems were ready for power transfer, the middeck would be powered off (if required), as would the GSE control box, then the GSE cable disconnected from the payload, the Orbiter flight cable connected and power turned on. Level-IV personnel would then power up the payload and perform an Interface Verification Test (IVT) to ensure all was working as planned.

Glenn Chin told the authors that his prime role was middeck payload processing and integration. "Back then, we really didn't have a middeck payload process *per se* and middeck payloads were rapidly becoming frequent flyers, so we had to be able to treat all PI customers with a user-friendly process so that each flight/mission would be a positive experience for them. My big deal was to develop and standardize a flow process for each PI customer per mission, and integrate so that each customer had KSC processing timeline from on-dock to PI experiment turn-over to KSC, where KSC would take over and deliver middeck payloads to OPF or pad and then install into middeck and IVT as required before middeck close-outs occurred. It was a very hectic and fast-paced rhythm, working multiple PI customers per mission and planning multiple missions on the radar and over the horizon, to ensure all customers were equally supported for mission success."

Closeout Operations

For most module missions, this would be the last time people would access the hardware while still on Earth. Closeouts included operations like final stowage, verification that all hardware was installed, all 'remove-before-flight' items were removed, all cabling was installed in the final launch configuration, all valves were in the proper position, and a final switch list verification performed to ensure every switch was set correctly for launch. As this was completed, photos for the final configuration would be taken and people would exit the module.

For pallet or MPESS missions, some closeouts would be performed in the OPF, but because these components could still be accessed at the pad, most closeout operations would occur a few days before launch at the pad before the Orbiter Payload Bay doors were closed for launch.

Once OPF operations were completed, the Orbiter was transferred to the Vehicle Assembly Building (VAB), stacked to the ET and SRBs mounted on a Mobile Launch Platform, and then the crawler rolled out the whole combination (now known as "The Stack") to the pad.

AT THE PAD

At the launch pad, operations would range from removing covers from experiments before launch, to rebuilding an experiment (as on USMP-1/STS-50 in 1992) only weeks before launch, to the first loading of a science payload with very hazardous hydrazine propellant (as on the ORS payload flown on STS-41G in 1984).

Mike Haddad: "I loved going to the pads, one of the most historic sites on the planet, and saying to myself, 'I'm standing here on the pad, doing awesome/fun work and getting paid'. Even though we were very busy at times, I would always take a moment to stop and look at the landscape that you could see from multiple hundreds of feet above ground. [There was also] the fact that you were up close and personal with an Orbiter, two SRBs and an ET, in launch configuration ready to leave the Earth."

"Once the Orbiter was installed on the launch pad, we attended the daily meetings to coordinate any experiment specific work required," recalled Lori Wilson, "We were also responsible for coordinating any experiment work required on the landing strip and were assigned to alternate landing sites and emergency sites in foreign countries. This required a passport, a physical and, of course, shots."

To do these operations required very long discussions early on in the planning phase by all parties involved, to say "Can this be done?" This was especially evident when animals were involved. Loading animals on the Shuttle at the launch pad before launch was a huge undertaking, explained Mike Kienlen. "The purpose of the mission was to do science. And you were sitting there and talking with people in the Shuttle program or wherever. You held all the aces. You'd say, 'This is what we need to do', and they'd come back and say, 'You can't do that'. You'd point out that 'If I can't do this, then we can't put animals on the mission, so what's the issue? Safety requirements? Okay. Well, we can probably talk about this for a week. We'll end up in the Center Director's office and then he'll be calling Headquarters and then we'll say the Safety engineer said that we can't load the animals, so we'll probably have to cancel the Spacelab mission. Your goal is to make the mission happen'. Nobody wants to stand in front of the senior management and say, 'Yeah, I'm the one who's saying we can't do it'."

According to Todd Corey, "Orbiter work at the launch pad was more predictable and presented less problems from a schedule change perspective. I did my best to stay abreast of the situation, and called our payload team out to the job site only when we had a reasonable chance of starting on time."

Mike Kienlen continued, "Crazy things have happened in the PCR; 99 percent of every payload that flies is super contamination-sensitive, and the PCR is a challenge to keep clean because it's a room attached to a Shuttle. Everything's moving. You've got air bags around the perimeter, you've got things inside the Shuttle bay doors, inside the room which are normally outside. So it's really hard to be clean, and as we were getting the room certified for the next mission, they were sandblasting and painting outside because it's a launch pad, a big, giant steel structure next to the ocean. They were always doing corrosion management, always sandblasting rusty areas and repainting it. [One time] It was the last couple of days [they could do the work], because all the painting would have to stop when the Shuttle and the payloads arrived. One of the workers took his sandblasting tool and stuck it in the pipe to clean the rust out on the inside of the pipe. He blew 30 pounds of sandy air inside the PCR, which completely covered the entire floor with sand dust. It was all over the place. No one knew that's what had happened, and they opened the PCR the next day to find the room covered in sand. They had to work around the clock for a week to clean the PCR."

To get to the PCR required a ride in the FSS elevator (the same one the astronauts used) to the 135-foot Level. The pad did not have many solid floors; it was all grating with a view through it revealing just how far it was down to the ground. Once out of the elevator, the path was over the Fixed Service Structure/Rotating Service Structure (FSS/RSS) transition, then around the outside of the PCR to get to the main entrance door to the suit-up room. Often, the experimenters and Level-IV personnel had to perform work in the PCR, so Level-IV personnel would be escorts for the experimenters while on the pad. Before going to the pad, a briefing would take place describing the pad configuration and walkways, as mentioned above, to see if anyone was afraid of heights and what would need to done if they were. Mike Haddad added that, "Most were OK, some were not, and a few never told us so we found out ourselves when we got out of the FSS elevator. I was escorting one PI, with a few team members, and as we stepped out of the elevator onto the 135-foot Level, the PI froze in place, one foot in the elevator one foot on the grating. NOT good. Obviously this person was scared to death of heights and hadn't told me. We kept the door open while we helped him move his one leg back into the elevator, and I told the rest of the team to proceed on while I took the PI back down to the ground. That poor man, I felt for him. He badly wanted to see his experiment for the last time before it went into space and tried to beat his fear of heights, but he just couldn't. He had to monitor the PCR work via the OIS [and sometimes we had video feeds] from one the trailers/boxcars at the base of the pad." It should be noted that due to the grating on all Levels, the Level-IV personnel would inform the PIs/PDs during the design phase of their GSE that anything on wheels to be used at the pad had to have dolly-size air-filled tires, rather than small caster wheels that could get hung up in the grating.

The payload sat inside the PCR, either on the Payload Ground Handling Mechanism (PGHM; operations will be discussed more in Chapter 6) waiting to be installed in the Orbiter, or already in the Orbiter. This would be the last location to access the payload and Payload Bay of the Orbiter before closing the Payload Bay doors for launch. A variety of operations would take place in the PCR. On the PGHM there was 360-degree access to the payload, so here most of the operations dealt with closing out the part of the payload where access would be lost once installed in the Orbiter.

Launch Readiness Review

The Launch Readiness Review (LRR) occurred at about L-3 weeks,[2] to verify the mission's readiness for launch. The meeting was used as an assessment of Shuttle and payload processing activities, including: launch commit criteria; GSE and facilities; payload and ground crew activities; training; launch and landing systems; landing and retrieval; scrub turnaround; and contingency plans. All organizations were responsible for identifying significant configuration or critical process changes that had been incorporated since the last launch, and which might impact launch, flight, or landing operations.

Flight Readiness Review

The Flight Readiness Review (FRR), was held approximately two weeks prior to launch. The FRR examined tests, demonstrations, analyses, and audits that determined the system's readiness for a safe and successful flight. It also ensured that all flight and ground hardware, software, personnel, and procedures were operationally ready.

---

[2] L-time is the actual time before launch. T-time is countdown time, so L-time = T-time + built-in holds

Late Stow of Middeck and Module Vertical Access Kit

Late stow refers to the task of installing and/or testing payload-related hardware late in the launch count very close to launch, usually within L-24 hours.

Middeck Late Stow Operations

Most late stow middecks were prepared in Hanger L, but some were also prepared in labs in the O&C. Wherever they were, Level-IV personnel would go to that location and do a turnover process from the PI. They would make sure it was clean, that it was the right part, and if it required continuous power they would put it on KSC battery power, making sure the batteries were all charged beforehand. It would then be taken out to the pad and transferred to Orbiter power. The Orbiter people would install it and then Level-IV would go in and transfer the power. Jeannie Ruiz said, "We didn't do any functional tests, just made sure the light came back on. The power was critical because some of those had maybe 15 minutes [maximum time allowed without power]. The tricky part was they couldn't be off power for very long, so it was nail biting sometimes because you had to get the Orbiter guys to flip their switches, and if they weren't timely then you had to hurry them along. That was fun, but there were some weird hours and you might get a phone call afterwards. 'Oh, a light went off on this experiment. What do we do?' And then you'd tell them what to do. We weren't always sitting in the Firing Room after that; we would go home or back to the office and we would go back [to the Firing Room] later."

Some middeck experiments, mostly dealing with life science samples, required very late stow (loading) onto the Shuttle. This could be as late as L-12 hours. The operation could be an entire middeck being stowed late, or just samples installed into an experiment previously loaded on the Shuttle middeck. Many of these had strict requirements to maintain the samples/experiments in pristine condition, requiring temperature and power support.

This would occur within the last hours before launch, so it was critical that those operations went smoothly and efficiently because any deviation could cause a launch scrub. The personnel performing these tasks were under a great deal of pressure to complete them as planned.

Module Vertical Access Kit

A very critical pad activity dealt with placing perishable science samples into a vertical Spacelab module very late in the launch countdown, about a day before launch. Access to the module was very complicated, and required the use of the Module Vertical Access Kit (MVAK). MVAK was a system of pulleys and motors which lowered a person in a harness down into the module on the end of a cable.

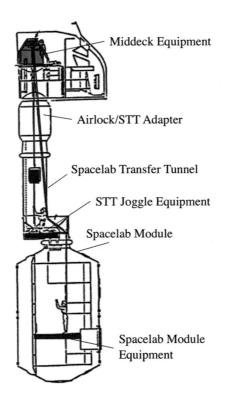

Middeck Equipment

Airlock/STT Adapter

Spacelab Transfer Tunnel

STT Joggle Equipment

Spacelab Module

Spacelab Module
Equipment

MVAK graphic. [Don Dolby.]

The original concept of the MVAK was to use it to change out any Spacelab module subsystem electronics and such inside the module that failed after the Orbiter and Spacelab arrived at the pad. This was to avoid having to roll the Shuttle stack back to the VAB, de-stack, roll the Orbiter back to the OPF, and then gain access to the module to perform a changeout of failed component. But it was soon realized that MVAK could be used to support Spacelab Science and Spacelab subsystems. Having late access to the module at the launch pad opened a huge door of opportunity, allowing very late loading of samples, animals and any other items that were very time-critical, based on launch time and when science would begin on-orbit.

The hardware used for MVAK was installed by Orbiter personnel, with support from Level-IV personnel, because the MVAK hardware interfaced directly with the Orbiter. Precise coordination was needed between the Orbiter personnel and Level-IV personnel to ensure that all the hardware was installed and tested properly before any personnel would ride the equipment.

MVAK was a totally optional task for the techs, with no extra pay even though it meant working on and below a live suspended load, usually in a confined space.

Had there been a fire, it would have needed total team work to get the MVAK tech through the joggle and up the tunnel in full smoke.

The team consisted of the following:

- *Experiment Transport* (up to 6 people): Technicians transported the experiments from the Payload Van up to the White Room.
- *Middeck Team* (3–4 people): Hoist Operator and two techs installed and removed packages from the air-sea-rescue hoist hooks. They also helped pass the experiment packages through the Orbiter hatch and onto the MVAK platform above the tunnel airlock.
- *Joggle Tech*: Passed the package through the joggle. Some of the weight would rest on the tech as the package was manhandled through joggle. From the joggle ring, the tech would guide a hoist line through guiding pulleys on the joggle ring.
- *Module Tech* (one or two persons): Inserted the experiment into the rack, performed the IVT, and verified that experiment operation was normal.

This system was used many times to load time-sensitive science samples, such as the primates and rodents for the Spacelab Life Science (SLS) missions.[3] In the event of a launch scrub, the activity would have to be repeated to remove the first set of samples and install fresh ones.

Closeout Operations

These were the final closeout operations before launch. Some of those tasks done in previous closeouts were repeated for those final flight elements that had not been closed out. The final closeout would take place following MVAK and middeck late stow tasks. After this, there would be no access by anyone until the crew began work on-orbit.

LAUNCH OPERATIONS

Depending on mission requirements, EI personnel also sometimes supported launch countdown operations from the payload consoles in the Firing Room located in the Launch Control Center (LCC). There, the team monitored and commanded payload systems, right up to the moment of launch. The Terminal Countdown Demonstration Test (TCDT) occurred about two weeks before launch. It began as a normal countdown operation and, as it approached T-0, the clock read 5, 4, 3... and then everything stopped. This was not an aborted liftoff, but TCDT. Prior to every launch, the astronauts and ground crew participated in

---

[3] These missions included Spacelab-3/STS-51B; SLS-1/STS-40 and SLS-2/STS-58

launch-related activities at NASA's KSC in Florida, for their final on-site preparations before liftoff. Other operations during TCDT dealt with how the crew would handle an emergency.

C-1 Payload Console Operations: TCDT and Launch

Payload personnel occupied the C-1 payload console during TCDT, to practice any launch day operations and support. Sometimes the payload was powered up, so personnel would make sure all information coming from the payload was as it should be. This would also be a time to practice late access items. Anything that did not work as planned would be fixed for the actual launch countdown.

Launch Countdown

Launch countdown would begin a few days before launch and was a repeat of TCDT, but hopefully with the countdown clock reaching T-0 and launch. This was the real thing; all the work, all the long hours, all the problem solving, all the friends made, came down to this moment of launch. No matter how many times you sat in the Firing Room on the C-1 console, every launch was unique and exciting, but also nerve-wracking. Have you done everything correctly? Did you miss something that means the launch and the mission is not successful? Mike Haddad recalled, "Those last 10 seconds for me really brought it home. We were launching! You tried to stay cool on the outside, but inside you were going crazy. T-0 and liftoff! Once the Shuttle cleared the tower, at that point Johnson Space Center would take control of the ascent."

Scrub Turnaround Operations

If there was an abort and the Shuttle did not launch, the system would proceed with scrub turnaround operations. This could be 24 hours, 48 hours, or longer before the next launch attempt, depending on what caused the abort. For many missions, this would cause a scramble by the Payload Organization, because all the late loading work would have to be repeated. Those items on the Shuttle would have to be removed and replaced with new items. It was very busy, very critical, and very labor and hardware intensive. "But that was fun," said Jeannie Ruiz. "Working with the Orbiter guys beforehand, you knew all the planning because it was critical. Part two was the scrub turnaround requirements for each of those, and the critical thing is [that] the PI might not really understand how the Orbiter was going to work and turn things around, and what that really meant to them and their sciences. I might have to talk to the Orbiter to guys and figure out what their plans were, and then interpret that to the science PIs so they knew when to be ready for a scrub, have the next samples ready, and when we'd be there to change them out. That was a lot of coordination, a lot of pre-planning, which was cool. You had to have all that ready lined up, with more time spent in pre-planning than doing the actual work for those. It was great; that was the most fun part."

One mission required unplanned MVAK operations due to length of the delay, which at one point seemed impossible due to all the tasks that needed to take place. This will be discussed later in the book.

Post-Launch

Right after T-0 and lift-off, the PAO Officer's call would typically be saying "…and the Shuttle has cleared the tower…Houston is now in control…" Mike Haddad recalls, "At that point our job was done, so everyone in the LCC would stand and look out the windows to see the Shuttle on its way to the heavens. Being right there, the sound was already hitting the LCC and shaking everything in sight. To me, the noise was like the Shuttle's way of saying 'goodbye [to us], see you at landing'. As the sound faded and the Shuttle ascended, you could finally catch a breath, but we all knew that first 2½ minutes [until booster separation] was a critical time, so you could not really relax just yet. After SRB separation, it was another six minutes to Main Engine Cut-off [MECO], ET separation, and then the Orbiter was in orbit. They may have needed to perform a few more burns to circularize the orbit, but the Orbiter had made it. So off came the headset, shaking hands everywhere and then it was time for beans." (see Callout: "The Beans Are Go.")

Go For Beans! [NASA/KSC.]

The Beans Are Go

A much-celebrated KSC tradition began with STS-1. After that first Shuttle liftoff in April 1981, launch controllers enjoyed beans and cornbread as an immediate payoff for a successful launch. Former NASA Test Director Chief Norm Carlson started the tradition with one small crock pot of northern beans for his hungry staff. The tradition grew in popularity and Carlson turned the cooking over to Kennedy's food service contractor. Hundreds of launch team members, managers and dignitaries would swarm the Launch Control Center lobby after each Space Shuttle liftoff, dipping into a dozen 18-quart cookers brimming with beans. This popular tradition wafted the aroma of launch success through the air ducts, into the elevators, out the automatic doors and into the parking lot. About 400–500 bowls of beans were served at the Launch Control Center.

Successful Launch Beans

Courtesy of Norm Carlson, former NASA Test Director Chief.
Put 6lbs. of dried great northern beans in an 18-quart electric cooker.
Cut 10lbs. of smoked ham into cubes.
Add ham and ham bones to beans.
Add ½ shaker of lemon pepper.
Add 3 lbs. of chopped onions.
Add 2 stalks of chopped celery.
Add 1tsp. liquid smoke.
Cover with water and cook for at least 8 hours.
Then Enjoy!

As time passed, workers enjoyed new customs and diversions from a demanding line of work. "We have a very professional and focused team here, but we don't have to wait for launch day to blow a little steam," Dave King, the Space Center's Shuttle Processing Director at the time, said in 2001. "We like to celebrate the smaller victories along the way as well."

"It was all fun. It was all fun," recalled Tia Ferguson, "I remember staying for way past the 12 hours because we had to get the hardware installed in the Orbiter before launch, doing things like MVAK. And I remember being so tired after 16 hours and just crashing. We'd worked so hard and it was a perfect place to be, when you're young and you don't have a family, and you have that time where you can just work extremely hard. And then after the launch the winds would blow. I remembered the days the wind would blow. And I had convinced several of the other engineers to learn windsurfing. And so if the wind was blowing after a launch, you wouldn't find anybody in the office because we would all be going out windsurfing."

REAL-TIME MISSION SUPPORT

"When this concept of Level-IV started, to the best of my knowledge there wasn't any plan," recalled Scott Vangen. "Certainly there may have been thoughts, but certainly not a plan that Experiment Engineers would follow through to the mission ops. Our job was integrate and test before launch, then the missions were controlled from Mission Control [in Houston] and the Payload Operations Command Center [after the first few Spacelab missions, it would be at MSFC]. They took over, and we just carried on with the next mission. In the back of the mind, there may have been [a plan] − and I think Bill Jewel [the 'Father of Level-IV'] knew this; he left the thread there because nothing had been done before. We didn't know what was going to happen. Well, what we learned on Spacelab-1 was that by the time you've processed the whole Level-IV, III/II, I and are good to go, the PIs [realize that] another person understands their experiment, sometimes as well if not better than them, because of the integration component. But the PIs also understand the NASA side of things with the Shuttle, and probably even mission ops. Even though we hadn't done mission ops, we probably had a more intuitive feel for what this Spacelab would do."

Understanding the experiment(s), the Spacelab subsystems that supported the experiments, and the Shuttle's capabilities to support the Spacelab, through weeks and weeks of testing, becoming familiar with the hardware, software, crew, PIs, timelines, failures, repairs, and such, the Level-IV personnel knew basically everything required to make the mission a success. Trusting that the Level-IV personnel knew what they were doing, and building that trust with all involved in the flight, would prove invaluable.

Scott Vangen continued, "When something was going on in orbit and they got into a malfunction situation and a Commander called down for advice, [a KSC person] Joe or Jane from experiment Y would say, 'This is probably the best execution step'. The crew would know who Joe or Jane was and sometimes they would want to talk to Joe or Jane, put them on the comm loop." Traditionally, air-to-ground was controlled very carefully through a Capcom (Capsule Communicator) in Mission Control, Houston, who was traditionally an astronaut. But the payload community had a different way of communicating with the crew once they were on orbit. The Alternate Payload Specialist (APS), also called the Crew Interface Coordinator, (CIC), focused upon supporting science-related communications, and how that happened was all carefully orchestrated. "You'd already established this repertoire with the Commander or the Payload Commander or whomever. There was a connection," Vangen observed.

Like most things in life, you can say you can handle a situation or understand how to work problems and fix things, but until that actually happened during real mission, there was still a question of "should I trust these people to do the right thing when called upon in the heat of the battle?"

"After Spacelab-1, they said we really should take advantage of this, starting with Spacelab-3 and Spacelab-2," recalled Scott Vangen. "I think nearly all of the experiments had their Level-IV engineers on Spacelab-2, which was nearly a dozen. But they looked at it as helping mission success. The mission managers, who were typically from Marshall [MSFC] for the Spacelab missions, were saying 'Why would I not take this person who's been with this payload team for a year or so and have them there for the last two weeks of the most important part of the whole thing?' And we loved it because we didn't have to monitor remotely or just wish them good luck and wave them goodbye. We followed them and we went to the sims. During the mission, there were no time outs. The Shuttle continued to orbit, the days and nights went by, and they were coming home on a set date. Every minute you lost your principal sciences [through a malfunction], it was not a minute you could recoup."

But on each of the missions, what was seen at the Cape and what was learned by Level-IV personnel brought value to the mission, because they could say, "This looks a lot similar to what we ran into at KSC." Scott Vangen observed, "Remember, we ran into this problem and we did this and we got out of it. That would be my recommendation. They would remember that situation and they would execute on your advice. Literally within seconds, they could have a possible answer and fix."

Those on the ground therefore became comfortable with Level-IV personnel supporting the missions. But how did the Level-IV personnel support the flights to make interacting with the crew during the mission as seamless as possible?

In answering this point, Scott Vangen said, "Experimental Engineers, along with their science team, executed multiple shifts. All Spacelabs were 24/7 all the way each mission. Some would do three shifts, some would do two shifts, mostly two shifts because eight hours would go by too fast; two 12-hour shifts was about right. Many of the other Shuttle missions, with few exceptions, were just a day the crew woke up, they had their timeline, they executed. That was their day's work. Then they'd go to sleep, get the proverbial wake up music next morning, and they'd do another day."

Problems encountered on-orbit could mimic ground problems, in which case the EI engineer would recognize that and then relay the solution to the crew on-orbit. For any new on-orbit problems, the EI engineer could pull from past experience to help isolate the problem and suggest solutions. One EI person's experience (Scott Vangen) was recognized by his selection by the Principal Investigators as an Alternate Payload Specialist for the Astro-2 mission (STS-67, 1995). Another was chosen to be one of the main interfaces to the crew (a Crew Interface Coordinator) for the Space Radar Topography Mission (SRTM). The astronauts flying these missions would also participate in many KSC EI activities. What better practice for the real thing than operating the flight hardware first on the ground before they saw it in space? As a result of this, the EI personnel, the PIs and the astronauts often became co-workers and good friends over the many months and years needed to prepare payloads for a mission.

POSTFLIGHT OPERATIONS

For each mission, EI personnel would be at the primary and back up landing sites to assist with removal of the middeck experiments. An enormous amount of planning would take place to ensure all the GSE, transportation and personnel would be ready for postflight landing operations. Mike Kienlen recalled, "The first time I remember we were out there, we needed a van but they didn't have one. So we went and rented a *U-Haul®* van and the Ops people said, 'You can't drive out on the runway with *U-Haul ®* written on the side of your van and take the experiments off'. We had to tape over the *U-Haul®* name on the side of the van. But it was hot in the desert and the scientists wanted climate control for their stuff. We hunted around for cargo vans that had air conditioning but we couldn't find one. And then one day, a bunch of scientists showed up in a 15-passenger cargo van full of seats, which had dual air conditioning and insulation, and the windows were tinted. So the next mission, we went to rent one of those." Mike explained how he went to the rental company to rent a 15-passenger cargo van. They opened the doors up and all the seats were locked into the van, so he asked them if they had any without locks because they might have to take the seats out. He was told the seats had to stay in, but Mike decided to rent one anyway. When they got out to Dryden, the technicians cut the locks and they took the seats out. Mike Kienlen continued: "We put the seats back in minus the locks, got back to the rental car center at LAX airport, and we gave them patches and pins and autographs, pictures of the astronauts. We cut the locks off every time and no one ever said a thing."

The schedule of events would be reviewed over and over to provide the most efficient order of operations. Shortly after the vehicle was safe to approach and the flight crew had exited the Orbiter, it was time to perform any necessary postflight operations.

To support landings out in California, the Level-IV team would fly to Los Angeles and rent what was needed to go out over the mountains into the desert to work the landings. The management was not there, just the workers. So the Shuttle would land, and the Level-IV team would be standing there, watching the Orbiter technician opening the hatch. The crew would come out, an ASP (Astronaut Support Person) would go in, then a Quality guy would go in, followed by an engineer and two techs. There was science instrument material and data that had to come off that was in Spacelab lockers, and an hour after landing (sometimes less, sometimes longer), Level-IV were pulling lockers off the Orbiter. Mike Kienlen: "Within an hour, an hour and a half, the experiments were in the scientists' hands, either doing what they need to do in the trailers that we had there, or we would have Learjets lined up and we'd drive them right to the Learjets. And those guys were going to get the hardware back to their labs to do the work they did. So it was so much fun, because every mission was different and we'd show up to do what we needed to do each time."

Tia Ferguson recalled, "We were the first ones on the Shuttle after the astronauts got off. Back then they would walk down those stairs. And I remember the smell, the smell of vomit, and it smelled like a foot locker where somebody had thrown up. That was what the Shuttle smelled like when you went in for the first time before it had aired out. And I remember the protein crystal growth [mission] had a fan that helped cool the insides. And so that fan was sucking cabin air in and stuff was collected on that fan. There were cockroaches. Yes, there have been roaches in space. There was popcorn, and Lord knows what else. The seeds that they used to eat, and hair, and all sorts of nasty stuff stuck on those fans that had just come down from space."

One of Todd Corey's favorite memories was being asked to join the recovery convoy crew on the dry lake bed at Edwards when the payload team was short-handed from a technician perspective. It was his first trip to Dryden, learning the new job under Mike Kienlen's guidance. Typically, they would remain in the Payloads trailer in the "office" area and stay connected via radio. Todd said, "Now I was able to witness the Orbiter landing on the lake bed strip from a close distance and provide hands-on support to carry the off-loaded middeck experiments from the Orbiter crew hatch to our Payload Van for transport to PIs waiting in the labs for their experiments. That was very exciting for a new NASA employee."

"Sometimes I drove the truck, but it was engineers that needed to get this task done," recalled Mike Kienlen. "Scientists were waiting in a lab ready to get the stuff and my job was how to help. How do I help those engineers and techs, and get it to them as fast as they wanted it? If the science was so complicated they couldn't bring the lab to Dryden, the only option was to bring the science back to the lab. That lab may have been set up for life sciences and if the lab was at the Hanger-L at Kennedy, and they needed to get the science back really fast, then they would rent a Learjet. And then we'd have to work with the Ops at Dryden to get approval to land a private jet at an Air Force Base. Only certain companies had the certifications to bring a private jet in. I remember one time, the jet landed but it was at the wrong spot at the Air Force Base. It was down at the Air Force side not up at the NASA side. I was standing there with a tech, Quality and the engineer and we were talking to the pilot of the Learjet. We said, 'Look, this is too far away. You need to be up at the other end'. The pilot came back 20 minutes later and said, 'I got approval to relocate my plane. I can carry three of you guys on the jet with me if you want to come with me'. All four of us needed to go, but I thought, 'Well, the technician needs to go, the Quality guy needs to go, and the engineer needs to go, so I'll drive the car back and meet them at the other place."

Sharolee Huet explained that, "The first time I went to DFRC, I got food poisoning the day of landing [but after landing had happened], so I was sick when we went back the next day to remove items at the Mate Demate Device (MDD). I had to stay in California until I could fly home. The next time, the lake bed was flooded and they had pool toys out by the road – a sea monster and an alligator. Not what I expected in California. They had a mud slide on the road between us and LA. Then there was Crazy Otto's. Everyone had to have breakfast there once. The thrill I had was going to Griffith Park and realizing the train depot was a favorite filming spot for TV shows. We were naïve and we saw a lot of stuff you don't see around here [KSC]."

Early destow using White Room (blue box with stairs) and shark cage. [NASA/KSC.]

Deintegration

All the work performed to assemble, integrate, service and test the flight hardware had to be reversed postflight. Taking everything apart was just as critical as putting it together. The same care would be taken to protect the hardware. The value of it had not changed, and in some instances it was probably more important than preflight. Mike Haddad recalled, "This was a responsibility that we took very seriously. But the work on preparing for future missions did not slow down, so many times we would be integrating one flight, while de-integrating another."

Preps for Transport Back to PI and Turnover

Once the hardware had been de-integrated, preparations were made to transport it back to the PI, or it would be turned over to the PI. EI personnel would help as much as needed to ensure the hardware was returned to them in the best possible condition.

After this generic description of operations, the following chapters detail how these operations were implemented for facility preparations and preflight, mission, and postflight tasks.

References

1. Moates, Deborah J. and Villamil, Ana M., *EASE/ACCESS Ground Processing At Kennedy Space*, Center Payload Management and Operations Directorate, National Aeronautics and Space Administration, Kennedy Space Center, Florida, August 6−7, 1986, Space Construction Conference Langley Research Center Hampton, Virginia, p.102.
2. Dollberg, J.C. and Kirkpatrick, P.D., NASA's Payload Ground Safety Process Review Process, at Kennedy Space Center Joint ESA-NASA Space-Flight Safety Conference. Eds: B. Battrick and C. Preyssi. European Space Agency, ESA SP-486, 2002. ISBN: 92-9092-785-2, p. 265, Bibliographic Code: 2002ESASP.486.265D.
3. *Young NASA Personnel Performing Hands-On Operations on Flight Hardware – A History of Experiment Integration*, Michael E. Haddad, NASA Kennedy Space Center, May 1−5, 2000, 37th Space Congress.

6

Towards that First Payload

*"It was very rewarding to those of us
that worked this first payload!
A new space program with only a
few flights under its belt, and we were a major player."*
Mike Haddad on witnessing
the launch of STS-7, June 1983

During the mid to late 1970s, many Kennedy Space Center (KSC) facilities were being converted from supporting the Apollo program to preparing for the Shuttle program. One of these included the modifications to what was originally known as the Manned Spacecraft Operations Building (MSOB) but had been renamed the Operations and Checkout (O&C) Building during the Apollo program. This chapter will also discuss many of the early negotiations and decisions that were made to create the Level-IV operation at KSC, as the work to prepare for Shuttle operations increased.

Level-IV, or Experiment Integration (EI) personnel were only minimally involved with flight processing for missions STS-1 thru STS-6. There were Spacelab hardware, pallets and Mission Peculiar Equipment Support Structures (MPESS) available that would carry experiments into space, and during the ground processing flow some of the team would work in the O&C Building on those payloads. When they were delivered to KSC, the unpressurized 'U'-shaped pallets that held the experiments carried on the Space Shuttle's second and third flights did not include the sub-systems that provided thermal control and electrical power, or which commanded and acquired data from the sophisticated experiments. That job, of outfitting the pallets with those essential sub-systems, checking them out to make sure they worked, and handling the hardware until integrated into the

© Springer Nature Switzerland AG 2022
M. E. Haddad, D. J. Shayler, *Spacelab Payloads*, Springer Praxis Books,
https://doi.org/10.1007/978-3-030-86775-1_6

Orbiter at the Orbiter Processing Facility (OPF), belonged to McDonnell Douglas Technical Services Company (MDTSCO) [1]. It should be noted that most, if not all of the 135 Shuttle missions contained experiments located in middeck lockers, but not all of the middeck work by EI personnel will be mentioned in this book.

BUILD-UP OF OPERATIONS

From the 1960s, some of the Gemini and Apollo spacecraft were integrated and tested in O&C. By the 1970s, the design of the facility featured modifications for the Space Shuttle, including work stands and Ground Support Equipment (GSE) to be provided by KSC. This work was begun in 1976, and the following year NASA contracted W&J Construction Corp of Florida to convert the O&C High Bay into a facility to support the processing of horizontal payloads in support of the Spacelab program. By 1978, the new facility was ready for use.

Looking west from Altitude Chambers at construction of Level-IV stands (background), Level-III/II stands (center) and CITE stand (foreground) in the O&C Low Bay, c 1978. Notice Apollo-era Altitude Chamber "M", middle right, is still in Low Bay at this time. [NASA/KSC.]

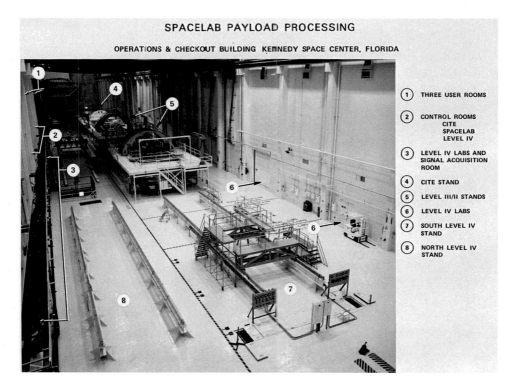

Looking east from ATM in completed O&C Low Bay c. 1980. [NASA/KSC.]

Amending the KSC Organizational Structure

As the time neared for Shuttle operations, a major change was made in the KSC organizational structure. One item of particular interest, reported in January 1979, was NASA's decision to perform virtually all Spacelab Level-IV integration (mounting of experiments to racks or pallets) at KSC. In early 1979, the Space Transportation System (STS) Projects Office and the Space Vehicle Operations Directorate were each divided into separate management organizations for the Shuttle and for Shuttle payloads. The STS Projects Office, managed by Dr. R. H. Gray, was divided into a Shuttle Projects Office under Gray and a Cargo Projects Office under John Neilon. The KSC Spacelab Project Office would be included in the latter organization and would continue its planning activities. Similarly, the Space Vehicle Operations Directorate under Walter Kapryan was divided into a Shuttle Operations Directorate under Kapryan, and a Cargo Operations Directorate under George F. Page. Isom (Ike) Rigel would be Page's deputy and also Acting Director of Spacelab Operations, responsible for hands-on activities with the Spacelab. With these changes, Neilon and Rigel would become the principal leaders and points of contact within the Spacelab program to ensure Spacelab's readiness for upcoming missions. In addition to these changes within KSC, Alex Madyda assumed his duties as the Marshall Space Flight Center (MSFC) Resident Spacelab Program Support Representative at KSC [2].

In the late Spring of 1980, the KSC Center Director asked Bill Jewel if he would put together an organization, an operating plan, and operating procedures for performing Level-IV. As Bill Jewel explained, "I had some discussions of how we would organize Level-IV at KSC and one of my requirements was I wanted to be able to select all the people that were there. I wanted to begin with 25 people who I picked to be my core people at NASA KSC. And I wanted it not to be a voluntary thing. We called them to the fourth floor, all 25. I gave a presentation that said, 'You're now in the Level-IV organization. This is not an option. If you have a problem with this, come see me and we'll discuss it'. Dick Smith, the Center Director was there, as were all the first line Directors and so I told Dick to begin with, 'This is not going to be an easy job. I'm not going to do this just walking along, I'm going to pound on it. If you want me to do it, I'll do it', and he agreed with it. He was always agreeable, and I'd known him since he was the project manager at Marshall for Skylab and when I was Instrument Unit [IU] section chief here at KSC for the Apollo-Saturn program. So we called all these people in on the fourth floor conference room and told them the story. I went through the organizations and said, 'I'm not going to tell you as a group where you belong, but I'm going to get with you individually and tell you where you're going to go."

Some of those individuals included Bill Mahoney and Dean Hunter for Mechanical Engineering, Herb Brown and Enoch Mosier for Electrical Engineering, Ed Oscar for Instrumentation, and Gary Powers for Test Director. Bill Jewel continued: "Ed Oscar was the only guy who came around, and he said, 'I can't do this because every February I volunteer to go teach skiing in Colorado'. So I said, 'Okay, here's what I'm going to do. I'm going to give you the agreement that every year I'm going to let you take your vacation in February and you can do that for two weeks. You can take your annual leave and I'll guarantee it, and I'll put it in writing'. So I put it in writing and gave him a copy. I said, 'Here's your guarantee'."

Gary Powers, NASA Level-IV Test Conductor, recalled: "Bill Jewel selecting me was an honor actually, but I was ill-prepared to be honest about people that didn't know what tests and procedures were. I could not imagine that, because I had lived in that world for so long; it was a wakeup call for me. I hope I recognized people for what they were – people – and that they were there and they wanted to learn. They wanted to do it."

Bill Mahoney was lead of Mechanical Engineering and needed people to begin work. Dean Hunter was his lead engineer, then he hired Tony Ornelas, Luis Moctezuma and Bob Ruiz. He even had a Co-op student, Barry Bowen, performing engineering functions. When Gene Krug was hired in, he had a unique role in

Level-IV. He was put on Bill Mahoney's staff, but he did not report to Joe Luckovich (recently added) or Dean Hunter "Mahoney was going to let me do the same work [as the engineers] but I couldn t do it. They wanted the engineering student slots and not me taking the engineering slot. They'd bring people out of college straight into the work, which had never been done before. That's how things got started. But if it hadn't been for Mahoney and Dean it wouldn't [have]." At first, Gene's main job was to pick up schedules and get the coffee pot going, but his authority increased. He would go to Lockheed, McDonnell Douglas or NASA to accomplish some task he was assigned to: "Then there were the times I went to Ace Hardware in Titusville [Florida] and got metric screws, so we didn't have any stoppage, where it would take McDonnell Douglas a week because it had to go through all this testing and everything. Things like that were different and had never been done before."

Training was also a critical part of getting Level-IV started, as Gary Powers recalled: "Some of those motivational classes were ordered, mandatory classes that we had, and thanks to Bill Jewel for that. I have to take my hat off to him for being successful in that field. The classes had nothing to do with technical classes, but all had to do with motivational and attitude classes."

Expanding the Team

These would be the managers, the leads, and some of the engineers, but what about getting more people to work Level-IV? Dick Smith gave Bill Jewel the option of first choice for all the KSC new hires, as Bill explained: "It wasn't pretty, [because] that didn't go well with some of the other directors, but it didn't bother me. I got the pick of the crop [and] that was great, that was the objective, to train young people. [It was] too late to train us old guys. That's how we got the system going and kept building it up by recruiting people." The group needed engineers to manage the operational aspect, and test conductors for major integrated testing, keeping up the schedules, coordinating any support for the systems engineers, etc. Bill Jewel assigned Gary Powers to be the lead for test and operations. He had assigned almost an all-female team to Gary, except for one or two men. According to Gary Powers, "They were real cold on what operations meant to them, so it was a real learning curve, throwing them into the midst of a battle. But the beauty about it was watching them develop and stand alone as confident engineers, and also being able to take a test procedure, exercise it and being able to communicate with NASA engineers and foreign national engineers."

Juan Rivera was one of the first to start working at Level-IV: "I reported to Bill Jewel and that's how I met Ed [Oscar]. He would be my supervisor. Ed told me, 'I want to put you working with the HITS [High Rate Multiplexer Input/Output Test System] system. You'll have to learn everything about the software on the computer'. The other person I interfaced with was Jim Cassidy – we all called him "Hoppy" Cassidy[1] – who was in charge of making sure that everything was working correctly."

With the people now committed to the work, the next question was what resources would have to be used to get the jobs done? Bill Jewel wanted a personal computer for every office: "So I bought two computers. I bought an Apple *Lisa*, the first one at the Cape, and they gave us a really good discount so I bought several more." At that time, scheduling was done on magnetic boards. "They had large 'Mag' boards, which required selecting a person to move the magnetic pucks on the thing," recalled Gary Powers. "Well, we built Mag boards to show the Level-IV activity, [but] it was all guesswork. We had Level-IV processing, Level-III/II, then CITE [Cargo Integration Test Equipment]. We did all of that before we went out and installed that Spacelab into the Shuttle Payload Bay. We had people from Headquarters come in and walk through the Mag board and they were quite impressed, so Bill Jewel got a lot of credit from headquarters."

IBM was in charge of all the software. Level-IV used a PDP-8 computer, which was called a mini computer at that time, then later upgraded to a PDP-11 computer. Then they upgraded again to the Data General computer and so on, but IBM was in charge of monitoring the software that was running in the High Bay.

But the job could be overwhelming. "Since I was the only physicist working there at the time, they assigned me all the experiments," recalled Juan Rivera. "So I looked at a very thick book and there were about 20 experiments, and I had to review all the procedures that had been written by somebody from somewhere else and put it all in the language that NASA Kennedy understood. I went to Bill Jewel and said, 'This is too much for me. I think what you should do is give some engineers a group of experiments, or one engineer an experiment, or one engineer a few experiments'. So I took maybe one or two and that's how the other people came in." Then the real trick was how to make it work, as Rivera continued: "I was part of the team, and what was required from us was that we needed to know machine language, the ones and zeros. I was one of the few that knew Boolean Algebra, and I was the only one that knew discrete mathematics; the stuff that I learned in school that I never expected I was going to use on a mission. And then we had to put it in electronic form. The beauty of this – discrete mathematics and Boolean Algebra – was that a series of commands that we were sending could be simplified. That reduced the need to spend too much memory, because memory at that time was at a high price. Today we talk about gigabytes, but in those days we were talking about a 64k memory. So instead of occupying a lot of memory, it would occupy less memory and accomplish the same thing."

---

[1] A parody of "Hop-along Cassidy", the fictional cowboy hero created in 1904 and later immortalized in film, radio and on TV.

What about contractors? How would they play with the whole Level-IV concept? Bill Jewel explained that process: "The contractor technician was another big fight. We had a lot of fights to get Level-IV working. I thought I was going to get technicians to work for the government without getting in a dog fight with a bunch of union problems. So I went to the Director of Procurement at KSC, and said, 'I want to write into the Source Evaluation Board [SEB] requirements, [which was another board he ran] the fact that the technicians would work directly for NASA engineers. They could be replaced or supplemented by engineers as I saw fit or as our schedule demanded. I will write all this down'. He [the Director of Procurement] said 'Who do you think's going to going to support that?' And I said, 'Well, I have two contractors who've come in and asked me what I needed'. So I told them, 'I need you to get your unions to agree that when you bid for this, that they understand what's going to happen. Because the first time that I have a union walk out, I'm going to terminate the contract. I'm not going to play that game because I've been in there with all the contractors'. In fact, during Saturn [the Apollo launch vehicle program] every one of them walked out at one time or another, so we either had the contractor engineers to take their place, or the NASA technicians and engineers took their place."

It was an interesting setup; the engineers could do the work of the technicians. There could be engineers pulling cables and doing what was necessary to put the Payload Checkout Unit (PCU) together in the Level-IV control room. Other engineers were trained for tube bending, cable terminations and to build cables, and all of the things that normally only a technician would perform. Bill Jewel would tell them that they had to be trained for that and they were going to take the training whether they liked it or not: "If someone disagreed and told me 'I'm an engineer', I would tell them to find another job. I wanted [them trained] for two reasons. Number one, they may be called on. And number two, I wanted them to know what the technician should be doing and how it should be done. If they didn't know that, then they couldn't manage a technician."

Level-IV at Marshall?

Level-IV was originally to be conducted at MSFC, so how did the decision to move it to KSC, happen? What details needed to be worked out? Bill Jewel recalled, "At least two of the science groups at headquarters were behind us. We had a good argument, because we didn't want to lose the chain of control for the hardware. If the hardware went all over the country, to this guy and that guy and the other guy to integrate, and then to Marshall, to the Cape, then back to Marshall, then back to the Cape, then they [KSC] had no control over it. That was one of the big arguments." Dick Smith got Bill to put together a pitch, and they got on a NASA plane and took it to Marshall to see Otha (O.C.) Jean of the MSFC Spacelab

Payload Project Office. Bill Jewel said, "We [told them we] wanted the hardware they'd already developed for the Level-IV works stand, and the PCU and stuff that hadn't been completed yet. All of it. The meeting went on for about an hour to an hour and a half, no problem. Only Dick Smith, O.C. Jean and I were in there. Only the three of us. And so that was it from there. He implemented his part of it and then we implemented ours."

Now that some really important decisions had been made, the GSE and flight hardware could start to arrive, but, as Bill Jewel explained, "It didn't work out near as good as Headquarters wanted because there were a lot of people that wanted to maintain control through a Program Office there. I had a lot of arguments with that Program Office. A couple of times in Europe, they [Headquarters] said, 'Well, you can go ahead and ship the hardware' and I said, 'No, you don't ship me the hardware without any documentation because there's nothing I can do with it. I don't even have enough documentation to know how to unpack it yet."

"It was that fall of 1980, where Marshall was onsite at KSC delivering the PCU and the HITS hardware and beginning the acceptance tests," recalled Donna Bartoe. "That was kind of my first mission, the whole handover of the hardware and software to KSC. It had already started by the time I reported for work. The Marshall gang was headed by Jim Lewis and they basically lived here for three months during this system handover. They were all living in condos on the beach, at Cape Canaveral, along with the main contractors and McDonnell Douglas, Dan Graves of IBM, and Lynn Haley who was the head software guy."

It had been decided early on to perform the Level-III/II and Level-I operations at KSC, so work had started on those integration areas. The Level-IV stands would be in the west end of the O&C Low Bay, but what about the rest of the integration locations around the Level-IV area? Bill Jewel had a hard time getting the Rack Rooms away from the Program Office: "They [the Orbiter program] didn't want to give us the rooms. But finally, after I had a big fight about it, I got hold of all the rooms."

There were other areas inside the O&C Low Bay/High Bay that could be used, but as with the Rack Rooms, Bill Jewel had to fight to get them operational. One such area housed the two altitude vacuum chambers that were originally used for the Apollo Command Modules, which had to be converted for use by the Level-IV Module Vertical Access Kit (MVAK). Bill Jewel recalled, "I had a big fight about that, because I had to put in stairs and a top and stuff like that to get into it, and take out all of the vacuum chamber equipment over there and the other side of the High Bay where the Tool Bay was. There was another big fight about the ATM [Apollo Telescope Mount] Clean Room back on the west end. It was partly in there, partly not." Level-IV also did not have anywhere for logistics, so they put logistics in the basement (the Tunnel under the Level-IV area).

Some of the organizational structure was still not in place, so changes needed to be made to better align the people with the work to be performed. The Program Office and the Spacelab Office were different entities. Spacelab organization had changed and they gave the Spacelab to Roger Gaskins, who had come out of spacecraft operations when they first changed over to payload and Shuttle operations. When Level-IV started, the Spacelab was a different program under a different division. "It was not working," explained Bill Jewel. "The contractor was getting in my hair, and so I wrote on my performance report for this particular year that 'I'm going to absorb the Spacelab division'. So I took all the people, some of which I wanted, some of which I didn't want, and some of them I got rid of one way or another. I finally put it all together and the contractor and tech was under control. McDonnell Douglas was under my control and so were the Boeing techs."

McDonnell Douglas held the overall Spacelab module and subsystems part of the deal while Boeing had the experiment part, but they were both under the same contract. ' That Spacelab contract was a joint contract, which was the only one I've ever seen," admitted Jewel, "It didn't work between Marshall and the Cape, and that was a single contract, and I said 'It [the joint contract] won't work'. They tried it for a little while and then finally they abandoned it and changed it, so there were two different contracts. And they did that mostly through contract negotiations." McDonnell Douglas had the MDTSCO contract used for overall Spacelab module and subsystems, while Boeing got the Spacelab Experiment Integration Support (SEIS) technician contract used for Level-IV/Experiment Integration. But still some of the contractor personnel were not quite the right fit for KSC operations. Bill Jewel said, "Get Skip Montagna down here, and let him help you to understand that he can give you access out to the Cape. So you can understand what goes on, and where and who, and when and how."

Getting Along

How would these contractors and the NASA Level-IV personnel interact? The NASA engineers were in charge of hands-on for Level-IV, and when things moved to Level III, the testing did not stop. That was the hard part, because NASA staff were now interfacing more closely with a contractor. Gary Powers admitted that, "They didn't appreciate NASA engineers being hands-on. There was a problem brewing there because of the cooperation with each other. Because we had the long test hours, sometimes longer hours than McDonnell Douglas did. Yet once the experiments were installed, and the racks were installed inside the Spacelab shell, the requirement for further testing continued. And we didn't know the end to it. We had no end until it was completed to our satisfaction, and we had to have Spacelab Level-III/II powered up in order for us to power up the experiments. So we had some conflicts of interest with a contractor in, and it wasn't particularly their fault. Bill Jewel was a strong leader, and unfortunately he was charged with the responsibility to manage the interface between the McDonnell Douglas contractor and NASA. He smoothed out the differences and everybody took in what they did. And everybody had that attitude of 'We will not fail'."

What about NASA Level-IV and foreign nationals? Gary Powers commented on this area as well: "When Bill told me to come over, he sort of threw me to the wolves, sitting in a roomful of people – foreign nationals as well as NASA people. I had to wing it, to be honest, because I realized quickly that the Germans didn't know what we used. They didn't know the paper system we used, the quality control, quality stamps. IPRs [Interim Problem Reports] and all of that consumed their time. IPRs were mind boggling, because we were in a development program that we had."

Maurice 'Mo' Lavoie admitted that he spoke a little French. "I understand some of the language. I can read and write and comprehend quite a bit of the French. So when PCU and all of the Mitra computer and I/O systems were installed, we had a variety of Europeans that came over with them. Mark Sanz was one of them from the French company SEMS [Société Européenne de Mini-Informatique et de Systèmes] that built the Mitra. It was very helpful and confirmed that we had the right documentation. There were several gentlemen from the computer manufacturer who came over and I told them, in French, *'Parle lentamente, s'il vous plait*; I [can] understand [you] if you speak a little bit slowly'. And so we got into a working relationship where they were teaching me French from their Parisian perspective. Mine was Canadian French from when I was a kid."

Maurice said he got his nickname because "They [the French] got tired of calling me Maurice." They were in their early sixties and he was in his early twenties, so to them he was the "child". They generally gave a person with the name Maurice the nickname 'Momo' during their childhood, "so Donna Dawson and a bunch of the other ones started calling me 'Momo'. I told them that wasn't going to work. I said, 'If you're going to call me a nickname, just cut it in half and make it Mo', so from then on I was known as 'Mo'."

Outfitting the hardware was another challenge, as this included both the GSE and flight hardware. The cabling under the Spacelab floor had never been integrated in Europe. One of the problems was the jigs that they used to drill the racks. The jig mounts – the panels that went into the racks – had to be custom fitted, re-drilled and annotated as to which racks they went on. This had to be specific for the racks because it was discovered that they had been hanging the jigs on the wall at MBB ERNO in Bremen, Germany and the jigs were being deformed slightly just by gravity. Bill Jewel explained the problem: "The panels would not universally fit because the racks were made one at a time. The racks were originally made in Italy, then went to ERNO to be finished. All those cables and piping and all that stuff had to be built there, but the Level-IV guys had to put it in."

Level-IV personnel made some three trips to Europe, to ERNO in Germany in particular where they were putting together Spacelab parts, to see how they assembled it, how they disassembled it and how the rack got installed. They were able to go through the factory and see Spacelab in its bare bones shape, take pictures and exchange information. There were two versions of the Spacelab: the Long Module, and the Short Module. At ERNO, they were separated and the Level-IV personnel were shown how they mated together. As Gary Powers observed, "We saw all that and took time to get pictures. [We] had very good exchanges in the meetings with the Germans. They explained [their procedures] very well and we explained very well what was expected of them when they came to the United States."

Level-IV used engineering model boxes called Remote Acquisition Units (RAU), and would run the experiment computer operating system to a point where it would display on the emulated data display systems. They would go down and simulate parts of the RAU signal with scopes, probes and power supplies just to see if they could get feedback. Mo Lavoie told the authors, "We would verify that the signals went from the operating system through the input/output unit and down the cables to the Low Bay. The network was not designed to go that distance on the Spacelab. It was only designed to go about 100 feet [30 meters], but we were driving it almost 1000 feet [300 meters] and it was working. At least, we had an engineering model, the RAU, to prove that we could drive that signal down there and get the response back. So we were set up ready to go by the time the real hardware showed up."

Juan Rivera added, "Everything that we did was recorded. The head [on the recording device] read 16 tracks, so we were able to do more with less effort. Every experiment data was recorded in those magnetic tapes, and we had an office that kept all those tapes stored. You had to go through a process so they knew who had what, where they were, and what time and date everything was recorded. If something malfunctioned then we had to tell Bill Jewel and explain why this thing was not working, so we had to go and retrieve the tape and look at the printout. We dumped memory, and then we looked at the piece of paper, and then we knew what went wrong."

So how would what was being learnt overseas be translated to setup back in the United States? Gary Powers explained that process: "After we made those trips, we came back and Ernie Reyes, who was our manager for the Level-III/II Spacelab, not level-IV, put together a model on his desk – out of beer cans – which showed our design for how the big stainless steel rails went along and would be placed in the O&C Building. A guy from McDonnell Douglas called Hugh Downs, Ernie Reyes and myself sat down with the guys from DE, the Design Directorate, and shared with them what systems we were going to use for Spacelab, where it came in, where to put the rails in, and also the Level-III stand, the Level-II stand, and the CITE stand. We set all that up and, thank God, it worked."

Mo Lavoie worked until November of 1980 in Eugene Nelson's office, doing nothing but studying electronics and studying the Spacelab computer systems, and having fun doing it. "[We were] Learning the Spacelab's systems, when Spacelab wasn't even at Kennedy Space Center yet; it was over in ERNO. Emmett Crooks was sent to ERNO to be involved in all the systems testing." At ERNO, Emmett was gathering documentation "under the table". Level-IV could get the specifications on the remote acquisition and the CIMSA MITRA125, but Emmett was getting the design-level documents. Maurice Lavoie said, "He was making copies of all the design reports that he could possibly get his hands on, and boxing them up and sending them to Eugene Nelson. Eugene Nelson was handing them to me. I had a lot of their design documents, copies of them. Some of it was in English, some of it was French, some it was hand-drawn, some of it regular – not quite computer, but typed documentation. So when I got into Level-IV in November of 1980, I had a lot of that documentation available to me, which was pretty darn good."

SEIS Level-IV Technician Bruce Burch added further insight into these early days. "Level-IV was just getting built up. The equipment was just coming in; all the fluid conditioning units, the water conditioning units and the rack conditioner. The first thing I [was assigned to] when I was there, was working with Roberto Tous [NASA Level-IV Mechanical Engineer] and Dean Hunter [NASA Level-IV Mechanical Engineering Lead] on the rack conditioning unit. They were trying to run it, but the pressure gauge was fluctuating so bad that they couldn't read it. They asked me how to fix that and I said, 'You just put in a needle valve in between and you can cut it down so you don't get the pulsation off the top of the compressor'. So we did that, made a change, and that's when I found out how much work you had to do to just make a simple change. When I came from the field, I would have just done that at somebody's house without even thinking about it; just cut the line, put a valve in it and be done. Here, we had to write paper, and Roberto had to get everything; the engineering stuff and stamped paper and all that. So I would say that was a learning experience for me then. That unit is still in use out there."

A Pivotal Year

George Veaudry started on Spacelab in 1981 in the Cargo Management and Operations Directorate, "checking out the Spacelab Ground Support Equipment [GSE] that was going to be utilized for the test, and checking out the Spacelab flight Environmental Control Life Support System [ECLSS]." He also did the GSE test and checkout to prepare for when the first Spacelab-1 module arrived. After he finished the test and checkout of the GSE, he again turned to the test and checkout of the ECLSS: "I supported the test and checkout of the ECLSS for the first three Spacelabs."

As the real hardware started to show up, no-one was sure that they were going to be able to test the experiments properly. Maurice Lavoie recalled: "We had a little bit of time to play with the flight hardware, and to me it *was* play. We really didn't have procedures in place. We had a lot of written steps that we had just written down on notepads. For example, I had a small boot sequence that I needed to prove, so that we could talk to the real data display systems. I came in early one morning and wrote my little boot sequence of what I wanted to do. I wanted to do flashing yellow, with my name on the flight data display system. So I toggled it into the Mitra computer, toggled the function into the input/output unit which had a bank of switches to address the display system, and executed. And I had a technician down on the floor who said, 'Something's wrong with the data display system, it's flashing *Maurice*, in yellow'. I quickly disabled it, but word got back to Enoch Mosier and some of the other guys, and I got my hand slapped for doing things to the data display system without a trace. It was kind of a test, but some of that activity that I did was for my education and it proved to be beneficial during tests, because when an experiment failed to operate properly with the RAU, I knew what the toggle into the input/output was to get a response back as to what the problem was; what the internal registry of the RAU was reading, where it was getting errors and where it looped back and told us it was not functioning properly."

Some, offsite to KSC, wanted to obtain hardware such as the pallets to perform some of the integration work, so there were continuous arguments about sending a pallet or sending a rack to Goddard or to Ames, for example. "Racks go nowhere. Pallets go nowhere. Spacelab hardware goes nowhere out of this facility. Period," said Bill Jewel. "I stuck to it. It never went anywhere. I said, 'No, sorry. We will send you a simulator, we'll send guys there to help you understand it, and we will send you drawings, but we're not sending hardware. We're not going to lose control of it. We don't know what you'd do with it. It's not going to happen'."

Bill Jewel went on to outline what processes would be used by Level-IV to get the work done. He had to invent how to make it work on a daily basis. The first thing he had to do was to come up with the standard operating procedures and plans, who reported to who, and when, why, and how it would actually relate at different levels of concern for problem reports or what the procedures looked like. Jewel said, "Our procedures in Level-IV had different names than the ones that they used in the Shuttle and what were used in Apollo. All had different names, and we always liked to name things here." Gary Powers explained that, "The best part was the software and the hardware coming together for our big job, which was to test flight software against flight hardware. We said it was somewhat like smoke testing, but our job was to find the problems before flight, and if we could do that then we were highly successful in flight."

Hermann Kurscheid told the authors: "I was actually there, at KSC in 1981, for the first Shuttle launch. We had already presented our configuration requirements and all we had planned to do at KSC for Spacelab-1 to the KSC people. I remember very well Gary Powers, who grabbed me and took me to where he had some other people from his group organized. He had me in to talk and present our payloads and activities, what we wanted to do integration wise and testing wise. I think Lori Wilson belonged to Gary's group there. I recall more people, but not specifically who, from that meeting."

NASA Level-IV Mechanical Engineer Bob Ruiz explained further: "[The] Configuration control before Spacelab-1 was much looser; I mean they actually let us red line for a while. The red lines would cover the whole page. They would go all over the place, and then scratch something out and write it in there with a red pen initially, and then do it again; put notes up the side of the paper. There were always procedure changes, what turned into Deviations. Thousands of those. Then eventually they stopped all that because it got out of control, they said no to Deviations." Hardware configuration control was a little looser too. Bob Ruiz recalled a funny event: "I needed some different bolts or screws. The screws were wrong, or wrong length or something. I called up and said I needed some different screws and the guy showed up with a baby food jar [full of screws], poured them on the table and said, 'Take one of these, whatever one works'."

Bruce Burch noted, "We probably did more GSE than flight hardware between missions; getting the maintenance and getting stuff to work, doing everything to get it to work for what we needed. For example, we had to put a bigger motor in the RCU. We tried to do rack flow balance and couldn't get the flow rate, so we ended up putting a bigger motor in and then put in a different pulley, a variable speed pulley. We could adjust the RPM and got this to work. After we proved that it worked, they went and bought a variable speed motor and put the controller on the outside, where you could adjust the speed of the motor to help get the flow rates you needed."

"I got to put the Spacelab computer systems together," said Maurice Lavoie. "I got to run the diagnostics on them to make sure they worked. It was neat that they had a switch panel tied into the Mitra computer, so that you could look into the memory and you could see in hexadecimal code lights, individual little lights. It was ones and zeros. It was all binary code. So when we came up with errors in testing some of the experiments and communicating to them via the RAUs, I'd go ahead and look at what the hardware was doing [and whether it was] a hardware problem or a software problem. I would find in the hardware where it stopped and find that in the code. I had to know the code, so I knew the hexadecimal version of the instructions that were being read on the paper and I would be able to say, 'To me that's not right'. I would alter it on the switches panel, to prove that it was a software problem by allowing the software to continue to run properly on the fly. We got to do a lot of things in Level-IV which were on the fly, which other people in NASA never got an opportunity to do. It was true hands-on. It was great fun."

The ground crew was essential, but was the flight crew involved at all in these processes? The astronauts would come to KSC with the same laptop that they were going to use on orbit to talk to the experiments on board Spacelab. Gary Powers explained: "They [the astronauts] spent quite a bit of time with us, and it got to be a friendly endeavor because we found out they were real human beings. They bled just like we did, in other words. And they communicated with us." He described how, from his background during the Apollo era, the astronauts were just a picture; they never got in personal contact with the ground crew. The only contact they had was when they got aboard a spacecraft and talked to the control room. But the ground crew admired the flight crew, and what they were giving up out of their lives and their job. There was a lot of respect for what they did, of course, and what they had accomplished. There have been other books published, primarily about the astronauts. It was maybe not intentional, but not a lot was attributed to the ground crew who put everything together, hands-on.

There were also many different labs back in the O&C. Different lab people could work on payloads, but as with Level-IV, they could not work there without the correct procedure, and it had to be approved. "End of story. I had a lot of people balk at that," said Bill Jewel, "and people used to laugh because I'd say, 'And that's the way it is, *Merry Christmas*'. I said '*Merry Christmas*' a lot back then, but it worked because we had no choice. There was no way we could implement everybody's idea of how to do a job. We had to have one way to do it. I had a sign that said, '*This is not Burger King. You get it my way or you don't get the SOB at all*' "

ORBITAL FLIGHT TEST (OFT) PROGRAM

For the first historic Shuttle flight, designated STS-1, *Columbia* launched on April 12, 1981 and landed at Edwards Air Force Base (AFB) on April 14, 1981. No Spacelab hardware was flown on this mission. Mike Haddad recalls, "I was Co-op in a group at KSC that monitored and controlled all the KSC facility ground systems, HVAC [Heat Ventilation & Air Conditioning], power, pneumatics, etc. This was done from the Complex Control Center [CCC] located on the first floor of the Launch Control Center [LCC] in a room just off the lobby. The firing rooms were located on the third floor. The full time engineers would take me with them when they traveled to the different facilities, to get a first-hand view of the systems we watched on CCC consoles. It was really more of a software job but we used our backgrounds – in my case mechanical – to understand the systems being monitored and controlled by the software. I learned the GOAL [Ground Operations Aerospace Language] software used in the LCC. I was a Co-op on the CCC console, in the LCC for STS-1. I'll tell you the excitement was everywhere. That first attempt on the Friday April 10 was scrubbed due to a computer problem on *Columbia*, but we successfully launched two days later on Sunday April 12. I was hooked more than ever, convinced that when I graduated was going to work at KSC and only KSC."

STS-2 *Columbia*, launched November 12, 1981, 2+ days
Payload: Office of Space and Terrestrial Applications (OSTA-1); one engineering Pallet (E002).

First use of a Spacelab pallet/Orbital Flight Test (OFT) pallet. The OSTA-1 payload consisted of a number of remote-sensing instruments mounted on the Spacelab pallet in the Payload Bay: Shuttle Imaging Radar-A (SIR-A); the Shuttle Multispectral Infrared Radiometer (SMIRR); the Feature Identification and Location Experiment (FILE); the Measurement of Air Pollution from Satellites (MAPS); and the Ocean Color Experiment (OCE). Three other payloads were carried but will not be discussed here.

At the KSC O&C Building, the NASA and MDTSCO team was delighted to finally have some hardware to prepare for flight, after years of facility modifications and paper studies of flow plans, schedules, analyses, and procedures. The electrical harness and avionics subsystems were mounted on the pallets, together with the active thermal control system consisting of the Freon® pump, cold plates, and associated tubing. The secondary structure for mounting a cold plate and several avionics boxes had defective welds. Substitute platforms were machined and retrofitted during the staging process. By November 1979, the pallet was ready for the next step: Level-IV (stage, not EI) integration of the payload. The pallet was turned over to the Office of Space and Terrestrial Applications for experiment installation. By this time, it was already apparent that the OSTA-1 mission would not be possible before March 1981. Nevertheless, it was decided to proceed with experiment installation and checkout, which was accomplished during the next several months, using the Level-IV Rails (work stand) for the first time. As Maurice Lavoie explained, "The first payload I worked was OSTA-1, with the PITS [Payload Integrated Test Set] and a very smart person called Polly Gardner. She had gone off and got all the education she could get on the PITS. That was not a computer, it was a sequence processor. The way it tested was that you started a sequence, and it would go down through the sequence, ask information from the Payload Flex MDM, and bring that information back. But when it got to the end of the sequence, it would stop. There were no loop arounds. There were no go-tos. There were no continued processes. It was just a straight line. This is all that was done. So Gerry Rivera and I spent a lot of time down at the OSTA-1 verifying that the flex MDM was communicating properly with the experiments on OSTA-1. Luis Moctezuma was down there quite a bit with us, because they had all these strain gauges and sensors and things on there that they needed to prove were getting the appropriate temperatures and things. So we did weird things like getting a can of Freon®, spraying sensors, and validating that the signal got back to the Flex MDM when the PITS was going through and looking for that signal response.

And then, hexadecimal wise, we could take that and convert it. It was X-degree assuming, and we saw that degree go up, that degree go down, while we were spraying it or not spraying it. We also had hot air guns, almost like your hairdryer, so we'd heat them up and things like that."

There was then a long wait for the first Shuttle flight, primarily due to problems with installation of the thermal protection tiles. The OSTA-1 pallet and its payload for STS-2 were temporarily stored in the former ATM clean room. In late 1980, however, the pallet was moved to the CITE stand (another first) to prepare for a simulated integration with the Orbiter. A symbolic turnover of OSTA-1 from Rockwell to JSC was accomplished on March 4, 1981, and a second turnover to KSC took place on March 10. Following the successful completion of the tests in the CITE stand, a Payload Certification Review was completed in July 1981 to certify that OSTA-1 was prepared to support the STS-2 Flight Readiness Review, certify that the OSTA-1 integrated payload and carrier were ready for testing with the Orbiter, and affirm the operational readiness of the supporting elements of the mission [2].

OSTA-1 installation into *Columbia*. [NASA/KSC.]

Following the launch of STS-2, OSTA-1 mission operations were conducted from the Payload Operations Control Center (POCC) in the Johnson Space Center (JSC) Mission Control Center [3].

The Shuttle landed at Dryden Flight Research Center (DFRC) on November 14, 1981. The early de-stow data materials were successfully removed, beginning one hour after landing with crew compartment items. Two days later, GSE was installed in the Payload Bay to access the OSTA-1 payload experiments for file and tape recorder work. Once removed, all this hardware was transferred to its proper destinations.

The Shuttle Orbiter with the OSTA-1 payload was ferried to KSC, arriving on November 25, 1981. At KSC, the OSTA-1 payload was removed from the Orbiter in the OPF, placed in the Payload Canister, and transported to the O&C Building for experiment removal, experiment laboratory testing, and OFT pallet refurbishment. The experiments were completely de-integrated and were returned to the Principal Investigators (PI). The pallet subsystems (the PCB, FMDM, etc.) were returned to the KSC hardware inventory.

STS-3, *Columbia,* launched March 22, 1982, 8+ days
Payload: Office of Space Science (OSS-1); one engineering Pallet (E003).

The Office of Space Science-1 (OSS-1) pallet consisted of: the Plant Lignification Experiment; the Plasma Diagnostic Package (PDP); the Vehicle Charging and Potential (VCAP) experiment; the Space Shuttle Induced Atmosphere experiment; the Thermal Canister experiment; the Solar Flare X-Ray Polarimeter; the Solar Ultraviolet and Spectral Irradiance Monitor (SUSIM); the Contamination Monitor Package; and the Foil Microabrasion Package. Two other payloads flew on this mission but will not be discussed in this chapter.

At the same time the STS-2 OSTA-1 pallet was being prepared for launch, the STS-3 OSS-1 pallet was being prepared at the KSC O&C Building by the NASA and MDTSCO team. As with OSTA-1, the OSS-1 pallet electrical harness and avionics subsystems were mounted on the pallets, together with the active thermal control system consisting of the Freon® pump, cold plates, and associated tubing. Like OSTA-1, the pallet was ready for experiment integration by November 1979. This time, however, the experiments would be of a scientific nature, with emphasis on deriving information on the Orbiter's environment that would be important for future scientific instruments to be mounted on this new breed of payload carrier. Another difference was that the OSS-1 payload would be mounted on the pallet at Goddard Space Flight Center (GSFC) in Greenbelt, Maryland. The pallet was turned over to the Office of Space Sciences, but instead of remaining in the O&C it was transported to Goddard for installation of its experiments. The pallet was transported from KSC to GSFC by road in late 1979, using the Payload

Environmental Transportation System (PETS), a truck-mounted container with environmental control that had been developed by KSC for just this purpose. Although transportation by road posed certain planning and scheduling problems, in the end it was a relatively straightforward operation. Experiment integration was completed after overcoming several problems, and then the payload was put into mothball status until it was transported back to KSC in late 1981 for final launch preparations.

OSS-1 in CITE. [NASA/KSC.]

Placed into the CITE stand in the O&C, the payload would go through final testing and all the required reviews before launch. From the Spacelab standpoint, the mission was another resounding success. A last-minute change of landing site to White Sands because of weather problems at Edwards added an extra degree of excitement to this mission, but the Spacelab pallet and support systems had performed flawlessly. In terms of the validity of the procedure for pallet processing off-site from KSC preflight, it appeared that this mission would probably prove to be an anomaly in the program. The die had already been cast, and major components of the Spacelab would never again leave the confines of KSC for payload preparation.

Mike Haddad recalled, "I came onto the scene at KSC Level-IV after this mission took place. I had graduated with a BS in Mechanical and Aerospace Engineering in April 1982 and started at KSC Level-IV only three weeks later in May. All my friends told me I was crazy, that I should take more time off before starting work because once you start you'll be at it for decades to come. I guess I could have taken more than three weeks off before starting, but I just wanted to get out there."

From Flight Tests to Flight Operations

STS-4, again using *Columbia*, was launched on June 27, 1982. Though this mission did not contain any Spacelab hardware, it was the first flight of the McDonnell Douglas-developed Continuous Flow Electrophoresis System (CFES) middeck experiment that would be worked preflight by Level-IV/EI personnel on future missions. *Columbia* had performed all four OFT missions very successfully. After the seven-day STS-4 mission, President Ronald Reagan publicly declared that the Shuttle would be classed as operational from STS-5, a bold statement considering the relative newness of the system, and one which would prove premature in light of subsequent events.

There was so much to learn about how things were done at KSC to make a payload ready for launch, as Mike Haddad recalled: "Talk about drinking from a fire hose! All that [the procedures] needed to be learned in a very short time frame. The payloads were lining up in the future so it was time to get working. Dean [Hunter] came to me one day shortly after I started and said, "Mikey, Bill [Mahoney] is doing a presentation to a group of payload people scheduled to fly on a future Shuttle flight. Why don't you go sit in on it, this will give you a good overview of our job'.

"So I went into this large conference room that must have had 50–60 people in there; the Mission Manager with all his payload people; and all the disciplines from KSC – managers, engineering, safety, quality, medical, security and on and on. I grabbed a seat in the back of the room and Bill went through his pitch. We did not have PowerPoint in those days, so the presentations were all on clear 8" x 11" [20cm x 28cm] overhead projection transparent foils that he placed on a viewgraph machine to project the foil onto a screen. Bill finished with his pitch and then asked if there were any questions. The Mission Manager, who later I found out was real arrogant, said to Bill, 'Sorry that just won't do. There is no way we will agree to any of that'. Bill, trying to be a diplomat, said 'Those are the processes we have established here to get the job done. We will help you as much as possible to get you through KSC and achieve a successful mission'. The Mission Manager continued to argue back and forth with Bill, and finally Bill said, 'Well, I guess you're not flying on the Shuttle' and sat down. Dead silence in the room. I thought to myself, 'What did he just say?' The silence went on it seemed like forever, then the Mission Manager spoke up: 'You can't say that!' and Bill responded, 'I sure as hell can'. Bill tried to be the nice guy, but he could hammer you hard at times, and this was one of them. The Mission manger then told Bill he would go over his head. Bill said 'My boss [Bill Jewel] is just down the hall. When you get into his office be sure to look for a sign he has on his wall. Also, you may be confused on why he will tell you *Merry Christmas* even though its summer 1982. Oh, and his boss is just a little farther down the hall. If you want to go all the way up, the KSC Center Director is on the top floor of the HQ building. It's that large building just west of us here'. Finally the Mission Manager backed off. I thought 'Wow, this is going to be a great place to work'."

STS BECOMES OPERATIONAL

Columbia's fifth flight, STS-5, launched on November 11, 1982, carrying a crew of four – the largest spacecraft crew up to that time. The payload was first two commercial communications satellites to be flown aboard a Shuttle. Level-IV had no involvement with this mission, which was classed as the first operational flight in the program.

Herb Rice told the authors that, "When I got there, I think there were 16 people in that electrical group and that was totally inadequate. So I basically did a funny thing. I put together a personnel spec on what I wanted and went to personnel about it. They didn't exactly like it, but they said okay, and they wound up handing me about 100 applications that I had to review. So I just took my spec, went through the applications, and tried to see the ones that matched it the most. I kind of had a little numerical grading system idea. When I got through, I was looking for around 20 people or something like that, so I just took the top 25 people because I knew I wouldn't get everybody I asked for. The only funny thing about it was that, at that time, almost all engineers were men, but out of those top 25, about 16 or 17 were women. Much to my amazement, I think I only got only accepted by one or two of the guys, but most of the women accepted. One had previously been Miss Florida, of all things. I mostly wound up with a high percentage of women and it became known as the women's branch out there too, among managers. But they were also extremely good because, at that time I guess, any woman who made it through engineering *had* to be pretty good. And they did [the job] for a decade or so. That was pretty cool."

STS-6, launched April 4, 1983, was the maiden flight of the *Challenger*. Designated OV-099, it had been upgraded from the Structural Test Article (STA-099) after OV-101 *Enterprise* had been deemed too heavy and expensive to be converted for spaceflights. Though this mission also did not carry any Spacelab hardware, back at the Cape preparations were escalating towards the first flight of the Spacelab module.

When Jim Dumoulin started, he noticed immediately that while the flight and checkout systems were state-of-the-art, there were few, if any, computer tools available to help individuals automate their workflow. The only Information Technology tools the office had was a secretary with a 1969 vintage IBM Electric II typewriter and a 1977 vintage 8K Commodore PET computer. Everything was typed on typewriters and copies were made with carbon paper. Jim Dumoulin recalled, "The only thing on my desk was a phone, a NASA acronyms book and an official, government issued Webster's Dictionary. On my first day of work, I remember thumbing thru the dictionary and looking up 'spacecraft'. The definition was 'An imaginary vehicle that may one day take people to the Moon'. I looked at the copyright date on the dictionary and it was 1958. I wasn't used to working in a paper-only environment and ended up bringing a computer to work that I'd built in school. Our Level-IV Electrical Branch began writing all of our work documents on this computer."

Level-IV Electrical Engineers were always making improvements to the payload checkout environment. The ground checkout computer the PCU was using was a Perkin Elmer 8/32 Real-Time computer, and during full up Mission Sequence Tests (MST) it was being pushed to its limits. When the real-time operating system overloaded, the system would crash hard and they would have to restart the test. The system was under configuration control, meaning Level-IV's hands were tied in trying to improve the hardware, so they had to settle for a way to push the system to its edge without causing the computer to crash, as Jim Dumoulin recalled: "Someone noticed that just before we would crash, the duty cycle on the CPU's activity light would go solid red. One of the engineers in the group built an impending crash alarm box, measuring the duty cycle of the CPU activity light. An alarm would sound if the system was so overloaded it was about to crash. We had time to checkpoint all files and back off some of the workload. Over time we got good at using this box and averting future crashes." The production MIP-O-METER (MIP is short for Millions of Instructions Per second) was a one-rack unit, 1.75 in (4.5 cm) tall device with a silver front plate in a blue BUD box (one manufacturer's name used for this kind of device).

The MIP-O-METER was inserted into an empty rack unit in the PCU. It was used to help justify getting the PCU Perkin Elmer 8/32 upgraded for later missions. As the ground computer got more loaded with software applications, terminal users, printing, etc. the Perkin-Elmer had an LED that indicated the duty-cycle of the load (full-off = no load; full on = full load; blinking 50% = 50% busy, etc.). Instead of guessing the duty cycle of the LED by eye, the MIP-O-METER would give a digital read-out from 0–100 percent CPU load. This was an easier way to see the PCU was nearing its limits and back off before it was overloaded

Scott Vangen mentioned how Enoch Mosier gave them a little tolerance in the PCU. When it had an error, the CID (Computer Interface Device) would ring a bell with a little ping, which would tell them that there was a fault. These faults were associated with one or more of the avionics models indicating a problem (e.g. RAU data bus error, Mitra computer fault, etc.), which would have to be investigated. People eventually got conditioned to this ping, which was not a good thing. Scott Vangen recalled: "Well, I don't know whose idea it was, I think it was probably a joint thing. Probably me and Craig [Jacobson] and Jim [Dumoulin] and Donna [Dawson]. We had this toy thing that was a gag Christmas gift, a little plastic toy penguin. It just looped, sending a penguin up a stair, then it slid down to the bottom and it did it over and over again."

"I connected this toy to that buzzer," said Craig Jacobson "So I modified GSE without paperwork, without authorization, to connect these penguins up. And so when the buzzer went off, the penguins would run. What I made was really benign technically, and made it so that if anything went wrong, there was no way the PCU or CID or anything could be disrupted or whatever. But still, the idea that I was in there wire wrapping stuff for something that was completely not authorized..." Scott Vangen continued: "The penguin toys we turned into what we affectionately call the Penguin Alert System [PAS] and we actually would write that into our procedures. 'If you hear a PAS...' We had the penguin toy up there for about a year before, for some reason, we had to take it down."

THE FIRST LEVEL-IV PAYLOAD

While STS-7 was not a mission in the Spacelab series, it was the first payload to fly into space that was assembled, integrated and tested by Level-IV/EI personnel. Mike Haddad recalls: "It was very rewarding to those of us that worked this first payload. A new space program with only a few flights under its belt, and we were a major player – well, one of four – in a mission that flew into space. I can't accurately describe the feeling of watching this mission take place. After placing my gloved hands on it and fitting the parts that held it in the Orbiter, observing the launch and then seeing it on TV, in space, was surreal."

STS-7 *Challenger*, launched June 18, 1983, 6+ days
Payload: Office of Space and Terrestrial Applications-2 (OSTA-2); one MPESS (F001).

STS-7 carried the OSTA-2 payload, a joint US-West German scientific Mission Peculiar Experiment Support Structure (MPESS) payload. The first payload to use an MPESS, OSTA-2 remained in the Payload Bay for the entire mission and was the first in a series of planned orbital investigations of materials processing in microgravity.

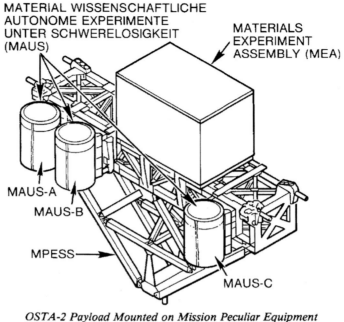

MATERIAL WISSENSCHAFTLICHE
AUTONOME EXPERIMENTE
UNTER SCHWERELOSIGKEIT
(MAUS)

MATERIALS
EXPERIMENT
ASSEMBLY (MEA)

MAUS-A

MAUS-B

MPESS

MAUS-C

OSTA-2 Payload Mounted on Mission Peculiar Equipment
Support Section (MPESS)

OSTA-2 payload graphic. [nasaspaceflight.com.]

The pallet (SN001), OSTA-2 flight configuration MPE Support Structure (MPESS F001) and structural/mechanical MPE were delivered to KSC on January 12, 1983 [4]. MPESS arrived from MSFC secured, on the side, to its yellow support structures. At this point it was EI personnel responsibility to install the mechanical interfaces, called Spacelab trunnions, that secured the MPESS to the Orbiter and all KSC GSE that supported the MPESS.

In total, five trunnions were used to fix the payload to the Orbiter. Those placed on top were designed to take loads in the X and Z directions; a pair were used in the primary (forward) position and another pair in the secondary (aft) locations. The final one was installed on the base, called a keel, and was designed to take loads in the Y direction, similar to the keel of a boat.

There were two lengths of keel trunnion; the 8-in (20.32-cm) normal keel forward of the Xo 1191 location, and the shorter 4.82-in (12.25-cm) keel aft of the Xo 1191 location. (The Xo location was a measurement in the "X" direction that started from Xo=0 at the front of the Orbiter and increased in number as it proceeded back over the vehicle. This number was used to define exactly where on the Orbiter items were located, in conjunction with Yo and Zo coordinates for the other two directions in the Orbiter's three-dimensional system.) The trunnions were made of titanium with a chrome polished plating, and were secured with 32 stainless steel bolts and one large titanium bolt for each primary trunnion, and ten stainless steel bolts and one large titanium bolt for each secondary trunnion. As Mike Haddad recalls: "The trunnions on the OSTA-2 were the first piece of flight hardware I was responsible for installing that would fly into space. A technician and myself performed the installation of the stainless steel bolts; real hands-on work." It was not a simple nut and bolt operation, however. The bolts were placed, head up, threads down, through the trunnion then through the MPESS structure and out the bottom of the top section, but still inside the main structure. Then a castellated nut would be installed, tightened down and torqued. A castellated nut, sometimes referred to as a castle nut, is a nut with slots (notches) cut into one end. The name comes from the nut's resemblance to the crenellated parapet of a medieval castle.

A cotter pin would then fit through the cut outs in the nut, the hole in the bolt and out the side of the nut, with the end bent back to secure it in place. With the nut being inside the structure, aligning the nut openings with the hole in the bolt was not easy, especially while requiring a wrench on both the bolt head and nut. This was done to keep the sharp edges of the cotter pins inside away from the flight crew, in case an Extra-Vehicular Activity (EVA) had to be performed. The sharp edges of the cotter pins would catch on the plastic gloves worn by EI personnel and tear them during fitting. Many dozens of gloves were needed, as well as small bandages when a finger got caught up on the cotter pins. The one large titanium bolt for each trunnion had be torqued to a very high level, several hundreds of foot-pounds (newton-meters) of torque, requiring very special tools. The keel was installed using the same type of process, but on this mission the stainless steel bolt was found not have enough lubricant, so when the stainless steel nuts were being installed, one of them seized up. When stainless-on-stainless seizes, it is like a welded connection and there was no way to loosen the nut, so the bolt had to be cut off. Due to this problem, checking for sufficient lubricant on stainless bolts and nuts before installation became mandatory.

When trunnion installation was completed, the MPESS could be lifted and placed into the Level-IV GSE stand to be prepared for experiment installation. EI personnel were also required to install the cabling. The cables and the connectors were inspected to ensure they were not damaged and were cleaned prior to integration onto the MPESS. The cables would allow the experiments to talk to the Orbiter. The experiment end of the cabling was secured near the experiment location via cable clamp, then routed outboard from the experiment location. The extra length was coiled up and temporarily secured to the outboard location on the MPESS. The extra length was required to allow the cable ends to reach where they would mate to the Orbiter's payload-dedicated Standard Interface Panel (SIP) following installation into the Orbiter. The connectors on both ends of the cables (experiment end and Orbiter end) were capped to prevent any contamination. The caps would be removed just prior to attachment to its designated experiment and Orbiter connection. All of this temporary hardware was non-flight, so they were mark with red "Remove Before Flight" tags and logged.

"We wanted to make sure nothing that was non-flight actually flew," Mike Haddad recalled. "The log began when the first non-flight item was installed, and followed the payload on its ground journey until the last item was removed, normally during final close-outs at the pad just before launch. The large red tags gave us a visual of what needed to be removed, the log was a recording of what was installed and removed, and the procedure listed in the log was the official document that installed and then removed the items. For multiple items, like non-flight tie wraps, one tag was installed on a tie-wrap and that tag listed the number of non-flight tie wraps installed. Then that same number of tie wraps listed on the tag had to be removed and bagged to ensure all had been accounted for. We knew the difference between flight and non-flight tie wraps, but the extra check and balance was in there for a reason. It sounds like overkill, but it was necessary to ensure all non-flight items were removed before launch."

Mounted to the top of the MPESS was the Materials Experiment Assembly (MEA). This was a large rectangular box about the size of a desk, and was the first experiment (that actually contained three different experiments) mounted onto the MPESS by Joe Lackovich, one of the Level-IV/EI Lead Mechanical Engineers: "I know that I was working on STS-7 payload OSTA-2, specifically the MEA experiment. At that time, although I was a lead engineer for OSTA-2, I did write the procedure for installing the experiment on the MPESS. I guess Dean [Hunter] signed the procedure so that I could perform it."

EI Mechanical Engineer Frank Valdez recalls Lori Wilson, one of the Operations persons, getting nervous and saying that that the scaffolding which was needed to remove the experiment lifting sling from the experiment was too high, and that she would prefer for the techs not to go up there. But it had to be done, and so the technicians did their job. Working at heights became a normal mode of operations. Starting at the floor level, there were the Rails that extended up off the floor. On top of those were the GSE supporting structures, and then the flight hardware was mounted to the top of the GSE support structures. This would place personnel at least some 10−15+ feet (3−4.5+ meters) above the floor.

When the installation of MEA was complete, an inspection was performed on the MPESS cable connectors that would mate to the experiment and the experiment connectors, to ensure there was no contamination and no damage. Once this was verified, the MPESS cables were mated to the experiment connectors.

A small German experiment contained in three Getaway Special (GAS) canisters, the Materialwissenschaftliche Autonome Experimente unter Schwerelogisgkeit (MAUS), was also flown on STS-7. After arrival at KSC, the three GAS cans went through offline operations by the PIs in an offline lab of the O&C. Offline meant work performed by the PI before being turned over to KSC for experiment integration.

Frank Valdez explained that, "The PIs needed a crane to do disassembly and reassembly of the GAS cans' outer covers, so myself and facility designers needed to figure a way to secure a crane in a room not designed to contain one. The result was a small crane attached to the ceiling supports."

STS-7 offline photo showing PI work and crane addition to room. [NASA/KSC.]

This was a good example of EI personnel, a PI team, and a facility designer working together to solve a problem, something that occurred hundreds of times during the Spacelab Level-IV/EI timeframe.

Once the offline processing of the GAS cans was complete, they were turned over to another EI mechanical engineer for weight (mass) and center of gravity (WT&CG) checks. With the mass of the GAS cans exceeding the maximum manual lifting capacity, a crane operation was required. The GAS cans were transported into the O&C Low Bay on their support GSE, and a control area was established. A crane was then lowered and a GAS can lifting sling attached to the crane, then two drop legs were connected to the top of the GAS can.

GAS can on end, sitting on WT&CG table. Technicians Bruce Burch (left, standing) Chris Madore (left, blue smock), Herbert Linhart (right, kneeing). Level-IV Engineers Don Tiffenbach (behind Herb, bending over), and Jeannie Ruiz (standing behind Don). [NASA/KSC.]

The CG X & Y components would be measured with the GAS can standing on end, so it was lifted directly off the GSE and placed on the CG measurement table. To get the CG Z component, the can had to be rotated 90 degrees and laid on its side in a special fixture mounted on the CG measurement table. With the measurements taken, the can had to be rotated 90 degrees back on its end and placed in its GSE support stand for next phase of integration. Following WT&CG, the cans were installed onto the MPESS under the supervision of Frank Valdes, the EI Lead for MAUS. A crane operation was once again necessary for each individual can. Access stands were configured to allow access to attachment locations between the GAS can and MPESS surface.

Frank Valdes (lower right) with Jay Smith (middle), installing MAUS-B GAS can. [NASA/KSC.]

A careful dance between the crane crew and technicians took place to align the GAS can fitting to the MPESS fittings, so that the bolts could be installed to secure the can to the MPESS. At this point, everything could have come to a screeching halt if there had been interference between the two. Luckily for this mission, no interference was encountered and the attachments were aligned for bolt installation. Another potential show stopper was if the bolts were too short or too long to allow proper securing, but again no problems were encountered. Once the bolts were installed, they were tightened down and then finally torqued to the proper value, and the sling was removed from the cans. Most attachment hardware then required lock wire to secure it in place. These same steps would be repeated for each GAS can. Cabling was then connected to the GAS cans, and the bonding straps installed and checked.

With all support structures added, cabling installed, and the MEA and MAUS experiments in place, Level-IV testing was completed. No servicing or final closeouts were done in the O&C for this mission. Instead, as with most missions, final closeouts would occur at the launch pad, although an inspection and cleaning of the payload was performed before it left the Level-IV area, using the appropriate cleaning solutions and materials that had been approved during the experiment planning phase.

Payload Canister Operations

CITE testing was not required for the STS-7 payload, due to it being a very simple interface to the Orbiter, so the large strongback (fabricated to the length of the Shuttle's Payload Bay) was used, requiring two overhead cranes to lift it. Featuring four drop links for each of the four OSTA-2 trunnions on top, the strongback was lowered down and the drop links attached. This operation was performed by MDAC with observers from EI. Then OSTA-2 was lifted from its GSE support stand in the Level-IV area and moved to the east, High Bay area of the O&C, to be installed into the large Payload Canister (PC).

Mike Haddad recalls the strange sight of seeing this huge strongback lifting only the very small OSTA-2 payload. Due to this very complex movement for such a small payload, a special lifting device called the Partial Payload Lifting Fixture (PPLF) was later developed. This was a smaller fixture, requiring only one crane instead of two, and resulted in huge operational savings, with counterweight adjustments to compensate for a wide range of payloads with very different weights and CGs. This device would be used in the future for lifting MPESS, individual pallets and special structure payloads, but was not used for handing the much larger Spacelab experiment rack complement, Spacelab pressurized modules or multi-pallet payloads.

Once OSTA-2 was installed horizontally in the canister, the canister doors were closed and special conditioned air was supplied internally to keep the payload at the proper temperature, humidity and contamination levels during its trip outside. With the proper environment established inside the PC, the large east door of the O&C was opened and the canister moved outside to be prepared for transfer to the Vehicle Assembly Building (VAB).

Processing Vertical Payloads

For operations that would take place in the Vertical Processing Facility (VPF), the Payload Canister had to be rotated to vertical. Vertical processing and horizontal processing often occurred in parallel, allowing as much time as possible for each process before coming together to make the full payload compliment for that mission. This mission was a good example, with the two coming together in the VPF when all vertical processing was complete, thus maximizing the horizontal processing time and then minimizing the time the horizontal payload was in the VPF.

STS-7 full payload complement in canister, vertical position, June 16, 1983. Forward (top) to aft (bottom), the payloads are: West Germany/Rbb's SPAS-01, or Shuttle PAllet Satellite; OSTA-2, the second Shuttle experiment sponsored by NASA's Office of Space and Terrestrial Applications; Anik C-2, a communications satellite owned and operated by Telesat of Canada; and Palapa B-1, a second-generation communications spacecraft designed and built for Indonesia. [NASA/KSC.]

The three payloads were transferred into the canister and then all four payloads were transported out of the VPF, reversing the previous steps to bring them in, and moved to Launch Pad 39A. Watching the large Payload Canister on its end moving down the road toward the pad was another unique sight.

Canister Operations at the Pad

Canister operations at the pad were tricky. The canister would be positioned under the Payload Changeout Room (PCR) and then a large crane would lift the canister off the transporter, keeping any services (ECS, purge, etc.) connected during the lift.

Lifting canister off canister transporter (red air conditioning hose between canister and transporter) and into position outside the PCR. RSS is the large gray structure housing the white PCR (recessed white area). FSS is the large gray structure in the center with the white lighting rod at the top (notice hinge for RSS in the middle of the FSS). The Shuttle stack is on top of the MLP to the right. The flame trench is shown between the white railings, bottom of photo. [NASA/KSC.]

The canister was properly aligned with the PCR and then secured using large latches. As mentioned previously, the canister represented the Orbiter payload, so it was placed in the same coordinates in the PCR as the Orbiter. Then the doors on the PCR were opened first, to allow the doors of the canister to open into the PCR.

The Payload Ground Handling Mechanism (PGHM) was moved into place so its large hooks would grab the trunnions on the payloads to remove them from the canister. The canister doors were closed followed by the PCR doors. The crane was then positioned to take the load of the now-empty canister, the seal depressurized and the latches released. The canister was then lowered back down vertically onto the transporter, the sling disconnected and the Payload Canister removed from the pad.

The Rotating Service Structure (RSS) was then rotated to mate up with the Orbiter. Then the doors on the PCR were opened first, this time to allow the doors of the Orbiter to open into the PCR. This would be the time any last minute operations were performed on the payload before installation into the Orbiter. This usually consisted of any closeouts, such as on the parts of the payload that could not be accessed once installed into the Orbiter. Then the payload was transferred from the PCR to the Orbiter.

Once secured, the coiled up cabling installed by EI personnel on OSTA-2 while in the O&C had the non-flight tie wraps removed, was un-coiled, and then the capped ends were placed near their Orbiter connections. Tie wraps were recorded in the Remove Before Flight Log (RBFL) to cross-check their removal and then the connectors were un-capped, with the caps also recorded in the RBFL. The cables were inspected to ensure no pins or sockets were damaged, cleaned if required, and then connected to the Orbiter SIP. Frank Valdez observed that they had found extra cable length − which was always better than not having enough cable length − so it, too, was secured to the MPESS using flight tie wraps. Orbiter-to-OSTA-2 testing followed, by performing payload on/off command switch activations.

OSTA-2 final closeouts by EI personnel consisted of removing covers from the OSTA-2 experiments, by accessing those locations from the PGHM's platform. The covers were recorded in the RBFL, which now showed all non-flight items as removed. When all payload closeouts were complete, the PGHM was backed away from the Orbiter and final closeout photos were taken. The Orbiter doors were then commanded closed, followed by the PCR doors. As Mike Haddad recalls: "I would always be there for final door closure, just to get one last look at the payload(s) before they were launched into space. It was kind of like watching your kid leave for college; you have done all you can to prepare them, now they're on their own, and you hope all the months and years of hard work you put into preparing them for that moment pays off. You would see the payloads again following the mission, though later in my career I worked on non-Level-IV payloads that were deployed never to return, so this would be the last time ever to see those payloads on Earth. Examples of such payloads were the Magellan spacecraft that went to Venus, the Galileo probe that went to Jupiter, the Ulysses that was placed in a polar orbit around the Sun, and the Hubble Space Telescope."

On the day before launch, L-1 day in space terms, the RSS was rotated back away from the Orbiter in preparation for launch. No EI-related late stow items were required for this mission, so the final EI task to perform was launch support at the C-1 (Payload) console, in the LCC Firing Room located next to the VAB. Because OSTA-2 was not powered up at launch or required any support during launch countdown, EI launch support at the C-1 console was minimal.

During the six-day STS-7 mission, no EI support was required. Landing occurred at DFRC on June 24, 1983. There was no early de-stow, and no EI support was required. The Orbiter was mated to the Space Shuttle Carrier Aircraft (SCA), a modified Boeing 747, and transported to the KSC Shuttle Landing

Facility (SLF); again no EI support was required. The Orbiter was then removed from the SCA via the Mate-Demate Device and moved to the OPF. OSTA-2 was remove from the Orbiter in OPF and installed in a Payload Canister by MDTSCO, with EI support, before being transported back to the O&C for postflight de-integration. De-integration in the O&C Building included removing MEA from the MPESS and putting it into a shipping container for transport to MSFC. The GAS canisters were taken to the offline labs and turned over to the PIs. EI personnel then started de-integration of the MPESS in preparation for its next mission.

STS-8 *Challenger* launched on August 30, 1983. EI had no direct involvement with the payloads flown in the Payload Bay, though Rey Diaz, whose first experience with the flight hardware came during post-processing of the STS-7 middeck experiment, did his initial processing of flight hardware to go into space on STS-8.

STS-8 was the first opportunity for Rey Diaz to see a launch and to be part of a mission processing, as he told the authors: "I had middeck payloads and I performed testing, offline testing, and the preparation of a hardware prior to taking it to the launch pad. There, I performed an Interface Verification Test (IVT), because these were the ones that required power. After that, we pretty much closed out the payloads for the mission. After the mission was completed, we removed those payloads a few hours after landing and did post-processing so that the scientists could have their late data for their experiments."

Number One on the Runway

As these first eight missions were being prepared and flown, EI processing of flight hardware at KSC was ramping up very fast to support the aggressive schedule of Spacelab and non-Spacelab missions. As Gary Powers explained, "Here comes a new boy named Spacelab and it came with a bunch of what ifs and newness. And I'll tell you something about the newness of it. We had life science missions come along where we were going to launch Spacelab in the Payload Bay with live animals. Well, we had to figure out a way to get to them into Spacelab late, because some life science samples had limited lifespans in which we had to keep them alive. Because we anticipated a long stay on the pad after we left the VAB, we had to install them late at the pad. But how could we get them through the Aft Flight Deck into the inside of the Spacelab? That was a real accomplishment."

By the closing months of 1983, after three years of preparing the hardware at the Cape, Spacelab-1 was now at the top of the manifest as the next mission to fly, and was the focus of attention both in the US and Europe. The pressure was now really on the Level-IV team to make sure all that could be done was done to ensure that this first launch of the Spacelab module and its 70+ experiments would have a successful mission. It was now time to put all that practice and theory into operation. It was time for Spacelab to fly and for those in Level-IV to hope that it did not fail because of their individual contributions.

References

1. *Spaceport News*, Vol. 19, No. 8, April 11, 1980, p. 3.
2. Lord, Douglas R.; **Spacelab, an International Success Story**, NASA SP 487; Washington D.C., 1987.
3. **OSTA-1 Scientific Payload Data Management Plan Addendum**, OAO/TR-82/0025, April 1982, National Aeronautics and Space Administration Office of Space Science and Applications Washington, D.C.
4. **PMIC-MA03-469-26 Data Requirement (DR) MA-03 Payload Missions Integration Progress Report**, November 16, 1982 thru January 15, 1983. Payload Missions Integration Division, Teledyne Brown Engineering, Building 4708 George C. Marshall Space Flight Center, Alabama 35812, January 28, 1983.

7

Path to a Quick Turnaround

> *"We got to do a lot of real engineering involving flight hardware.*
> *That was a great education and great experience for me…*
> *All of these things, that you really don't learn in college,*
> *became commonplace for us."*
> Eli Naffah, NASA Level-IV Co-op
> while attending University of Central Florida.

A great deal of work across scores of operations was performed by Level-IV personnel, far too much in the early years to detail in the pages of this book. Indeed, there was so much that went into each mission that a book could literally be written for each of those flights. What follows are just a few examples from each mission. In short, many tasks had to be performed on each payload, with new employees being hired all the time. There were far too many activities to detail in depth here, but what follows under each mission is a selection of some of those activities that took place to make the launch date. Each entry focuses upon a specific mission, or had direct influences for that mission.

As the flight rate continued to increase, it soon became apparent that teams would be working multiple missions at the same time, given the limited number of Level-IV personnel available. Their involvement encompassed planning, document and drawing reviews, procedure writing, assembly, integration, servicing, testing, and more – and all preflight. Then there was launch support, mission support, landing operations, and all the postflight work that was required for each mission. Mike Haddad remembered just stopping to take a break one day in late 1985 to list all the payloads (Spacelab, non-Spacelab and middeck) he was working on at the same time. There were 13.

M. E. Haddad, D. J. Shayler, *Spacelab Payloads*, Springer Praxis Books,
https://doi.org/10.1007/978-3-030-86775-1_7

SPACELAB FLIES

From the initial creation of the Level-IV organization, it became apparent more personnel would be required to meet the flight rate, even with everyone working multiple missions at the same time. By 1983, due to problems discovered with the Shuttle Solid Rocket Boosters (SRB) following the STS-8 launch, STS-9 and Spacelab-1 would be delayed from its planned date of September 30, 1983. While those problems were resolved, the Level-IV personnel continued pushing ahead at full speed to ensure the each of the Spacelab series of payloads would be ready to launch on time.

STS-9, *Columbia*, launched November 28, 1983, 10+ days
Payload: Spacelab-1, Long Module [MD001); Tunnel (MD001); Floor (MD001); Rack [see Appendix 2]; and one Pallet (F001).

The Spacelab-1 mission had over 70 scientific experiments that were conducted in a variety of fields, including astronomy, solar physics, space plasma physics, Earth observation, material science, technology, and life sciences. Canada, Japan, the United States and 11 different European countries were involved with the mission. This would be the first true international effort in space and Level-IV was a main player in making this and other missions a success.

NASA Co-op Eli Naffah remembered, "Obviously there were engineers involved that issued us the work, so we could go back to the engineer and consult. They gave us a lot of responsibility and autonomy to go work with Teledyne Brown Engineering, or whoever was providing the experiment, whoever built it, to get the drawings and all the particular information on what we were going to be getting. When the hardware came in, it didn't necessarily match the drawings or look like what it was supposed to look like. We modified the procedures or the hardware in order to make it fit, so we got to do a lot of real engineering involving flight hardware. That was a great education and great experience for me. In fact, so much so that I just wanted to hurry up and get through school and get back to work. It was that much fun. So all of these things, that you really don't learn in college, became commonplace for us."

Donna Bartoe recalled that "While the Spacelab-1 experiments were still in Europe, there was a gang at Marshall [Space Flight Center, MSFC] that were assigned to the astronaut training. They were developing pretty elaborate experiments and simulator models. So we tapped those engineers to help us liaise between the Teledyne Brown engineer and the experiment engineer from MBB/ERNO [Germany]. That's how we pulled together to do what we needed to write our experiments and checkout procedures for that first full Spacelab mission."

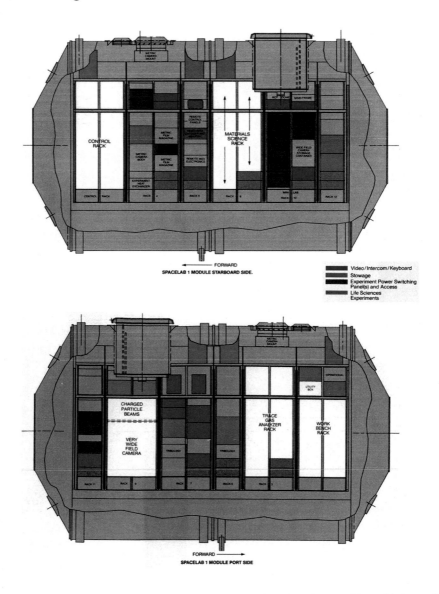

The Spacelab-1 module interior configurations (starboard and port side, with legend key). [NASA]

A KSC Way of Life

All the new people of Level-IV were thrust right into the action. Trying to learn the Kennedy Space Center (KSC) way of life was not easy for anyone. Comprehending all of the government requirements placed on civil servants, and

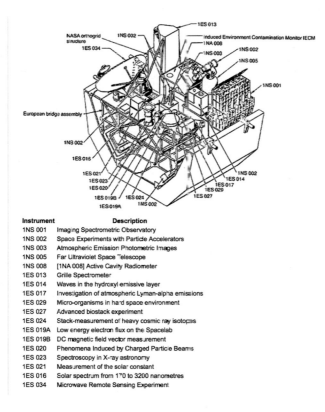

Instrument	Description
1NS 001	Imaging Spectrometric Observatory
1NS 002	Space Experiments with Particle Accelerators
1NS 003	Atmospheric Emission Photometric Images
1NS 005	Far Ultraviolet Space Telescope
1NS 008	[1NA 008] Active Cavity Radiometer
1ES 013	Grille Spectrometer
1ES 014	Waves in the hydroxyl emissive layer
1ES 017	Investigation of atmospheric Lyman-alpha emissions
1ES 029	Micro-organisms in hard space environment
1ES 027	Advanced biostack experiment
1ES 024	Stack-measurement of heavy cosmic ray isotopes
1ES 019A	Low energy electron flux on the Spacelab
1ES 019B	DC magnetic field vector measurement
1ES 020	Phenomena Induced by Charged Particle Beams
1ES 023	Spectroscopy in X-ray astronomy
1ES 021	Measurement of the solar constant
1ES 016	Solar spectrum from 170 to 3200 nanometres
1ES 034	Microwave Remote Sensing Experiment

The Spacelab-1 pallet configuration diagram. [NASA]

then trying to abide by them all, as quickly as possible, came on top of becoming familiar with the facilities, the ground systems, and the processes used to prepare those facilities and systems for the workload ahead. It was a challenge to take their experiences of the past, and then apply them to this unique, hands-on opportunity to work on this very special flight hardware to ensure a successful mission. But it got even more complicated with the fact that Level-IV personnel needed to travel offsite to teach those who would travel to KSC the "KSC way of life".

Gerry Rivera recalled that, "Travel was a fairly common thing in Level-IV. I remember traveling to Huntsville [MSFC in Alabama] on a few occasions to meet with the designers of the Spacelab, or meetings with the PIs, the Principal Investigators. I remember flying the NASA-4, a Gulfstream 1. You would take off from the Shuttle Landing Facility [SLF] at the Cape and go straight to land at the Huntsville Space Center."

Hardware was arriving all the time, and the Spacelab Experiment Integration Support (SEIS) technicians were called upon to assist with the offloading operations, as Bruce Burch explained: "One of the things we did on Spacelab-1 was to physically help unload the hardware from the plane. They had a 747 come in to the SLF and we went out there and unloaded it. We had semis [trailers] of stuff coming to the back of the O&C [Operations and Checkout Building] and we were [there till] two or three o'clock in the morning unloading all their experiments, but the PIs also had bicycles, kayaks, and all their personal stuff."

A European Perspective

Hermann Kurscheid, Spacelab-1 European Space Agency-First Spacelab Payload (ESA-FSLP) Head of Ground Operations at NASA KSC, added the European point of view to these early days: "The first flight, we had part of the payloads coming from the USA and part from the Europeans. We had the European Bridge Assembly [EBA], a structure-type framework that was to support European experiments. We were responsible for one part of the Spacelab-1 payload. The NASA mission manager was Harry Kraft from Marshall, but we had our own test equipment in Bremen, Germany, at ERNO, so they had already completed the Spacelab integration. We could assemble our hardware there, test it and then come over to KSC and assemble the two parts of the payload for the first mission there. We could then do the Mission Sequence Test."

Lori Wilson did not recall any big problems with any of the experiment developers, especially the Germans. "In fact, Hermann [Kurscheid] made everything go smoothly on his end. Any of the 'overall' problems that may have existed with the Germans were handled by Gary Powers. I occasionally had to negotiate a minor schedule item with Hermann, but it was always equitable."

Donna Bartoe added that, "We had a little bit of good thing going for us, because the ESA MBB/ERNO team came with the experiments. They had developed the experiment test procedures, at least the ones they used in Bremen, Germany, and we took those and then had to adapt them to fit our system. We had these dual roles, and in retrospect there should have been a whole other team being Experiment engineers and Systems engineers. We were doing both jobs. An experiment that I drew was called Experiment 13 and it was a Grill Spectrometer from Belgium. Dirk Frimout was the PI, who later flew on STS-45 Atmospheric Laboratory for Applications and Science-1 [ATLAS-1]. But it was a bitch, a pointing experiment [that] didn't point [on its own] like other telescopes. It was just sat in the Orbiter Payload Bay, but [it was necessary] to point the Orbiter as it was trying to look at the Earth's atmospheric limb during sunrise and sunset. So it was a very complicated procedure to write, because you were interacting with the Orbiter's state vectors and we had to get the Orbiter numbers plugged in there correctly."

With all the foreign equipment that was coming onsite to support Spacelab-1, there was an issue over where should all be housed, as Hermann Kurscheid recalled: "Bob Sturm had a lot of accreditation business to do there. I mean, we had the containers behind the O&C Building and we had a permanent office over the time we were there. So we had our little city there behind the O&C Building in the containers." Housing for them would be in the O&C office area.

Donna Bartoe continued: "We had already completed the acceptance testing and were kind of messing around with the system, trying to understand what roles everybody had prior to the arrival of the Europeans. I was assigned to the PCU [Payload Checkout Unit] hardware and in particular I specialized in the computer interface device. I backed up Mo [Maurice Lavoie] on the Mitra flight computer, and Ann Bolton was software. So we were just kind of messing around before the s**t hit the fan when the experiments and the European contingent showed up. They descended on us. I remember, they filled up the east side fourth or fifth floor offices in the O&C that had no windows. I remember from day one when they showed up they were furious, it was an international incident. Bill Jewel and the Mission Manager got involved. I can't tell you how mad they were [because of where their offices were located]."

Finally, the flight hardware started arriving, mostly in pieces. The offsite Ground Support Equipment (GSE) for the flight hardware arrived, then the offsite personnel responsible for the flight hardware up to that point began arriving, and then the fun began. For Spacelab-1, the experiments came in separately from the racks.

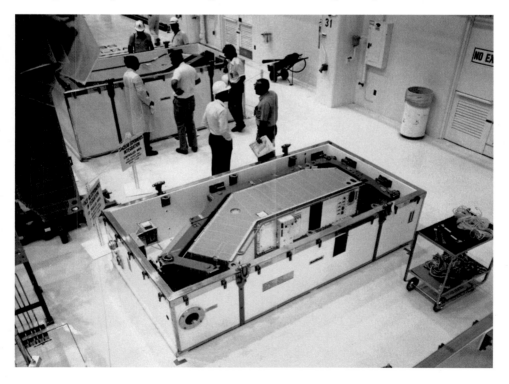

Spacelab racks in open shipping container, double rack at the top, single rack at the bottom, Level-IV area KSC, May 1982. [NASA/KSC.]

"The first taste of flight hardware for me was rack six of Spacelab-1," said Bruce Burch. "I was the mechanical [technician] and Andy Petro was the electrical. We decided to wipe some panels off, so I went and got some soap and some regular squirt bottles, but Bob [Ruiz] said 'No, it can't be that. It's got to be [suitable] for flight'. So everything was cleaned with Freon® 113."

"When I first came on board, most of the GSE had already been accepted," said Gerry Rivera. "It was operational, but prior to every mission, even though the GSE was already operational with acceptance tests, we double checked everything before we hooked it up to the next payload, for voltage checks, polarity checks. We would put load banks and make sure that it would provide the amount of power that we knew the payload would require. That was done on every mission."

Racking It Up

Assembling the racks for the first time was quite a challenge, because the actual flight racks differed from the drawings. As Bruce Burch recalled, "They had some stuff that slid in and out on rails, and we had a hard time getting the rails parallel to each other. We had a parallelogram to measure them, but they had actually cocked sometime and we were trying to get them straight again. I remember trying to get a pair of four-foot calipers in there trying to straighten it up. Also, the PI was there with the old baby food jar in his pocket, with all kinds of nuts and bolts, who dumped them on the tables and asked, 'What size do you need?' But it was the same hardware. That was back in the early days when we didn't know any better."

As the work began, new problems or situations would occur, even though as much as possible was done ahead of time to plan and avoid anything that may hold up the activities. Jay Smith recalled working on the racks: "We didn't have a way to get to the back of those racks. I remember talking to Dean [Hunter] and he said, 'Can you guys build us something that would get us in there?' We basically designed these platforms to walk up, or lean on, or sit on, so you could actually work in there, because there was no way to do it. Dean said, 'You know, that's perfect'. Dean was such a nice guy and he said, 'You need put this into the suggestion program'. Dean signed off on it, and I think I got a check for $1,500, or $2,000 or something, for a suggestion."

Everyone had their part in getting tasks done with whatever they had available. Gene Krug described how he could have a piece of scrap paper with some design written on it, and take it to McDonnell or NASA's machine shop and they could do it for him: "I could get anything I wanted. [Bill] Mahoney gave me a charge account so I could go get anything in federal stock. He was like a straight arrow; you couldn't operate that way today."

Bruce Burch recalled helping Walt Preston as he was working the JSC Rack 10. "I was done with my rack, so I'd go help Walt on his. I remember there were a lot of little screws and we finally figured out we needed to use hemostats [surgeon tools] to actually hold the nuts and stuff in different places to try to put that together. We couldn't use standard tools. It was a lot of work."

Hop and Drop

Bob Ruiz worked an experiment called "Hop and Drop". It was a device for which the astronaut grabbed a handle connected to brackets at the top of the rack, and a number of bungee cords pulled from below. Bob recalled: "I was watching the video and I didn't know how it [the experiment] worked. He was hanging and all of a sudden he went 'bang', he went down. I said, 'Oh my God, the bolts broke off'. My stomach was knotted up, I thought I was going to throw up. Finally, somebody said, 'Nah. That's how it works' "

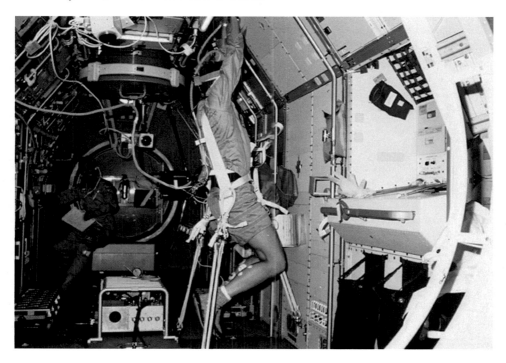

Spacelab-1, Vestibular Hop & Drop Experiment, performed by Byron Lichtenberg during the Spacelab-1 mission. A new D-ring on front of rack enabled attachment of bottom right bungy straps to Lichtenberg. [NASA/JSC.]

The Hop and Drop was also one that Mike Haddad worked. "We found that the bottom support ring that attached to the front of the rack interfered with the floor panel when you tried to raise it. Testing was about to begin, so we left it in place so they could perform the testing required. After testing, we tried to figure what could be placed in that location, to do the job but not interfere with the floor panel. Frank Valdez [Level-IV Mechanical Engineer], who was an avid boater, said, 'Why not use a Schaefer Spinnaker Pole Slider, Welded/ Lined [D-ring on a slider] like I use on my boat?' We suggested it to the designers and, knowing the materials and loads it would see, they thought it should work fine. So we obtained a D-ring, got it approved for flight, and I installed it and we ran a short test to make sure you could rotate it up to use, then stow it and the floor panel could be raised. It all worked."

The racks contained all this hardware, but how would it be cooled if needed? There were two main cooling subsystems for the racks – water cooling and air cooling. For the air, the subsystem could only supply a limited flow, so each rack and experiment was allocated so much flow, but how would the flow be measured and adjusted as needed? Rack flow balance was the answer.

Bob Ruiz was asked about the rack flow balance: "I thought it was a pretty neat job, because you got the real flow curve, flow versus pressure. I drew it by hand back in those days, though; no computer stuff. Each experiment had a cooling tube that you hooked in and measured the vacuum and the flow through each tube. You had to adjust one, check the other; adjust one, check another. And then finally we got close enough to balance. Okay, good enough." Bruce Burch added, "We spent months doing flow balance, trying to figure out how to make it work. We tried using Magnehelic® gauges [used to measure performance of pressure in filter units and ductwork], using a big dial gauge that measured inches of water, and then we finally got the digital gauges."

This was done for each rack, but then all the racks would be installed into a "Rack Train" during the next step in rack integration. As Mike Haddad noted, "As each rack integration was completed, we would roll them on their GSE stands out of the Rack Room and out into the Low Bay floor, to prepare them to be installed on the flight floor located in the South Rails of the Level-IV area." This operation would be performed for each rack, thus creating the "Rack Train" of experiment racks that would then require integrated operations in preparation for integrated testing.

Not all was perfect at the start. There were some people that maybe just did not fit the standard of excellence that was demanded to do the Level-IV job. Jay Smith remembered that at the time their tech lead ran things differently, and he was just fast and loose and not nearly as professional as some of the others. After one of the crane operations was completed, he threw a torque wrench down from the top of the scaffolding towards a tech, Ken Huber, who was below him. Ken was a sharp guy, older than then the other techs, as Jay Smith recalled: "I learned stuff from Ken. Ken was smart enough to not try and catch it [the wrench], so he just stepped out of the way and that torque wrench hit the ground, bounced across the floor, and that was the end. I think that they fired the tech lead right after that. He was gone and Herb Linhart took his spot."

As well as the racks, there were subsystems that supported the experiments in the racks, one of which was the vacuum vent system for the Spacelab module that connected to the front of the racks that Level-IV assembled and tested. The critical thing was that it had to be assembled and leak checks performed for all the connections, so there were no leaks. Any leaks during the mission would pull the atmosphere out of the module and into space, which was not desirable. So each rack, and any experiment on that rack that needed vacuum, had be accounted for and adjusted as needed. Frank Valdes explained: "The other thing is we had to make sure that one experiment wasn't pulling too much from the other experiment; basically, that they had enough capacity so both of them could operate at the same time. We also had to provide the vacuum GSE. We had two turbo molecular pumps, a big one and a little one, so during testing we were pulling vacuum on that system to simulate what it would be like in space."

The racks, Rack Train, vacuum systems, and all the other components mentioned so far, all dealt with the Spacelab-1 module. This had a single pallet that flew behind the pressurized module, which required build up and assembly of the support structure with a platform on top, called an Orthogrid. The Orthogrid was used to mount the experiments that would fly external to the module, exposing them to the vacuum of space.

Breaking a Hard Point

The basic Spacelab pallets came to KSC stripped down, with only very minimal hardware installed. It was the responsibility of the Level-IV team to install the underlying attachment hardware that would be used to build up the support structures that held the experiments. The main support locations for pallets were called "hard points" and they were the load-bearing interfaces.

"We actually broke a hard point," admitted Tony Ornelas. "[Those things were] strong and sturdy. We had one big, strong technician pulling on it and the torque bar was supposed to click when it got to the correct torque. He was putting on the 450 foot-pounds (610 Nm) or whatever it was, and it broke the damn pin on the hard point. Talk about being in some kind of s**t. Well now what did I do? I screwed this up bad. So of course we have to write a PR [Problem Report]. Dean said, 'They want you to write up a 'lessons learned' on how this happened and what happened'. I was feeling really low and mad about it, but Dean said, 'No, you're being too hard on yourself. Don't worry about that. I mean, if you're not breaking something or doing something, you're not doing anything'. The fact is that he was willing to just brush it off by saying, 'We understand that what you did was not intentional. The torque wrench failed, it didn't click, but he kept pulling and pulling on it and everybody was expecting it to click'. As a young engineer who was feeling really responsible for something like that, I should have been able see it and say, 'Wait a minute, you're pulling too much on it, something's wrong'. So we changed the operating procedures to torque in increments."

After the hard points were installed, the support structure would be added to the pallet. For Spacelab, the two main structures mounted on the pallet for experiments were the EBA and Orthogrid.

EBA removed from its shipping container in the O&C Level-IV area. [NASA/KSC.]

Hermann Kurscheid, explained that, "We also flew a carbon fiber structure on Spacelab-1, the EBA, but had to do a late change because of a potential emergency return. The engineers, the Safety people, said there is a possibility of a cold case emergency return [this was a situation where the Orbiter was in the shade of the Earth on-orbit, creating a very cold environment, and needed to perform an emergency landing while cold]. So several covers were put on the struts, and all that – the thermal blanket, the cutting and tailoring – was manufactured in the O&C building by a guy from ERNO in Germany. Actually, there were two more changes on Spacelab-1. The microwave antenna was planned to swivel moving the dish, but also due to safety reasons they finally decided to fix it in place. [We] also planned to fly the sled on Spacelab-1, but then it was decided not to put it on Spacelab-1 and [instead] fly it on [Spacelab] D1."

"We assembled that Orthogrid piece by piece, that whole thing bolt by bolt," recalled Jay Smith. "We laid the Orthogrid on the floor in the Clean Room and there must have been thousands of bolts in that thing, because each section was about two and a half by two and a half feet (3/4 of a meter) or something. And then it was bolted to the next section, and bolted to the next section; it was a Lego® set."

Once the Orthogrid build up and installation was complete, the experiment attachment hardware, such as cables and fluid lines, was integrated onto the pallet and Orthogrid. One of the most important tasks was the installation of the experiments themselves onto the Orthogrid. Due to the weight of the experiments, and the need for very precise movement and positioning, most of the installation procedures were done via crane operations.

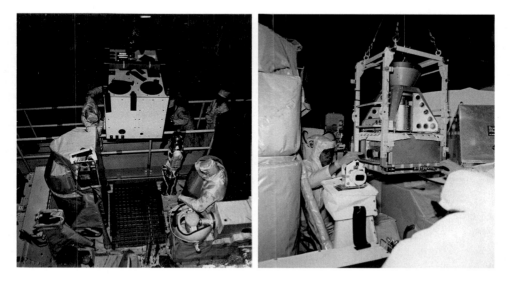

Left: Induced Environment Contamination Monitor (IECM) installation, October 22, 1982. Right: Experiment 2 Electron Beam Accelerator (EBA) installation, June 23, 1983, with Jay Smith in the center. Note the close spacing of the experiments and the "Remove Before Flight" red tags. [Jay Smith.]

Jay Smith noted that, "The lifting operation was a slow operation. Those crane operators were amazing guys, to be able to lower that million-plus-dollar experiment onto that Orthorgrid carefully and very smooth. You couldn't replace that stuff. I mean, can you imagine if we screwed someone's experiment up? You couldn't recover from that."

The integration of the pallet containing so many experiments in such a small area was a huge challenge, but just gaining access to the pallet was a task of its own. "None of us had built scaffolding before. It was all aluminum, but the deck plates were all wood. You couldn't have wood in the High Bay," explained Bruce Burch, "so we had to go and drill out the rivets and remove the wood. Then we got diamond plate and had the machine shop cut diamond plate sections, hand-filed them to get all the sharp edges off, drilled them, and then pop-riveted them. We'd go out and build towers and stuff, just training and getting familiar with it, knowing how you could interlock them and build the bigger platforms." Sometimes, special scaffolding would be used for specific operations, as Bruce Burch explained: "For Luis's [Moctezuma] alignments, they bought tower scaffolding that had a center section open to put the theodolites in, but eventually they quit using them. It was easier to just build standard scaffolding that would fit without all this other fancy stuff. We could actually build the other stuff quicker and it was better."

There was a very strict alignment requirement for the GSE. The alignment work for this mission, as well as all other missions, started with the installation and alignment of the fittings used to attach the Spacelab carriers to the Trunnion Support Assemblies (the big blue structures used to support the Spacelab carriers in the Level-IV area). That was followed by the installation of a system called the "Zero-G System", a set of two jacks installed at the forward and aft frames of the Spacelab pallets. The purpose of this system was to minimize the sagging of the pallet as experiments were being installed and their alignment measured. As experiments were installed, their alignment was measured relative to the Spacelab pallet, or to other experiments. This information was shared with MSFC, since they were responsible for managing this mission. There were also very strict alignment requirements for the flight hardware. As part of operations to verify the alignment, a very solid location was needed to setup a reference point from which to work. A large inside support column used to support the roof of the O&C Low Bay was thought to be the perfect spot to setup the reference target, as Luis Moctezuma explained: "For Spacelab-1, since we were expecting movement or changes of the pallet system, I wanted to monitor those changes relative to a baseline system that I thought would be solid. I had not realized that part of the column we used to install the baseline system would be affected by outside temperature. Measurements taken at the same time of the day were very close, within arc seconds, yet measurements taken throughout the day changed; arc minute differences. Also, the change was gradual." Essentially, the temperature during the day would affect the column, thus changing the reference mark and disturbing the alignment requirements.

AS TIME GOES BY

As time went on, the Level-IV personnel would learn what worked and what things should not be repeated due to the ramifications caused by those actions, as Eli Naffah explained: "You're growing up, maturing as an individual, but you make mistakes. I was told to go make sure all the hoses calibrated for the Ground Support Equipment, for example, and then they wanted to do a test, but they couldn't find any hoses because they were all sent out. There were a lot of life lessons."

Many experiments would first go through offline operations, to give the PI their final opportunity to complete any last minute work, or fix any problems, before going online and being installed for flight in a rack or on the Orthogrid.

One of the experiments to be installed on the Orthogrid was a Japanese experiment called SEPAC (Space Experiments with Particle Accelerators). But before it was placed on the Orthogrid, offline work had to take place. Scott Vangen recalled that, "SEPAC flew on Spacelab-1. We were troubleshooting offline, and the Japanese kept asking me for plus and minus screwdrivers while they were disassembling something.

I said, 'What do you mean by plus screwdrivers? And they looked at me, and signed 'plus' [placing their fingers in + configuration] for a 'plus' screwdriver. They were trying to signal a Phillips [cross-head], and a normal [flat-head] screwdriver was a 'minus'. They didn't know what the equivalent word was in English from their Japanese [term] for Phillips, so they used plus and minus; [there were] little things like that." Following the offline work, SEPAC, or any other experiment, would then be transported from the offline location to the online Level-IV Low Bay area.

Jay Smith could not remember how many experiments were supposed to be test-fitted before they received them: "There was no way they test-fitted it. They said that stuff worked, but it didn't and we had to modify many experiments. I remember doing lots of mods to modify all kinds of stuff. Re-drill and enlarge holes, having to grind stuff off, or file it, or whatever, to make it fit. I mean, nothing fit; holes didn't line up. I just remember we did lots of mods [modifications] to get these things [the experiments] on there. Just did whatever we'd got to do."

Have you ever heard of using an electronic meter to torque bolts? Bill Jewel explained that The Raymond Extension Meter was first used on Spacelab-1, on the pallet, for mounting several experiments. Very expensive bolts that were manufactured in the US were sent to the UK for fitting and calibration of the sensor, then returned to the US for heat treating. The bolts were stretched to ensure clamping of two parts, where vibration and load stresses were critical. "Torquing is really only a thread friction indication, and does not ensure any movement of parts and stress fatigue, in one or both parts, at an attached point in a high loads environment," added Bill Jewel

Jay Smith explained, "[That was] the first time I ever used a 'fluke' (electrical) gauge to torque a fitting. Those bolts had the threaded centerpiece and we'd put them on, put the wrench over it, thread them on and torque. I mean, that's precision torquing. I knew when you torqued something, the bolts stretched. That was fascinating."

Keeping all the necessary experiment attachment hardware could be a challenge, as Mike Haddad recalled: "I remember many times going down into the Tunnel [a location under the O&C Low Bay floor where flight hardware was stored] to get the flight hardware I'd kitted out, and a few times found that bolts, nuts and/or washers would be missing. The total I had at the end of the day before did not now match. I eventually discovered that Tony Ornelas would get there early and basically take the hardware he was short from our kitted hardware. That's how he sometimes finished his integration work before us."

In his defense, Tony Ornelas explained that, "We had to go in there and get hardware whenever we needed it, but I was doing Spacelab-1 or -2, and it required swapping from Peter to pay Paul. I would just get enough of the hardware needed. I could just say, 'We got some extra bolts over here, the same type and size, and you can take those now so we can continue with the integration work'."

Sometimes keeping the hardware was not the problem; just getting the kind of parts you needed in the first place could be a challenge. But those in Level-IV always stepped up to the challenge. The "can-do" attitude was just part of doing whatever was required to get the job completed and/or solve a problem.

This New World of Spaceflight

Maurice (Mo) Lavoie, revealed that, "My first experiment, crystal growth in microgravity, was the Iodine Crystal Growth Experiment, which, the first time we connected up the French experiment to a Remote Acquisition Unit [RAU], it failed royally in test. We found out that it had a bad chip inside. When we took that experiment offline, Gary Owens and Dick Scaltsas [NASA Quality] said, 'You're going to have to document anything you make changes to', but the PI said, 'That's my box. I do to it what I want'. So we actually signed documentation saying it was no longer a NASA Spacelab property. We opened the box and we found a chip that was black, toasty black. The PI said, 'We don't have another military spare like that that we could take the conformal coating off, un-solder it, pull it out, put a new chip back in'. I talked to the French people and said, 'Well wait a minute, so you want a military-grade chip? What's the part number?'. So I came back to Orlando in the evening and went to Sky Craft Electronics. I walked in there and looked in their parts bins. I asked how much they wanted for a military chip and they said '69 cents'. So I handed him a dollar and walked out with a chip in my pocket. I went back to Kennedy Space Center the following day and handed it to the French people. We unsoldered the failed chip and removed it, put the new one in and soldered it into place, conformal coated it, tested the box, put the box back into the Spacelab and flew it. No documentation needed."

In this new world of spaceflight, Level-IV personnel would encounter unexpected problems and also discovered design flaws in the flight hardware. Some of that hardware had been used for years in ground applications, but thanks to the work of Level-IV, a potentially dangerous situation was found that could arise from a design flaw.

There was an issue with the type of insulation on some of the cables, as Gerry Rivera explained: "Kapton was an insulation that is no longer used on designs at the space center, but Kapton was pretty much widely used for the onboard cabling on the Spacelab, on the under the floor cabling. Kapton has a very, very unique quality if you have a short on the wire that generates any kind of heat. Some people called it a smart short, which is where you short out two wires, but the amount of current is not really enough to trip your breaker or your fuse at the source. However, with the wire now carrying current higher than perhaps it was designed for, the heat would cause Kapton to ignite. And the thing that made it unique was that Kapton would keep burning as long as you had that heat source. It would actually burn under water. It generated its own oxygen in the process. So, we had an instance when we were doing testing and one of our personnel pressed on a Kapton cable – apparently he pressed it against one of the metal brackets under the floor and it caused a Kapton short, which of course generated smoke and generated heat on the spot. A PRACA [Problem Reporting And Corrective Action] was opened against it, and of course I took care of it right away, but it was sent out to failure analysis at the O&C Building. They came up with a full report that Kapton ignition was a problem. Kapton is no longer used on spacecraft."

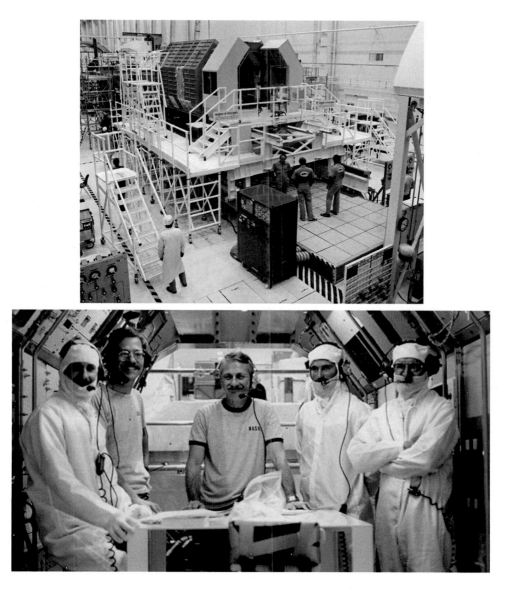

Top: Spacelab-1 Rack Train in Level-IV South Rails. Experiment rack in gray, GSE Racks 1 & 2 are tan. Blue Rack Conditioning Unit (RCU) supplied air to the Rack Train. (Jay Smith.) Bottom: Level-IV personnel with crew members inside Spacelab Rack Train during Spacelab-1 testing at KSC, c 1983. L to R: Michael Grovier (Boeing technician), Michael Lampton (Backup Payload Specialist-2), Owen K. Garriott (Mission Specialist-1), Scott Vangen (NASA Electrical Engineer) , George Bors (NASA Quality Assurance). [NASA/KSC.]

Testing the Rack Train

The Spacelab-1 Rack Train was now together, and all the experiments installed on the pallet and EBA. With the preparation completed, testing would begin to see if all this flight hardware would operate together in a flight-like scenario. Testing involved some very long hours, as Jay Smith recalled: "Really long hours of that. I remember we stayed over on more than one night and just crashed in the tech shop. We were working on stuff that had to be monitored, and so we had to be there 24 hours a day."

Donna Bartoe recalled that, "We didn't know how screwed up the system was till we started Spacelab-1 integrated testing. I remembered that the PCU was the system that was particularly broke, compared to HITS [High-rate Input/output Test System] which was pretty well in good shape. There were a lot of people they ended up having to bring back. A lot of the McDonnell Douglas people, the hardware people on PCU, and the IBM software people for sure. They had to just come back in and live with us, trying to get the system up and running. I mean, it was just awful. Nothing worked. That's when the overtime hours started piling up."

Scott Vangen revealed that, "On Spacelab-1 [and Spacelab-3], part of our challenge in addition to keeping a PCU running was the reoccurring argument about whether the RAU serial interface problem was on the experiments side or the RAU side. We ended up being called RAU defenders because in the end the RAUs were pretty robust. The experimenters got it to work with their RAU emulators, but it turned out that wasn't a sufficient enough fidelity test."

As the testing began, it led to some really long days, and many times the work would carry the personnel right through meals and into the night.

Mike Haddad remembered working Spacelab-1 integrated testing. "Everybody was hungry, so we were going to run out and get sandwiches, for all 105 of them. By mistake, we placed the order from a phone that could not be called back in on, so of course the Titusville restaurant called backed thinking the order was fake and could not get through. We got the orders and money from everybody who was working late, then Jay Smith and I jumped in a van and ran out to the place in Titusville. We got there and I went to the takeout person and said, 'We're here for the 105 sandwich order', and she said they thought that was a crank phone call so they weren't made. I reached into my pocket, pulled out this wad of money and said 'Can you make them now? We have people back there [at KSC] running testing right now that are hungry'. They almost shut down the restaurant and brought all hands on deck, you know, servers, dishwasher, anybody who could make a sandwich, and they cranked them out. It took about 45 minutes, but they cranked out the sandwiches and we brought them back to KSC and all were happy."

Offline experiment integration for Spacelab-1 started in April 1982, followed by Rack Train integration in August 1982. The Level-IV Mission Sequence Tests were completed on December 10, 1982, after which Spacelab-1 went to the Level-III Test Stand 2 in January 1983.

The astronauts would come to KSC to learn about the flight hardware they would work with in orbit, and testing was one of the operations for the crew to observe and participate in to get that knowledge of the flight hardware. Spacelab-1 Payload Specialist (PS) Byron Lichtenberg said, "Once we got into the Level-IV thing, it was really good. We sat down and had the pre-brief to understand the scope of the test and the desired outcome of things. We had the simulators, the Payload Crew Training Complex up at Marshall Space Flight Center, and that was fairly representative, but working with the flight equipment really only happened down at the Cape, there at the O&C Building. That was the first time that we had a chance to really look and see and feel and touch the flight equipment, participate in testing, and look at the stowage and things when you were inside the module. That definitely helped a lot because it gave us some confidence that we actually saw the real flight deal, working with the flight computers and everything."

Jim Dumoulin said that the first couple of Spacelab missions were the fascinating ones, especially testing the first module mission and the first pallet, the Instrument Pointing mission: "Basically, Spacelab-1 and -2, because the main thing was trying to figure out [that] when something didn't go right, where was the problem? It was a brand new checkout system, brand new ground software and brand new flight software. And for the experimenters, it was their first attempt. So, the best part of the challenge was just to be able to figure out the problem, like a detective. And then, as an electrical engineer, you could solve it with software, you could solve it by changing pins on a cable. Once you knew what the problem was, you could fix it in any number of ways. You always had a flight assignment and then you had a ground assignment. You had a GSE thing you were responsible for. Mine was the Payload Checkout Unit, and so that meant all the hardware models and simulators and everything, the parts of the checkout environment that kind of tweaked the payload and put it through its paces. Then, in Level-IV, there was always a third thing. It was always either looking forward or doing some tweak to the system, looking at a special project or whatever. So there were three things going on and there was never a dull moment."

Jim Dumoulin building a prototype in 1983 to replace the PCU Control Room Orbit Simulator Punch Card Reader. [Scott Vangen.]

To prepare for testing, there could be several tasks that needed to be performed before the actual "powering on" of the experiments. In Level-IV terms, these were known at Pre-Operations Setups and/or Operations Support Setups, which could include cooling support to electronics, sample testing of ground gases to be used during the test, or even something as simple as opening a valve on a flight component.

It's a Gas

Mike Haddad recalled that, "One of experiments inside the Spacelab module had a flight bottle that contained a pressurized gas. I can't remember what kind of gas, but it required the shutoff valve to be opened for testing then closed after testing so no more gas was used than necessary. And that was my job in support of testing; very simple. Well, after one of the days of testing I closed the bottle like before and went home. I came in the next day and opened the valve, and the bottle was empty. The German experiment engineer got very upset, stating I hadn't closed the valve properly and all the gas had leaked out overnight, so we couldn't do the test. Well, we got into an argument and I said, 'Let's take this out of the module,' so we didn't disturb all the other work that was going on. Dean Hunter heard us arguing and came over and asked me what was going on. I told him that I'd closed the valve the same way as before. It was documented, and now the experiment engineer was saying I did not close it correctly. Dean did not question me, but turned to the experiment engineer and said, 'There is a problem with your bottle or the valve'. The engineer started yelling at Dean in German, and as this was happening Dean's bottom lip started to quiver. I was told I would know when Dean was really mad because his bottom lip quivered. So I saw this and backed away, knowing it was about to get ugly. The experiment engineer finally stopped yelling at Dean in German and Dean looked at him, got right in his face and said, 'Don't you ever talk to me or yell at me in German ever again. You want to say anything to me, say it in English! Because I'll have your ass on the next plane out of here and you'll never set foot in Kennedy Space Center ever again'. The engineer said, 'You can't do that!' and Dean said, 'I sure as hell can. I approved your access to KSC and I'll take it away in a heartbeat'. That was it, conversation over, it was done. We took the bottle out and replaced it with a spare to continue testing. Working on the empty bottle, we found the valve had failed, so Dean was right. It was fortunate that the bottle valve had failed on the ground. If that had occurred during the mission, the experiment would have had to be shut down."

Testing, like almost everything else for the first few Spacelab missions, was a real learning curve. For example, the hierarchy of who was controlling the test, the "Test Director", and how the other people involved fell in this testing hierarchy; who would do what, who would be responsible, was it NASA or a contractor? All of these questions needed to be addressed and answered before the first aspect of the test was started.

"We tried to grease the skids," explained NASA Level-IV Test Director Gary Powers, "but, the McDonnell Douglas people were a hurdle that we had to cross. It took a little time, but we did it. We came to a valuable relationship where they accepted who we were after they saw mission success. They didn't like it [Level-IV] particularly, because they always thought that hands-on was theirs. McDonnell Douglas thought it was their responsibility and Level-IV was not letting that happen. After the mission successes, that sort of mellowed out. That was the beauty about it; everybody played their part and role, and everybody cared. Doing it right."

Jim Dumoulin said that, "On the Mission Sequence Test on Spacelab-1, we already had a working system, and I was really just surprised that they just let me go in and change the assembly language code of a working system. And there weren't any people stopping us from doing so. We used an API [Application Program Interface] that was built into the designs that came from Marshall, from IBM. They had assumed that maybe somebody in the future would want to do this, so they built an API, but they didn't expect anybody in NASA to write assembly code and use it. I read everything when I got there. I read every manual that they had given us. The guys that built it were old Apollo guys who'd figured that we would need this capability in the future. It helped us solve a lot of problems by having that visibility. Other than that, you could just crash and you didn't know why. And then I remembered that occasionally, and during the flight, they used the rationale we had for the paper we closed to explain anomalies on board the flight. Usually we were onto the next payload by then, but they did have them. Some folks, like Scott Vangen, got heavily involved in the flight mission stuff."

Hermann Kurscheid just wanted to make one remark regarding Spacelab-1, which was that "After the MST, KSC organized a party or get together in the recreation area, also with the families, and there were a lot of contacts made and also the people got an understanding of who was who, who was doing what. And it was very interesting to learn all that, and I think quite some friendships did exist from that time. Very important for the success that the people understood each other and got along with each other."

Some of the Hardest Workers

In those days, not many women were involved, at least on the technician side. It was just at the start of more women engineers coming through. On one occasion, Eli Naffah remembered working with one of the female technicians, Marcia Siddons, on something inside the Spacelab that was being a problem: "Nothing was matching up the way we had expected. It was a real problem trying to get into the area to install some hardware. [It was] long hours, and the technicians at that time had a union agreement where they had to take breaks, and you had to factor all that in. But this woman just really worked her tail off, and she stayed with me inside the module doing things that needed to get done to help us stay on schedule. I remember that some of the big techs, like Jay [Smith] and some of these other guys, couldn't even get their hands into a certain area to tighten something down or to install it. I remember I asked her if she would go in there and do that, and she went in there and she did such an excellent job. She was one of the hardest workers. And you know, I think I learned a lot at that point about judging people, you know by their looks, or their gender, or whatever, From that point on, I never underestimated anyone on my team and basically tried to make sure that I gave everybody a fair shot."

Mike Haddad remembered that, "Marcia was also very strong for her size, around 5 feet (1.5m) tall. I remember one time we were unloading a set of scaffolding that, for whatever reason, was stacked on-end instead of flat. Someone pulled loose the wrong tie downs and the whole stack of about dozen sections started to fall towards one of the other techs. Well, she jumped into action and was able to temporarily hold up the stack long enough for other techs to come in to help with the load. It must have been a couple hundred pounds she held up by herself."

Craig Jacobson added that "Susan [Jacobson] was both an electrical and a mechanical technician, she could do both She was one of the small ones, like Marcia. Those were the two that they always sent under the Orthogrid and stuff like that, because they were small. Somebody dropped another bolt or something down in Orthogrid, they were the ones who had to go in and get it out. So those are the kinds of things she did."

The Smarty Tour

At this point, it should be noted that Spacelab-1 was not the only payload being worked on There was also a group of flight hardware that was small in size and less complete, called Partial Payloads. "When I started in Level-IV on my first day down here at KSC, my mentor was Eddie Oscar, the world's oldest teenager," recalled Jim Sudermann. "I ended up working on a lot of work called Partial Payloads, where we put a different kind of computer on it, like the Sperry, a smart Flight Multiplexer De-Multiplexer [FMDM]. When we fired them up they literally blew up, and we had to send them back to Sperry. We made a t-shirt about it, '*the smarty, smartfignugen tour*,' where the SFMDM went from Marshall to KSC, to Marshall to KSC, to Marshall to KSC; the '*Smart-Aid '92 tour*'."

By this point, Spacelab-1 was still in the Level-IV area undergoing individual independence experiment testing before that floor would move down to Level-III/ II. At that time, there were some transient problems when some of the experiments were powered on. On looking in to it, it turned out to be an induction problem with the GSE cables that were running from the ground GSE supplies to the actual Level-IV Spacelab interface, which were the Experiment Power Distribution Boxes (EPDB). These were not run out straight, but coiled in order to cover the distance between the power supply and the physical location of the EPDBs. As Gerry Rivera explained, "The coil cable gave you an inductance effect when you first turned on the power supply, which was actually creating a transient spike that took out some of the experiments, as they were designed to do. So, we laid them down on the floor in a linear way and that certainly took care of the problem. As young engineer, probably only about 26 years old, it was my first troubleshooting effort, and boy that was a lot of lessons learned. I really enjoy that kind of work."

There were a few serious issues, and most of these were not purely electrical. A French experimenter had misinterpreted the specification on how that experiment had to interface with the basic module. All the other experiments had been interpreted the proper way, but he wanted to make a legal issue out of it and claim that his interpretation was right. Herb Rice said, "It made no sense to ask the other experimenters to change, and we had to tell him, 'Look, if you won't do that [change the interface], our only choice is to not fly your experiment'. I don't know if it was me or someone else who said, 'Well, I'll tell you what, you've got a week to see if you can make the change'. So reluctantly, they made the change, and less than a week later we were ready to test the interface again and it worked. So we flew it.

"Another serious issue dealt with the Atmospheric Emission Photometric Imaging (AEPI, 1NS 003)," Herb continued. "There was a concern that it might not come down and lock, so they wanted to be able to strap it down using an eye-bolt. Bruce Burch asked, 'Why don't you use, a 'U'-bolt? It gives you a way bigger area and then it puts the load over to different parts'. So they ended up designing a 'U'-bolt and putting it on there."

AN INTEGRATED PAYLOAD

As discussed earlier, experiment processing and Spacelab module processing were done by two different groups in two different locations, but eventually they had to come together as an integrated payload. In January 1983, the Spacelab-1 experiment Rack Train was moved from Level-IV to the Level-III/II test stands, to be integrated with the flight Spacelab module that contained all the subsystems used to support the experiment Rack Train. This being the first time a fully integrated Train was installed in the module, there were the inevitable structural distortions and some cable rerouting was necessary. Then the aft end cone was mounted to close up the module, and the pallet was put into place to complete the flight configuration.

Some (Spacelab) people in Level III/II worked the Spacelab module but not the experiments, and thought they would be powering up the experiments and taking it from there once integrated with the rest of the flight floor. Bill Jewel said, "No, no. I don't know where you got that impression." Scott Vangen remembered, "getting phone calls from people saying, 'We need to look at your experiment procedures for Spacelab-1, because we're going to be running them in III/II to test some things'. No, I don't think so. This was an example of where the Level-IV personnel role went beyond the Level-IV area, one of the reasons why the name got changed from Level-IV to Experiment Integration."

This Mission Sequence Test for Spacelab-1 was performed during March and April 1983, and simulated about 79 hours of the planned 215-hour flight. The Orbiter was simulated by GSE, with the high-data-rate recording and playback being demonstrated. The test was remarkably successful, with more than 70 experiments, the Spacelab subsystems, and the control software all working well together.

Once integrated testing was complete, the whole Spacelab-1 payload was moved from the Level-III/II stands to the Cargo Integration Test Equipment (CITE) stand on May 18, 1983, for a higher-fidelity simulation of the Orbiter interface and use of the KSC launch processing system. This was done to fix any problems between the Spacelab-1 payload and the Orbiter systems used to support the payload. During this CITE test, the data link to the Payload Operations Control Center (POCC) at Johnson Space Center (JSC) was simulated using a domestic satellite in place of the Tracking and Data Relay Satellite System (TDRSS). The CITE test was remarkably problem-free and the POCC-to-Spacelab link test-verified the command link from JSC to the Spacelab and the correct reception of telemetry from the Spacelab to JSC. The data display unit again had to be replaced, while the six-hour off-gassing test under continuous power revealed traces of a contaminant. This was solved by operating the tunnel fan and scrubber to remove the contaminant. A subsequent module leak test demonstrated a pressure decay of only one-eighth of the allowable rate.

Bruce Morris said that, "The first hands-on activity was by Betty Valentine, to remove all the non-flight hardware from the Spacelab module. That belonged to Level-IV for the outgassing tests. It was really good training to be able to do those small activities, because the hardest thing of course was learning how to write a procedure and get it signed off. All this stuff you had to know, about how to really do hands-on work versus just watching somebody else."

This prepared Spacelab-1 for the next big step, the installation of Spacelab-1 into the Payload Bay of Space Shuttle *Columbia*, which was located in the Orbiter Processing Facility (OPF). But before that move could take place, stowage and other final O&C operations were performed. The stowage was not final stowage for all the items. The ones required to be stowed as late as possible in the Spacelab-1 module would be performed during final stowage for flight while in the OPF.

Hermann Kurscheid recalled that, "O&C activity was stowage. We had to do a lot of stowage and testing. There was a lot of fit checking and final stowage. Ron Woods was there from JSC, and he needed to get all the crew stuff together. From our side it was Rudy Vogel, and he was also very busy there to get it all together."

Closeouts complete, the Spacelab-1 payload was moved into the Payload Canister and then transported to the OPF on August 15, 1983.

In the Orbiter Processing Facility

After arriving at the OPF, the Spacelab-1 module and pallet were removed from the canister and installed into *Columbia* on August 16.

Finally, the Spacelab-1 payload and Orbiter were now together. Preparations were conducted for the final testing, Three specific and important tests were conducted during the next month:

1) The Spacelab/Orbiter interface test to verify power, signal, computer-to-computer, hardware/software and fluid/gas interfaces.
2) The Spacelab/tunnel/Orbiter interface test to verify tunnel lighting, air flow, and VFI sensors; and
3) The end-to-end command/data link test of the Spacelab/Orbiter/TDRSS/White Sands/Domsat/JSC/GSFC link.

A few problems encountered were quickly solved. With this being the first Spacelab module to arrive at the OPF, there were some learning curves and other tasks that had to be worked. Lori Wilson recalled that, "The OPF workers weren't used to us at first, but they came around real fast. Remember, they'd never had a special group come in to work on a payload *in the Orbiter*. When they saw the professionalism of our people, and the unbelievable complexity of the experiments, they backed off and supported us with whatever we needed. In turn, we showed them that we understood the dos and don'ts of their Orbiter requirements. I really don't recall any specific problems with anybody there."

As testing began, a problem was found with RAU 21 on the Spacelab pallet. This required access to get in and replace it, but the problem there was no access; even the special access equipment at the OPF would not work. What could be done? RAU 21 was a McDonnell Douglas box they had put on in Level-III/II, but the blanket was the European blanket that Level-IV had put on. Bruce Burch explained: "We put a web strap on the big overhead crane, I put a fall protection harness on, and they actually picked me up with a web strap and the big crane and lowered me over in the pallet near the Orthogrid. They lowered me between the Orthogrid and the Orbiter aft bulkhead, [where there was] just enough room to fit. I wouldn't fit now, but there was just enough room to get down there. Then I got unhooked from the crane and got on the pallet, going over to take that blanket off so that they could come in and do some repairs on it. Bill Mahoney was there, and the Rockwell guy that controlled the OPF said, 'You're not doing that in my building'. Bill Mahoney told him we were, and called the head of NASA Safety for KSC. He was out eating with his wife at a restaurant. Bill told him what he wanted to do and they let us do it. I've never seen anybody else do that before, with the regular fall protection harness being lifted up by crane and a web strap."

"I was a task leader for the lifts," added Frank Valdes. "We actually tested it out in the [old] ATM [Apollo Telescope Mount] room with a crane, and we actually lifted Bruce up in the harness. We tested that, and then we had to cut the bunny suit to let the hook through the bunny suit for the harness. Then Bruce was complaining about his leg straps being too tight, but it was the safest way to lift him. He did a real good job."

Another problem that had to be taken care of was that the overhead bucket that was used to gain access to the payload was leaking oil. Because of its location, it could not be fully placed outside the Payload Bay. Frank Valdes explained that "Vicky Johnson and Lori Wilson were the Ops people, and they were there whenever the payload doors were open, of course. But when the bucket mechanism started leaking hydraulic oil onto the Spacelab-1 pallet, we had to go in there and build a plastic tent over the pallet." Bruce Burch took up the story: "So they got me involved, [Walt] Preston and I think Jay [Smith]. We laid out that pink [plastic] stuff and taped it all together, because at the time that was the only anti-static plastic we had. We had a rope going across, and then we had to tie a rope down from the 13 platforms, and then we put ropes on the side to keep it up off the payload." Frank Valdes added, "The oil leaked onto some of the flight blankets, so we had to clean that up with distilled water and laundry rags. I remember we washed the rags, then used a Laminar flow bench on a room right outside the door. I thought 'This would be a great place to dry the rags'."

Maurice (Mo) Lavoie recalled that for OPF work, they wanted to verify that all of the Spacelab hardware would communicate properly with the Orbiter through the Pulse Code Modulator Master Unit, so that they could get telemetry information to and from the S-Band system for command purposes. They also did End-to-End testing. They got JSC to send information through their S-Band system to the Orbiter while it was in the OPF, to command Spacelab to do certain functions, and received proper responses through the S-Band. They also did Ku-Band testing to where experiments were driving their data stream through the high rate multiplexer into the Ku-Band system. Mo Lavoie said, "I got to suit up and be inside the Space Shuttle, crawl around through the tunnel and into the Spacelab, operating the display systems while we were doing some testing. For between 18–20 hours, where you worked through the night, you were out there on the Orbiter schedule. You could take a break and you could go out to the car and sip coffee or something. It's amazing how Coca Cola® and peanuts can keep you awake; between the caffeine and the stimulant and the protein, it would keep you going."

From OPF to Pad via the VAB

Once OPF operations were complete, the Orbiter would then be taken to the Vehicle Assembly Building (VAB) to get rotated to vertical and integrated with the External Tank (ET) and Solid Rocket Boosters (SRB). Finally, the whole assembly would head out to the launch pad. The Orbiter was moved to the VAB on September 23, and the assembly rolled out to the launch pad on September 28. Unfortunately, it was rolled back in October, de-stacked and sent back to the OPF due to booster concerns. The revised flight date finally chosen was November 28. The Orbiter was returned to the VAB for a second time on November 4, and the Shuttle was rolled out to the pad again on November 8.

Michael Stelzer revealed that, "There was no planned access for the Spacelab-1 (STS-9) module at the launch pad. However, we needed to ensure that the MVAK [Module Vertical Access Kit] equipment worked should any issues have arisen with the Spacelab-1 module systems during testing at the pad. Fortunately, no issues appeared in testing, and MVAK was not needed for the Spacelab-1 mission."

With no internal closeouts required, attention turned to external closeouts on the pallet. Level-IV personnel performed final closeout operations at the pad in the

PCR before door closure for flight. Everything was now ready to fly. All that could be done had been done to prepare the Spacelab for its first journey into space. When the Orbiter Payload Bay doors closed, this would be the last time the payload would be seen until it reached orbit. It was a very emotional event.

Level-IV personnel supported launch operations preparation from the C-1 Payload Console in the Firing Room at Launch Control Center (LCC). Not just anyone could get into the Firing Room for launch. Those who would be "on console" for launch needed to participate in the Terminal Countdown Demonstration Test (TCDT), S00017 (*Sue-17*, as pronounced at KSC). This was a dry run of launch day operations, where the crew suited up, went to the pad, climbed into the Orbiter and were strapped in, the pad was cleared, and the clocks counted down to just a few seconds before launch and then stopped. The crew then got out, went through an emergency egress operation and, if all went well, the launch team and crew were ready for launch.

As the launch date approached, articles in various scientific, aerospace, and popular magazines gave the upcoming mission mixed reviews. Certainly it was recognized as the most ambitious Space Shuttle mission to date, characterized by the following "firsts":

- First flight of the European-built Spacelab (development cost approaching $1 billion)
- First flight of non-career astronauts (Payload Specialists Merbold and Lichtenberg)
- First flight on a NASA mission of a non-American (Merbold, a German)
- Longest Shuttle mission (planned for 9 days)
- Largest Shuttle crew (Commander, Pilot, 2 Mission Specialists, 2 Payload Specialists)
- Two shifts for 24-hour operations
- First operational use of the TDRSS

Time Magazine described the Spacelab as "an outsized thermos bottle" and the mission as the "first real marriage of space engineering to fundamental scientific research in the manned space effort." It recognized the Spacelab as the first true scientific research station in orbit and a major advance over Skylab – and probably over Salyut, the Russian space station in orbit at the time [1].

As launch day neared, Dean Hunter wanted to make sure everything was as perfect as possible. He approached Tony Ornelas and asked if there was anything on his mind that he was not comfortable with. Tony responded that when they installed the ISO experiment, Jay [Smith] thought maybe he had not torqued one of the 24 bolts all the way, even though the documentation showed all was good. Tony remembered: "Dean did some quick analysis and thought the remaining hardware would support [it]. We were just going to have to let it fly. But that weight on me… thinking, 'What if this damn thing failed, the whole Shuttle could blow up'. [But] We flew it and everything worked." Postflight checks of the bolt showed it had been fully torqued, but this highlighted why all the work needed to be documented and no corners cut, even though the pressure to keep on schedule was huge.

LAUNCH DAY

November 28, 1983 was launch day. Very little had be done other than to monitor Spacelab and the experiment systems during launch countdown. The Shuttle launch team was going through the final steps of the launch procedure, S0007 (pronounced *Sue-7*). The reference tool for conducting a Shuttle launch countdown was a five-volume manual (in 1983 that is; a sixth volume was added later) encompassing some 5,000 pages of instructions. The Shuttle Countdown Manual was the lengthiest Operations and Maintenance Instructions (OMI) used at KSC to document procedures for assembly, processing, testing and launching of the Shuttle. The S0007 procedure was reviewed and updated prior to each mission.

Jim Dumoulin said that, "During missions, of course, we always had an audio loop to listen in and monitor the mission. You wanted to see how your payload was doing. So if somebody perked up and had mentioned a problem, we were kind of researching it before we got a phone call from JSC or from Marshall, so that we could know how to answer it."

The First Spacelab Flies

As the clock continued to count down toward T-0 (liftoff), the world was listening to NASA FAO and launch commentator George Diller calling out the last few words before the Shuttle left Earth for its voyage into space. The tension built as he stated, "T-6 seconds… go for main engine start… 3…2…1… solid motor ignition and liftoff, liftoff of *Columbia* and the first flight of the European Space Agency Spacelab… The Shuttle has cleared the tower." Co-author Mike Haddad remembered, "Yay! Our job of preparing it for launch was now complete. As *Columbia* climbed to orbit all was well, with SRB separation at T+2:07 (minutes: seconds), main engine cutoff (MECO) at T+8:29 and ET separation at T+8:47. After a couple of Orbital Maneuvering System (OMS) burns, Spacelab-1 was on orbit ready for activation. Hopefully, all the hard work would now pay off as the experiments were powered up for on-orbit operations."

Maurice (Mo) Lavoie said, "RAU 21 was the one that my wife got a call about at 2:30 in the morning. She said, 'Some Air-to-Ground guy wants to talk to you. It's Houston Capcom, Mission, and do you mind talking to some of the astronauts about RAU 21?' It was toward the back of the Spacelab-1 on the pallet, and it appeared that whenever they rotated into the sunlight, they would get dropouts from RAU 21, bit drops, and then they'd rotate back into the cold and the unit would work fine. It was not getting enough cooling from the cold plate, so we had to do some modifications of procedures after Spacelab-1 to make sure that we had a nice clean surface, a nice contact-to-contact."

For the flight crew, it was important to discover whether anything that had happened during the pre-fight activities at Kennedy, or things they had learned during that interaction, would help during the mission. Spacelab-1 PS Byron Lichtenberg reflected, "Thinking back, I think I could say that just [having an] understanding how Spacelab's systems, computers, CDMS and everything worked, and actually seeing the real displays and understanding the time lags and things, when you would roll in a subroutine or different things, or a sub timeline, that gave us pretty good confidence and knowledge on board. So I think that [the work during Level-IV] was helpful for sure. Clearly, any time you can interact with the flight hardware and interact with the folks that are intimately involved in it really helps your confidence level. It makes a big difference in your ability to kind of think ahead in flight. I would really just like to give Kudos to all the ground folks that worked there at KSC. I mean, sometimes it would take hours to go through a series of steps that, if you did it real-time just following crew procedures, it was done in minutes. Everybody was really professional and I made some, really, really good friends there while we were doing that."

Hermann Kurscheid told the authors that, "Some of our guys went to JSC, but I completed preparations for the return and the Baseline Data Collection [BDC] or hardware receiving at Dryden. I was only a short time at JSC, and then I traveled to Dryden Air Force Base in California, to make sure the PIs got their experiments back as early as possible, because there were also the live specimens in there."

Most of the previous discussion dealt with the engineering aspect and technician work, but just briefly mentioned here is the work of Level-IV Quality. As Bob Raymond, NASA Level-IV Quality, explained further: "QA [Quality Assurance] was a hands-on inspection, working hand-in-hand with engineering and technician staff to make sure the product was put out correctly. The engineers and techs did the work; my job was to go in and make sure it was done per specification and stamp [the paperwork]. There would be a technician stamp and there would be a QA stamp on critical items. Not everything was on critical items."

Eli Naffah added, "There was a Quality guy that I used to work with that would stay late with me inside the Spacelab module, working real late trying to get something done long after the techs had gone home. I remember that we had the authority to go in and do the work, even if the technicians weren't there. We all worked hard together and we played hard together. We were very cohesive unit."

Popcorn, Paper Fish, Beer and *Monty Python*

What follows is some general information regarding things that occurred during Spacelab-1 processing, but then carried over into most of the missions worked by Level-IV.

Taking breaks was important to keep from being overwhelmed, and there were many different types of breaks that the Level-IV organization performed. G e n e Krug recalled, "We got popcorn every afternoon because [Bill] Mahoney believed in the breaks and everything, but he liked the money. So I knew an old boy in NASA Quality at the VAB from Titusville. He could get 50-pound bags (of corn) and I would get the peanut oil you can get for deep frying. I got my paper bags out of supply; they cost a penny, I charged them a quarter. People would show up about two o'clock [in the afternoon]. One thing Bill always said was, 'Your work's got to stay on schedule. Keep on schedule, you do whatever you want to'."

There were also traditions or routines that were implemented, and each country had their own which made for a real variety of events that would occur at KSC. Scott Vangen recalled: "On Spacelab-1, the User Room, which was on the fourth floor of the O&C, had all of the experiment teams, the PI and their respective support team. So, on holidays for their country, they would do things and then invite people to participate. In any given test week, there was probably one nation or one payload or two that was celebrating some oddity that many of us domestic, young engineers probably didn't know about. For example, the French had something they called the 'Fish Day'. They suddenly showed up at the User Room and there were these little cutout paper fish that were appearing everywhere. The Germans weren't shy about their beer, nor the other Europeans about their likes and dislikes with American cuisine. The English would get into some *Monty Python* routines and the next thing you know, they'd be putting on Gumby hats."

Jim Dumoulin recalled that, "After the successful STS-9/Spacelab-1 launch, ESA threw the NASA team a spectacular launch party at a huge Spanish mansion on Merritt Island called Hacienda del Sol. ESA flew in a large transport aircraft onto the Cape Canaveral skid, which was full of shipping containers to load up all the prelaunch test equipment. As it turns out, the 'empty' Spacelab shipping containers were full of gourmet food, wine, cheeses and Becks Beer direct from Europe."[1]

On January 19, 1984, Spacelab Program Manager John Thomas presented a preliminary report on the Spacelab-1 mission results to Administrator James "Jim" Beggs, followed by a preliminary science report from Mission Scientist Charles "Rick" Chappell. From the Spacelab viewpoint, it was hard to visualize a more successful mission report. All objectives were successfully achieved, including 17 detailed test objectives, 81 functional objectives, and 361 verification measurements.

The success of this first Spacelab mission was a crowning achievement for the Level-IV team, resulting from all the hard work, long hours, tough decisions, daily problem solving, and establishing new processes and procedures.

---

[1] Years later, the Hacienda del Sol mansion and the surrounding ten acres of land on Merritt Island was sold. The new owner was none other than Dave Thomas, the founder of Wendy's.

ALL CHANGE AT THE CAPE

NASA changed its flight designation system starting in 1984, in response to the growing complexity of its launch manifest. The new system of designation tried to pack more information into the flight number than a simple sequential ordering could. The "STS" in STS-41B, for example, still stood for Space Transportation System. The first number represented the Fiscal Year (FY) in which the Shuttle launched ("4" for 1984 in this case). The second number was originally intended to designate the location from which the Shuttle would launch. The number "1" would stand for Kennedy Space Center, while number "2" would have denoted Vandenberg Air Force Base (AFB), although this was never used. Finally, the letter in the designation marked the Shuttle's launch sequence for that Fiscal Year – "B" in this case denoting that it was the second planned launch for FY 1984.[2] This confusing system would last until 1986. When flights resumed in 1988, NASA reverted to simple flight numbers.

STS-41D, *Discovery*, launched August 30, 1984, 6+ days
Payload: OAST-1 (Office of Aeronautics and Space Technology-1); and MPESS (F004).

Primary payloads for this mission were three commercial communications satellites that were deployed successfully and became operational but did not have any Level-IV involvement. Another payload onboard, which at that time fell under the Level-IV heading of Partial Payloads, was the OAST-1. The Solar Array Flight Experiment (SAFE), was a device 13 feet (4.0 m) wide and 102 feet (31 m) high (when extended), which folded up for launch into a package 7 inches (180 mm) deep. The array carried a number of different types of experimental solar cells, and at the time it was the largest structure ever extended from a crewed spacecraft. It demonstrated large, lightweight solar arrays intended for use in building large facilities in space, such as a space station. Middeck payloads that flew and had little or no Level-IV involvement were: Continuous Flow Electrophoresis System (CFES) III; Radiation Monitoring Equipment (RME); Shuttle Student Involvement Program (SSIP) experiment; IMAX camera, being flown for a second time; and an Air Force experiment, Cloud Logic to Optimize Use of Defense Systems (CLOUDS).

The OAST-1 MPESS and structural/mechanical MPE were delivered to KSC on January 10, 1984. The solar array hardware arrived at KSC on February 1, 1984, for checkout and installation for a launch.

---

[2] The United States Fiscal Year runs from October 1 to September 30.

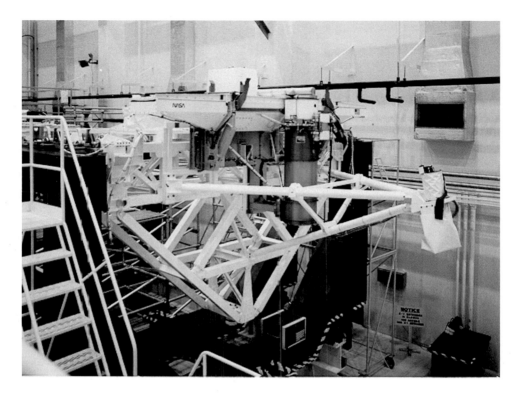

Office of Aeronautics and Space Technology – 1 (OAST-1) payload in Level-IV area of O&C. [NASA/KSC.]

OAST-1 was basically a box containing dummy solar arrays made out of Mylar sheets, which would open up and extend in space to about 102 feet (31 m). Each Mylar sheet bore round, white, disk targets attached to each of the panels. A laser at the base of the OAST-1 experiment was pointed to scan those targets, so that as they extended it would provide an indication of torsion and flexing levels. Once the panels were extended, the thrusters on the Orbiter would be fired to induce motion on that solar array, and again the laser would map how much torsion was being induced. The actual physical extension and retraction of the panel was done in California. Gerry Rivera: "At KSC, we integrated the panel to the carrier, to the MPESS, and we ran an interface test, which was like a loop-through test where we would put a dummy plug that would simulate what was coming from the extended arrays, and verify our controls from the standard switch panel."

Maynette Smith said, "We inherited a system known as PITS, or Payload Integration Test System, which was really a very primitive kind of system. But we used it to process OAST-1, Large Format Camera, OSTA-3, ORS – all those Partial Payloads went through PITS. PITS was finally decommissioned and replaced with the Partial Payload Checkout Unit.

Testing Solar Array Experiment (SAE), OAST-1 (STS-41D) with Payload Integration Test System, aka PITS. Sitting at the terminal is Juan Rivera (left), with Maynette Smith (middle, hands on keyboard), and Ed Oscar (right). [Maynette Smith.]

"The first experiment I tested was the Solar Array Experiment and the Dynamic Augmentation Experiment [SAE/DAE] on OAST-1," Maynette Smith continued. "For a brand new engineer, they did not have any training for this. They just said, 'Okay, you're assigned, you go'. I went out to Sunnyvale, California, to watch them unfold it in the zero-g test stand. Judy Resnik was the astronaut that was going to be working it, so I got to meet her out there. Funny story is, I went out very dressed up, and I had longer hair and bangs, and I walked in and they said, 'Are you Judy Resnik?' I said no, but I wish I was because I looked like her. She was fantastic, because it was her first flight on STS-41D, and she treated me like I knew everything, worked with me, and called me all the time checking on procedures. We had to go through all kinds of approvals to activate the laser because it was in a controlled environment."

Mike Haddad recalls, "I met Judy Resnik during some of her trips to KSC for OAST-1 testing. We got to meet many crew members even though we weren't working their specific mission, because we had so much hardware being assembled in the Level-IV area for different missions. It should be noted here that the Level-IV group worked very closely with the astronauts, but the unwritten rule was to stay professional with the crew. Do not try to get pictures with them or ask for autographs, etc. In other words, don't be a tourist; they get that all the time. They came to Level-IV for work and a little fun. Of course if they offered, it was ok to accept.

The Schlierf Box and NASA's Standard Initiators

"We had a simulated universal time clock and it was glitching, not very consistently, so it was hard for them to get the trigger on when it was a problem," Roland Schlierf remembered. "Craig got me in there, explained to me how the Spacelab time clock worked, and he and I put together this box. Craig was the brains behind this, I was a Co-op. It took three Co-op terms to design, fabricate and use this box. Basically [it was] plugged into our time clock to get it to trigger off in an O-scope, so that when the error happened, we had captured the time and we could look at it and understand what was going on. Scott and Craig were able to understand what it meant; I wasn't smart enough to understand what that meant. I think you'd be hard pressed to find another place in NASA where a young engineer could come in and do something like that. That was meaningful. That was my introduction to Level-IV."

When something was deployed in flight beyond the envelope of the Payload Bay doors, there was always the risk that something might fail, preventing the Payload Bay doors from being closed to come home. Such payloads were equipped with NASA Standard Initiators (NSI) which were explosive devices. The crew could blow these NSIs, grab the payload with the manipulator arm of the Shuttle, and then release the payload in space. Gerry Rivera explained: "The NSI was a very unique device. In Level-IV, we would verify that we had the right amount of power at the NSI connector with it de-mated from the NSI. We put a Faraday Cap on the NSI to make it impervious to any kind of ambient RF energy. Once it got installed in the Orbiter, once that cable was mated to the NSI, we had to run a circuit resistance test through that NSI. The NSI testing was the most critical and the most meaningful, because of the safety concern for the whole [flight] crew."

An Attempted Launch

The launch attempt for STS-41D on June 25 was scrubbed during the T-9 minute hold due to a failure of the Orbiter's back-up General Purpose Computer (GPC). The launch attempt the following day was aborted at T-6 seconds when the GPC detected an anomaly in one of the Orbiter's three main engines. Two of the main engines – No. 2 and No. 3 closest to the aft body flap – had blazed to life, but the No. 1 engine, directly at the "top" of the pyramid, had failed to ignite.

Mike Haddad recalled: "While sitting at the C-1 console for launch, you couldn't see the pad, so each console had small TV monitors to view different areas of the pad from different cameras located on the pad structure and around the base of the pad. During those last few seconds before launch, your eyes were glued to those monitors and you could hear the countdown over the headset you were wearing. Well we heard, 'Go for main engine start' and we could see what looked like a normal three-engine start. But all of a sudden we heard 'We have a cut off... NTD we have a RSLS [Redundant Set Launch Sequencer] abort'. The two engines running came to a screeching halt. I will never ever forget that sound, it was worse than fingernails on a chalk board. We were all stunned. What just happened? But we could see on the monitors all looked ok, the Shuttle was still there, the engines were off, or at least we could see they were not running even though on-board they did not have an indication Engine No. 1 had shutdown. Then we heard that pad sensors had detected a fire at the base of the Orbiter near the three main engines. The engine engineer requested to turn on the heat shield fire water, spraying water up in the vicinity of the engine bells of *Discovery's* three main engines. A hydrogen fire burns clear; you can't see it. So now we had an invisible fire burning near or on the Shuttle. The launch team with the crew worked real hard to get everything under control. The fire was extinguished and then the crew was safely removed from the Orbiter and off the pad. One day in my NASA career I will never forget."

Discovery returned to the OPF and the No. 1 main engine was replaced. To preserve the launch schedule of future missions, the 41D cargo was remanifested to include payload elements from both the 41D and 41F flights, and the 41F mission was cancelled. With the Shuttle restacked and returned to pad, the Launch finally occurred on August 30, 1984.

Once in orbit, there was a planned series of operations to test the Solar Array: extensions to 70 percent and 100 percent; retractions to 70 percent and 0 percent; extension back to 70 percent; then the final retraction back into its container for the ride home. Extensive Level-IV testing of the payload before launch once again contributed to successful operations on orbit.

STS-41G, *Challenger*, launched October 5, 1984, 8+ days
Payload: The Office of Space and Terrestrial Applications-3 (OSTA-3), Large Format Camera (LFC), Orbital Refueling System (ORS); one Pallet (F006), one MPESS (F002); and a special deployable Earth Radiation Budget Satellite (ERBS).

The OSTA-3 payload was designed to conduct experiments in Earth remote sensing. This experiment payload consisted of: a Shuttle Imaging Radar (SIR-B) for studies of the Earth's surface; a Large Format Camera (LFC) for cartographic mappings of the Earth; a Measurement of Air Pollution from Satellite (MAPS) experiment to determine the distribution of CO in the atmosphere; and a Feature Identification and Location Experiment (FILE) for classification of surface materials. The SIR-B was an upgraded version of the SIR-A flown on the OSTA-1 payload during the STS-2 mission. The MAPS and FILE sensors were re-flights of those same instruments on the OSTA-1 payload. Components of the Orbital Refueling System (ORS) were also connected, demonstrating it was possible to refuel satellites in orbit.

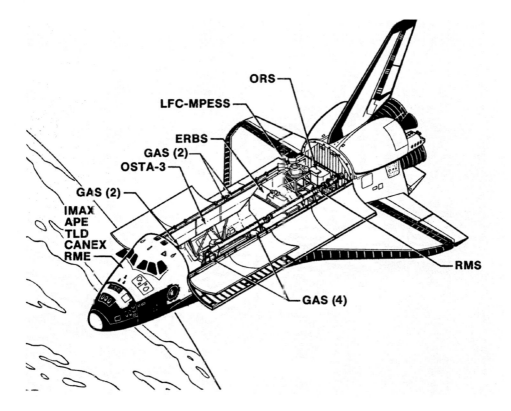

Diagram of STS-41G payloads. [NASA]

Integration for this flight consisted of both a pallet and MPESS being assembled, integrated and tested in parallel. Those two large structures basically came in stripped down, and had to assembled and integrated from the bottom up.

Mike Haddad noted, "I had the responsibility to install a portion of the OSTA-3 pallet Freon® cooling loop. Freon® 21 was to be used in Spacelab systems but was very toxic, so they switched to the more benign Freon® 114. This hardware needed to be installed first and leak checked before mounting the remaining components. The cold plates used to cool electronics and such had been installed already, and my job was to connect the fluid lines and fittings that tied all that hardware into one long cooling loop.

"At that time, we were using the cheaper, lower tolerance AN type fittings that required a metal seal between the lines and fittings. Because they were AN fittings, some just would not stop leaking, so Walt Preston, the SEIS tech that was helping me, devised a solution. He said, 'Let's just coat the seal with flight Bracote 601', a grease used in vacuum systems. Well, he used his finger to apply a precise amount to the sealing surface, then re-assembled the fittings, and by gosh there were no leaks. The Bracote worked. Now this would not be used for long-term systems, but with Spacelab being in space for a maximum of two weeks, there was no concern sealing the surfaces this way."

With the flight system together, there was the question of what GSE would supply the Freon® to the flight system to support testing. Bruce Burch added: "There was an old cooling unit that originally used a poisonous Freon® 21 gas, and they converted it to run on Freon® 114. Gary [Bitner], Debbie [Bitner], and I think [Don] Doby, helped rebuild this whole thing. We had trouble with it when we put it on OSTA-3 [41G], because we couldn't get a flow rate. So we started taking things apart, and found that we had a piece of tape over one of the inlets to one of the flight cold plates, which was acting as a filter and it had kind of clogged and backed up."

Green Freon® cooling unit, with (l to r) Gary Bitner, Debbie Bitner and Bruce Burch. [NASA/KSC.]

Mike Haddad recalled the difficulties that having a young family played on the workers at the Cape: "George Veaudry was one of the few people that was married and had kids at that time, and I remember we were down in the High Bay and some problem came up, and he just got so frustrated. Finally he said, 'Damn it, I haven't seen my kids in two weeks. I come here early in the morning, they're asleep. When I leave and go home late, they're asleep. I want to go home to see my kids!' I thought it was the coolest thing, that he finally had enough. We were just working nonstop. So, with me being single and having nobody to go home to, I said, 'George, I'll take over for you. Go home, see your kids'."

SEIS Technician Gary Bitner recalled: "I got put on the fluid stuff. We had a little problem, every night about 5:30pm or six o'clock, in that the temperature would go up on our cooling units. I remember we did a little bit of investigation, and figured out that they had two [facility] chillers running and they all backed off in the evening. So we talked to the chiller people for them to notify us whenever they were going to do it, and then we opened up the valves to get a little more water."

OSTA-3 installation of SIR-B antenna by Luis Moctezuma (on right, with beard). [NASA/KSC.]

Preparing SIR-B

For OSTA-3, Luis Moctezuma did the installation and alignment of the SIR-B antenna, as well as the SIR-B electronic boxes: "The installation and alignment of the antenna to the pallet turned out to be rather difficult and very tricky. The plan put together by the Payload Developers involved the use of an optical flat installed by them on the antenna, and already measured to the antenna frame during the assembly. This would have made the KSC work very simple, since our job at KSC would have only required taking measurements to the optical flat. As we started taking measurements at KSC, we noticed that the quality of the optical flat was not very good, and that we weren't getting good enough repeatability on our measurements. We ended up having to take measurements to the antenna structure, a very difficult task given the bulkiness of the antenna and the lack of visibility of the targets that defined the antenna's longitudinal axis with a single instrument."

Tony Ornelas added, "Luis was doing an integration, where he had a bolt on there and then he torqued it up, but couldn't get the holes align to install the cotter pin. I had the idea to just change out the nut and bolt. [We] Tried two different combinations and it worked. That was really a high satisfaction moment for me."

"Ed Karo and his team were there for SIR-B," said Maynette Smith." He was the PI and he worked SIR, which was on OSTA-1 (STS-2), and then worked SRTM (STS-99). They actually did not broadcast the antenna, but they had somebody checking to make sure it wasn't radiating, We didn't actually test that we could move the antenna itself, and that was one of the things that we found. I think the history was that nobody thought about the design on-orbit; they didn't think about the impact of me having to integrate, test, and maintain it in a one-g environment. And this was another one where you had to make allowances for that."

Michael Stelzer was hired by NASA as a permanent employee, and was assigned to perform the Weight and Center of Gravity (WT&CG) measurements for the OSTA-3 payload of the STS-41G mission. The WT&CG of payloads installed into the Shuttle Orbiter affected the way the Orbiter flew, so it was important to get this correct. Michael Stelzer added that, "Jeanie Ruiz had performed Weight and Center of Gravity determination for a previous payload and helped me with the procedure. The effort involved hoisting the fully integrated and tested payload a multiple number of times to get the measurements, and then performing a complex set of calculations with those measurements. Hoisting was one of those operations that could go well or, if something was missed, go very bad very quickly. As an engineer a few months out of college, having the responsibility for hoisting a very expensive payload that a lot of people had worked on for a very long time was huge. I was fortunate to lean on the experience of Jeanie and a very competent set of technicians that made the effort successful. However, my 'pucker factor' was up there. OSTA-3 and the STS-41G mission went on to fly a successful mission in October 1984."

The Orbital Refueling System (ORS) would contain the very hazardous hydrazine, so special training would be required to load this fluid before launch, as explained by Bruce Burch: "Me and Jay went out to Houston and one of the things we observed was they did not handle their Self-Contained Atmospheric Protection Ensemble [SCAPE] suits very well.[3] They were kind of thrown in the corner, unlike at KSC where they were properly stored and maintained."

---

[3] SCAPE were protective suits that covered the person's entire body and were used to protect people during extremely hazardous operations.

Damon Nelson explained that he was hired on full time with NASA KSC, as an Orbiter Auxiliary Power Unit (APU) systems engineer. "And then, when I moved into the Level-IV group, they had just announced this payload called the Orbital Refueling System that just happened to use excess APU parts. The idea was they were going to use APU tanks to demonstrate this hypergolic fuel transfer on orbit. It seemed kind of a natural fit that I had that APU background, so that was my first assignment, working the Orbital Refueling System. It was the STS-41G mission that it was going to fly on, and it was going to utilize an EVA, a spacewalk, the first to involve a female astronaut, Kathy Sullivan. So the experiment was headed out of Johnson Space Center, and again, interestingly enough, the people that were responsible for the experiment and the Ground Servicing Equipment were some of the same guys that I had worked with back in my Co-op days out at the thermal chemical test area. I did travel three or four times out to JSC as a Systems engineer to participate in their fueling operations, which they did on site. I got familiar with the Ground Support Equipment and worked very closely with the development engineers. I was going to be the sole source for KSC expertise on this particular experiment."

With the arrival of the flight hardware, an unplanned event occurred, as Damon Nelson explained: "We had a plan to have this equipment decontaminated as much as possible, because part of the effort was to briefly bring this equipment into the non-hazardous Operations and Checkout Building." Jay Smith remembered asking, "Does this still have hydrazine in it? I mean [I need] no face shield, gloves or anything?' They said, 'No, it's fine', so I said 'Okay'. I remember taking the wrench and loosening the fitting, and the first one that I pulled off, hydrazine poured out of it. I just walked away."

Damon Nelson continued: "Needless to say, our world got very complicated. We were not in SCAPE suits at the time because we had not anticipated getting any of this gas like we saw come out of the system. So we immediately shut everything down for eight hours, working with Safety, jumping through numerous hoops. Finally, we were allowed to bring the guys back in, with some protection and with the proper fans and such to vent the area, and we were able to make that connection. But that was a very crazy day."

Integrating the Payload

While Damon was responsible for the fueling of ORS, Mike Haddad was responsible for all the integration of the subsystem hardware that supported ORS, as well as the Large Format Camera (LFC) and its subsystem hardware, which was another experiment that would fly next to ORS on an MPESS.

Mike Haddad recalled, "About 6−8 months before payload arrival, it was my job to review the drawing, figure how best to integrate all the ORS and LFC flight hardware onto the MPESS, then write the procedure that would perform the work. The MPESS had arrived weeks earlier, basically empty, and was placed into a support stand in the Level-IV area of the O&C. For ORS, the main support structure fit snug around the box section of the MPESS, and the ORS would be placed directly on top. The LFC was little trickier. Due to its large size, a special cantilevered support structure would be mounted to the forward of the MPESS that LFC was placed in. All that integration went smoothly. Next came all the cables, brackets, electronic boxes and the LFC GN2 pressure system. All but the cabling went as planned. I really should not say this, but during integration of LFC/ORS cabling onto the MPESS, some of the cabling had been previously installed, but I noticed it was not per print. We were running up against a wall on time, and it was a very simple fix of moving the cable to the other side of the clamp. Normally, we would have to stop what we were doing and go open up a PR. It would have taken half day. The QA guy I was working with at the time was one of the best, Ed Grabowski, and I said, 'Ed, if you turn your back I can get this fixed in five minutes'. And because we were returning to print, it was not a change, just a fix to make it right. So he turned his back to review the upcoming steps of the procedure and I fixed the problem. I think after that we suggested to some of the designers to put a footnote on the drawing that it was ok to flip the clamps. That way, we could do the operation without taking paper on it."

Mike Haddad (left), leading ORS installation onto MPESS. The structure that would hold the LFC is in the foreground. [NASA/KSC.]

The ORS could now be lifted into place onto its support structure. The forward section of the ORS contained all the EVA tools and other hardware that would be required for on-orbit operations. To provide a secure location for the astronaut to stand on while accessing the area, a special device called an Adjustable Portable Foot Restraint (APFR) needed to be installed, but because there was no way to calculate the exact position of all the APFR hardware, the last support to be installed had to be match drilled. Mike Haddad explained: "Match drilling seems straightforward, but we needed to make sure the particles coming off the drill bit did not contaminate the rest of the hardware around that location. So I had to tape off the area and, using sheets of plastic, isolate the drilling location from the rest of the flight hardware. My main concern was I really had only one shot at getting this correct, and if I screwed it up major de-integration of the payload would occur. But all went according to plan." With ORS installed, the LFC followed next, and it, too, was successfully installed.

Not all went well with testing, however, as Gerry Rivera explained: "When we went in to hookup our GSE cabling to the LFC, we found that some of the pins on the LFC side − it was one of the data connectors − were either recessed or bent. The whole section of the payload had to be disassembled so that connector could be accessed and replaced. That was a huge effort."

Working Long Hours

Mike Haddad recalled: "I remember we were testing, and sometimes the door on LFC [to expose the lens] would open, sometimes it would not, and sometimes it would open halfway. So we spent the weekend taking the camera on and off, and electrical engineer Gerry Rivera was responsible for trying to find the problem. I think in the end it was a grounding problem or something, in one of the flight electrical boxes on the payload. I remember we had to pull off their electronic box from the bottom of the MPESS, which was very tricky because you were basically working upside down. Once off, it was turned over to the PI and they had to go tear the box apart. They fixed the box and then I reinstalled the box back on the MPESS. It was commanded three or four times, and every time the door worked fine. It was fixed and ready to fly."

Maynette Smith continued: "I went to JSC for some testing with it loaded with hydrazine. For Level-IV testing it wasn't loaded, but you had a lot of software and we had residuals and so we kept three valves closed at any one time to ensure that we weren't releasing any hydrazine. Basically, three valves would have to fail for hydrazine to be released. For OSTA-3, there were timing issues between the flex MDM and their computer system and the ICDs [Interface Control Documents]. That would be either a non-starter on orbit or it would be a tremendous cost on orbit to try to fix, and for a short-term mission, you just couldn't afford that downtime."

Mike Haddad commented that, "Everything was completed on the LFC/ORS payload. The last task to do in the O&C was final closeout of the EVA tools, located on a shelf at the front of the payload. We had completed that earlier in the day, but astronaut David Leestma, who would use the tools during the mission, wanted to see them before we did the final closeout. I remember it was late at night and the whole O&C Low Bay was dark except for a few lights in the Level-IV area over the payload. The only people there were David, a Quality person, a security person and me. David looked everything over and said he was happy. If he was happy, I was happy. I said, as we were closing up the area, 'No other human will touch these until you do on-orbit'."

When testing was completed, the LFC/ORS payload was placed into the Payload Canister and sent to the Vertical Processing Facility (VPF) to mate up with the rest of the payloads that would fly on that mission. Then all that was transported to the launch pad and placed into the PCR. This would be the location for loading the hydrazine for launch. Because of the hazardous nature of this operation, it was done before the Shuttle was rolled to the pad, so there was total pad clear. Damon Nelson added some detail to that operation: "I believe it had about 210 pounds of monopropellant hydrazine, but we got out to the launch pad and again, the idea was that someone else would run the procedure – that I had developed – from the Firing Room, and we would be on headsets. I was going to be the 25-year-old lead system engineer on site in the Payload Changeout Room at the launch complex.

"We had two technicians and myself, and then we would have a Quality Assurance [person] because the idea was you would have a buddy system, so if anything went wrong we would go out in pairs, and we would go in in pairs as well. Then we had Mike [Haddad] as our backup system engineer, along with two additional technicians and an additional Quality. The initial idea was we would have one NASA, Boeing KSC technician and one JSC-provided technician because they were the ones that had the hands-on experience out in Houston. Unfortunately, about two days before the operation, the two JSC technicians failed their KSC physicals. Physicals were very demanding, because again you would potentially have to exit the pad going down multiple flights of stairs, and you could not just be out there and not be in fairly decent physical shape. All of our personnel were relatively young, but the JSC technicians were much older. So that was a big twist, just a couple of days out, that we were going to go with a total KSC team out at the launch pad. We started the operation and we had a couple of snafus so that, after about four hours, when we were ready to do the personnel change, we still had not started the actual fueling operation. So a decision was made that we would change out the technicians and the Quality person, but since I was the only KSC engineer that had actually seen the fueling operation, I would stay on. That's how I ended up being in a SCAPE suit for such a long time, more than 10 hours. It was about twice the maximum length of time that you're supposed to be in a suit. We had to get the approvals, the Safety clearance to let that happen, but we got it done. I talked to both of the astronauts postflight and they were very positive."[4]

---

[4] It should be noted that the IMAX® camera flew on this mission, and a number of scenes were shot that are shown on the IMAX® Movie, *'The Dream is Alive.'*

Jay Smith said, "When we filled [the hydrazine] at the pad, that was wild because they cleared that whole pad and we went up. There were only four of us, and we went up in our SCAPE suits. We had our oxygen bottles going in that same elevator, just like the astronauts, and went up inside the PCR and switched from bottle air to facility air. Then we transferred all that monomethyl hydrazine into the unit."

On-orbit operations went well, but even then Level-IV personnel had a piece of the flight and, like all flights, what happened in the Level-IV processing had a direct effect on mission operations. Maynette Smith continued: "For the OSTA-3 payload, I can remember this guy from JSC kept cranking [while closing] the SIR-B antenna while on the ground, to stow it for flight. When they deployed that thing [in space during the mission] it kind of sprang out. There was too much pre-loading on it, and it was a little surprising on orbit for [Shuttle Commander] Bob Crippen and company."

Mike Haddad said that, "LFC was also very successful on orbit. In fact, I am in a photo that was taken as the Orbiter flew over KSC, because just before the pass over KSC, the PI called me and said to go outside because they were going to take a photo of KSC. So if you were to zoom *way* in on the photo, you would see me standing out in front of the O&C Building looking up. I still remember that while I was out there, many colleagues walked by and gave me a strange expression, until I told them why I was there looking up. The camera had that kind of resolution. A single negative frame of the camera was 9 in x 18 in (23 cm x 46 cm). I remember talking to the PI after the flight regarding some of the other photos it took of the Earth. One such photo was over Boston Harbor, and he noticed what looked like a scratch on the image. He thought, 'Well that really destroyed what was an excellent photo of that area of the US', but zooming in on the end of the scratch showed it was not a scratch at all. It was a jet, and the 'scratch' was the contrail of the jet flying over Boston Harbor"

Removing Hazardous Payloads

After the mission, some postflight operations needed to occur in the OPF as soon as possible, to prepare the payload for ORS-only removal, due its hazardous nature, while it was still in the Shuttle's Payload Bay. This was a very rare occurrence. Once ORS was removed, the rest of the entire payload was lifted out, placed in the Payload Canister and transported back to the O&C for de-integration.

Damon Nelson continued: "The ORS had to go back to a hazardous facility for de-servicing, because none of the hydrazine was ever consumed on orbit. I'll never forget that after we got the experiment off the Shuttle and onto the transporter, and we took it over to this hazardous facility in a convoy that was kind of an interesting story in and of itself, they gave us a hard time when we got to the facility. We were in the parking lot and we had covered this experiment as best we could, but the manager hadn't opened the facility doors. Here we were with this piece of flight hardware that's just been in outer space, sitting out in the loading dock while we were waiting for them to open the facility. So that was kind of ironic. But when we got into the facility, that was a really neat part of the job. We saved everything; got the hypergolics off the experiment. And, again, hats off to everybody involved: a NASA contractor team working directly with a NASA JSC team and pulling off this really important mission."

Mike Haddad explained that, "The SCAPE suits we were using for de-servicing were CAT 6, like a rubber suit that completely covered our bodies. These suits had the vent out the back. During the offload, one of techs sat down to take a break on a step-up stool that had handles, not realizing he had just crimped the vent on his suit. He began to inflate like the *Pillsbury Dough Boy* and he basically wedged himself between the handles. What to do? He could not move, but he had to maintain a positive pressure in his suit, so I told him just to disconnect his airline, let the suit deflate long enough for him to get up, then reconnect the airline, so there was always a positive pressure on his suit and no concern about breathing in toxic atmosphere." The de-servicing continued successfully, and then ORS was transported back to JSC.

Breaks were still important, because work was increasing, the flight rate was increasing and the pressure to perform at a high level could weigh on the mental and physical condition of the staff. Debbie Bitner revealed that, "We used to have 'John Cougar Mellencamp hour'. I was bringing watermelon in as a snack, and then people started flocking over to my desk wanting watermelon, so I started bringing it in every day at the same time." Mike Haddad added: "It was like a break because we were working, working, working, and then it was melon hour. I can never thank Debbie enough for taking the time every day to do that for us."

SATELLITE RETRIEVAL

STS-51A, *Discovery*, Launched November 8, 1984, 7+ days
Payload: Satellite Retrieval Mission (SRM); two Pallets (F007 and F008).

The mission included deploying two communications satellites, Anik D2 for Telesat of Canada, and Leasat 1 (also known as Syncom IV-1) for the US Navy. The Level-IV part of this mission dealt with assembling hardware on two Spacelab pallets to recover two other satellites that had failed to achieve their proper orbit after being deployed by a previous Shuttle mission. This was not a mission that had taken many years to plan. The failures had happened a matter of months earlier, so this was definitely what you might call "shoot from the hip", trying to recover satellites that were not designed to be retrieved. It was basically creating a mission from scratch, designing and manufacturing the hardware, then assembling it into flight configuration and flying it, all with in a number of months, not years.

The satellites, Palapa B2 and Westar 6, had been deployed during the STS-41B mission earlier in 1984, but had been placed into incorrect orbits when their kick motors malfunctioned. The original STS-51A mission was redesigned to include retrieving those satellites, and Level-IV personnel were selected to build up the flight hardware that would be used in an attempt to bring them back home on the Shuttle.

Bruce Burch remembered: "We built a lot of drills and stuff. We had to do a ton of match drilling and building special drill bits to try to line all this stuff up that they sent us. This was being done on the fly; it was not a planned mission. So we ended up with a whole bunch of machinist tools. We were going to Palm Bay [to get tools and such] and having drill bits specially made."

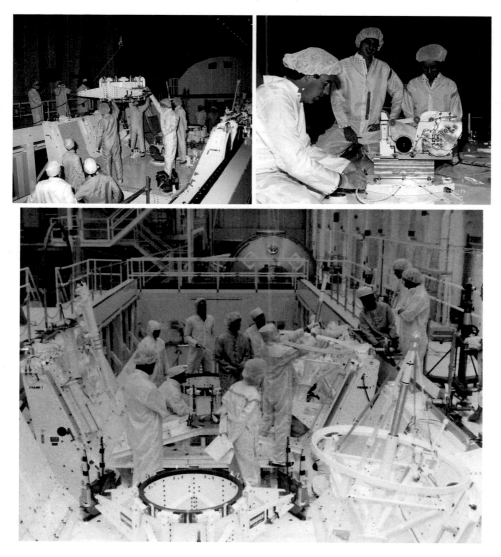

Top left: Satellite Berthing adapter installation. Top right: capture latch work, with Gary Bitner (left), Chris Talon (center), and Debbie Bitner. Bottom: Crew Equipment Interface Test. All work performed in O&C CITE stand. [NASA/KSC & Gary/Debbie Bitner.]

The Coolest Thing

NASA Level-IV Mechanical Engineer Debbie Bitner said that the satellite retrieval mission was her first flight hardware assignment, because they had kept her working on the GSE. She thought it was the coolest thing she ever did: "The time from the day that hardware came in the door to the day we put it in the canister was 30 days. We were on a timeline. We had to get that up so they could reach the satellites in orbit, and so they had a specific schedule to do that, and I was responsible for the electrical wiring on the pallets. The funny thing was that they had some goofy design where they were gluing down wire clamps." Boeing SEIS technician Gary Bitner looked at the design of some of the securing flight hardware, which had spikes sticking up. That was something very hazardous for any crew during an EVA, and it was not secured very well. Garry said, "There was no way that was going to hold on a torque. I said, 'I can just hit it and knock it off'. So then they asked, 'What would you do?' I said, 'I'd go buy one of those devices from Black and Decker had this little plastic fitting that you could put a drill bit in, and it sat down flat and drove straight home. I'd use that to put in helicoils as support'. We got a vacuum cleaner [to collect the shavings from drilling] and that's what we did. We used two clamps in one hole [butterfly orientation], put the wires together, and it looked nice and neat. It worked out really good."

"We had the designers there at KSC while we were doing all the integration, so if we ran into a problem, they could make a change right away," Gary continued. "That's how we were able to get it processed so quickly. Astronauts Joe Allen and Dale Gardner would come down after most of the shifts we'd done, look around, walk around and look at it, and go back upstairs to talk to you. It was kind of neat."

Don Dolby remembers doing a sharp edge inspection with the EVA crew, where the crew would point out anything they thought could possibly catch on their spacesuits. Level-IV would then put pieces of flight tape over it: "I remember that was pretty neat, getting to work with a crew doing that."

All that integration took place in the CITE stand, because the Level-IV and Level-III/II stands were full of other flight hardware. The team had to figure out where they were going to put it together, determine what access would be required and build it, figure out the kind of tools required, and work out the logistics of getting the hardware there.

The small, tight team created all worked together very well to pull this off. That was one of the good things about Level-IV; they could move ahead pretty quickly. The Satellite Retrieval Mission (SRM) was launched on STS-51A on November 8, 1984, just nine months after the Westar and Palapa satellites were left stranded. Both satellites were successfully retrieved, returned to Earth, refurbished and later relaunched. Michael Stelzer recalled: "Being a part of a NASA team that could rapidly develop and timely implement a daring and technically challenging mission was an incredibly rewarding experience."

SUMMARY

Level-IV had a great year in 1984. With the success of some high profile missions, the team was clicking on all cylinders. As the year came to a close, Level-IV continued to bring in new talent to try to keep up with the increasing flight rate of Spacelab-related missions planned for 1985.

Reference

1. Lord, Dcuglas R.; Spacelab, an International Success Story, NASA SP 487; Washington D.C., 1987, p. 343. This book was a valuable reference source throughout the compilation of this chapter.

8

Ramping Up the Flight Rate

> *Spacelab-3 was the first and only*
> *Shuttle/Spacelab mission to fly primates…*
> *Our instructors told us: "Don't stare at the monkeys."*
> *We had to change into sterile-type gowns with undies only.*
> *Those gowns were see-through, so not only*
> *were we trying not to stare at the monkeys,*
> *we were trying not to stare at each other!"*
> Bob Ruiz, NASA Level-IV
> Lead Mechanical Engineer
> Research Animal Holding Facility STS-51B.

By the start of 1985 the flight rate for the Shuttle missions had increased, and of course so had the workload to prepare these missions for launch. There were three planned Spacelab missions for 1985, two with the Long Module and the first pallet-only flight. Several other non-Spacelab missions were also manifested, including the first two fully classified missions for the Department of Defense (DOD).[1]

THE UNSPOKEN ONES

Level-IV personnel had to pass security clearance to work the classified missions. These officially began with the launch of STS-51C in January 1985, but of course they cannot be discussed here. Those who worked the primary payloads have

---

[1] Ten classified DOD missions were flown between January 1985 and December 1992, with many others planned but ultimately cancelled.

© Springer Nature Switzerland AG 2022
M. E. Haddad, D. J. Shayler, *Spacelab Payloads*, Springer Praxis Books,
https://doi.org/10.1007/978-3-030-86775-1_8

stories of those missions, but these personal memories may never be able to be told to the outside world. One can only imagine what occurred to ensure those flights were successful. One day, perhaps, those stories might be told, but not for some time to come.

Mike Haddad noted that, "I did work a few of those missions, but mostly on unclassified middeck payloads not on the primary payloads. And while each of those have a story on their own, I feel it would just not be right to talk those payloads here. I was asked many times if there was anything I could reveal about working those flights. Nope. All I can really say is it was very interesting for those of us working together on a classified mission, [with] what had to be put in place to allow us to do our job, not interfere with classified operations, and still have a joint set of classified and unclassified hardware fly into space."

STS-51B *Challenger*, launched April 29, 1985, 7+ days
Payload: Spacelab-3 Long Module (MD001); Floor Assembly (MD001); Racks (see Appendix 2); Tunnel (MD001); plus an MPESS (F003).

The investigations selected for the Spacelab-3 mission originated in the United States, France, and India, and covered several disciplines. For the materials processing discipline, higher-quality crystals were grown by two methods: seed crystal growth in a saturated solution, and condensation from the vapor phase. The two fluid physics experiments studied the dynamic behavior of rotating and oscillating liquid drops, and the convection processes found in planetary atmospheres and in stellar interiors. The performance of equipment and facilities specially designed for investigations on the Spacelab Life Sciences mission series was evaluated. Two of the investigations, one in material science and one in astronomy, had previously flown aboard Spacelab-1. Some of the experiments were located in the module, some on the pallet, and one in the middeck.

"Let's start with the first item assembly servicing and test," Bruce Morris told the authors. "I did two stowage racks, and then I did the rack with the big Earth-sensing weather conditioning dome experiment. It was a single rack fully filled up, and then I did the rack that had video equipment that was an upgrade for the video system inside Spacelab. The big experiment for modeling weather patterns had a PI [Principal Investigator], Fred Leslie, from Marshall [Space Flight Center, MSFC] on it. It was probably the hardest integration, because it had [what] seemed like hundreds of cables running in the back. It was stuffed inside a single rack, not much room Actually, that rack went together really, really easily. We had some stowage racks that were a lot more trouble, for the FTS [Fourier Transform Spectrometer] experiment, because they were trying to repurpose some old stowage containers from Skylab, which was a good idea except they didn't fit. And we had to do a lot of real-time machining, which happens with almost everything you do in Mechanical, Also, the first political issue I ever got into was with the racks, because we called a designer down from Marshall. He came down and he grabbed the long arm on the back end of this storage container, and he literally hung on it with his feet off the ground until he got the holes lined up. And he said, 'Quick, push the bolts in now'. I said, 'No, we're not doing that', and he got real annoyed because I wouldn't do what he said. That was the first time I got to go up to Dean's office and say, 'Boss, I don't think we should do this'. He said, 'You think?'"

Cheryl McPhillips explained to the authors that, "I got my first experiment assignment on Spacelab-3 General Purpose Work Station (GPWS), which was a pretty easy one to test because it didn't have a lot of complex interfaces. It did have air. It was basically the dissection chamber from AMES."

Bruce Morris described how Anna Villamil (a NASA Level-IV Mechanical Engineer) basically put the entire Drop Dynamics Module from JPL together on a Problem Report (PR) because they intended to fly it as a Class D payload (low priority, low to medium national significance, medium to low complexity). Once it got through the management team, all their planning and budget and drawings showed it to be a Class A payload (high priority, very high national significance, very high to high complexity). "She had these drawings that came in the day the hardware showed up. They were all just pictures with handwritten lines to the part numbers."

Gerry Rivera revealed that, "During Spacelab-3, I was still working in the Experiment Integration group and on ATMOS [Atmospheric Trace Molecule Spectrometer]. I remember some troubleshooting that had to be performed, because a lot of times it was about the data that they had." The PI was not sure if he was putting out the correct data, or getting the right data in front of the subsystem from Input/Output (I/O). Gerry frequently had to break the data lines and put in the breakout box and adapter cable to scope the lines for them. "I do remember the experiment was a full success during the mission and they were very, very happy with the KSC support they received," Gerry continued. "So that one is always in my memory, as one of my favorite moments also in Spacelab and Level-IV."

Jim Dumoulin recalled that there was one noise problem in the High Rate Demultiplexer (HRDM) that took weeks to track down, because it never showed up during simulations. HRDM was responsible for separating different data rates from Spacelab's downlink telemetry and routing it to the High-rate Input/output Test System (HITS) for display to the User Room. The problem they were having *only* occurred when the entire test team was on station supporting a full-up test. They looked for random noise everywhere, including the increased use of the elevator and restrooms during tests, and the increased use of the water fountain and refrigerator in the break room. During one of the many days searching for this elusive source of noise, the Electrical team was down on the High Bay floor with the input cable of the HRDM routed thru a Break-out-Box (BOB) to monitor the signal. A Spectrum Analyzer and Oscilloscope were displaying the noise. Jim Dumoulin recalled: "Someone noticed that each time the noise pulse appeared, a yellow light was flashing in the distance in the High Bay. It seems that Safety had a yellow light illuminating a sign that said 'TEST in PROGRESS', which they turned on only during full-up tests. The strobe light was plugged into an outlet below the test stand and was injecting noise into the ground plane. We switched the strobe light to one that was battery powered and the problem went away."

Work on Spacelab missions continued to increase, and Level-IV management knew more good people would be required to support the upcoming workload, so more personnel were brought on board to those branches that were still short. Level-IV also had an upper management change.

NASA Payloads Director John Conway added, "When I first went over to Payloads, the Level-IV people were there. And of course they thought they walked on water; basically, in their own manner and in their own mind, there was nothing else in the world but them. So their emissary, Scott Vangen, called on me to determine whether I was going to have the right level of support for Level-IV."

OF MONKEYS AND MEN... AND MVAK

Bob Ruiz told the authors that he worked on the Research Animal Holding Facility (RAHF) for Spacelab-3, which held two squirrel monkeys and 24 rodents. These racks came in as Ground Support Equipment (GSE) racks. "They came in as GSE racks but with flight hardware installed, and we had to swap out all of their equipment and then put it in the flight rack, which was quite a job. Flow balance and us, we did the same thing. Flow balance of each individual rack and the Rack Train."

The PI from Ames that was in charge of the experiment wrote a long briefing that basically said it was absolutely impossible to transmit herpes from that strain of primate to a human being. It had never happened before. But a naval technician up in one of the big bases in Florida was handling similar monkeys (Navy and Air Force pilot research used a lot of monkeys for testing things that were going to be exposed to humans) and he died from herpes virus, the very thing that the PI said could not happen. Level-IV personnel had to be given additional medical clearances so they could trace any exposure, but the bigger thing that impacted the mission was that a colony of these monkeys had to be found that were herpes virus free and could be trained in time. Bruce Morris remembered, "It turned out it was so hard to do that, they actually couldn't find four monkeys. They could only find two that were free of the herpes virus, so we had to lose two of the specimens right off, before we ever got out to the pad. We only got two monkeys to take out there, not four."

Bob Ruiz told the authors that Spacelab-3 was the first and only Shuttle/Spacelab mission to fly primates, the two small squirrel monkeys (*Saimiri sciureus*). He was Lead Mechanical Engineer for the RAHF, which meant he installed the RAHF hardware into the Spacelab and serviced the water/coolant system. The

RAHF was complex. In addition to the cages, it had a coolant system, an air circulation system and a food and water supply system. Bob Ruiz recalled, "One problem we ran into was the pull force for the cages. The flight crew had a hard time pulling out the cages; it turned out to be the electrical connectors."

Bruce Morris added that RAHF was pretty quirky. They had to do a lot of specialized, critical tasks that did not operate in the way they expected; things discovered in the final assembly. About once a month, they completely drained all the water out of the drinking water tank, then serviced it, cleaned it with alcohol and then reloaded it with fresh water. It could not remain empty because of microbial growth. Samples had to be taken once a week, or whenever a major problem with a drinking water system was discovered. The plating came off and was pitting on each one of the servos that controlled the drinking water that went down to the monkeys, so the servos and the little drinking caps down in the cages were clogging up. Bruce Morris said, "The mission was in real trouble and we found that late in the flow. [After] analyzing the drinking water and tracking the failures, I had to tear out almost all the drinking water systems, the upper third of the rack, return it to Ames for them to replace it, then reinstall it and retest all of that massive rework that we weren't expecting. But it was a good thing that we found it, because if it got on orbit like that we would have had a lot of thirsty rats and monkeys."

Cindy Martin-Brennan recalled: "I remember RAHF, remember big-time troubleshooting [and] the experiments struggling with the RAHF, and the way it was communicating with the Remote Acquisition Units [RAU]. I remember some major problems and really having to dig into some of the software. We could not figure it out, but working with Sharon Walchessen, and I think I might've had the Teledyne Brown people [as well], we finally figured the problem out."

The bane of every test engineer was documenting an Unexplained Anomaly (UA) for problems that no one could explain during Level-IV testing, but which could possibly show up during flight. Jim Dumoulin explained that, "Noise or random spikes on data lines were a constant problem. In one instance, we were tracking down noise on the Spacelab module internal intercom system. Someone managed to isolate the noise and send it to a speaker. It turned out to be a local radio station. It seems that the ESA designers of the Spacelab intercom didn't worry about USA FCC spectrum space, because they figured there would not be any radio stations in space."

Mike Kienlen started in Operations and his first job was loading the animals on Spacelab-3 with Bob Ruiz: "I walked into the first meeting [and was told] 'This is Bob, he's the engineer. He's responsible. You're the Ops guy. You two are in charge, handle it'. Then management walked out the room. It was incredible. It was just phenomenal for a group of engineers to have [such] hands-on responsibility."

Bob Ruiz explained the Module Vertical Access Kit (MVAK) operations with the RAHF. As Lead Engineer, he was responsible for RAHF late and early access operations. Late access involved loading animal cages into the RAHF in the Shuttle on the launch pad around 24 hours before launch (initial requirements from the PIs were between 18 and 12 hours). The only catch was that the Spacelab would then be vertical with its floor now upright and the RAHF sideways. Bob Ruiz recalled: "The only way in or out was by a hoist mechanism called the MVAK."

"Planning for the late access, we had two jobs," explained Bruce Morris. "The biggest issue, though, was to develop the procedures and the hardware to do the specialized late access for the animal servicing." This included refilling water tanks for the research animals, activating and making sure that the system came up online properly, installing the rats and the monkeys, and then installing all of the time-critical supplies that went along with them, such as food, test kits, and sampling gear. "None of that had ever been done before."

Bruce Burch expanded further: "I'm not sure how I got selected for MVAK [operations], but we went to training. A lot of us went to training for the MVAK, the actual training. But then I was selected to be the guy in the module."

Bruce Burch during MVAK training, removing the rodent cage from the transportation container to install it into the simulated RAHF enclosures to the left. Using radios for communication did not work, so they went to the OIS system with a hot mic. [NASA/KSC.]

Michael Stelzer concurred that the Spacelab Module Vertical Access Simulator proved instrumental in helping train the technicians and flagging deficiencies in the existing MVAK hardware. For example, the technician hanging from the hoist found it difficult to translate just a few feet from the center of the module over to the edge of the racks where they needed to perform the stowage. Mike Stelzer said, "To fix this, we added a web to an existing safety net below them that they could hook their feet into to help them maneuver. Padding was added to most everything that needed to be lowered, to prevent damage to items in the airlock or transfer tunnel." Over time there were a number of failures with the hoists, which were proving unreliable. A type of aircraft grade hoist used in helicopters was looked into and it was decided to upgrade. Bruce Burch was the lead technician for the installation, and was key in helping identify changes to the equipment and processes that facilitated the stowage operations.

Bruce Morris made it clear that, "The monkeys definitely were not pets. They'd scream and yell and spit, bite, all those kinds of things. And so we had to train in having to deal with that in case one of them got out of the cage."

Bob Ruiz added: "Before we actually worked with the animals, we had an orientation session with them at the animal facility. The technicians and engineers all got to meet the primates and rodents, but our instructors told us 'Don't stare at the monkeys', so we didn't. One surprise for the monkey meet-and-greet was that we had to change into sterile-type gowns with undies only. Those gowns were see-through, so not only were we trying not to stare at the monkeys, we were trying not to stare at each other!"

"We couldn't recycle the same team," explained Bruce Morris, "so we had to have a two-shift capability that we could extend day over day in case it [the launch] was scrubbed. And then of course it turns out these rats, when they give them to you, the first thing they do is urinate. And so now you've got this big old wet spot going down a bunny suit."

Level-IV personnel actually went out to California on Spacelab-1 to do a mockup of unloading the animals through the tunnel, using a platform to set the cages on that ran on tracks, with ropes to move it. Bruce Burch explained that, "We went and tried that on Spacelab-1 to see if it would work, and we found we could actually just pick the cage up and carry it out faster than we could using this little contraption."

Mike Kienlen added that, "Loading the animals was a challenging task. How do you get management approval to run a procedure without Quality verification of tasks? The Safety engineer said their position was they had to lower a Safety engineer into the Spacelab, to certify it was safe. Quality said, 'We'll have to lower a technician down there and a Quality engineer to verify every task he's doing'. And they physically couldn't do that. The neat thing about Level-IV was we worked as a team and came up with an answer that made everybody happy. Some of it was camera verification, some of was all the training with the Quality guys, [where it] got to the point where they trusted the tech. When the tech said, 'I threw the switch', he actually threw the switch and the Quality people could buy it off. The module technician was on a headset hanging in a bosun's chair for hours. That's when you get into things like what happens if the crane breaks? What happens if the technician passes out? What happens if there is an emergency at the pad and we have to evacuate? We had to develop all those procedures. And if there was a scrub, we had to take [the animals] all out, we had to clean it all up, start all over again and put them back in"

Besides all the hardware challenges with Spacelab-3, all the different countries involved brought other aspects of how Level-IV did their jobs into play.

Monkeying Around at the Pad

Don Dolby said, "We were the early crew in the middeck area, where we were loading all the animal food and things like that. And then the other crew came in and put all the animals and the cages in. I was right there in the middeck hooking up the packages and then sending them down on the crane, and then there's somebody about halfway down that was help with guiding things through that little joggle area into the Spacelab. They had little stick figures [showing] where each person was at and what they were doing.

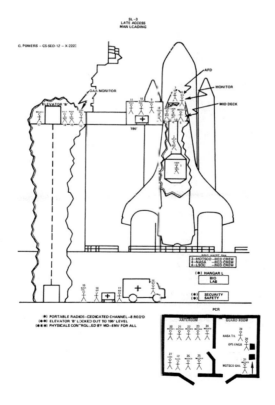

Spacelab-3 Module Vertical Access Kit (MVAK) late access graphic. [Don Dolby.]

Bruce Burch, recalled the event: "I was hanging on a crane and swaying back and forth. MVAK actually had three cables that would come down. One cable was holding yourself and then they could pull a cage up and lower a cage down using the other two. They'd go back and forth and they had a little bit of flexibility with the joggle too, with pulleys to move you over to one side or the other that could help try to move it. But once you got in the module down to the bottom, you had enough angle on your swing that it was easier just to swing across and grab the other handrail. If you were going to be there for a while, you had straps on your bosun's chair. When I was loading the animals, I would strap myself to the rack front of the handrails and, as the stuff came down I could push them; otherwise they would push you away as you tried to slide these in."

Bruce Morris added: "We ended up not flying a full complement of rodents, and the same with the monkeys. We had a couple of [rodent] cages that failed very late in the flow. They had to be custom fitted for their particular slot so that the pull and push load on it was not so high when an astronaut was working with the cage. The timelines and everything worked fine, but we had a failure in the water system when we got vertical and did the final water tests, so we actually had to go down and tap the water lines to see if there was an air bubble in it, to clear it."

While the Level-IV personnel thought all this was a just another part of their job, those on the outside had a totally different perspective, as Bruce Burch recalled: "I had to go get a new badge, so I walked over to NASA security and they were all talking: '…and we were watching, they had some guy on a crane down there.' I didn't know everybody was watching NASA Select, and had been watching the whole time. I thought it was just Bob up there, the engineer and just our little crew. I was glad I didn't know that; I would have been so nervous."

Bob Ruiz commented that, "We had practiced so much that the actual loading of the animals into the Shuttle went perfectly, and luckily the Shuttle launched so we didn't have to get them back out and reloaded for a second launch attempt." The final timeline on loading the animals was between 22 and 17 hours before launch.

After months of training and numerous iterations of the timelines and procedures, the Spacelab-3 late stowage was successful, and *Challenger* launched as STS-51B.

On-orbit operations with the RAHF encountered a few problems. There was the famous feces issue, where they found out that they had miscalculated in the design. "The way they were going to trap the waste products was having negative pressure inside the rack," explained Bruce Morris, "so cabin air was pulled through the cages through a filter system that trapped the waste and then that filter was closed out several times during the mission. The modeling for how it was going to work in zero-g didn't quite line up with what they planned, and they actually found out they had positive pressure inside the rack. So instead of sucking the waste into the trays, it blew it all out in the cabin. All the loose feces and urine and all that other stuff was all spread all over inside the experiment."

Co-author Dave Shayler added: "With monkey feces floating in the cabin, the astronauts were certainly less than pleased, especially since Commander Bob Overmyer and Mission Specialist Bill Thornton, both of whom had flown before, had expressed concerns before the flight that the cages would leak. They tried to convince the RAHF designers and engineers that this would happen but to no avail. As Bob Overmyer wrote in 1987, two years after the mission, 'On about the third day, all of a sudden, particles of feces and food particles suspended in droplets of urine were leaking out of the cages'." [1] Indeed, the Astronaut Office had expressed their concerns for some years but to no avail. There were no plans for direct crew handling of the animals, although both medically-qualified crewmembers, Norman Thagard and Bill Thornton, had been given additional veterinary training to care for their animal passengers in the event of illness or injury.

None of the crew wanted to bring home a sick, injured or dead animal. In fact, Thornton spent a considerable amount of time caring for them during the flight, ensuring they were fed and watered by the automated system. But on the second flight day the feeding system played up, and when one of the monkeys became sick, Thornton had to revert to hand-feeding it after a two-day fast. Wary of past experiences, when one of the monkeys had bitten one of the trainer's fingers pre-flight, the situation was not helped when the ground suggested the debris was probably food crumbs and not animal waste leaking into the cabin. This suggestion upset the crew, as Bill Thornton clearly recorded in the postflight crew report: "I can absolutely tell you that, as a boy that grew up on a farm with chickens, pigs and so forth, I knew at the age of three years old [the difference between] feces and food pellets." The postflight report also stated: "A large volume could be written on the problems with this system, and that serious reviews were needed before the device flew again." But part of the objectives of flying new equipment in space was to evaluate both success and failure, though hopefully not to such a degree as experienced on STS-51B [2].

While one reason to fly the hardware is to determine if the design will work on-orbit, some experiments just have failures that cannot be anticipated; thus the need for humans in space. Mike Haddad said, "It used to get me mad when all you would hear on the news was what failed during the mission. No it was not a complete failure! You learned that that design would not work in space!"

When the Drop Dynamics Module payload reached orbit, it had a power supply failure. Taylor G. Wang, the PI that flew as the Payload Specialist (PS), opened up the experiment and rewired the power supply, which some thought was bad Public Relations (PR). Bruce Morris disagreed: "I was in the POCC [Payload Operations Control Center] and to me it was great PR humans in space can fix problems. They got it to work and it operated flawlessly."

Extensive planning also had to take place for landing operations, as Bruce Burch explained: "We actually had to go out before launch. We would do both places [primary and backup landing sites] in case they made up an abort landing. You still had to get the animals off, so you had to be there ready. We supported all three places; KSC, Dryden and White Sands."

Mike Kienlen added that, "When it landed in California, we had to have a team out there to get the animals off, and get them back to the scientists to return them to their labs. Lear jets were lined up, [and] all kinds of stuff were put in place to maximize the science. Whatever the requirement was, you had to meet the requirement. What would make this the best it can be? We figured out a way to get there. That's what made Level-IV so cool."

Bruce Morris recalled: "I also did all the procedures for recovery, because for the first time we were going to do a Spacelab early recovery. We had to get all the animals off while the Orbiter was on the runway and none of that had ever been done before. Bob [Ruiz] and I split that duty. Bob took the team that went out to Edwards. I took the team at KSC, and so we were on the runway waiting for the landing. And of course it landed at Edwards, at the primary site."

"Early access involved retrieval of the animals, after the Shuttle with Spacelab landed at Edwards Air Force Base," Bob Ruiz told the authors. "We were really surprised when we entered the module. Rodent droppings were everywhere [and clearly] the crew wasn't too happy having those little nuggets floating around. We picked up the droppings and bagged them. When we looked in the primate cages, the little squirrel monkeys seemed glad to see us and chirped happily, at least that's what it sounded like to me. We grabbed the little monkeys [in their cages] and took them out. Later on, the techs enclosed the droppings in clear resin, with a label 'Made in Space', and gave them to those of us that worked the mission."

Bruce Burch remembered: "We went in and everything kind of worked. We got them out in a timely fashion. We actually took gas respirators, thinking it was going to smell bad, but it didn't smell that bad."

Gary Bitner had a particular memory: "We went in and we were not supposed to look the monkeys in the eye, that were still in a sealed cage. But that monkey just put his hands out and jumped towards me, [as if to say], 'Get me the hell out of here'."

"When I started in summer of 1985," said Johnny Mathis, "my first job on the floor was to go in with people postflight and take pictures of everywhere that the feces had managed to get into the module. For a 19-year-old kid, that was amazing, to actually get to go into a module that had flown in space. [I was] wide-eyed; Wow. When Spacelab came back, we had to take it apart, of course, and that was when I found the value of having Co-ops on our team."

"It was Spacelab-3 and they had the rats and the monkeys. When they came back we had to de-integrate it and it was a mess inside," recalled Eli Naffah. "Of course, I was still Co-oping. The engineers didn't want to deal with it [the mess], so they told me to go into the Spacelab, to clean and take samples so we could begin the process of de-integration of the racks. I remember it was disgusting. There was feces everywhere, it smelled to the high heavens, and it wasn't very good ventilation. You had to go in with a bunny suit that was hot. I felt like passing out a number of times from the asphyxiation."

Other activities besides hardware and software occurred for very high profile missions like Spacelab-3, as explained by Bruce Morris: "The other cool thing about getting this involved in a mission like this is that you also learn how to brief management about this stuff, because everything we were doing was so critical. It had to be briefed all the way up to the Center Director about our plans."[2]

ABORT TO ORBIT

The 19th mission in the program finally saw the second verification flight for the Pallet-only configuration with the Igloo and Instrument Pointing System (IPS), flying as the third official Spacelab mission. But STS-51F was not without its issues in trying to get off the Pad and into orbit.

STS-51F, *Challenger*, Launched July 29, 1985, 7+ days
Payload: SPACELAB-2; three Pallets (#1 Pallet F004, #2 Pallet F005 and
#3 Pallet F003); an Igloo and Instrument Pointing System (IPS).

The flight objective was to demonstrate Spacelab's capabilities through a multidisciplinary research program, and to verify system performance. Investigations in the fields of astrophysics and solar astronomy included a sky survey for extended infrared sources, X-ray imaging of cluster galaxies, cosmic ray measurements, studies of small-scale structures on the Sun's surface, and a measurement of the coronal helium abundance In addition, there were measurements of: the solar ultraviolet flux; the plasma environment and plasma processes near the Orbiter; and zero-gravity effects on technology processes and on the behavior of liquid helium. Life sciences problems investigated included bone demineralization in humans and lignin formation in plants.

---

[2] STS-51G in June 1985 deployed three commercial satellites from the Payload Bay of *Discovery* and carried the deployable/retrievable SPARTAN (Shuttle Pointed Autonomous Research Tool For Astronomy) free-flyer. Though not a Spacelab mission, part of the payload support equipment was an MPESS structure (designation unknown).

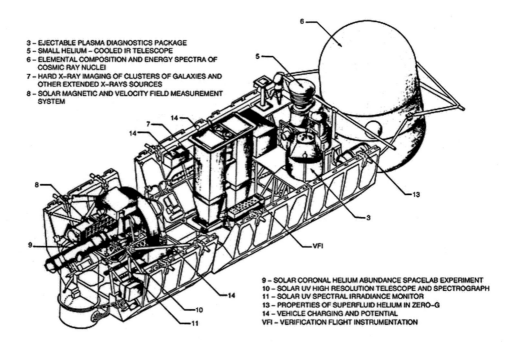

3 – EJECTABLE PLASMA DIAGNOSTICS PACKAGE
5 – SMALL HELIUM – COOLED IR TELESCOPE
6 – ELEMENTAL COMPOSITION AND ENERGY SPECTRA OF
    COSMIC RAY NUCLEI
7 – HARD X–RAY IMAGING OF CLUSTERS OF GALAXIES AND
    OTHER EXTENDED X–RAYS SOURCES
8 – SOLAR MAGNETIC AND VELOCITY FIELD MEASUREMENT
    SYSTEM

9 – SOLAR CORONAL HELIUM ABUNDANCE SPACELAB EXPERIMENT
10 – SOLAR UV HIGH RESOLUTION TELESCOPE AND SPECTROGRAPH
11 – SOLAR UV SPECTRAL IRRADIANCE MONITOR
13 – PROPERTIES OF SUPERFLUID HELIUM IN ZERO–G
14 – VEHICLE CHARGING AND POTENTIAL
VFI – VERIFICATION FLIGHT INSTRUMENTATION

Graphic of Spacelab-2 in Level-IV North Rails, c 1984. [NASA/KSC.]

This was the first multi-pallet-only mission of the Spacelab program, carrying three Spacelab pallets. One was located on its own in the forward section of the Payload Bay, the other two were fixed together behind, with a large egg-like cosmic-ray experiment in the very aft location. The Igloo structure and IPS also flew for the first time on this mission. The GSE used to support the flight hardware was positioned in the North Rails on the west end of the Level-IV area, where it remained for months on end as the team worked to learn, install, assemble, integrate and test the flight hardware to the point of getting it ready for its mission.

Maurice (Mo) Lavoie explained that, "The CHASE experiment was a British experiment for which Craig Jacobson and I had to travel over to England, spending two weeks out near Oxford with the developer. [We] wrote all the procedures, getting all the specs and drawings we could from them, then brought it all that back to the US to write up all the experiments that we could do on the ground to confirm that the experiment was working properly."

Bill Jewel recalled that, "The Europeans had the Hard X-Ray Imaging of Clusters of Galaxies and Other Extended X-Ray Sources telescope. Big deal; big, big mess. [So] I gave Jim Harrington [NASA Spacelab Program Director] an ultimatum, in September of 1984: 'Either you ship it and get it here by December, or keep the damn thing, or ship it to Washington, because after that, it's not going to fly on Spacelab-2. It can't happen because I don't have time for it. We'll just take it off the program, and Merry Christmas'."

At this point, Level-IV was still attracting new people, because the Shuttle manifest at that time (1984) called for multiple Spacelab missions over several years.[3]

Cheryl McPhillips said, "[Everyone was] just so passionate about what they were doing. [When] I took the job at NASA, my first job was making cables, and I burnt my hand on the soldering iron because I was overeducated and didn't have any hands-on skills."

NASA Level-IV Mechanical Engineer Angel Otero added that, "I always consider myself a tech, so every time I had a job that required two techs, I would only schedule one so that I could do tech work. If you look at the cabling procedures from Spacelab-2, most of the signatures are mine."

Luis Delgado told the authors that, "My first job was a Freon® loop servicing with Jeannie Moates [Ruiz]. She had other things to do, so she gave me that. I finished the leak checks and I did the servicing on Spacelab-2. But the Freon® loop was very interesting, I had super techs. Jim Nail, Walt Preston, Jay Smith and Gary Bitner showed me how to be an engineer. And if you weren't clear, or they didn't like what you were doing, they would ask you and challenge you."

Personnel that had been in Level-IV for a while, even some of the youngest, were able to perform many important functions. Most think all that Co-op students did was make copies and fetch coffee or food for the engineers. Not in Level-IV. Eli Naffah, explained that he, too, "did some work on Spacelab-2, some cabling work. I was involved with maintaining the GSE that we use to do the checkout activities, basically as a Co-op, where for the first few years I was assisting engineers where they had larger tasks trying to carry out the integration, and then testing it to make sure it worked."

The Spacelab program was still in its infancy and some of the subsystem hardware used to support Spacelab missions had not been flown yet. The Igloo and IPS were a two such items. "Spacelab-2 was the first Spacelab that utilized the Igloo," explained Gerry Rivera. "Spacelab had a series of subsystems, including computers and I/O units, which were all pressurized modules for Spacelab-2. There were the three big pallets, but in the front of the first one was this cylindrical vertical structure they called the Igloo. The IPS was part of a subsystem that the experiment would be mounted to. It was a big ring with actuators and a high pointing accuracy."

---

[3] The authorization to build a space station "within a decade" came in January 1984 from President Ronald Reagan's State of the Union address. At that time, it was foreseen that the Spacelab missions would continue in parallel while the Space Station was being constructed, but by the mid-1990s plans for both programs had changed considerably. For Spacelab, irreversibly so.

Even with the flights of Spacelab-1 and Spacelab-3, people were still learning the GSE and how information went from one location in the Operations and Checkout (O&C) Building to another. "We had the high density recorders in Room 1263, which I use to call 'the dungeon'. It was on the first floor," explained Cheryl McPhillips. "All the cables from Level-IV, Level-III/II and Level-I, CITE in the High Bay came into 1263 and then went up to the User Rooms from there. So I spent a lot of time in the dungeon trying to figure out where the signals were, and the signals would come back even from the OPF [Orbiter Processing Facility] and pad up to the users in the User Room."

"The other thing that was part of that was you also had video and audio coming off Spacelab," explained Craig Jacobson, "so that had to be patched through 1263 to upstairs. But also you had the display module within Spacelab. It was not a video display, it was called a vector display, so you couldn't get a signal from that directly. We had to point a camera at the display within Spacelab so that people could see what was on the display, what was happening with their payload data. And then, when commands were typed, they could see the commands. Prior to a test, I would write all that stuff down on my paper and then I didn't have to write a WAD [Work Authorization Document] to do my job, because it didn't interface to flight and it was only to monitor. It wasn't sending commands, so you couldn't hurt any flight stuff. I'd be the only one there, and I would power up the PCU [Payload Checkout Unit] all the way to the point to where one more switch throw would power up flight. Then of course I would power up 1263 and all that equipment, and then I would go up in the User Rooms and make sure all the HITS terminal data was coming through."

The PCUs required software support, and here Jim Dumoulin was one of the primary people. When experiments came through preparation for space flight, the interfaces between those experiments and the Spacelab had to be tested. Level-IV would write the software and then test the interfaces out between, say, an experiment from Germany, or from the United States. Cindy Martin-Brennan explained that, "Jim Dumoulin trained me on writing some of that software. The funny thing, I remember, was that I changed my degree. My worst class, my most hated class, was learning assembly language. Hated it. I never wanted to do it again. Well, never say never, because when I got to NASA, guess what I programmed with Jim Dumoulin? Assembly language. At the time, the contractor up in Huntsville, Alabama, provided support on the software, the operating systems, and all that stuff, to allow us to test the interfaces. They provided the backbone and frequently they would send us updates to their operating system, another tweak here or tweak there. So Jim and I weren't writing the main backbone, but we spent a lot of time troubleshooting some of the problems that came along. They would make an upgrade with software, but it sometimes introduced another problem. That's always the way software is; you think you've fixed the one bug and you create another. A lot of his stuff ran on things called Perkin Elmers, a type of computer. We would travel to classes and learn about what Perkin Elmers were doing, and some of the new functions that we could maybe incorporate into our software that we used to test out some of our Level-IV, for payloads. I traveled to Huntsville a lot."

Sue Sitko (left) and Cindy Martin-Brennan troubleshooting a software program in preparation for Level-IV testing. [NASA/KSC.]

STS-51F Payload Specialist (PS) John-David Bartoe explained that before coming to KSC, the actual High Resolution Telescope and Spectrograph (HRTS) flight instrument was taken to White Sands (New Mexico). "They set it up and then we brought the whole crew out and had them operate the experiment; the Sun actually pouring into the instrument. A great way to find all the problems in the software for your experiment is to let the crew touch it."

Donna Bartoe told the authors that there were actually two Naval Research Lab experiments on the IPS. "One was the HRTS, which was a telescope and had a film camera. They did a lot of pointing and taking pictures during the mission but they weren't getting [real-time] data. The other was SUSIM [Solar UV Spectral Irradiance Monitor]. They already had their ground checkout procedures for those kinds of environments, so I had already met the development team when I did my trip to NRL and had started looking at their procedures."

John David Bartoe was from the National Research Laboratory and was working on one of the three solar instruments when picked as one of the Spacelab-2 Payload Specialists. He spent a lot of time down at KSC learning about the new IPS and the experiments. There were solar telescopes to observe the Sun, and experiments designed to look at deep space; conflicting requirements flying on the same flight. After Spacelab-2, for a short time the two were going to be split up to fly again on two separate flights. SunLab would just fly the solar instruments, while Dark Sky was going to fly the deep space instruments. Scott Vangen told the authors that, "Mission Manager Roy Leicester invited me to join him when they did preliminary discussions with the Europeans on that, and I flew to the United Kingdom and France with him when we were thinking there might be both SunLab and Dark Sky missions."

Cindy Martin-Brennan recalled that, "One of my first experiments was called the Solar Spectrometer, and it was a French pallet experiment that also flew later on ATLAS-1. We were basically test engineers, testing the experiment to make sure the interfaces would work, and the way we tested the interfaces was trying to break them. So, often I would go to Germany, Japan and France, to work with these folks whose experiments were on the Spacelab and see how their stuff operated in their home laboratories."

John-David Bartoe recalled: "One thing that we did, that I was insistent on very early when we were first developing the flight computer before the experiment, was that I told my software engineer I wanted all the pages he used to be the flight pages; not to go create several ground pages to run some things that were hidden somewhere that I didn't know anything about. So from day one, we learned the instrument connected with what it was actually going to look like during the mission, which actually turned out to be a big help for the software team."

"We were trying to learn this artificial intelligence, to figure out how to incorporate that into the way we were testing some of the payloads in Level-IV," explained Cindy Martin-Brennan, "so with that training, again, I got training on the PCU but also training on these Symbolic machines, where Jim [Dumoulin] was the lead."

The Level-IV Electrical and Software people clearly had a number of issues and tasks to work on Spacelab-2, but so did the Level-IV Mechanical and Fluids people

Angel Otero recalled that, for the Hard X-Ray Imaging of Clusters of Galaxies and Other Extended X-Ray Sources (XRT, Experiment 7) on Spacelab-2, "There were some gas bottles that we needed to use to refill the tanks of Experiment 7. In the offline lab, they had five bottles, but they couldn't tell me which one was full or empty. I wanted to make sure the bottle that I plugged in was full and the panel that we were going to use before the test had a gauge where I could plug in the bottle and open a valve so that I could see the pressure. I closed the valve, [and] put the bottle back in. It was a very specific gas mixture, of xenon, helium, methane, something like that. One of the PIs saw me, or I told them I did it, and he went mad at me saying I had contaminated the whole bottle because I'd opened it, and oxygen had gotten inside the bottle. Oxygen could not be let into the experiment because it would deteriorate the detectors. There was 900 psi inside this bottle, and he was going crazy, saying that because I opened it for a second to let the gauge read it and then closed it again, I had just contaminated the whole bottle. I remember talking to Robbie Brown [Teledyne Brown Engineering] about it, and even going down to talk to the chemist in the lab. I remember buying chemistry books talking about osmosis, to figure if there was any possible way that I could actually have screwed this up. The guy was adamant that I screwed up. The main PI was in England, so of course they couldn't convince anybody that I damaged anything, but for the next two weeks or so, I had to take samples of the gas coming out of the experiment every day to verify that nothing had deteriorated."

On Spacelab-2, Luis Moctezuma performed the alignment/coalignment of the IPS Solar Telescopes, the X-Ray, and the infra-red telescopes. He explained to the authors that the installation and alignment of the Solar Telescopes that flew on the IPS was interesting, since it had to be performed with the cruciform in a vertical orientation in order to minimize gravity effects on the alignment measurements as the telescopes were installed. The assembly and alignment was performed in the Apollo Telescope Mount (ATM) cleanroom and required the removal of the center section of the ATM floor. Three pedestals were installed to a "seismic pad" at the center of the ATM and the cruciform was installed vertically onto the pedestals. Felix Joe was responsible for the installation of the telescopes to the cruciform, as Luis recalled: "He would do the work on first shift, and I would do the alignment on second shift. Alignment was typically done during or right after installation. It required quiet time, so any delays had the potential to impact schedules. At the beginning of the work, we had a software problem and could not get the system going. To avoid any schedule impacts, we collected the data manually and I ended up doing the calculations by hand."

Scott Vangen recalled that, "The very first thing they had to do was put the IPS with the instruments hanging over the Igloo, which violated the very first rule − never pinch off the Igloo − because if you think about it, once you got it integrated you could never get back in there. You'd have to remove the whole payload assembly and you would have to go back to square one. Fortunately they never had to go into the Igloo, but that was a really bad decision."

Safety was always number one at KSC, so Safety personnel were always involved with any operation that could be dangerous or hazardous. John-David Bartoe recalled a couple of comments about safety. "When I was down there, I actually found the Safety crew to be very cooperative whenever we had to interface with them. We kept our instrument under nitrogen purge all the time, and of course nitrogen is a really big deal at KSC. But the Safety guy was there for 20 minutes and that's all it took for him to be totally comfortable with what we were doing."

Gaseous helium (GHe) and liquid helium (LHe) were all used a great deal at KSC. GHe was a pressurant for those experiments that used cryogenic fluids because its freezing point was so low, about −452.2 °F (−269 °C), so the gas would not freeze in the very cold systems. Different forms of LHe were used for the experiments themselves. While not cheap, like GN2/LN2, it was still a commodity in high use at KSC.

Frank Valdes recalled that, "For the time, we were the highest user of helium in the country. We got the helium from the National Bureau of Mines, probably from somewhere in Texas.[4] So we were getting five 500-liter Dewars[5] of liquid helium delivered a week, and we converted that into superfluid helium. We had to do that offline, between me and Luis [Delgado]. We actually salvaged some of the Apollo portable Dewars as well"

Luis Delgado explained that, "For the superfluid helium we had to get liquid helium shipped from Amarillo, Texas into a big commercial Dewar. We put them in the ATM Clean Room and we had to use cranes to pick them up to figure out how much weight we had. Maintaining the Ground Support Equipment for the helium experiments was a job on its own. The biggest problem in the world for liquid helium and superfluid helium is that you're working below atmosphere pressure. So anything that leaks, it's going to leak into your tank. And once it leaks inside the tank, it freezes and things blow up.

"We had to service the Dewars by pumping down. We lowered the pressure on that Dewar, and that eventually lowered the temperature of that helium. It went from 4.2 Kelvin to 2.2 Kelvin. That becomes the superfluid for all experiments, so we had to fill the experiment with helium that was very close to that superfluid helium temperature, and then immediately start pumping on it to keep it at 2.2 Kelvin."

Frank Valdes added, "We would do a top off. Luis Delgado and me had to practice, which we did on North Rails [Level-IV area] a couple of times, just to get the timeline to see how long it was going to take. It was a long process, to convert, and then that's when we ran into a problem by blowing one of them up. [That happened] over another long holiday weekend. We had just converted three superfluid helium Dewars. One of them had been superfluid for a week and we just topped it off and reconverted. We serviced it as the other ones because they transferred from the Dewars that we got from the Bureau of Mines. They were beasts, those

---

[4] Created in 1910, the United States Bureau of Mines (USBM), part of the Department of the Interior, served as the primary US government agency for conducting scientific research and the distribution of information on the extraction, processing, use and conservation of national mineral resources. The USBM was abolished in 1996.

[5] A cryogenic storage Dewar (named after Scottish chemist and physicist Sir James Dewar, FRS, FRSE (1842−1923) is a specialized type of vacuum flask used for storing cryogens such as liquid nitrogen or liquid helium, whose boiling points are much lower than room temperature.

lightweight, easy portable, commercial Dewars. We were just going to top off the Dewars because if they warmed up, it took a while for them to cool back down. And you had a lot of loss, a lot of boil-offs."

Cryogenics was not the only hazardous element of Spacelab-2. It also had NASA Standard Initiators (NSI), as Gerry Rivera explained: "There were two panels on the aft flight deck of the Orbiter that we mostly worked with, a Standard switch panel and the R7 panel, which was the big Spacelab subsystems panel. But there was a third panel called the IPS. It would control the NSI and it had the primary and the secondary NSI circuitry. The panel had a test port to hook up a special meter designed for that. It was an Ohm meter that was on a very low current scale, and it was set up permanently that way with a metal overlay, with a NASA part number and everything. There was no way you could overdrive that one Ohm resister inside the NSI; more than five amps and it would blow the NSI. Just by hooking up an Ohm meter on the wrong scale, you could blow the NSI."

Lori Wilson remembered: "The OPF personnel were extremely helpful with our requirements for one of our Spacelab-2 telescopes, which required continuous power with a two- or three-hour window in emergencies. A lost power anomaly occurred once about 2:00 a.m., and they called me at home. I, in turn, called one of the Electrical engineers to go in and fix the problem. All went well."

Focusing on two experiments, John Bartoe's HRTS and the X-Ray Telescope XRT, the Level-IV End-to-End (ETE) testing proved valuable. John Bartoe explained to the authors, "I was always a big believer in End-to-End tests."

One ETE test involved shining a bright light into HRTS and moving it up-down-right-left. Then the currents in the IPS gimbals were measured to be sure the IPS was also trying to move in the correct direction. The second ETE test also occurred in the Orbiter. The camera inside HRTS sent a signal from HRTS to IPS, to Spacelab, to the Orbiter, and finally to a monitor in the Aft Flight Deck. Both tests were successful.

Another ETE test ended in mystery, as Scott Vangen explained: "The PCU sent state vectors for attitude and velocity, all those quaternions, all that stuff. We couldn't explain some actions of the XRT. We suspected that the Orbiter state vector coming from the Orbiter to Spacelab had a sign error or something like that, and it turned out it was an old problem that had been there since STS-1. It explained a problem they had with the KU-band antenna pointing." John-David Bartoe noted, "I understood that two of the angles were reversed; it was beta-alpha instead of alpha-beta." Scott Vangen thought this had occurred "when we were doing all that collision avoidance mess." According to Donna Bartoe, the Orbiter ICD was wrong, which John-David Bartoe concurred with: "Because all the ground testing was against the ICD. My recollection is that it was only solved once we got on orbit. I can remember those air-to-ground calls where they said, 'We think we figured out the problem.' I think the storyline there, once again, was the importance of the fidelity of testing. Even with the limitations we had, it caught those things that just pure simulation would never find."

Operations at the pad were always a huge undertaking, as explained by Angel Otero. On Experiment 6 (Element Composition and Energy Spectra of Cosmic Ray Nuclei) for Spacelab-2, he had to keep a constant purge while the Orbiter Payload Bay doors were open, and then disconnect when the doors closed. He had to enter the PCR and change the bottles a couple of times because it was a constant purge, 24/7, so those bottles would not last forever: "I hooked it up to the experiment and I had a computer that I had to program to make sure that it opened the valves. There were a bunch of valves inside that would keep track of the pressure, and then adjust the pressure and open the valves to flow gas into the experiment depending on what the conditions were. The second bottle was there so that you could switch over to it while you were replacing the first one when it emptied, so that you could have continuous flow. There was never a break."

Lori Wilson observed: "One of the jokes that I remember was about an experiment on Spacelab-2 while at the big Shuttle scheduling meeting. It required servicing with superfluid helium at strictly scheduled intervals. The Periodic Table sign for helium is 'He', with a capital 'S' for superfluid. It appeared on our schedules as 'SHe servicing!'"

The High Crew were a very special group of people that could provide access to anywhere that needed to be put up in the PCR (or any other location at KSC). Using scaffolding boards, rope and c-clamps, they could erect any possible configuration to get personnel into places most thought were impossible. Due to their very specialized talent, they were the only group at KSC that could work at heights without a safety harness. Their expertise was used to support all the planned pad work but became essential in helping the Level-IV team solve what sometimes seemed to be insurmountable problems that occurred later during pad operations.

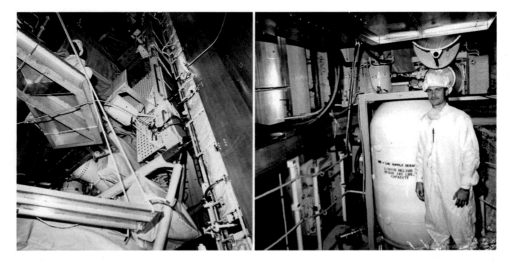

Left: Gary Bitner on access platform performing servicing of Spacelab-2 experiments. The view is looking up as the Shuttle (yellow handrails) sits vertically at the pad. Right: Luis Delgado at the pad, servicing Experiment 13 with liquid helium. [NASA/KSC.]

"Jay [Smith] and I were transferring superfluid into Experiment 5," recalled Don Dolby. "We had to pump all that down and I think we worked multiple shifts. Everybody coming in, working all night, pumping the helium down for days, I guess, to get it at as cold a temperature [as possible]."

Pad operations could have some very interesting aspects, especially when the weather turned bad, as Don Dolby remembered when he and Jay were working out there on a stormy night. They could see and hear the lightning popping and they could actually observe the Orbiter moving a little bit from all the wind, but they were stationary inside the Payload Changeout Room (PCR) on the Payload Ground Handling Mechanism (PGHM) that enabled access into the Payload Bay. Don recalled: "Working up on those boards, we had to be really careful because we had all our tools tethered and we certainly didn't want to drop something down inside the Payload Bay and hit another experiment below us, or bang into the Orbiter."

Lori Wilson explained that, "Pad work proceeded without conflict. We understood the Orbiter requirements and they, in turn, supported ours. On Spacelab-2, an issue came up about the scheduling of some experiment servicing, but once I explained that we had a window in which this MUST be done, and that I wanted to schedule it as late as possible in the window for a reason, they worked with us on it." All in all, the one thing that impressed Lori with working at KSC was the level of cooperation among groups. Whereas most teams in industry consist of maybe 10 or 20 people, all the different groups in Level-IV worked together with each other and with the Orbiter people as one team: "I say it again: No better place to work."

Jim Dumoulin was the prime Level-IV Experiment Engineer on Spacelab-2 Experiment 14. This was a plasma physics experiment from the University of Utah that worked in conjunction with the deployable Plasma Diagnostics Package (PDP) from the University of Iowa. Experiment 14 had a Fast Pulse Electron Generator (FPEG) and a Langmuir probe that was mounted high on a Spacelab pallet, as close to the sill as it could get without hitting the Payload Bay doors. As Jim explained: "I joined my PI and tech on the sill of the Orbiter to perform pre-launch electrical bonding measurements and to clean the surface of the Langmuir probe. As my tech gently wiped the tip of the probe, the silver sphere came off in his gloved hand. Due to the launch schedule, we rolled out to the pad with the probe broken and unsure if we had the time or access to do a repair at the launch pad. I spent the next two weeks working the paperwork so we could relocate access platforms and mix a special conductive epoxy in the PCR at the launch pad. Due to outgassing, the fast-setting epoxy had to be mixed on a grate outside the only bathroom on the Rotating Service Structure [RSS] and we had only minutes to run up two flights of stairs and pass it to a bunny-suited technician in the back door of the Clean Room. The tech quickly relayed it up to another tech on the payload handling fixture, who attached the silver sphere to the top of the probe before the epoxy set."

Working with cryogenics involved a constant battle to keep the flight hardware and GSE from freezing. Don Dolby continued: "Jay and I worked with Luis quite a bit on all that cryo stuff. And I remember we had heat guns that we'd have to point at the relief valve while we were pumping that down, just to keep them from over pressurizing. I think they'd have a couple of us come in and work different shifts while we were pumping all that stuff down. It took days, if I remember correctly, and we had to keep a heat gun on there to keep those relief valves from freezing up."

With all the GSE required, all the operations taking place at the pad, and all the timeline crunches needed to get the work completed and not impact the launch, eventually something was going to go wrong. As Frank Valdes recalled: "We had the 'Fire at the Pad' incident. JPL [the Jet Propulsion Laboratory] had obtained these brand spanking new vacuum pumps and they wanted to take two of them. They only needed one, but they wanted to take both. I really didn't want to haul out all that equipment; we didn't want more than we needed."

Luis Delgado added that they were pumping down the experiments and the Dewars, and Frank Valdez had Steve Williamson, the technician, build a totally illegal electrical extension cord: "They were overloading this extension cord. The PI was looking at the pressure gauge and the reading was showing more pressure was needed. The pumps were running hot, and the cables were running hot, they just didn't know it. When they shut down the pump and reconnected all the pumps and hit the start button, everything went south. That surge exceeded the limit of the cord, sparks flew, and we had a fire on the extension cord."

Frank Valdes continued: "[Mike] Kienlen and I were standing there and talking and there was this orange glow on the [Orbiter] doors. And we just looked at it and said, 'That doesn't look good. It doesn't look right. Something's not right now'. Sure enough, the adapter cable had started to smoke." Luis Delgado added, "Then, Mike Kienlen, one of the Operations folks, grabbed one of the fire extinguishers and hit it when it was almost out. We shut down the power breakers."

"It had just a little three-inch flame on it," explained Mike Kienlen, "and the engineer, being a smart engineer, ran to turn the power off. Me being an Ops guy, I ran straight to the fire, not knowing what the heck I was going to do when I got there. It was shorting out to the diamond plate and sparks were flying all over the place, so I grabbed the fire extinguisher and we put it out. After reconfiguring the pumps to their own circuits, we were sitting there and the technician said, 'Well, we can turn it all on now, or what do you want us to do?' and we looked at each other and said, 'Well, if you call the Firing Room and ask them if we can turn it on, they'll probably have to talk about it for a couple of hours'. The science had to have it turned on, and the scientists were screaming we had to get the power back on. So we said, 'All right, turn it on, turn the cryo pumps back on and once it's up and running we'll call the Firing Room and tell them we've turned the pumps back on.

"The incident got back to the Test Conductor, and of course got back to the Fire Marshal. He came out there; I guess Frank and the team just didn't report it fast enough for them, for their liking, and we kept on going. [Onsite] cops and everybody, the fire department, they showed up in full force." Luis Delgado recalled, "Lockheed people came over and everybody had to inspect and look at the burning cable. We had to wake up Herb Brown. He had to come in the middle of the night to look at the cable." Frank Valdes added that, "[Bill] Mahoney then wrote out a letter saying that there would be no electrical work done by Mechanical personnel unless blessed by Herb Brown."

The fire was now behind the Level-IV group, the cryo servicing was complete, and final operations were underway to close the payload pay doors for launch.

T-24 Hours and Leaking

Damon Nelson explained that, "Twenty-four hours prior to launch, we were backing out of this equipment, they were getting ready to do the Payload Bay door closing, and we had our experiment running; it was actually pumping to help keep the helium conditioned for pre-launch. The PCR is a super-clean facility and we were all in our Clean Room garments. As we were walking out, a Shuttle guy tapped me on the shoulder and said, 'Wanted you to see something here'. The hardware was in a vertical orientation, so everything was 90 degrees straight up. Then you had the Orbiter cargo bay, and this guy showed me some oil that was on the protective liner of the cargo bay. They were getting ready to remove this liner to close the Payload Bay doors and he pointed up and said, 'You got an experiment that's leaking oil onto our Payload Bay'."

Frank Valdes recalled: "We had worked 26 hours or whatever, then Damon was finishing up the closeouts, taking all the GSE out and closing out stuff. Then they found oil leaking out of one of the flight boxes, the VMA [Vacuum Maintenance Assembly]."

Luis Delgado remembered it distinctly because it was the second problem that Experiment 13 had after the servicing: "The VMA was a pump system that was on the payload and it was fed through T-0 power and would keep this pump running for the two days before we launched. The mission managers got together and decided to shut down the pump early, so the Experiment 13 PI was crushed. He was walking around for that last day like a family member had died. The experiment, he said, would slowly warm up and pretty soon burst disk number one was going to blow, and then pretty soon burst disk number two was going to blow and it was going to be dead."

Jim Dumoulin said, "I was on the C-1 Console during the third shift for the Spacelab-2 launch attempt on July 11, 1985 when one of the flight boxes for the JPL experiment on superfluid liquid helium [SL-2/Experiment 13] started to overheat. I woke up the JPL PIs at their hotel and they said that without the pump, the experiment would be ruined. For the rest of the night, I monitored the temperature of the superfluid helium tank. Without the added vacuum the tank began to go critical, and just when the Sun came up it ruptured its burst disk, flooding the launch pad with helium. All throughout the launch pad there were little tubes connected to a mass spectrometer in the MLP [Mobile Launcher Platform] sniffing for helium. When the tank ruptured, every alarm on the launch pad went off, signaling a massive hydrogen leak. The guys at the C-3/C-4 [Cryo and MPS consoles] went scrambling and spent most of the morning resetting their systems. Their day went from bad to worse when, at T-3 seconds after main engine ignition, a malfunction in a coolant valve caused an RSLS [Redundant Set Launch Sequencer] launch abort, halting the launch."

John-David Bartoe remembered: "Our pad abort was the second one. The first one [STS-41D] was the one that had the hydrogen fire underneath, which could not be seen because they didn't have any IR sensors at that time. Ours was the second one, and that was the one and only time during the mission that my heart rate went through the roof. My first thought was a fire like what happened before.[6] But it turned out to be okay."

Because of the pad abort, the launch would not occur for a week or so. With the delay, Level-IV personnel had a window to try and remove the failed pump, get it fixed and install it back on the payload in time to make the launch. Of course this made the PI very happy because there was chance his payload would not fly dead. The original installation had been performed by Level-IV engineer Tony Ornelas, but that had been while the payload was horizonal. Now it was vertical, so the problem was how to remove something at the pad in an orientation never tried before. Tony figured it out, as he explained: "I remember going in there late one night. The access, as configured by the High Crew, was really, really tight and hard. We had to go in there and pull it out and people were just concerned that we really had to be careful because if we dropped something there was going to be trouble. We had to make sure not to drop anything and always tethered tools and stuff like that."

In preparation for the repair, the pump company was contacted and sent soft-goods to KSC. The pump was sent to the tech shop in the O&C, where it was taken apart to figure out why it leaked oil, then rebuilt and thoroughly sealed, putting RTV all around the case. Luis Delgado added, "That way, if it leaked into the case it was still going to be contained and it wasn't going to come out and leak on the cargo bay." Once repaired, the pump was sent back to the pad, installed, and it

---

[6] John Bartoe refers to the immediate events following the June 26, 1984, STS-51D pad abort that raised concerns from the crew when reports of residual hydrogen gas catching fire outside the Shuttle were passed on to them.

worked. Then Frank Valdes and Luis Moctezuma had to go through all the helium servicing steps again to get it fully operational.

The delay helped one experiment, but for another it caused a major problem. Experiment 3 (PDP – Plasma Diagnostics Package) had a concern that the batteries were going to die, so Level-IV had to change them out even though that was never meant to happen. As with the VMA, the experiment was now vertical. Jay Smith recalled: "The Principal Investigators were all freaked, and so Tony [Ornelas], Gary Bitner and myself went out there and took that thing apart. Of course if we dropped something, it was going to hit everything on the way down to the bottom through all these other experiments, but we got those batteries out, replaced them, put it all back together and didn't screw anything up. Those Principal Investigators were so happy that whole thing worked, I think they bought us tequila."

Battery replacement for Experiment 3 Plasma Diagnostics Package (PDP), with (l to r) Tony Ornelas, Gary Bitner, Jay Smith. [NASA/KSC and Jay Smith.]

Tony Ornelas admitted, "Removing the payload batteries was a last-minute-type deal. That's another important thing to show, that things like that would come up and we didn't just perform work in the O&C. We worked at the pad as well, getting ready to launch, and it was still a Level-IV activity." Level-IV personnel were responsible for going in there and changing out those batteries, even though it was in the Orbiter at the launch pad getting ready to launch. That was something Bill Jewel and the upper management fought over, because the Vehicle Manager said, "No, that's our job. That's our vehicle, we're going to do that." So Level-IV management went to bat for Tony and his team, stating, "No, you're not. Level-IV has the experience, the knowledge, the engineer. We will do that." Tony continued, "So that, I think, was a point of contention for a while before it was actually resolved. When we were done, I went to sleep in the O&C Building on my desk."

The superfluid helium had to be in its cryogenic state at launch. The vacuum pump that was serviced before it launched was connected into room 10A in the MLP to maintain the superfluid helium. Everything was going fine. Then the people sitting at the C-1 Console monitoring the temperature of the experiment stated that something was not right. The temperature was varying, gradually, in ranges so something was wrong. The team knew it was still cryogenic, and set about trying to figure it out, as Craig Jacobson recalled: "Finally, I used some of the post-processing tools on LPS [Launch Processing System] and found out that there was a 60 Hertz signal imposed on the temperature signal. So I went down to PCR with my toolkit, with my soldering iron and everything in it. I wrote a TPS [Test Preparation Sheet] and got a burn permit. I was in room 10A in the LPS, taking the GSE control box out of the rack and soldering a capacitor to filter out the 60 Hertz signal across the temperature lines. While I was doing this, the Lockheed guys were standing over in another part of 10A looking at me, wondering what this NASA guy was doing there. I was waiting for them to call somebody and complain, or be concerned or whatever, but they didn't do it. So, soldering the capacitor on it took care of the problem and the thing flew. The temperature came in the way it was supposed to and they actually left that capacitor in the GSE. It's been there ever since."

Finally, after all the work before the pad abort, and the fixes following the pad abort, Spacelab-2 launched. Mike Haddad recalled the event clearly: "I remember a number of us were at a local Cocoa Beach restaurant watching the launch from there. We had been there quite a while before launch consuming alcoholic beverages, and as [the Shuttle] ascended and thinking all was good, I made a trip to the restroom. Coming out, everyone had a worried look on their faces. I asked what was going on and they told me that one of *Challenger's* main engines had shut down and they may not make it to orbit – and if they lost a second engine, be prepared for a bad day. The failure resulted in an Abort to Orbit [ATO] trajectory, whereby the Shuttle achieved orbital altitude, but lower than planned."

From his perspective on the flight crew, John-David Bartoe added that, "The on-orbit problem, with the lower orbit, [meant] all the timing was off for all of the documents that we had on board. All the timing was incorrect. So Mission Control and the POCC had to take over and generate a new set of pointing plans for every single orbit for the entire mission. And then they would send up the procedures for the IPS on that teletype machine on board. We had paper everywhere. It was the first time that they ever had to change the roll on orbit because the first roll ran out. [There was] so much paper on that. There's a great picture of Roy [Bridges] with paper all around him, but it was really the ground team that was just hustling every 90 minutes to come up with the next 90 minutes."

All of the GSE that was in the PCR had to remain there during launch, because there was not enough time to remove it all between completing the servicing and launch. As Luis Delgado explained, "After launch, we had to take those Dewars out of the PCR. That was a process. They would bring a crane, we grabbed the Dewar, we picked it up off the extended platform and lowered it to PCR floor. Once rolled out of the PCR onto the deck of the pad, we had a hammerhead crane outside that we used to hoist these Dewars 200 feet up [and then down to the pad surface]."

Eli Naffah added, "I think the ultimate was seeing something that you put together actually fly. The really cool thing about Level-IV experiment processing was that you got to see it from beginning to end. Obviously we didn't design the experiment, that was done before it came [to KSC]. But to go from drawings, to hardware, to integration, to test, then see it operate on orbit with the astronauts there, and then have it come back and then you remove it, that was a full cycle. In NASA, you didn't often see that in a relatively short period of time. I thought that was really cool."

Craig Jacobson remembered one particular issue: "The SOUP payload [Solar, Optical, Universal Polarimeter] was on the Spacelab-2 cruciform. Ann Bolton was the lead and I was back up. They had a problem powering up [on orbit] and they spent half the mission trying to get power up. And then suddenly, somebody just happened to notice that it was powered up. So they did get to do some observations, and we tried to figure out what was going on but didn't ever sort it out. But in hindsight, we had learned later that because of the wiring going into the cruciform, to the Orthogrid on the IPS, the 28-volt power dropped in voltage to 24 volts or something like that. Over the years, the more I've thought about it, the more I kind of think all of us didn't quite handle that right, because in Level-IV we cranked up the power supply voltage to overcome that. Well, you can't do that on orbit. So we tested in a non-flight-like environment by cranking the voltage up."

The following is a continuation of the exchange that took place during a gathering of Level-IV personnel and crew members:

John-David Bartoe added, "Another thing I wanted to mention was the extreme value of having KSC people in the POCC during the mission. Donna [Bartoe] was there, Scott [Vangen] was there." Some in management did not want KSC personnel at the POCC, but KSC management and the PIs really pushed for it. They understood that having the experience of the Level-IV people on-hand during the flight was essential for mission success. KSC management was willing to take their people out of the flow for the next Level-IV mission they were doing. For Spacelab-2, Scott Vangen had to get out of the Astro-1 flow, which had already started at KSC. Scott Vangen recalled: "It wasn't just the mission itself. You went up for the POCC Sims [simulations]. Multiple flows doing all of those things. That was the best time of our lives!"

As the Spacelab-2 mission was coming to a close, the teams on the ground were preparing for post-landing operations. One of the telescopes used film to record some of the data and it was a requirement to remove that film within one hour of landing. The experiment was NRL's Solar Ultraviolet High Resolution Telescope and Spectrograph (HRTS). Felix Joe said, "Gene Krug and myself, and I believe Steve Bigos and Walt Preston, were the techs that went to recover the film at Edwards."

John-David Bartoe wanted to explain about Spacelab-2 from the point of view of what he saw as the value of Level-IV: "First off, for the crew to be able to control the actual instrument in Level-IV was a huge advantage! At an STS-51F Technical Crew Debriefing – these were the debriefings where you sat down with George [Abbey] and talked about the mission – we were having a conversation about the IPS and the simulator at JSC, which was a piece of garbage. Loren Acton said, in retrospect, that we should have made another trip to Dornier and used their engineering unit, but Tony England didn't think it would have all that capability, and that the only facility that really modeled the Orbiter/IPS interface was at the Cape. Tony continued to say that integration testing at KSC was where state vector losses were giving us a really funny reading. That was real, and that's what we found on orbit. Tony thought that part of the solution was that somehow the users had to have more time during hardware integration at the Cape to exercise different modes. That was only place you could have gotten the experience. Tony said that the unofficial training that we got the Cape was really necessary and extremely valuable. He could not imagine flying the mission without all those hours he'd had to do that at the Cape."

THE CALM BEFORE THE STORM

STS-61A, *Challenger,* launched October 30, 1985, 7+ days
Payload: SPACELAB D1 Long Module (MD001); Tunnel (MD001); Floor (MD002); Racks (see Appendix 2); plus German unique support structure.

The primary objective of the Spacelab D1 mission was to conduct basic and applied materials processing research in the Spacelab module, funded by DFVLR (Deutsche Forschungs- und Versuchsanstalt für Luft- und Raumfahrt, or German

Test and Research Institute for Aviation and Space Flight), and in the Materials Experiment Assembly (MEA) funded by NASA. Life sciences investigations were also conducted in the microgravity environment of space. Approximately 70 investigations were defined, covering fluid physics, solidification of crystals and metals, and human and plant cell responses to weightlessness. Most investigators were German; however, there were researchers from France, Italy, the Netherlands, Spain, Switzerland, and the United States. Many of the life sciences investigations were performed using two facilities, and some included a cooler/freezer combination, one incubator for the 18–20 degrees C range, a second incubator for the 30–40 degrees C range, and a glove box. There was also the Vestibular Sled which included the sled unit, the human vestibular system experiment, and the space motion sickness experiment The Long Module contained three Spacelab double racks for the materials science facilities and three single racks for life sciences investigations. The crew included NASA Mission Specialists and two foreign Payload Specialists. Useful data were collected from almost all investigations. The spacecraft was gravity-gradient stabilized.

"For D1," Angel Otero explained, "Donna Dawson, Scott and I, and one other person, went to Bremen about a year before the D1 payloads were supposed to get to KSC. We went up there to see how it was going, because for D1 we were not supposed to do any work. All the Level-IV work was supposed to be done in Germany."

As the time approached for transporting the Spacelab D1 hardware from Europe to KSC, it became apparent that not all the Level-IV-type work would be completed before arrival in the United States.

Mike Kienlen worked both Spacelab D1 (STS-61A), and D2 (STS-55). At that time, he was about to become the Level-IV Test Conductor. "They [the Germans] showed up at Kennedy Space Center six months before flight, with two 747s full of equipment [the experiment Rack Train came to KSC in a USAF Lockheed C-5A]. They were good engineers and working with people from another country was just so neat. They were so happy to finally get to KSC and do that mission. Their launch party was the best! They flew in a private charter Lufthansa flight from Germany to bring the food for their party. So all those containers were empty, or mostly empty, when it was time to go home. They loaded all their luggage and the beach stuff they bought, because they were here for a year, and then they insulated one of them and filled it with dry ice because they probably took back 1000 pounds (453 kg) of beef that they'd bought because it was so much cheaper in the US. One guy bought a Harley Davidson, took it apart and shipped it back in pieces. So when the mission was over, we spent a whole bunch of time helping them load everything up that no one knew about in these containers."

As the hardware was being unpacked, certain assembly and integration activities fell to the KSC Level-IV team. With the fact that there really was no plan to do a lot of Level-IV work on D1, most of the engineers already had commitments to other flights. Some of the engineers could take on the extra duties, but some operations fell on others – mainly the Co-ops – to perform those tasks required.

"One of the big ones I was involved with was Spacelab D1, which was the first German Spacelab," Eli Naffah told the authors. "There was a venting system inside the Spacelab that I was responsible for putting together, and I had to work with the Germans in doing so. Nothing fitted at all, the drawings were garbage, so we had to do a lot of modification in order to get it to fit in the Spacelab. I remember when we were done our procedure [paperwork] was all red; like we'd bled all over it. But it was good, it ended up working quite well on orbit. It was kind of cool getting to work with them."

One of the main experiments was the Vestibular Sled, with a seat that ran along rails mounted to the center of the Spacelab module floor. Angel Otero remembered, "The first thing I did was throw away Mission Specialist Guy Bluford's back support. It was between the backrest and the seat on the sled. The backrest was folded and there was a piece of foam, and foam needed to be wrapped in plastic so it didn't shed. This one was shedding, you could see chunks of foam falling off it. We couldn't keep that on the flight hardware, so I threw it away. The next day, one of the Germans came by asking me if I'd seen Guy Bluford's back support. I asked what it looked like and when they explained I said to them 'It's in the trash'. I said, 'You guys can't have something sitting on flight hardware that's not wrapped in plastic because it was shedding. Not good; it could contaminate the hardware'. So they had to make another one for him."

Other problems occurred that needed to be documented, but the way they were documented could sometimes get the Level-IV person into trouble. Mike Haddad remembered, "They were having some kind of problem with the sled, and to show the people offsite that could help fix it, video had to be taken of the sled in operation. At that time, the photo support was a contractor who did still and video, and could be called up at short notice to support us due to the nature of our work. But they said they could not support with this video until the next day. The sled engineer needed it right away because they would soon be heading to the airport. We had a mini-video camera in our office that used the small VHS tapes, so I got suited up in the bunny suit, went in the module and videotaped the operation. We gave the tape to the sled engineer and off he went to the airport. There was a photo support contractor taking still images in the module at the same time and he knew I did not work for his contractor, so he wrote a grievance against me for doing what should have been their work. I did not know this till days later, when Dean Hunter called me into his office and told me about the grievance. I said, 'Dean we needed the video', and that was the last I heard of it. Not sure what Dean had to do, but he covered our backsides so many times."

A ground version of the sled was also brought to KSC, as Angel Otero explained: "We had the flight version and then they had a ground version, and they needed to build up some ground data, some subject data on the ground. They were asking people to volunteer to ride the ground sled and then they would put the helmet on you and whatever. So I figured that if I'm the test engineer it would be cool, then at least I'd know what it felt like to ride on the thing. So I volunteered to be one of the test subjects and I rode on the ground sled."

While the sled was a high visibility experiment, with it running down the center of the module, there were plenty of other experiments that were worked on by Level-IV personnel, as well as other problems to deal with, not all of them only hardware related. Angel Otero elaborated: "Hector (Borrero) and I were doing a D1 job. We were going to do a test on one of the experiments and we needed vacuum to make sure that the sensors were reading right in the rack. I was inside the module with a German guy and a Quality guy, and Hector was outside with a vacuum pump and another German guy. When we started testing, we told Hector to start the pump. He cranked the pump up – and nothing happened. We weren't getting any vacuum inside. So I stuck my head out of the hatch and asked what was going on. And then the German guy that was with me started talking German to the other German guy, and they were going back and forth. So I started talking in Spanish to Hector, and then the German guy stopped and looked at me. I said, 'Either we're going to understand each other, or nobody's going understand anybody'. The Germans liked to switch to German when they didn't want to tell you they'd screwed up. So I told them I could play that game too."

As mentioned earlier, some operations had to be accomplished by other individuals, because the group normally doing certain operations (for example, the Electrical group was mostly responsible for writing and running the experiment test procedures) were already too busy working future missions. Members of the Mechanical group had begun to write and run testing procedures, and one of those fell into the lap of Mike Haddad: "Holography was used in the Spacelab D1 mission to study bubble motions in liquids and density distributions near the critical point, under reduced gravity. If I remember correctly, it was in Rack 10, located as part of the payload element 'Prozeßkammer' [process chamber]. It contained a Hasselblad camera that used film to record the data on orbit. The test procedure was very simple, which was probably by design. They did not want a rookie performing a complicated testing procedure his first time. It was basically a re-run of a procedure they had performed in Germany. I took their data and created the correct KSC-type procedure to perform the test. This was to ensure the equipment was still functioning properly after all the transportation and other work done at KSC. It involved flipping a few switches in the proper sequence and then getting information from the experiment to ensure all was up and running properly. The test went smoothly, no problems, and the German engineer I worked with was a very pleasant fellow. My first and only full-blown experiment testing procedure. Unfortunately, during the mission, something on the Hasselblad camera failed and the data was lost."

Often, especially after Spacelab-1, Level-IV had two, three and sometimes four different missions going on in parallel. Preflight, postflight, one was coming, one going; they would get overwhelmed. At one of the peak periods, they were so busy and at such capacity that partners were housed in trailers in the O&C parking lot;

there was just so much going on. The O&C was bursting at the seams. The partners were asking why they were out in these trailers; they were our partners, so why were they not in the building? "I don't know what came over me," said Scott Vangen, "but it was just kind of a spontaneous thing. I said, 'Well if you think about it, The Germans, the Italians and the Japanese, all in trailers. We won that war. We get the air conditioning'. And they burst a seam laughing and laughing. Then we stopped and we looked at each other and said, 'Isn't it interesting that the three axis powers of World War II, Japan, Italy and Germany, are now our three biggest international partners in space flight?'"

Every once in a while, a late requirement was established that Level-IV personnel had to try and implement, regardless of how bizarre it might seem. As co-author Mike Haddad recalled, "We were getting ready to closeout D1 when a letter came down from NASA HQ stating that all logos inside the module needed to be removed. It seems someone up there noticed that the German companies that were flying experiments had placed logos on the front of their experiments in the racks. NASA felt that if those appeared on TV, it would appear like NASA was promoting those companies. Why did they wait so long to come up with this requirement? We had been working this mission for years, and only now did they made a decision like this. I was assigned to come up with a solution and just flipped out. It was kind of crazy to begin with, and there was no way to remove these logos. The Germans were very upset, but something had to be done. Most of the logos were absolutely beautiful; some were stickers some painted on. So what should I do? We noticed that the crew really liked the Velcro that was placed in certain locations to aid their on-orbit operations and we had rolls of the flight Velcro left, so I went and covered up all the logos with Velcro. That way, nobody would see them and we didn't have to destroy part of the rack to remove them. That was all fine and good until about two or three days into the mission. The onboard cameras were taking video of the Spacelab operations, and we noticed some of the logos could be seen. My big boss, John Conway [Director of Payloads, KSC] came over to me really mad and said, 'I thought you covered them up'. I told him I had, and showed him the paperwork and the closeout photos. So what happened? I guess the crew got on-orbit, thought covering up the logos was wrong, and pulled the Velcro off. To tell you the truth, I was kind of happy about it."

As with other Spacelab missions, members of the KSC team were requested to support the flight. For Spacelab D1, the mission support was at the German Support Operations Center (GSOC) in Oberpfaffenhofen, Germany. As Angel Otero explained, "I came into work one day and the Dean [Hunter] called me over and said that the Germans, the Vestibular Sled team, had sent a letter to NASA asking for me to be sent over to Germany to support the mission. I worked the installation of the sled and I actually was the Test Conductor for the sled. But then when I got to Germany, Byron Lichtenberg, who flew on Spacelab-1 and who never showed up at KSC for Level-IV during D1, was running the sled for the ground team and he didn't feel he needed me. So I attached myself to the MEDA team, which was a materials double rack, and I ended up writing an inflight maintenance procedure to fix it because it quit working the first day. So I was pretty helpful to the team in that sense. During the mission, the pressure sensor that released the latch on the door failed. We were joking, because there was no way that you would be able to open a 19-inch (48-cm) diameter aluminum door if there was vacuum on the other side. The procedure was to take half the rack off and then find the cables connected to the sensor, and we actually had the crew show the cable on the camera. Then we could snip the correct cables, which would release the sensor so they could open the door and keep doing the experiment.

"I have a chilling picture of me arguing with [Alternate Payload Specialist] Ulf Merbold on console when I wrote the procedure. Actually I was panicking, because it was in my handwriting and they sent it to Houston first to get reviewed." Ulf was the Capcom [Capsule Communicator], and Angel was sitting next to him talking. Unbeknown to him, the lady whose house Angel was living in during that time in Germany had gone to the viewing room, seen him there and started taking pictures of him. So she has some pictures of Angel arguing with Ulf Merbold, because he had started deleting a bunch of steps from Angel's procedure: "I was arguing with him and then Air-to-Ground One, Houston, called back and said that they liked the procedure the way it was and they had faxed it to the Shuttle the way it was. I spent two and a half weeks in Germany supporting the mission on console."

Angel Otero, right, describing the MEDA inflight maintenance procedure to Ulf Merbold at the German Support Operations Center (GSOC) in Oberpfaffenhofen, Germany, during the Spacelab D1 mission, November 1985. [Barbra Backhaus.]

Hermann Kurscheid told the authors: "I was down for Spacelab-1 until our plane left with all the hardware back to Germany, and they had already prepared there the D1 mission. I was not involved in the D1 mission directly, except during the mission itself in the Payload Operation Center, but when I came back from Florida after the Spacelab-1 mission, I started with some other guys to propose the D2 mission. After that I was then in Germany for quite a while, getting hardware and experiment proposals ready to form a mission from them."

Following the flight of D1, the hardware remained at KSC in preparation for the next German Spacelab mission, but due to the *Challenger* accident just three months later, everything changed.

TAKING STOCK

STS-61B, *Atlantis*, launched November 26, 1985, 6+ days
Payload: Experimental Assembly of Structures through EVA (EASE) / Assembly Concept for Construction of Erectable Space Structures (ACCESS); MPESS (F004); IMAX Cargo Bay Camera.

This mission demonstrated construction in space by manually assembling the EASE and ACCESS experiments. The experiments required two spacewalks, by Sherwood "Woody" Spring and Jerry Ross, of 5 hours 32 minutes, and 6 hours 38 minutes, respectively. Also located in the Payload Bay were the Get Away Special (GAS) canisters and the IMAX Cargo Bay Camera (ICBC).

One of the many unusual payloads processed at KSC consisted of two space construction experiments: the Experimental Assembly of Structures in Extravehicular Activity (EASE) and the Assembly Concept for Construction of Erectable Space Structures (ACCESS). Approximately a year before the EASE/ACCESS launch, KSC participated in the Integrated Payload Final Design and Operations Review at Marshall (MSFC, Huntsville, Alabama). KSC NASA engineers participated in the integration of mock-up EASE and ACCESS experiments and Mission Peculiar Equipment (MPE) onto a non-flight MPESS at Marshall, and participated in the integration and testing of the flight ACCESS experiment at LaRC (Langley Research Center, Hampton, Virginia). These activities enabled KSC engineers to obtain detailed information from the investigators about the experiments. This participation also enabled KSC to estimate the integration time and the resources that would be required at the launch site more accurately. Using this information, NASA generated an EASE/ACCESS processing schedule for the entire payload flow at KSC. Procedures included EASE installation, MPE installation, and ACCESS functional testing.

No functional testing of EASE was required at KSC. The ACCESS experiment, on the other hand, required a thorough checkout. Since raising the ACCESS assembly fixture required the use of the crane, this procedure was hazardous.

Testing began by raising the fixture and verifying that it could be properly secured in the vertical position. A number of other tasks were performed, and finally two bays were assembled and then stowed. At the end of the functional test, the non-flight lifting hardware was removed from the ACCESS experiment.

With many payloads being processed in parallel, and KSC resources (including personnel) being shared by all payloads, the schedule usually did not allow for a flight crew training session, but this was worked in to the processing flow. This test enabled the crew to check out the EASE and ACCESS experiments and the MPE. The Mission Specialists assembled two bays of the ACCESS experiment and also fit-checked the contingency tools.

Payload sharp edge inspection was particularly important for Shuttle missions that included a planned EVA (Extra-Vehicular Activity). The ACCESS experiment and associated MPE were checked for sharp edges immediately following the functional test. Two sharp edges were found on the MPE and were rounded. ACCESS was then closed out for flight.

Just prior to EASE/ACCESS closeout operations, a last-minute modification was received from Marshall to install 32 logos onto the payload. A few logos could not be installed due to hardware interference. Don Dolby recalled: "I remember we had difficulty trying to get the letters on straight, but that's the only thing that really stands out in my mind on that."

Another closeout operation that was normally scheduled was clearance checks. During envelope checks for EASE/ACCESS, the outboard handrail for the top foot restraint was found to be outside the payload envelope by four inches. The MPE handrail had to be removed from the carrier the day before payload transfer to the canister, and Weight and Center of Gravity (WT&CG) measurements, had been performed. Since EASE/ACCESS required no interface verification testing and had no electrical interfaces with the Orbiter that needed verification, the payload was transferred directly to the Orbiter after Level-IV closeout.

First it was installed in the Payload Canister at the O&C, then the canister traveled to the Vehicle Assembly Building (VAB) and was rotated to the vertical position. It then traveled to the Vertical Processing Facility (VPF) in order to pick up the three satellites that were flown on the same mission. Following transfer of the three satellites from the VPF into the canister, the payload traveled to the pad PCR using the PGHM. Then it was transferred into the Orbiter. A final inspection was performed and the Orbiter doors closed for flight.

After the mission, the payload was brought back into the O&C. De-integration of the EASE and ACCESS experiments was accomplished in three days, while de-integration of all MPE hardware (including the foot restraints) took approximately one week.

Level-IV personnel were also involved with other payloads that would fly in the Orbiter, but what made these different was that they were attached to the side walls of the Payload Bay (called side mounted) instead of spanning across the entire width of the Payload Bay, like those on an MPESS, pallet or Spacelab module. Mike Wright recalled, "I was assigned as the EI [Experiment Integration] engineer for this first flight of the ICBC [IMAX Cargo Bay Camera]. As the primary test engineer at KSC, I had to be trained to use the crew hand controller, which at that time was a rather archaic piece of equipment that looked like an oversized calculator. I exposed a short stretch of film in the OPF as part of the Orbiter IVT [Integrated Verification Test], which I later had the privilege of viewing during a screening of raw footage for '*The Dream Is Alive*'."

It should be noted that there was another IMAX camera, called the In-Cabin Camera, which flew 12 missions between 1984 and 1986 in the crew compartment of the Shuttle and would be mounted to one of the windows to take video of activities occurring outside the Shuttle. Level-IV personnel also worked with the in PI preparing this camera for missions, by providing offline support, transport to the Orbiter's middeck, installation for IVT, and then final stowage in the middeck for launch [3].

STS-61C, *Columbia,* launched January 12, 1986, 6+ days
PAYLOAD: Material Science Laboratory (MSL-2) on MPESS (F006);
plus Hitchhiker G-1.

MSL-2 was a structure for experiments involving liquid bubble suspension by sound waves, melting and re-solidification of metallic samples, and containerless melting and solidification of electrically conductive specimens. Other Payload Bay payloads were: Hitchhiker G-1; 13 Get Away Specials (GAS), 12 of which were mounted on a special GAS Bridge Assembly; Infrared Imaging Experiment (IR-IE); Initial Blood Storage Experiment (IBSE); Hand-held Protein Crystal Growth (HPCG) experiment; and three Shuttle Student Involvement Program (SSIP) experiments.

Michael Stelzer, explained that, "As it matured, the Space Shuttle program developed a number of accommodations within the Orbiter Payload Bay for small payloads." Almost every mission had some variation of these types of payloads, so Level-IV personnel supported many missions besides the Spacelab-specific flights. Support included offline preparations for Orbiter installation, transport to the Orbiter, participation in installation and testing of the payload, and finally closeouts – whether in the OPF or at the pad.

Hitchhiker G-1 (HG-1) was developed by the Goddard Space Flight Center (GSFC, Greenbelt, Maryland) and was a carrier that could accommodate almost 1000 pounds (453kg) of payload. Hitchhiker G-1 was designed to fly on the forward starboard side of the Payload Bay. At KSC, preflight processing went pretty smoothly. Alan Posey (GSFC) was one of the primary contacts from the HG-1 Project, and kept Level-IV personnel tied in closely with the GSFC team. Under the leadership of Jerrace Mack, the HG-1 was checked using the Cargo Integration Test Equipment (CITE) Space Shuttle Orbiter simulator, which revealed an error in the command system interface design that required changes in the HG-1 ground system software to resolve. After a successful retest, the HG-1 was ready for installation into the Space Shuttle. Mike Stelzer told the authors that, "As the Mechanical engineer assigned to HG-1, I was responsible for the move of the HG-1 to the Orbiter Processing Facility and installation into the Orbiter."

Installation and testing with the Orbiter went well, and HG-1 was *eventually* launched after several delays on STS-61C on January 12, 1986, with Level-IV supporting from the C-1 Console in the Launch Control Center (LCC). All three of the payloads operated well during the mission, returning valuable data. The GSFC Project Manager, Ted Goldsmith, was pleased with the results and the demonstration of this new carrier for small payloads. Following the flight, all that remained was removal of the HG-1 from the Space Shuttle Orbiter for return to the developers. Normally this would be pretty standard, except this time the removal was scheduled for the morning of January 28, 1986.

Mike Wright was the lead Electrical EI engineer for this payload. "Although a simple side-mounted 'payload-of-opportunity', Hitchhiker G-1 holds three distinctions for me in my career: First, it was the first payload for which I participated in the launch, in this case in the Firing Room (and after three scrubs). Second was my collaborative relationship with the HG-1 project manager from Goddard Space Flight Center (GSFC), and third, that ended up being key to my transferring to GSFC three years later."

The Hubble Telescope Upgrade Mission

Another non-Spacelab mission planned at the time that would use Spacelab hardware was the series of Hubble upgrade missions. These were planned *before* the launch of Hubble, but that all changed when Hubble got on orbit in 1990 and they discovered the primary mirror problem. During this early work, however, a special portable room needed to be created in the O&C Low Bay that would be used to process the very sensitive Hubble experiments.

As Bruce Burch recalled: "When we were developing the tower, originally we said we didn't need anything. Don [Tiffenbach] designed the towers, so we named them *Tiffy Towers*. They were covered with plastic and we thought that being in the High Bay, the air temperatures were already cool so we wouldn't need anything. But once we started working in there, especially dressed up in full bunny suits with very little exposed, which you needed to be for the telescope, people were getting hot, so we decided we needed to get some air blowing. Gary Bitner, myself and Don Dolby knew it had to be filtered, so we got the spares for the Rack Conditioning Unit. We just took the spare HEPA filter out of it and the prefilter, ordered a fan motor to go on top of it, and just built the filter boxes. It worked so well we actually built two of them." Bruce described how the three of them worked with Gene Krug, and would go the machine shop, have them cut the metal, poprivet it together and cut the holes out, and then they painted it. They did everything for this cooling box. The suit-up room in the front contained scaffolding benches and the Clean Room clothes. It was like an ante room, where personnel could enter and dress out. One other feature of Tiffy Towers was that the front would also swing out, so the pallet could be floated out to do lifts and then floated back in when installation of hardware was complete.

Main: Tiffy Towers. Insert: Air conditioning unit. [Gary and Debbie Bitner.]

O&C GSE upgrades were required, as explained by Frank Valdes: "The flight rate was going to increase. We had the Level-IV rails and then they decided that they needed more rails, so the Level-IV North Rails were extended. There was also going to be an Aft Flight Deck simulator built for the North Rails, but it was decided we didn't really need it."

1986 SPACELAB MANIFEST BEFORE *CHALLENGER*

Naturally, due to the time required to process payloads for each mission, there were a number of missions that were being worked when the *Challenger* accident occurred. Some of the work will be mentioned here, but more detail of the Level-IV involvements on those flights that would eventually fly after Return-to-Flight, or were permanently canceled, will be mentioned later in the book. Some special projects Level-IV personnel began working on before the accident are also mentioned here.

Astro-Zero

Work on this payload was completed all the way up to the CITE stand, and it was getting ready to go out the door when the *Challenger* accident occurred. The original mission should have flown on *Columbia* in March 1986, but did not fly until December 1990. The work that was done to prepare it for its first flight (formerly termed Astro-0, or Astro-Zero) is mentioned in this section of the book and, for convenience, will be referred to as Astro-1.

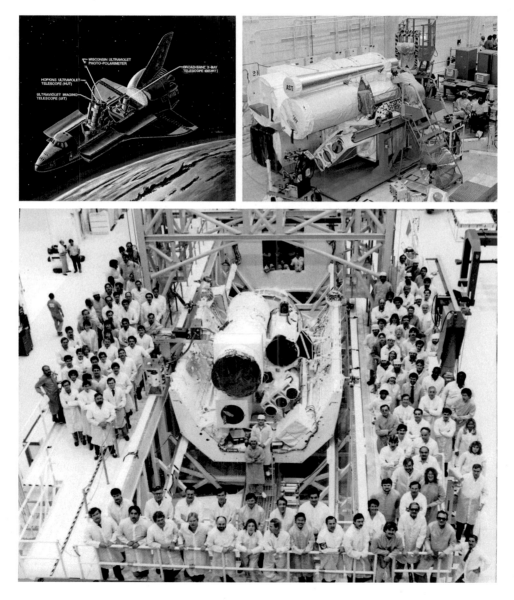

Top left: Astro-1 payload. Top right: Telescope package in Level-IV. Bottom: Astro-1 in CITE, c 1985. [Graphic: nasaspaceflight.com. Photos: NASA/KSC.]

Astro-1 Payload Specialist Sam Durrance noted that, "[They] delivered the telescope to Kennedy in the beginning of January 1985." One of the roles he had in the development of HUT was the alignment, because of the construction and the alignment and testing of the optical system. By the time the telescope was delivered, Sam had been selected as a Payload Specialist, along with Ron Parise and Ken Nordsieck, but they still had duties associated with the telescopes themselves. "I was at KSC when it was unloaded and put into the offline labs, where we had some pre-discussions about what was needed," Sam Durrance told the authors. "We performed some testing in the labs, and then there was a sort of a formal turnover, We'd been working on this telescope for four years, and we had to sort of give up control over it, which was kind of a strange thing. We had confidence in the people that had been working on it so far, but we still participated in the assembly and construction."

While KSC now had control, the PIs and their teams still were heavily involved with KSC operations. Sam Durrance recalled, "The first thing was to test each of the experiments themselves, each of the telescopes, with the hardware available for the testing, the supplies being used for all of that and the interface with the computer. That went fairly smoothly."

Bruce Burch added that, "They were going to rotate the Astro [cruciform] and put it in a horizontal position, because it sat on three pedestals in the vertical. We hooked up the lifting fixture and started to lift, and you could see that this thing was going to turn. It started to go over, so then the whole experiment was likely to have flipped. They were picking it up below the center of gravity instead of above it. One of the techs came out with the idea to add in two bars that bolted to the trunnion inside the trunnion fixtures and then lift from that, which then put it above the center of gravity. I think he got a couple hundred bucks for coming up with that one."

"The mounts on the Optical Sensor Package [OSP] were made out of Inconel®, which is a soft metal that has a very low percentage of expansion, and correction and contraction, with the temperatures that we were going to see on orbit," explained Bob Raymond. "The mounts that came from the factory were built at Marshall and they were built wrong. The mount of the Optical Sensor Package, as just one example, had to have square mount plates with four bolts in. When the mounts came out of Marshall, they had a round pad to interface with the Optical Sensor Package, so then we got to spend all night at the machine shop re-machining these pads so that we could mount that Optical Sensor Package."

Other instruments that flew on the Astro-1 cruciform also had their share of problems according to Luis Delgado: "I got the Hopkins Ultraviolet Telescope, the first one to go on, and right off the bat we went in and put the first of three shoulder bolts that connected the three feet on HUT to the cruciform. The one that was hard to reach was the one inside the center and [despite] the technician [having to] stretch, we couldn't get it in without additional lubrication. Apparently, we had the lowest tolerance on that hole, and they had a worst-case condition where the bolt was one of the thickest and the monorail was one of the smallest. And so they came with Emery cloth and we started try to polish the bolt shank, where that shoulder had to go. But we also had to be careful that it had to be pressed fit; it could not be a loose fit. That was a critical path too, because HUT was first and then WUPPE."

"The PI. Doctor Arthur Code, or 'Art Code', was an incredible, world famous astrophysicist, who created a fun group of scientists and engineers on that experiment," explained Scott Vangen, "Think about the experiment name – WUPPE, or Wisconsin Ultraviolet Photo Polarimeter Experiment. The nomenclature typically for experiments were three letter acronyms, and Astro was going to be no exception. [There was] UIT for Ultraviolet Imaging Telescope, and HUT for Hopkins Ultraviolet Telescope. WUPPE had five letters. You can't have five letters. You can't have WUPPE, but Art said, 'Well, that's just what we call it and we're not going to change it'. And that was the end of that."

Luis Moctezuma was the Mechanical lead for the Astro payload. As such, he oversaw the installation and alignment of all the Astro instruments to the Astro cruciform. Coalignment of the telescopes was crucial for the success of the mission, so a team was put together with personnel from MSFC, the Payload Developers and KSC, in order to track the alignment of the telescopes from fabrication all the way to installation onto the cruciform at KSC. As each of the telescopes was installed, Luis would measure the alignment of the optical cube and compare it to numbers provided by MSFC. MSFC had developed a model to predict the deflections and effect on the telescope's alignment as the cruciform was being loaded. The plan put together by the team was a good one, but as work started at KSC they learned the hard way that there needed to be somebody to follow the work performed at the different sites more closely. Apparently, last-minute work or changes were done after developer's alignment measurements were completed at their sites, as Luis explained: "This ended up causing discrepancies between MSFC predictions and KSC measurements. As we installed the telescopes and compared to MSFC's model predicted numbers, the numbers did not match. We checked and re-checked everything we did. MSFC verified the numbers, but could not determine what was going on. Everybody's eyes were on us since we were the last ones on the whole process. Because of the discrepancies, a decision was made to take measurements of the telescopes. This turned out to be a very costly decision for the HUT telescope, as Murphy played one on us."

Sam Durrance continued: "There was also some activity over doing the alignment of the actual instrument for HUT, and that was kind of a bad day. We had an accident. The spectrometer was down inside the telescope itself, and we had to use a theodolite to reflect it off the aperture itself so that we could get the alignment done. It was in a vertical orientation with the doors open, and it had never been like that before as far as I know."

A light bulb being used to illuminate the interior of the telescope exploded, and part of the filament ended up on the primary mirror of the telescope. They had failed to notice that the light system used was not provided with a shatterproof cover.

Sam Durrance added: "That, of course, had implications for how long it would take to fix. Of course, we were always under a schedule pressure of some kind but I thought it was going to be six months or so to do something like that. We did have a spare mirror, but then that mirror hadn't been calibrated, and actually hadn't even been coded yet. So they put together a Tiger Team to figure out how to do it."[7]

Sam Durrance oversaw the repair operation. The question was whether it could be fixed in time for it to fly. The repair was being done in the Level-IV area, with the Level-IV engineers and technicians helping Sam with all of that. Sam Durrance continued, "It was decided that we would essentially take it apart far enough to get to the mirror itself." The telescope had to be disassembled in order to clean the mirror up. The ATM room was activated in order to provide the clean environment required during the whole cleanup process.

Bob Raymond added, "We removed the HUT and put it in there [the ATM room], disassembled it, cleaned that mirror, and Sam Durrance oversaw all of that. We basically had to disassemble the whole experiment, open it enough to get to the mirror, fix it and reassemble it. The mirror was contaminated with the dirt and debris from the broken light bulb. That was a real eye-opener. And again, it was still a young program at the time [1985]."

Spacelab Experiment Integration Support (SEIS) Level-IV Technician Don Dolby recalled, "It was stacked up in sections, where we had to remove the bolts for the length of the telescope. It had a lot of MLI [Multi-Layer Insulation] blankets on it. I had to remove all that and then go into it piece by piece, just completely clean it out and disassemble it, and then put it all back together."

Sam Durrance continued: "They had some equipment that could be used to measure the reflectivity of the mirror. It turned out that it was okay everywhere other than the area right where the filament had hit the mirror, which left a little divot. A couple of people from the HUT team and people from Level-IV got on these harnesses and were suspended above it and lowered down into the telescope itself. With a little, very soft brush, they brushed the material off of the primary mirror itself and tried to clean everything up. They got it pretty clean with the exception of the filament. It actually tore a little bit more of the coding around that. So then they took it out and assessed it, and it did have the reflectivity, so we consulted a specialist in crack propagation. They said that if we just polished this divot so that it was smooth and had no edges on it, then it would be safe. It wouldn't just crack or wouldn't spread or anything like that. So it was decided that we

---

[7] A Tiger Team is a team of specialists formed to work on specific goals.

would use that mirror, just clean it up. And we put it back together, but then we had to realign it and refocus it."

Bob Raymond said, "Sam Durrance oversaw that and it was quite detailed. It was just remarkable to watch this guy work and put it all back together. And he did the alignment; very, very detailed, very, very precise. When he was happy, we were happy."

Sam Durrance explained that, "We had all of those activities completed within less than three months. It was amazing. It was one of the things that was kind of the essence of the NASA 'Can-Do', environment, It was not 'You can't do it', or 'It's impossible', or anything like that, but 'We can do it'. So we then installed it on the cruciform, verified alignment, and started doing the integrated tests."

As before, alignment was critical, as Sam Durrance explained: "That was what Luis [Moctezuma] did. He verified it, and if there was anything off we made whatever adjustments that we had to do. We had to match the pitch. Anyway, the whole of this was the requirements for the alignment, in other words, how it was all determined what the pointing requirements were, like observatory pointing accuracy requirements. Delta, which is right ascension, [had to be] less than or equal to plus or minus seven arc seconds, and Delta roll less than or equal to one degree. Those were the essential requirements. If they were not met, it would be detrimental to the performance."

Sam Durrance's logbook, showing the alignment requirements that needed to be met during Level-IV preflight operations. [Mike Haddad]

Luis Moctezuma explained that, "There was a requirement to provide AST's geodetic location as well as its measurements relative to true north. Through the Air Force's geodetic group, I obtained the geodetic data of three points and the azimuth of the lines between the three points." The three points were: the center of a water tower in the Industrial Area; the centerline of an antenna in the Industrial Area, and a point at the west side of the O&C Building. By setting up a theodolite over the target outside the O&C, Luis transferred the azimuth to a line inside the High Bay, which he then used to get the AST measurements. "We installed a target in the hallway, by the tech shop entrance, so that we could see inside the Low Bay through the Low Bay west entrance doors."

"Astro was to have flown in March of 1986," explained Scott Vangen. "That was its target date, because it wanted to be there for the Halley's Comet flyby, so it had the Astro complement of UIT, HUT and WUPPE, and it had the Wide Field Camera [WFC]. It was going to take super cool pictures of Halley's, basically. But because of *Challenger*, this did not happen. Fortunately, the WFC wasn't needed by the time we flew in 1990, so it was removed, but when we were preparing it we called it Astro-0 because we were pressing to fly in 1986."

There were lots of hazards that Level-IV personnel had work around, and therefore lots of potential ways of getting hurt. During installation of the WFC, an incident occurred that changed the way photographers documented the Level-IV work, as Mike Haddad explained: "I was Task Leader for doing a lift of WFC, which required scaffolding to be placed about 20 feet high around the Astro cruciform. To get good photos, the photographers climbed up to the top of the scaffolding as we had started the preps on the floor for the lift. All of a sudden, I saw one of the techs, mouth dropped open like they were scared, pointing at something behind me. As I turned, I could see the photographer had somehow manage to get himself wedged between the upper rails of the scaffolding. He was a very heavy man, his feet were dangling on the outside of the scaffolding, and the upper rail was under his neck and he was choking. I remember the young crane stop guy and myself were down on the floor, and he handed his controller to one of the other crane guys. Then he and I climbed on the outside of the scaffolding. The photographer was very heavy, there was no way one person could do it [free him]. The crane guy grabbed one leg and I grabbed the other leg, and that was only the way we were able to lift him up and shove him back on the scaffolding, to stop him from choking. We climbed on the outside of the scaffold, with no safety harness or anything because there was no time to put that stuff on. Of course that was illegal as hell, but that was the fastest way we could figure to stop him from choking. We climbed back down, and I immediately called Safety and told them what had happened and that I was there, so if they heard any other version from someone else, it was not the whole truth. Normally, the Safety person would be there for the entire operation but sometimes, due to the limited number of Safety people and the huge work load, the Safety person would ensure all the proper safety steps were in place for the operation and then go to their next assignment, trusting us to maintain all the safety protocol and notify them immediately if something happened, which is what I did. We also sent the photographer to the Occupational Health Facility to get checked out. He ended up being fine. After that, we placed rules about how the photographers documented our work to keep them safe."

Luis Delgado became the Mechanical and Fluids engineer for the HUT, which was built by John Hopkins University. They had an evacuated chamber for the detector that measured the ultraviolet light. From the day it was built and coated, this chamber had to be evacuated for the rest of its life, because that coating was continuously off-gassing and if it was not removed, the off-gassing would clutter the detector. It was a vacuum ion pump, an electrical pump, with a big magnet, working at a very low vacuum. Before the vacuum ion pump was started, a roughing pump had to be started to get it down to a certain level. Once it reached a laminar or a molecular flow on the vacuum, the ion pump had a chance to start, because it could get overwhelmed. If there was too much pressure, it could not be started. There was an electrical current on the pump, which was how it could be determined what the pump was doing by seeing what the current draw was. It was either starting, or it could not start. Luis Delgado noted, "We had to continuously have an eye on this 28 volts from the time this thing was built. They had this pump built into the experiment that was flying. So you fed it 28 volts from any source from the ground once you got in flight and you could open up, but you know, when we were in Level-IV, we were connected straight to the telescope. As we moved down to Level-III/II and further in the flow we needed to be more outboard to make it easier to connect power.

With this pump having to run 24/7, someone had to monitor it the whole time, or some other process had to be used to ensure it was always on and running properly. Luis Delgado described an alarm system with a phone dialer (this was back in the early 1980s) in case something happened to the vacuum. It dialed the KSC duty officer and the duty officer had a list of names and phone numbers. Luis was the first one on the list, and many times he had to go in at three in the morning because the alarm went off: "If it was no big deal, I could just reset it then that was it. But we had a contingency plan that if I couldn't reset it, I would call Herb Linhart, the tech supervisor, to get techs and Quality [in] and we would hook up a pump and get it started again with the roughing pump." Mostly, it was a false alarm or a glitch, but they had no cell phones back then. Information was also being recorded, so three times a day they would make sure that the pump was running and record how many milliamps it was drawing, because that allowed them to see whether the pump was reaching the end of its life, as Luis explained: "If the current draw kept rising, it was a sign that the pump might be failing, so we kept track of it."

There were two RAUs mounted to individual brackets, to support a plate attached to the outboard edge of the cruciform. Mike Haddad added that, "One of the many flight items I installed for Astro-1 was the pair of experiment RAUs. The mounting plate had a very smooth surface cut-out at the bottom that the RAU's brackets sat flush against. Well, they were supposed to sit flush but I noticed screws sticking out from the bottom of the RAU brackets. With these screws sticking out, there was no way for the RAU bracket to sit flush with mounting plate. We couldn't remove the screws and we couldn't cut them off or file them flush. What should we do? I got with the main TBE [Teledyne Brown Engineer] engineer Robbie Brown [no relation to the organization] and said, 'What if we put spacers under the RAUs, just thick enough so the screws don't impact the mounting plate?' But the spacers could cause another problem; if they were too thick, the bolts would be too short and would need to be changed out. A simple calculation of the numbers determined that the bolts were long enough. He got back with the designers to ensure the structure could take the different loading and that the offset of the CG for the RAUs did not cause any problems. Both concerns came back as ok. The next step was making the spacers. There were no spacers in existence, so we were able to get some Aluminum 6061 bar stock to make our own spacers. I conferred with Robbie and we determined the proper dimensions for the spacers, and where the holes would be drilled to match the RAU bracket mounting screws to hold the spacers in place. We then had our machine shop in the O&C make flight spacers, one of my first pieces of flight hardware made to fix flight hardware problems. We installed the spacers, then the RAUs. It all fit, and it flew that way."

There were many times during the Level-IV processing of flight hardware where the drawings did not match the hardware that showed up. Mike Haddad had one particular experience of this. In fact for this example, the cables that connected to the AST (Astro Star Tracker) mounted on top of the UIT, there *were* no drawings. Mike Haddad remembers, "The UIT cabling drawing showed routing of these cables as referenced by the AST installation drawing, and when I got the AST installation drawings, they referenced the UIT cabling drawings. Basically, both referenced the other, thus none were made. This was another piece of flight hardware integrated per PR. If I remember this correctly, I was able to secure the cables via nearby clamps and zip-tie them to existing cable harnesses. It did not look perfect but it worked."

Testing

Sam Durrance explained that, "The three of us [Durrance, Ron Parise and Ken Nordsieck] were also involved in almost all of that testing, because it was sort of agreed upon that since we would be operating [the equipment] when we were in orbit, the more time we had operating it on the ground, the better off everything would be. So pretty much most of the time when it was turned on, one or all three of us were there. And of course there were quite a few folks from Level-IV to help us.

"When we first started, the Level-IV engineers and technicians would do the actual commanding, but that didn't last too long and it was mostly Ron and myself and Ken that were doing commanding, with the Level-IV folks there to help. So we went through all the testing there. With another one of the issues of interfaces, we could just could not figure out what to do, but the Level-IV folks did some very good troubleshooting on this thing. They found out that there was a problem in the ASCII character that was used for either the *space* or *the* or *d*, I can't remember which it was, but there was an ASCII character for that. The ASCII character that we had on our computer that had been given to us as the interface was different from the one that was on the actual flight computer. Eventually, primarily Level-IV engineers and technicians found that and were then able to fix it, because they put the right character in the HUT."

(L to R) Astro Payload Specialists Ron Parise, Ken Nordsieck and Sam Durrance, with Mission Specialist-1 Jeffrey Hoffman (blue t-shirt) in the Astro-1 Level-IV control room, c summer 1985. [Scott Vangen.]

Craig Jacobson explained: "I supported data services for all payloads. WUPPE actually was having a data problem, I remember that, and that was Scott's payload. I helped troubleshoot that. There were those kinds of things, where they were getting telemetry up in the Control Room. I had programmed this box that could read the telemetry. I would program it for their format and I could show them that it [the telemetry] worked or didn't work depending on what was going on. If it worked, they knew that all the patching was correct, because they were always wondering if they were getting the right signal. You could program the sync words and the format of the sync words, frame counter and length, and bytes, and all that junk. That's where you learned to be able to punch in binary. That was always my test, but to do that you had to be able to program to decimal and you had to punch in the bits to do that. So you had to be able to convert in your head from one number format to another."

"Another thing that we participated in, and Level-IV essentially set up and did everything for, was of course when it got into orbit," explained Sam Durrance. "The telemetry and the data was going to come to the team in Marshall, but we had to simulate that at KSC. We had to make sure that all of those interfaces were working. So each of the teams got a control room, essentially, and had connectivity to the experiments, which should have been the same as it was going to be from the Payload Operations Control Center at Marshall. There were always problems with interfaces but, for example, there was an Ethernet that wasn't there that needed to be. All we did was ask Level-IV, and lo and behold it showed up, so we didn't have to worry too much about paperwork and things like that. The Level-IV guys did all that and made sure that everything was as close as we could make it to what it would be like [on orbit]. And then testing, the End-to-End test, and ready to do the final Mission Sequence Test, where we ran a series of the mission plan for 48 hours or something like that. We would exercise all of the interfaces and make sure everything was working, including all the commands that we had put together. That was one of the other aspects that we did with the help from Level-IV folks and people from Marshall and Johnson putting together the actual commands, the Data File that we were going to carry to orbit with us. In those days, setting that up was not that easy. I'd have hard lines for all the high resolution, high rate data, stuff like that. Today it is trivial. You could probably do it with your phone. That's the difference. So that was my first experience with Level-IV. It was clear that we were dealing with people that were qualified for what they were doing."

Astro-0 was in the CITE stand ready to go into the Orbiter *Columbia* and fly after the STS-51L mission, but it never left the O&C. There were other payloads in the flow, including the SLS-1, Spacelab-J, that ended up flying many years later, and there were some in the planning phase, such as Star Lab or SunLab, that would ultimately be cancelled.

CHROMEX Middeck Experiment

The CHROMEX program was developed to study the effects of microgravity and space flight on plants. The CHROMEX acronym was derived from an early focus on chromosome integrity and the cell division process. John Conway explained that, "[It was] decided to have the Level-IV people design and build CHROMEX. We did it for one reason, not because we needed another experiment, but so people would learn all the different phases that the PIs had to go through – the certification of this, the outgassing requirements of that, and so forth. That was the reason we did that, to train the Level-IV guys to be better supporters of the process."

Many NASA Level-IV engineers and Boeing SEIS technicians worked on different aspects of design, development, fabrication and testing of the payload. Just one example of the ingenious work of the team was using smoke from a standard cigarette to see air flow through a component of the experiment, thus being able to determine if the flow was correct or design adjustments needed to be made to achieve proper flow. CHROMEX flew on a number of flights following the *Challenger* accident but just a few details of the work will be discussed in this book. Each flight had its own successes and problems, but the team learned from each flight to ensure better operations the next time it ventured into space.

Military Spacelab Processing at Vandenberg

Bill Jewel said, "We were supposed to be able to shoot Spacelabs into polar orbit out of the Western Test Range out at Vandenberg [Air Force Base, VAFB], California. I went out to look at it and reviewed it, and two things showed up. Number one was that the Orbiter could not go across a small bridge there, because the railings were too high for the wings and it would've taken the wings off. Then I went out to look at the launch pad, and the Launch Control Center was very close to the pad. I said, 'I wouldn't be in that Launch Control Center if you paid me millions of dollars. If something happens at launch, it will blow the hell out [of that LCC] and you'll die'. The last and crowning blow was that Spacelab *could not* launch from there, because if they had a processing facility that had a crane that had to come up and go over a place, to go into a hole where they put it, the crane couldn't lift the Spacelab because it was too shallow. There was no space. The whole damn building would have to be torn down and restarted, or a new building put there. So I said forget it, it was impossible. I just came back and said, 'Well there is no Spacelab program at WTR. Forget it'. So we just terminated all that."

Jim Dumoulin expanded on the development of Level-IV at Vandenberg: "We did a number of meetings over the Air Force side, looking at how we would do Level-IV checkout at KSC in Air Force facilities. They didn't want to use the Payload Checkout Unit. They wanted to build their own at one point, so we had evaluated what would it take to build a PCU facility on the Air Force side. That was one of those special projects where we did a lot of different kind of 'what ifs', but things didn't ever materialize."

Mike Haddad recalls, "I was part of a team that went to VAFB, as the Mechanical lead on how to process Spacelab from the west coast. We made a number of trips and came up with suggestions regarding the processing flow, and what was needed as far as GSE to support the flight hardware and flight software. So, lots of challenges. We went to a variety of facilities as well as Space Launch Complex-6 [SLC-6, pronounced "slick-six"]. As Bill Jewel mentioned, the LCC was very close, I think about 1200 feet [365 m] from the base of the pad. We talked about flipping coins, with the loser sitting on console in the LCC for launch. Other local problems included many foggy days, a train that ran right next to SLC-6, and the mating habits of a local animal, which all prevented any chance of launching. After the 1986 *Challenger* accident, the military decided not to have any more military missions onboard the Space Shuttle, and the idea of using the 'Slick-Six' Launch Complex at VAFB was scrapped, at least for the Shuttle."

Jim Dumoulin elaborated on the early plans. "Prior to *Challenger*, the idea was we were going to launch from both coasts. A third of them were going to be military missions and maybe a third of those military missions were going to be Spacelab military missions. We knew where they were going to bring their own experiments, and because there was no threat, they wanted to make sure that we were secure. I remember we lost a third of our PCU control room because the military decided to build a vault next to the Astronaut Quarters. That PCU control room was right next to the astronauts' control room. They decided they needed to review flight procedures in a vault, so they took off the back of our control room, poured massive amounts of concrete in and put a bank vault back there. We lost part of our control room so they could do military missions, and it's now the astronaut kitchen. One Monday, I ran up the stairs and I said, 'Something's wrong. This hallway's not right'. Over the weekend, they had made the Astronaut Quarters bigger and they had moved the door. First of all, they grabbed that room to make it a secure room, but they had to leave the Astronaut Quarters and go into the hallway to enter it. But they didn't realize that our comm closet was also on the other side of that hallway, so we couldn't get into a comm closet. That was where all the cables came into the rooms, so then we had to always ask permission to get into the Astronaut Quarters to pull comm cables."

SUMMARY

All of the missions worked by Level-IV in 1985 were completed, and the first of the 15 missions planned for launch in 1986 had ended. Looking ahead, there were 19 missions planned for 1987. The flight rate was increasing, the pressure to make launches was increasing, and the teams at KSC preparing the flight hardware were working as hard as possible to maintain the excellence required. The flight rate was also having an effect on the crews. Astronaut Hank Hartsfield would remark: "We were going to be up against a wall [in terms of having crews properly trained]." The crews for missions STS-61H and STS-61K would have averaged no more than 33 hours in the simulator, according to some training predictions. "That was ridiculous. For the first time, somebody was going to have to stand up and say, 'We have got to slip the launch, because we are not going to have the crew trained'."

Those in Level-IV were driving ahead to make sure all was being done that could be done to make the planned launches. But what many did not understand was the workload on Level-IV. Not only were they preparing for the next mission real-time, they also had to de-integrate the previous mission or missions at the same time, while performing the planning for future missions, again at the same time. On top of that, Space Station planning was getting more attention, which took some of their time. Now they would be working two major NASA programs at once, the Shuttle program and the Space Station program. There were not enough hours in the day, as Mike Haddad recalls: "By the beginning of 1986, I was starting to lose my edge. I could not put the time into future planning of missions to help find problems early. I could not be preparing procedures and such for the upcoming mission(s) to the highest level needed to ensure total success, and the de-integration of previous missions to prepare them for their next flight was being somewhat ignored, which did not help those needing that hardware for the next mission. I really could not see how we could keep this pace up and maintain the highest quality required for spaceflight. The truth was… we couldn't. Then suddenly, everything changed on the morning of January 28, 1986."

References

1. Shayler, David J., and Burgess, Colin, **The Last of NASA's Original Pilot Astronauts**, Springer, 2017, p. 347.
2. Shayler, David J., and Burgess, Colin, **NASA's Scientist-Astronauts**, Springer, 2007, pp. 410–412.
3. Moates, Deborah J., and Villamil, Ana M., EASE/ACCESS Ground Processing at Kennedy Space Center, Payload Management and Operations Directorate National Aeronautics and Space Administration, Kennedy Space Center, Florida, Space Construction Conference August 6–7, 1986 Langley Research Center Hampton, Virginia.

9

The Shutdown Years

"Flight controllers are looking very carefully at the situation.
Obviously a major malfunction. We have no downlink.
We have a report from the Flight Dynamics Officer
that the vehicle has exploded."
PAO launch commentary STS-51L,
Mission Control Houston,
January 28, 1986.

Icicles on Launch Pad 39A, at the Kennedy Space Center (KSC) in Florida? When was the last time that happened? Never!

TUESDAY, JANUARY 28, 1986

The launch of STS-51L was scrubbed the day before due to a troublesome Orbiter hatch handle that could not be removed, and a few other reasons. A cold front passed through central Florida and the temperatures dropped sharply, to as low as 22 degrees F (-8 degrees C) overnight, with icicles forming on the pad. As the Sun came up, the temperature started to increase but remained below freezing. Most thought there was no way *Challenger* would launch. Icicles could break off during launch and damage underside tiles, and then a disaster would happen during re-entry. Many of the payloads personnel were unaware of the O-ring discussions taking place.

© Springer Nature Switzerland AG 2022
M. E. Haddad, D. J. Shayler, *Spacelab Payloads*, Springer Praxis Books,
https://doi.org/10.1007/978-3-030-86775-1_9

Ice on the pad, Launch Day, STS-51L. [NASA/KSC.]

"We were working like madmen before that," explained Luis Delgado. "I remember being in the O&C [Operations and Checkout Building] and we had one Spacelab coming, another one going, running maybe three shifts. People were, like myself I think, going to snap. We were going too fast, trying to get too many things done. People were stressed out. And obviously, when the truth came out, I mean we were driven to the bone."

Rey Diaz recalled, "I returned from [STS-] 61C in California. I was supposed to be in the contingency team at California and we became the prime because of weather problems. Now I was getting ready for 51L, my prime mission. We had to scrub the launch attempt on the 25th and again on the 26th. We would have to try it on Tuesday, the 28th. I returned to KSC on the 27th and I don't remember working on board the *Challenger* the night before the accident. I was tired, but not only that, [there were] all the things that were happening that night; the temperatures were going down. I pretty much got maybe six hours sleep. I was waiting for the launch and it was posted on until 11:38."

Michael Stelzer remembered: "It was wickedly cold on the drive to the OPF [Orbiter Processing Facility] that morning, and it felt good to get inside the Bay where the temperature and humidity were tightly controlled. We worked throughout the morning to get the cabling and hardware disconnected, the lifting frame reattached and the HG-1 [Hitchhiker G-1] hoisted from the Orbiter. Following the re-installation of HG-1 onto its transporter, we took a break and went outside to watch the launch of the STS-51L mission. A little over a minute into flight, an expanding ball of smoke let us know that something was off. I watched for the Orbiter to emerge and hopefully loop back for a Return To Launch Site [RTLS] abort. It was not meant to be. Professionally, but with heavy hearts, we completed our HG-1 tasks that day, and over the following days returned the hardware to the GSFC [Goddard Space Flight Center] team."

"I was the Ops guy for that *Challenger* mission," said Mike Kienlen. "There was an IUS/TDRSS [Inertial Upper Stage/Tracking and Data Relay Satellite System] and a Spartan, so it was probably the first mission that I was getting to where I was meeting the crew and starting to know who they were. And so when we lost the crew, that meant something to me, meant it more, made a bigger impact to me. When it first happened, you were going through what could be the possible reasons this thing failed. Maybe Spartan fell off the MPESS and landed on the aft bulkhead and into the engine compartment and caused the loss of it, or the IUS came loose and exploded. So [there was] the fear that you were part of the reason for the loss of it. There was always a thought about 'Do I even want to go watch this Shuttle launch, because do I want to see if it goes bad? Do I want to be there?' Those were tough days."

Slipping the Surly Bonds

Mike Haddad recalled that tragic day clearly: "On launch day, we would always try to take a break from whatever we working on to go out and watch the launch. Usually for me, when the countdown came out of the T-9 minute hold, I would terminate our operation, secure the area and go outside the O&C to watch the launch. Many times, we would go on the outside fire escape stairs that led from the fourth floor to the fifth floor of the O&C office area. From there, you could get a good look at the pad. Plus they had the countdown audio on the loud public address speakers throughout KSC, so everyone could hear how the launch count was progressing. We would also play it in the control rooms and such, so no matter where you were or what you were doing, you could hear it. Well, we all thought they were never going to launch that day, so I began the day thinking we would try tomorrow or the next day, so let's get back to working payload operations. Later that morning, we heard them come out of the T-9 minute hold. I thought to myself, 'What? They are going to try and launch today?' So I shut down the operations, went back to the office to get my jacket on, and headed to the fire escape stairway.

"I remember the day of *Challenger*, how cold it was on that fire escape. Beautiful clear day, blue skies but very cold. Temperature of 36 degrees F (2 degrees C) at launch. We were standing there and off it went. It was ascending and someone next to me took my attention away from the Shuttle just for a second. Then I heard the person next to me gasp and I looked back up and saw the cloud. After watching dozens of launches we knew something went wrong, and all of us started to say, 'RTLS, RTLS, Return to Launch Site'. RTLS was an option for the Orbiter to return to KSC if a problem occurred during launch. It could only happen during a certain span of time on ascent. If the JSC PAO [Johnson Space Center Public Affairs Officer] person stated, 'Negative Return', that meant they were beyond an RTLS. There was no 'Negative Return', so we kept looking for the Orbiter to come out from behind the cloud. We waited, and as the pieces were falling we heard the JSC PAO person say, 'We have no downlink'. Right then it hit us; they were not coming back home. The pieces we saw falling were what was left of the ET [External Tank], *Challenger* and the crew. The boosters were still flying, so I thought maybe a main engine let loose. We ran back to the office to see what the hell was going on, and we could see on the TV the parts splashing down in the ocean."

Luis Delgado recalled that a friend of his from college, Jose Arench, had been working Shuttle that night. They lived close together in Cape Canaveral. He came to Luis's house, and Rey Diaz and himself and their wives tried to get Jose to tell them what he knew, but he said, "No, no, no." He said they had had a meeting, but he could not say anything because if something got out, people would know it was one of them. Luis said, "So Jose held his word. He didn't say anything. The next day we heard it was a leak on the SRB [Solid Rocket Booster], probably from the same meeting that he was at. We started looking at a video; he said it was obvious."

Jim Sudermann said, "They had the feed from Houston, and the guy was talking and the thing blew up. It just came apart and we were watching it come apart. We were going, 'God dammit, that's not right'. The guy on the speaker was just chattering away as if everything was okay. We started seeing the stuff come out of the cloud, and it sort of dawned on us we'd just watched seven people die right there. That was a bad day, and there were a lot of people just kind of walking around in a daze. Me included."

Bill Jewel told the authors, "I didn't go to watch the launch. I had a bad feeling about it or something. I stayed in the building, but I heard it and I went out then and I could see it. I had a good friend, Ellison Onizuka from Hawaii, who was an astronaut on that."

Angel Otero said, "That day is burned in my memory till the day I die. We were de-integrating Spacelab-2 and we always quit whatever we were doing to walk outside and watch a launch. Because I was on the pallet, I had the bunny suit on and it took me a while to get the damn thing off, so I didn't get a chance to go up to my office to get my coat because it was freaking cold. So I was standing outside without my coat on, freezing my ass off, and then it happened. And I can tell you what this guy on my left did, what this lady to my right was screaming. Then I remember a bunch of us ended up sitting in a circle in the office, just staring at the floor like we didn't know what else to do. Every once in a while we would look up and make eye contact with somebody, shake our heads and go back to looking at the ground. Because we had no idea what was going to happen now."

Cheryl McPhillips remembered thinking *Challenger* was behind the plume, so she ran down the stairs of the LCC (Launch Control Center) and ran to the other side of the parking lot because she thought it would be coming back to the Shuttle Landing Facility (SLF) to land. But no, it was gone. She said, "I went by the O&C cafeteria later that day and it was full of people. And there wasn't one sound coming out of that cafeteria. Not one sound."

Donna Bartoe said, "I remember leaving my place, and the sprinkler water had frozen all over the plants. This was beach side, so that never happened. That's how cold it was. And I guess we kept thinking, 'You're not going to launch, right?' We had not launched with this temperature before. It was a complete shock."

Craig Jacobson observed that, "It was going up and going up, and then we were trying to figure out what was going on. Something was not right. It was like a huge mental disconnect. It made no sense, and the brain was just kind of trying to make sense out of it. I remember Donna started crying and people were coming out of the LCC, going under the stairs to go down or up, and they were crying. Everybody was just in shock and reacting to it in different ways."

Cindy Martin-Brennan told the authors that, "At the time, I didn't think they were going to [launch], This is probably one of the clearest memories I have. I thought, 'Well, there's no way in hell they're going to fly'. We had to do training every year, to stay up on things. I walked over from the O&C Building to a [training] room in Headquarters, and I remember loading in the training, putting on my headset, starting the training. Then somebody came into the training room and said the *Challenger* just blew up. My first reaction was that I yelled at this, 'That's not even a funny joke!' I was so angry at the person because I thought it was the sickest joke in the world. So I took off my headset and I just walked out into the hall. I went to Ladies room and there were people crying. It really had happened. And I remember going outside and looking at the cloud."

Damon Nelson stated he would never forget the day before, as he watched the technicians trying to resolve the hatch anomaly with a tool that didn't have a battery charge, which was so disheartening. Then he saw the accident and, quite frankly, didn't understand what was happening. Damon said, "We were in our mid-twenties, and all we had known was working for NASA. We had a meeting that afternoon, where our boss got us together and talked about how we would get through this."

Jay Smith remembered, "SPARTAN had some 'remove before flight' aspect, so I went out to the Pad and took the 'remove before flight' stuff off it before it flew. I still have that PCR [Payload Changeout Room] badge for 51L."

STS-51L PCR-VPF badge. [Jay Smith.]

Jay remembers that they were so busy, they didn't have time to go over near the LCC to watch the launch. "So me and Walt Preston just walked out front of the O&C building and saw *Challenger* take off, and then the parts kind of falling through the air. I remember Walt saying, 'My God', and that they wouldn't have a job in a year. He was right. It was almost a year to the day. McDonnell Douglas won that contract somehow, and they laid off almost all of the Level-IV technicians. Maybe Walt and Bruce Burch were the only ones left."

The Boeing Spacelab Experiment Integration Support (SEIS) technician room was just across the hall from the elevator that would be bring the crew down to enter the Astro Van for their trip to the Pad. They watched every crew come out of there, and did so for 51L. Some of the crew's families had taken their cars out to see the launch, but after the accident Jay Smith recalled, "A lot of our guys went out and got their vehicles and drove them back, took them to their homes from here and ferried stuff, because there was nobody to do that. These family members had just lost their loved one. They were not going to get in their car and drive it somewhere." Bruce Burch added that, "For a lot of the family members and visitors, we drove them in our vans back to their motels right after the incident. That was tough, with the families. We were right there. They came in and just asked us if we would take them."

Right after it happened, and before all the phones and communications were shut down as part of the normal set of procedures that were supposed to occur in case of an accident, the phone rang in Dean Hunter's office. Angel Otero recalled: "It was a radio station in Argentina. They'd called Dean's office because I think, if I remember well, our office number was 2401. The KSC Public Affairs Office was 2410 and the guy had dialed the last two digits wrong and got Dean. Dean yelled that he needed someone who spoke Spanish in his office, so I walked in and I ended up speaking live on a radio station in Argentina. I basically told the dude that if he was watching CNN, he probably knew more than I did."

Gary Bitner recalled that, "They cut all communications, shut the phones off. Everything got shut off, and they used our techs to run handwritten messages from the Astronaut Quarters over to the fourth floor of the Headquarters building, where the KSC Center Director and his staff were located."

Bob Raymond explained that, "The day *Challenger* happened, I was in the control room trying to close out paper, because we always had a major paper close out process before a launch. Our mission was next [STS-61E/Astro-1], and we were supposed to go into the [Payload] Canister the day that *Challenger* happened, so we were headed for the OPF. There was a young Mechanical engineer having an argument, a discussion [with me]. We were trying to close this paper, the integration of the IPS on STS-61E, and he had made engineering changes without the approval of the design agency. I told him I was not signing his paper off until we got engineering orders, some sort of proper approval from the design agency to say it was okay to do what he had done. He was adamant that we should be able to do this and I should sign it off, and I was adamant I was not signing that off. It was downright [wrong], you could go to jail for stuff like this. I told him people died doing things like this. We watched the launch and we watched *Challenger* blow up and I'd just got done telling him that people died doing stuff like this. I looked at him and all color had left his face, and he walked back upstairs. Several weeks later he resigned, and left instructions with Joe Lackovich to say, 'You need to tear that IPS down and rebuild it to print'. So that was a learning experience for us all, I think. I'm glad I stuck to my guns on that one. Who knows what would have happened."

Afterwards

Scott Vangen said, "I do remember going home in the morning, back to the beach house. And there was still residual in the sky, whether that was the smoke or the residue. This would have been a couple hours after the accident. I lived with Jim [Dumoulin] and Craig [Jacobson]. Eventually they showed up, and I think Donna [Dawson] came over too, and we tried to console each other, watching the news to figure out what was happening."

Bill Jewel said, "Later in the afternoon [of the 28th], I went out to someone I knew and said 'You can't win this contract now. NASA will never select a new contractor replacement for McDonnell Douglas'. It was going to be probably a year or more delay for the next launch and I said, 'There's not going to be any Spacelab activities, there's nothing flying, and it's going to be a real big mess'. They went ahead and bid it anyway, and they lost it of course."

"I was at the O&C when *Challenger* happened," said Roland Schlierf, "I was a Co-op and I didn't know the astronauts, but all my coworkers that knew them just really, really felt the weight of what had just happened."

BEYOND THE DAY AFTER

"I'll never forget the day after *Challenger*," recalled Damon Nelson. "I had a lifting operation involving flight hardware the day before the accident, so I went back to continue the operation and you could have heard a pin drop in that High Bay while we were doing that lifting operation. Everybody was so on edge with what we had just experienced, but we kept pushing. we kept working, kept integrating, and it was not for probably another four to six months that things started to shake out as far as what was going to happen." Earth Observation Mission (EOM) 1 and 2 would be canceled, but some of the payload eventually flew as ATLAS. Spacelab Life Sciences survived, but it got pushed years down in the manifest. It was about six months after *Challenger* that people started to get shuffled around a little, because the Shuttle was obviously not going to be flying for a while.

Debbie Bitner explained that EOM 1/2 were combined from two missions. They were a Mission Peculiar Equipment Support Structure (MPESS), a pallet and a Short Module. "I had assignments on the pallet and the MPESS, and I mostly did the integration on the MPESS, but then *Challenger* happened and that brought everything to a halt. At first, it was frustrating because they kept us doing the integration, because of course we didn't know what was going to happen with the program. So we'd go downstairs for days after the accident, feeling the after effects of the accident and not having any motivation or anything like that, and had to keep integrating. We felt like 'Why are we doing this?' I was one of the few people down there working at the time. One interesting thing about it was that we had a French experiment, and we had two of the PIs [Principal Investigators] come out to watch us integrate it."

Sharolee Huet admitted that she was afraid of heights, and still was when it came to the pad: "I was assigned to remove MLI [Multi-Layer Insulation] from the EOM 1/2 pallet/Orthogrid, but I was scared to climb out on the Orthogrid. Later [for ATLAS], it was no problem. Also, [there was] repairing reused and misfitting MLI. The MLI had been used for EOM 1/2 and then folded for storage, and the MLI interior metal surfaces were flaking and tearing for being stored flat. We repaired the interior layers the best we could with metal tape and then stored the MLI on foam forms to keep their shape. We also had many repairs due to cabling and hoses not lining up with the cutouts. We did lots of hand sewing to make the proper accommodations. It was months of work."

Gary Bitner told the authors, "I remember Skip [Montagna] said Seattle [Washington State] called and wanted to know who worked on the 51L mission, as part of the post-*Challenger* investigation. And then the Boeing boss at KSC called me into a room. They went and pored over everything that we did and looked at all the QA [Quality Assurance] and all the engineers and all that stuff. That was interesting, because it made you feel like you were a crook or something. They really questioned you, but I can see why they did it."

Scott Vangen recalled that, "Up to *Challenger*, we had launched Spacelab-1, -3, -2 and D-1, and we were just ready to kick the Astro-1 payload out from the CITE [Cargo Integration Test Equipment] stand into the Orbiter. [After 51L, the question was] Does everything just stop? And the answer came relatively quickly within the next day or two: 'No, we are not moving to the OPF, stand down. Do whatever preventative maintenance sustaining you need to do on your equipment'. And the payload? We had to talk to the PIs, who knew the world had changed. Astro-1 was on a rendezvous launch date for Halley's comet [which was] coming by that year, So we knew we were probably gonna miss that. What was going to be Astro-1, we simply now referred to as Astro-0."

Mike Haddad, added "The WFC [Wide Field Camera] was the only Astro-0 experiment that I was responsible for installing and removing postflight, which of course never happened. So I just used the de-integration procedure to remove it following the *Challenger* accident. That area was then closed out for WFC and didn't require any further work if and/or when we began flying again."

"There were so many things happening at the time and not happening," recalled Scott Vangen. "That's the strange duality of our whole space program. I think we knew the decision at least six months in that we were not going to fly an Orbiter within a few years. And the reality had hit the payloads folks. And rightly so, mission managers for the Astro payload at Marshall [Space Flight Center, MSFC] said, 'Let's remove the cruciform, get that back to Level-IV, so the science team can take advantage of the downtime for whatever enhancements'."

"After *Challenger*," Don Dolby continued, "everybody basically was shut down. They were loaning us out for doing all kinds of stuff, one of them being the lightning studies out north of the main road coming in towards Playalinda Beach. We had to go out there and work on doing some grounding on it, because they used it every summer. They had scientists and people coming from different parts of the world and the country to launch these little rockets that had wire attached, so that when they launched it would unspool and trigger lightning, and then they would do measurements on it."

"In the weeks following the accident, a few members of the Level-IV team were assigned to support the investigation but the majority of our jobs were on hold," Jim Dumoulin told the authors. When management began to realize that the investigation and shutdown was likely to last several years, they started to entertain suggestions about how to keep the workforce employed. Level-IV not only had a number of multi-disciplined civil servant engineers (whose jobs were not in jeopardy), it also had a support staff of highly trained techs (as part of the SEIS contract). Those techs were likely to be laid off unless they could propose new work. A number of engineers proposed (and management approved) multiple tasks aimed at keeping the workforce employed and engaged. The Electrical group used the downtime to improve the operational flexibility of Level-IV. In the two years following *Challenger*, they upgraded all the offices to have at least one computer for every four engineers, a file server and printer in every office, and an extensive computer network and cable TV system in all the payload processing facilities. Electrical techs went from building flight payload cables one day, to installing ethernet drops in every office and becoming cable TV installers. The team also expanded the payload closed circuit video switch and built a second Payload Checkout Unit (PCU) control room.

Jim Dumoulin recalled that, "Other Level-IV personnel provided their experiment processing expertise to the early Space Station planning activities that laid the ground work for the construction of the Space Station Processing Facility. These tasks proved crucial to handling the surge in payload processing complexity when launches resumed. Without the time to regroup and upgrade KSC's facilities during the *Challenger* shutdown years, it is unclear how Level-IV would have been able to ramp up to the intensity of the decade of science missions in the 1990s." (see Callout: *Halloween Parties*.)

Craig Jacobson told the authors, "We would look at what systems need to be updated and what was coming on, what we needed to do. It turned into that kind of care and feeding of stuff that didn't get done because we were too busy launching."

Frank Valdes explained that he was part of the Ground Equipment Review Team (GERT): "Basically, that's what took my time. We were actually busier after *Challenger*, during that downtime, than we were when we were working missions and they had these ground rules. Tony Ornelas was doing the Mechanical and basically everybody was assigned at Kennedy to review any equipment that interfaced with the flight hardware to determine that it was safe or not, hazardous or whatever; no single point failures, all that kind of stuff. Tony argued that his interface was only a shim or spacer. He argued that between that Blue Stand, there was a spacer and then there was a trunnion fitting. That was the shim. He only had a five-minute presentation to the to the review board. Ted Sasseen was the chair person. We had the vacuum system, the water server, Freon® servicers, and the Rack Air Condition Unit that interfaced with the flight hardware, so all those had to be reviewed for safety and operational concerns."

Scott Vangen explained that during the stand down period, there was a real push by the Level-IV Electrical team to review everything, check everything. What were they doing right? What were they doing wrong? What could they do better top to bottom? Scott Vangen said, "We in Level-IV saw an opportunity, in that our control rooms hadn't changed since they were first installed in the early 1980s. Keep in mind it was 1970s technology, so from a work environment, for the long hours you spent in there, it was a pretty loud environment and it was not necessarily very comfortable. If you were in there for 12-hour-plus shifts, day after day, it could get a bit tough. Dave Sollberger was our project manager. We repositioned all of the equipment for PCU-1 and HITS [High-rate Input/output Test System] 1 and 2 – the loudest equipment like the computers – behind a barrier wall, with a lot of windows so you could look into it. The terminals and the seats were put in a quieter part of the control room. We carpeted the walls with an acoustic kind of covering. They used it in certain work environments for suppressing sound. So when we started doing testing again, it really enhanced the work environment. It was quiet. We had more space. It was very efficient. So during the down time we improved our procedures, our control work environment. We took care of our customers or the payloads folks. We really looked forward to return to flight."

Halloween Parties

With the processing of all the flight hardware basically coming to a screeching halt, it was difficult because everyone had been geared up to work hard in 1986 to get all the payloads off the ground. All this talent, but no real plan yet on how flight hardware would be processed, or what could be done to keep their motivation, their "edge" for what was required for the Level-IV job. Mike Haddad said, "Scott Vangen came up with a great idea. The Payload Beach House, where everyone used to meet for social work and play while we were flying, now became the new focal point for our work. Not space-flight work, but Halloween Party work. Scott thought, 'Why not put all the pent-up energy into creating one of the best Halloween parties on the Space Coast?' That was the beginning of the famous Level-IV Halloween Party. The basic idea was to turn his three-car garage into something very unique, fix up the inside of the house along the same theme, and carry that into the back yard and maybe onto the beach."

Scott Vangen explained, "We were young, we were busy, we were happy, and we were on top of our game doing experiment payload processing. And the manifest was just going to increase. So as crazy as it was during the 1982 to 1985 [period], we thought 1986 to 1990 and beyond was going to be even bigger. And it was, but when *Challenger* happened, all that momentum came to a halt, at least from a flight activity perspective. Level-IV built a pretty healthy team of young engineers, mechanicals, electricals, techs, Ops, QA, into this whole team that congealed well together, worked together, played together, knew each other. What could we do with this energy? We lived right on the beach and our house was always open. People felt free to hang out, which was fun. So whether it was volleyball, just laying out on the beach, boogie boarding, barbecues, or whatever, it was just a fun rendezvous spot. So we were comfortable with people coming and going.

"So in 1986, in the first of what would become an annual tradition, we started the Level-IV Halloween Party. The first one was relatively humble, but do something once and it becomes a tradition. The Shuttle had still not returned to flight in 1987 and the first party was fun, so the idea was to make it bigger and it started to be themed from 1987. There was about a month's worth of preparation for an evening's party and maybe over 100 people attended that party – in costume. It was great fun."

In September 1988, the Shuttle returned to flight, but the party still took place that year and expanded to a whole new level. The house had a three-stall open garage that was converted into a theme-based location. In 1988, the theme was the movie *Aliens*, which took months of preparation. Special effects were built that would rival Disney, including *Alien* eggs made of paper mache. Some of them would open up and squirt silly string everywhere. The Level-IV team built a full-size Alien, made a space base, made the face-huggers and bio tanks. Everyone got into it, and the people participating had a lot of fun as well.

Top: Halloween team taking a pizza break. Bottom: John-David Bartoe (astronaut) and Donna Bartoe prepping pumpkins. [Scott Vangen.]

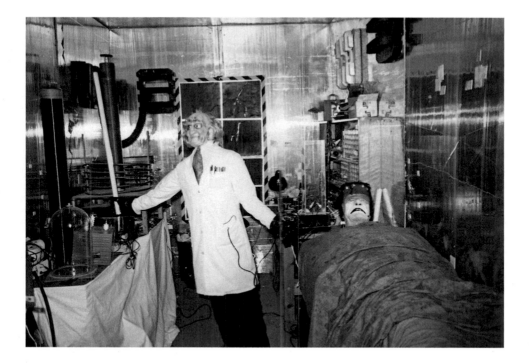

The mad scientist lab. [Scott Vangen.]

Scott Vangen continued, "We look back at that camaraderie extending into something as simple as a Halloween Party. You can laugh that off, but it was just a continuation of the camaraderie and working together, and the creative energy. In 1989, even though we'd returned to flight, Astro-1 was now back in flow, and we were getting ready for her to fly and other things hot on its heels, we still had the party that October." This time, they did a Swamp theme. "[We] flooded the garage in about four to six inches of water and had things come out of the water and out of the vegetation. It was crazy, and now hundreds of people were attending the party, all with costumes, amazing costumes."

The last Halloween Party took place in 1990 as *Tales from the Crypt*, with the garage converted into a mausoleum with various crypts, and things sliding out, things popping up, things surprising people. There was also a full-size latex model of Craig Jacobson, being embalmed by an embalming machine on an operating table. "To this day, the pictures that we have from the Level-IV parties bring back a lot of memories. Our management fully supported it, or they learned to support it. There was a 'no show award' presented to upper management that went as high as KSC's Center Director." Astronauts, payload developers, PIs, contractors, NASA techs, engineers, QA, Ops people, you name it. And friends and family. All would show up." By the end, the party was attended by some 300 to 400 people.

Mike Wright added that, "The most widely memorable Level-IV off-hours event was the annual Halloween Party. Elsewhere in the house were electronic arcade games, a rope-light illuminated dance room, and a 'séance' room. In fact, the amount of power required to operate all the special effects and other equipment necessitated connecting dozens of extension cords across to neighbors' houses. It was speculated that Florida Power operators would gather to watch their gauges curiously for the Cocoa Beach jump each year around the same time."

The séance room included a full-size casket (borrowed from a local cemetery), low-level lighting, a plasma glass ball, whispering sounds, and a "deceased" dummy with a rubber life-mask modeled after one of the young engineers. One story involved a visitor who was so impressed with the dummy's face in the casket that he went to find his wife elsewhere at the party to show her. Meanwhile, the *real* engineer laid in place of his dummy alter-ego in the casket. When the man returned with his wife, he was even more impressed with the life-likeness than before. When the wife leaned over the casket to take a look, the engineer opened his eyes. The woman screamed, leaping back across the room. Whether she was able to continue wearing her costume at the party remained unreported.

Mike Haddad explained that "Anything you could do, you'd do to build teamwork. We were basically working wide open when *Challenger* happened, doing what we needed to do for Level-IV, and all of a sudden everything completely stopped. We had all these new engineers but what were they going to do? [The Halloween Party was a chance to use their talents.] Scott would say, 'OK, you're young engineers. We need some kind of electric thing that, when you step on a pressure pad, this thing pops up out of a wall or something. You guys go make that happen'. It was the idea of taking that talent and putting it to use."

"I keep saying that I probably learned more about electronics as a Co-op student at the Halloween Party than I did with all my formal Electrical Engineering training at Auburn University," admitted Riley Duren. "I attribute that to Scott Vangen, Craig Jacobson and Jim Dumoulin, who were really incredible electronic engineers. And of course what I learned as a Co-op at KSC on the job."

Craig Jacobson remembered that, "Practically every inch of the whole property [was used]. The driveway was the graveyard, the garage was a crypt, swamp and Alien egg place, and then he had stuff on the roof of the house. He had stuff going down the sides of the house, in the yard, out by the beach. They rolled out carpet all over the yard, upside down carpet because of the sand spurs, and people were putting up big tent awnings out there, and speakers were set up for music, and there were dummies in the grass leading down to the beach. People just got more and more into it."

Mike Haddad concludes the *Tales from the Crypt*. "One last story about the Halloween Party. Because we had so many effects and stuff, smoke coming out all over the place, noise, lights, etc., someone driving by and not knowing about the party would think the house was on fire. So before the party we met with the Cocoa Beach Fire Department and told them about the party, the smoke, etc., and if there was a real fire, one of three people would call so they would know to respond. If anyone else called, they were well-intentioned, but there was no need to come to the house."

It should be noted that the payload beach house was *not* a place where people came to get drunk and do drugs; those were not in the picture. The house was a place, in a good location, where great friends would go just to have fun and enjoy each other's company. Astronauts, PIs, VIPs, domestic and foreign engineers, etc., all visited the beach house. Mike Haddad said, "Sure, every once in a while someone would have a beer or a glass of wine, but that was the exception, not the rule. Also, all the years I went to activities at the beach house, as well as in the period of time I lived there, I *never* saw anyone doing drugs." One person in Level-IV did have a family member who was dealing with an alcohol problem, and they used to bring them to the beach house to show them how young (and older) people could have a good time without the need for alcohol or drugs.

Expendable Launch Vehicles

The Shuttle may not have been not flying but automated Expendable Launch Vehicles (ELV) were moving forward, so a few people from Level-IV transferred over to ELV on a six-month detail assignment to help out. Mike Haddad recalled that, "Angel Otero, Mike Wright and myself went over to ELV and worked the Atlas program at Complex 36 on the Cape Canaveral Air Force Station. This first thing I was assigned to was reviewing the procedures to get familiar with the GSE [Ground Support Equipment] and flight hardware. I remember looking at these and some still had the original dates on the cover page from the 1960s. The same version of the procedures were used for decades because they were good and they worked; why fix something that was not broke? They also had us involved in the Centaur upper stage that was to be launched in the Space Shuttle." [1]

Left to right: Chuck Gay (former ELV launch director), Angel Otero, Mike Wright, Mike Haddad, and Jim Womack (head of Atlas-Centaur at the time) in ELV. [NASA/KSC.]

Angel Otero told the authors that he completed a six-month tour over with the ELVs. "When I was over there in the ELVs, I got assigned to be one of the NASA engineers for the de-tanking of the hydrazine out of the Galileo spacecraft, because I went over there to work Atlas-Centaur and Shuttle-Centaur, and for a while after *Challenger* the Shuttle-Centaur program kept working. They finally decided it was not going to work; putting a Centaur Upper Stage inside the Shuttle was just not realistic. So they decided to cancel it, but by then the Galileo Spacecraft was at KSC mated to a Centaur stage, so they needed to de-mate it so that they could return the Galileo to JPL and figure out what to do."

Mike Wright remembers, "With flights on hold for an extended time, Level-IV engineers were afforded opportunities to support other non-Spacelab efforts. For example, some of us were assigned to support Shuttle-Centaur preparations and testing at the Atlas-Centaur [AC] Launch Complex [LC] 36. Unfortunately, soon after the three-month detail, the Shuttle-Centaur program was cancelled due to post-*Challenger* safety concerns. Although at that point we were only involved with assisting the AC team, it was rather interesting to work at such a remote and historic location at the Cape. It really made you feel like you were 'back in time', when the space race was only just beginning. Other assignments for me included supporting the Delta-178 accident investigation in May 1986, only two weeks after a Titan launch failure and the third in four months involving every major US launch vehicle. Around that same time, about a dozen of us Level-IV engineers were assigned to design, build, qualify, integrate and test the Chromosome Experiment (CHROMEX), a middeck plant-growth experiment manifested on the first Shuttle mission after *Challenger* [which became STS-26]. I was the Lead Electrical Engineer, responsible for electrical design, qualification, and testing of the Atmosphere Exchange Unit [AES]. The experiment was a success, and flew again on subsequent missions."

Luis Delgado said that, "Chris Talon and I went over to Complex 36. Steve Francois was the head of Mechanical in that group. If there had been an RTLS, there was a requirement on Shuttle-Centaur to dump LOX [liquid oxygen] overboard on one side, hydrogen on the other in 300 seconds, so by the time we landed it would be empty [it would have been too heavy to land when fully loaded]. There was something with the timing of these valves, the Fairchild Company valves, that were made in LA. So we went to all these meetings at Fairchild with Steve Friends, and they started telling us about these valves and timing and closing. It was a very precise thing. The valves were starting to exhibit some kind of weird dynamics and there was a fear that they could slam shut and then that could blow the whole thing up. The water hammer effect, just rip the whole place. We had all these other risks that we identified that we'd got to work before we flew. So I remember Jim Womack, who was Steve's boss, called us into the trailer and said, 'Program's cancelled. It's too risky'."

Bob Raymond explained that, "We started looking to the QE [Quality Engineering] role. Are we going to stay in the chairman role of the Material Review Board [MRB]? Is it going to get transferred out of the Safety and Mission Assurance organization over to the Engineering organization, or somewhere else? Keeping us busy instead of laying us off. We needed to review everything and be sure that we could operate safely in the future, and be sure that the right people were involved to ensure the right decisions got made and we could go forward so this sort of thing never happened again, even though it really didn't fall on KSC. We needed to look at the whole picture, and that's what we did."

John Conway told the authors, "We had taken the technicians and the Quality inspectors and trained them in terms of what to inspect and what to write in a report and so forth, but we never told them why they were doing this, why they were doing this inspection, what they were looking for. We found that the reason people were not filling out reports, on hardware and stuff and software, was because they didn't realize the importance of it and how it was used. So we had big reviews. We did lots of stuff to talk about how we were changing, what we'd done, what we'd reviewed. We went through everything. We went through all of our processes and we found a lot of things to strengthen and improve."

Debbie Bitner recalled that, "Procedure reviews were awful. You know, it was hard to stay motivated and it was the first time I actually thought about leaving NASA. It was just we had a bad reputation at the time and we didn't know what our future was, when we were going to fly again, or what we were going to fly. They were reorganizing the manifest so much and the work was boring. It was just procedure review, Standard Operating Procedures [SOP], Standard Practice and Procedures [SPP]. We rewrote all the procedures and stuff like that. It wasn't a fun time at all."

"I think at the same time we also had to go through our Standard Operating Procedures," said Frank Valdes, "Well, all the procedures were unique except for the multi-run. I think we only had one multi-run. We had all these runs, and the titles were insane, and of course they were never closed out. So I had 98 runs that we were going to in the open items list, that I needed to close out. One of my things was telling people to please put a title that makes sense – not 'lifting this into that' or whatever – for tracking purposes. It took me two weeks working with Rita Wilcoxon, because she volunteered to help go through each run to figure out what they were lifting, where were they lifting it, what date they did it and stuff like that. So that was part of the stuff I did."

"We had to rewrite all our Standard Operating Procedures," explained Cheryl McPhillips. "That was when I went back to school and got my Masters in Space Technology, because I had the time, really, at work. We redid a lot of our planning and our procedures to try to do things better."

Maynette Smith remembered that things were pretty dead at the time. "We spent a lot of time doing catch up, or make work, like the HITS. During the shutdown time period, a lot of work on the ground systems involved just trying to get ahead. Things that probably needed to be done or improved, just doing things to make things crisper, if you will. I remember being busy, not crazy busy, but I remember busy enough to get this stuff done."

"We brought a number of new engineers into Level-IV in the time that we were getting ready for return to flight," recalled Rey Diaz. "When the *Challenger* took place we had to monitor the Astro hardware complement. During the time that we were waiting for the mission to be reassigned, which became STS-35, I think it was 90 days or 60 days, we had to do a power up and do some continued testing." At that time, Rey became a Test Conductor for payloads. Some of the work was primarily ensuring that the hardware they had at KSC was not going to deteriorate, so that it would always be in a flight-ready condition.

Scott Vangen knew they had to exercise the Instrument Pointing System (IPS) because of some requirement that the European Space Agency (ESA) Spacelab had on the gyros or the gimbals. "The cruciform was removed, and they hooked all the counterbalances up the IPS and sent the commands to release the payload restraints. They started commanding the IPS to tilt and it bumped up against the bumper ring that protected the IPS from going into the 'keep out' zone so that it wouldn't hit the sill or the sides of the Orbiter. Just prior to the *Challenger* accident, late in the flow, someone made the decision that the bumper ring had the wrong envelope as the 'keep out' zones. So they swapped out the original one that ESA provided and a new engineering CADCAM [Computer-Aided Design/ Computer-Aided Manufacturing] machine that NASA made was put in. The idea was that this was a better safety 'keep out' zone and met the requirements. When they did the IPS motion tests to test the gimbals, surprise, surprise, the IPS came up and interfered with the bumper ring. It couldn't even get out of the stowed deployed position. The bottom line was that had Astro-1 flown in 1986, if the *Challenger* accident hadn't happened and they had flown on the original manifest time that year, then when they got on orbit and they started deploying the IPS, it would have interfered with the bump ring and it would have been a complete loss of mission. Not life, but loss of science mission. There would be nothing they could have done about it at that point. Nothing. There was no contingency. They would have had to stow − assuming they *could* stow and it didn't jam into the bumper ring − and come home.

John-David Bartoe explained that, "I had put together a speech for the 25th anniversary of the first IPS launch at Dornier. I had every single crew member whoever had flown with an IPS send me some words. My speech was that I just read everything that everybody sent me. Jeff Hoffman had included the bumper ring business. And after the speech, this Dornier engineer came up to me and said, 'That story's not right. That's not actually how it went'. He explained it to me and I said, 'Would you actually send me that story?' So he sent me an email. He claims that was the accurate story, about how he suspected something was wrong when he was there working on an upgrade of the clamps on IPS at KSC during the down time. When he looked up, he had this gut feeling that it was not going to fit, that we were not going to be able to swing that thing up. So they convinced them to take that bumper ring assembly and put it on the offline IPS and try it there. And in fact, it did not fit, and it was all because the Astro was bigger. It wasn't intended to tip as much, so to constrain it more they took the bumper ring and moved it up. However, the ring on the movable part was a cone shape and part of the cone [diameter] would not go through. That's the story that he tells. His name was Manfred Bader."

Bruce Burch admitted, "I know we did a lot of rail polishing; Herb liked to polish rails. If you didn't have anything to do, he'd find some. We'd go out there with polishing compound and polish those stainless steel rails that rusted. Someone came up with an idea to paint the rails with a varnish so they didn't rust anymore. It worked great for a year or so, but then once you started rolling those big wheels over them, all that varnish started popping up. That was not good for a Clean Room, so we had to strip them all down and started having to polish them again."

"What I remember from after *Challenger*," Cindy Martin-Brennan recalled, "was having that uncertainty about what was going to happen to the space program. I just remember that time as being 'very busy' work, not productive time. [We were] working a little bit on the evolving Space Station program, but I just felt like it was busy work and not contributing to anything big, like I did when we started working the Spacelab stuff. You felt part of something. But I would have the same recurring dream of a Shuttle launching, me standing there, and this thing coming back down and seeing it crash right before my eyes, I had that dream for so many years. I haven't had it for a while. It would just jolt me out of my sleep but it's just stuck with me. I think a lot of us suffered. I'm nervous [it may be] post-traumatic, but I had the dream for a long time."

Bruce Burch recalled going through and rewriting all the SPPs at the time. "We went through and redid all of them and brought them up to date. We put new processes in, we did work on the offline labs we took in, and we built the experiments."

Jim Sudermann remembered that, "I wasn't sure that I was going to still be employed, but the management started telling us, 'No, we're not going to lay anybody off. Maybe some contractors, but not any of the NASA people'. So we started the Artificial Intelligence lab. They started dealing with a bunch of symbolic machines, the symbolic computers, and it was about pushing. So we started going up to Boston in Cambridge, where they were making this in volume. That's where we took a bunch of classes and we learned how to program in the LISP language [The name *LISP* derives from 'LISt Processor']. LISP was invented by John McCarthy in 1958 while he was at the Massachusetts Institute of Technology [MIT]. Parentheses; parentheses within parentheses. It's a very strange language."

While some of the Electrical and Software personnel were learning about AI to enhance support for future work, the Mechanical personnel were creating mechanical GSE to help with experiment operations once the Shuttle returned to flight. They started working on The (Experiment) Van to be used for middeck experiments. A break trunk was gutted and they placed a generator on it, built a wall behind the driver, framed it up out of aluminum and put stainless steel sheeting on the floors, the sides and the top.

Some of the Electrical personnel were also enhancing GSE that would help with experiment operations.

Powering Up the Middeck

Boeing decided they wanted to use the same middeck battery power control box design that Level-IV was using to power the middeck, as Rey Diaz recalls. "I did the initial design around that time and then I continued to work that throughout the time after the *Challenger* accident. We had a set of boxes ready by our Return-To-Flight STS-26 mission. It had a timer, a volt meter, some switches, and the battery was outside of the box. The first step was to load first and check the voltage. If you got less than 24 volts, then we were going to switch to a different battery because it was not giving enough charge in the battery to support the transfer operation."

"Everybody was looking at everything we had done on the payload side, to see how we could improve our systems," Damon Nelson told the authors. He became heavily involved in work associated with a development of a new facility, called the Space Station Processing Facility (SSPF). This was a great opportunity as a young engineer to be hands-on with the design engineering group that worked the facilities at KSC. He took a lot of the lessons learned from the Experiment Integration days and rolled those into this new facility, and for two years they got a really great opportunity to perfect this new facility. Damon Nelson recalled, "It was about that time that I started shifting to a different role. In Level-IV, I had previously been doing fluids and mechanical integration, and now I moved into a project lead role. It was about the summer of 1988 when I started this new role, plus I had a Test Conductor responsibility. My first responsibility was with a DOD [Department of Defense] mission called StarLab, a non-classified mission using a pressurized module and a laser system that flew on an MPESS. That laser system was out of Ball Aerospace in Boulder [Colorado], so that was my first real opportunity to start doing some travel out to some of the large contractors. We worked it for about six months and then we went into a telecon one morning and the army announced it was canceled just like that. I hadn't really seen just how quickly things could come and go. I had the experience with the EOM, but that was a gradual thing. This was just one morning we were a mission and the next morning we were canceled."

After *Challenger*, the Level-IV team was doing a variety of different tasks in various locations. That included building a model of the Space Station Processing Facility. Walt Preston headed up that project for the technicians. They brought in a whole room of woodworking equipment to build that model. It remained on display for some of the open houses. Don Dolby, one of the modelers, told the authors, "We made little tiny pallets and everything to try to match up what was going to be the Space Station Building. I think mostly, it was because they were trying to find something for us to do. They needed that done because I think they were trying to get that built, and I think maybe somebody wanted some ridiculous amount of money to build it. I think Herb went to management and said, 'Our guys are not doing anything, they can build it'."

Astro Modal Survey Test

Bruce Morris explained that he ended up working one modal survey test on the integrated Astro payload. Little 3D sensors and wires were placed all over the integrated Astro payload, with the telescopes and everything else on. "NASA was willing to send such unique requests to KSC because we had a group of people down there that had such a 'can-do' mentality. Our group was just designed towards 'Can you do this? Yeah, we'll find a way to do it'."

Mike Haddad recalled, "I was the Mechanical lead on preparing the GSE that would support the test. I remember, because I was working with the support and all the cables that came off those sensors. We provided stingers off the scaffolding that supported all the cables, so the weight of the cables wouldn't affect it." Bruce Birch added that, "The sensors were held on with dental stuff, so you could get the sensors on and off without damaging things. They had a chart, showed us where to put them, and we put them there and then moved on and did another section. We took Astro and placed it in front of the altitude chambers for the test. They did it for weeks."

Sam Durrance explained further that, "In order to do that accurately, you had to have a model of motion of the cruciform itself. We suspended the Instrument Pointing System and the cruciform and all of the instruments – everything on it as it was essentially supposed to be on orbit. To do these tests, they had these calibrated hammers and decided which hammer they should use to hit which place, then measured the feedback of that so that they could refine the Kalman filter, which refined the pointing accuracy or the pointing stability of the IPS. It was clear that we had a little time so we could do it, and they really wanted to do it. It was a fairly involved and expensive test, but the filter worked great and we actually used it on orbit."

OTHER ROLES

The Level-IV environment was changing, with new people coming in and some of the original Level-IV personnel leaving for other jobs within NASA; some at KSC, some at other NASA centers.

Tony Ornelas explained that, "When *Challenger* blew up, that was a turning point for a lot of people in KSC. Level-IV guys were really at full force. We had people to do the job and we had the resources to get the job done. We had proven our value, our capability, and then all work came to halt. I guess the management was looking to cover heads. I mean, they said, 'You got too many people, you're going to have to cut back or send them to different places'. And that's when Space Station was starting to build up over here at Johnson [Space Center, JSC]. Dean [Hunter] asked me if I wanted to go to JSC and maybe get some programmatic experience. We weren't doing any integration, so I said, 'Why not'. I was going to get to spend a year there, it was 1987. I came to JSC and once I got there I kind of enjoyed it and I wanted to stay. That's how I got signed up and have been here [at JSC] ever since. I still got friends here, but not to the same level. It was right after *Challenger*, and lots of people went different ways."

Josie Burnett admitted, "I didn't know what it meant to be a government employee. I never once felt like a government employee until the last 20 years into my career. The first job I had was on Astro and it was a Weight and Center of Gravity [WT&CG] job. And again, [it was] the uniqueness of this job, this position or these jobs that you were getting as an entry level engineer. You were being given a lot of responsibility. I remember Joe telling me, and Dean telling me, 'Look, we want you to do this work. We want you to assemble these experiments. It's your work, and we want you to build relationships with our customers, which are PIs and scientists. But don't worry if something goes wrong, if you break something'. But the sense of accountability was just built in, because he said, 'You can make a mistake, [it] might cost $1 million, but we'll cover you. That's all right. You won't get fired. We'll cover you'. You learned from your mistakes, [but made sure not to] make that same mistake twice. I was assigned to CHROMEX and my job at the time was very simple. Angel Otero designed a filter housing for a tiny little air exchange system, so the idea was to give fresh air to a different compartment where the plants were. I took his drawings and I went to the machine shop. They would machine it, and I would go back and forth to the machines and describe any design changes. We couldn't compromise crew safety, so that was why we went to the Johnson Space Center."

Level-IV Enters the Experiment World

As mentioned, the decision was taken to have Level-IV personnel build the CHROMEX flight experiment, so that they would get the chance to understand everything the PIs had to go through to get their experiments flight ready.

"That's the beauty of this Level-IV, right?" explained Josie Burnett. "CHROMEX was my first big screw up. In this case, it was a piece of flight hardware and it had to do with this machinist called Virgil. He was a crotchety old guy. Angel had this design for the filter housing and brought it back there. It was a very fine thread housing and it required Krytox®, a lubricant. I was just doing a random fit check of the housing and I crossed the threads. They had to cut it out and it ruined the piece of hardware. It was a piece of aluminum that cost I don't know how much. It would've been flight hardware, but I messed it up. And so they cut it out and we salvaged it. We made another one, but basically I still have that filter housing with me. I've been at NASA for 32 years now. I've carried my mistake around with me my whole career."

Sally Ride Task Group 1A

Mike Haddad said, "About a year or so after I started Level-IV, a few representatives from NASA HQ came down to KSC. I'm not sure what their goal was, but they asked a number of people in Level-IV a series of questions. Dean called me into his office and told me say whatever I wanted to these people: 'Don't hold back, spill your guts, nothing will be held against you'. So I dumped my brain about the whole Level-IV environment, what an awesome place it was to work, with incredible people, doing amazing operations on these multi-million-dollar payloads that were flying into space. Well, evidently what I said stuck with someone at NASA HQ, because in 1987 I was approached by HQ people who said that I had been selected to be the sole representative from KSC to a special group, Task Group 1A. It was being formed out of NASA HQ and led by astronaut Dr. Sally Ride. The group would select one person from each NASA center to make up the core team."

The Rogers Commission, in its concluding thoughts, stated that NASA "constitutes a national resource that plays a critical role in space exploration and development. It also provides a symbol of national pride and technological leadership. The Commission applauds NASA's spectacular achievements of the past and anticipates impressive achievements to come. Only with a clear strategy in place, and its goals for the future defined and developed, will the country be able to regain and retain leadership in space." In response to growing concern over the posture and long-term direction of the US civilian space program, NASA Administrator Dr. James Fletcher formed a task group to define potential US space initiatives, and to evaluate them in light of the current space program and the nation's desire to regain and retain space leadership. The objectives of the study were to energize a discussion of the long-range goals of the civilian space program and to begin to investigate overall strategies to direct that program to a position of leadership. Dr. Sally Ride was assigned to NASA Headquarters in Washington, D.C., where she led NASA's first strategic planning effort. [2]

Mike Haddad continued: "Shortly after being selected to the Task Group, Dr. Ride came to KSC to meet me, part of her trips to see each representative at each of their centers. Humble as she was, she walked in the room and introduced herself as Sally Ride. Then we talked for a while, discussing the role of the group. It was incredible. This took up most of my time around the end of 1986 and most of 1987, even though the group continued until 1989. After about a year, they wanted new personnel to cycle into the group, so I was asked to select a co-team member from KSC, and I picked Scott Vangen." The group's work was highlighted by Dr. Ride authoring a report titled *NASA Leadership and America's Future in Space*. She founded and became the first director of NASA's Office of Exploration.

When you are on the inside of something, or at it day to day, the pace of change seems slow, and sometimes you do not see how much change has really occurred, unlike coming back to something after an interval. This was true for the post-*Challenger* Level-IV environment, as Mike Hill explained. "I think the biggest change when I got back was post-*Challenger* and that was maybe a couple months before the next flight, so I missed all of the pain and suffering. I was there when the accident happened, then I took time off to get my Master's. KSC went through a period where essentially QE took over and rewrote a lot of the rules. And I was shocked to find that where we used to be able to do red lines down on the floor and had a lot of leeway on getting things done, much of that had been eroded and taken away. And there was the mood. I noticed the mood of the people was very different. It wasn't part of the accident, but I guess politically the QE folks took a lot of power away from us. It just made the work ten times more difficult to achieve, even just reviewing procedures and stuff. You remember how much of a power grab they had, and there was no real reason for it. It didn't add anything."

Scott Vangen summarized the period for the authors "The support of the country and our Administration at that time was so strong that we came back better and stronger. The Astronaut Corps, which I went on to serve years later, felt confident, fully confident in NASA doing the right thing for the crews. And that continues to this day. We know space is risky business, but it's well worth doing for the gains that are there to be had."

SUMMARY

By the mid-1980s, Level-IV had changed from the original concept. Some of those who worked Level-IV pre-*Challenger* went on to other NASA organizations, some left NASA all together, and NASA awarded the Payload Ground Operations Contract (PGOC) to McDonnell Douglas Astronautics Company on January 1, 1987. Contractors would now perform some of the Level-IV operations that used to be done by NASA personnel. As the Shuttle system returned to flight, Level-IV was different to what it had been before *Challenger*… and so was NASA.

References

1. For further details on the astronauts' involvement in flying Centaur as an upper stage on the Shuttle, see Shayler, David J., and Burgess, Colin, **NASA's First Space Shuttle Astronaut Selection, Redefining the Right Stuff**, Springer-Praxis, 2020 pp. 340−41 & 363−65.
2. *Leadership and America's Future in Space,* A Report to the Administrator, Dr. Sally K. Ride, August 1987.

10

Pallet and MPESS Missions

"As soon as you get [in space]
and see the view, it just blows you away.
But at that time you're pretty busy.
You've got to do a lot of setting up.
You've got to set up the Spacelab computers,
you have to set up [a] number of things
that need to be set up there."
Sam Durrance, STS-35 (Astro-1) Payload Specialist
recalling his first minutes in orbit.

As the Shuttle was being prepared for Return-to-Flight, the manifest was getting firmed up. For those in Level-IV, it was clear how the future missions were lining up. Payload operations were already swinging into action in preparation for the new manifest, including which Spacelab missions from before *Challenger* were still in planning and which ones were to be delayed or sadly cancelled. There were a number of tasks that were intrinsic to Level-IV and these are explained here generically, though some parts of the chapter do address specific missions.

AFTER *CHALLENGER*

"When PGOC [Payload Operations Ground Contract] was established I was hired as a tech, but within six months I was made a lead and then a manager with PGOC," Bruce Burch told the authors. "I managed both Mechanical and Electrical

M. E. Haddad, D. J. Shayler, *Spacelab Payloads*, Springer Praxis Books,
https://doi.org/10.1007/978-3-030-86775-1_10

[technician] areas, so I started out as Mechanical and then I progressed to senior manager. Then, when Maynette Smith became the Section Chief of that PPCU [Partial Payload Checkout Unit], there was a mixed PGOC/NASA team and they were essentially badgeless, so we all worked and it was a great team that worked really, really well."

When NASA asked for McDonnell Douglas people to join the Level-IV team, they enjoyed the same protection and status as NASA engineers. They could do whatever job needed doing without fear of union members (technicians) filing grievances. It is hard to exaggerate the magnitude of this concession, as McDonnell Douglas Computer Design and Science Engineer Dan Kovalchik explained: "We got spoiled by the cooperation we enjoyed from our control room techs. On the floor, we engineers were commonly exposed to what I refer to as 'the technician's attitude'. This attitude consisted of a variety of uncooperative activities, body language, and remarks that let us know in no uncertain terms that we were the enemy."

Rich Jasnocha started out as an RS-232 cable tester. "Then I moved on to software projects and it was doing these projects in this group that really made me change my career path from hardware to software, because my degree is in Electrical Engineering. My other Co-op stuff was the design of a differential single edit signal container with Rick Coats, so after that, I did the RTP, the Re-Transmission Processor. But that was when I was full time. RTP basically interleaves a window of data from a major stream of synchronous data, and then the captured window's output in the new synchronous data string. We used that successfully during payload testing and then after that did a data capture system. It was part of the RTP project that archived the payload data onto the hard drive of the PC. And you could de-manipulate, to capture data."

Dan Kovalchik explained that, "In Level-IV, the NASA engineers sketched examples of how they wanted commands and measurements to be arranged on their computer screens. We McDonnell Douglas PPCU engineers took these lists of commands and measurements and taught the PPCU how to interpret them via a command and telemetry database. Our most basic task, however, was to keep the PPCU running, ensuring that the dozens of computers were all running their correct operating systems, their specialized PPCU applications software, and their mission databases."

Before *Challenger*, Level-IV personnel had to work problems with the Ground Support Equipment (GSE). Interaction with the Principal Investigators (PIs), experiment people and others caused its own problems. Part of the job involved fixing the hardware/software problems, fixing the personnel problems, testing the flight hardware to stay on schedule and keeping up with the increasing flight rate. These tasks would again be performed going forward, but now with the new manifest, updated processes, and new people in Level-IV.

Going Forward

From this point forward, not every mission will be discussed. Only those under the Spacelab series or worked by Level-IV personnel will be included here. A few missions contained only small experiments, but are included here to show the variety of tasks performed by Level-IV personnel. Also included are the proposed missions in each series, to reflect on the ongoing work Level-IV would have been involved with had the plans reached fruition. Sadly, most did not and were cancelled, which contributed to the early demise of the highly successful and reliable Spacelab program.

THE LONG-AWAITED RETURN-TO-FLIGHT

For the Return-to-Flight mission (STS-26, *Discovery*, launched September 29, 1988), Level-IV worked mostly middeck issues because the major payload was not assigned to Level-IV. Saying that, some Level-IV personnel did support the IUS/TDRSS (Inertial Upper Stage/Tracking Data and Relay Satellite System) that was carried and deployed on that mission. Bob Raymond revealed to the authors that, "The TDRS-C satellite on the STS-26 mission had been contaminated while in the VPF [Vertical Processing Facility]. I was a discipline lead for QE [Quality Engineering], and we had to go have the satellite re-cleaned and inspections done again. But because it was the Return-to-Flight after the *Challenger* accident, they wanted additional eyes on the equipment once we got to the pad, to be sure everything was okay."

The next flight, three months later, was the classified DOD (Department of Defense) mission STS-27, but the first mission of 1989 (STS-29, *Discovery*, launched March 13, 1989) included the Space Station Heat Pipe Advanced Radiator Element (SHARE). This was a test of a new cooling system under development for the Space Station. Also flown was CHIX-in-Space, a student experiment designed to fly 32 chicken eggs to determine the effects of spaceflight on fertilized chicken embryos. Crystals were obtained from all the proteins in the Protein Crystal Growth experiment. The Chromosomes and Plant Cell Division in Space (CHROMEX) life sciences experiment was designed to show the effects of microgravity on root development. Finally, an IMAX 70 mm camera was used to film a variety of scenes for the 1990 IMAX film *Blue Planet*, including the effects of floods, hurricanes, wildfires and volcanic eruptions on Earth.

NASA Level-IV Mechanical Engineer Michael Stelzer expanded on the story behind the SHARE and SHARE II: "It's hard to imagine squeezing a 50-foot-long experiment into an already almost full Space Shuttle, but it was possible. Steve Glenn was the JSC [Johnson Space Center] Project Engineer and was instrumental in keeping the project on track and helping us at KSC [Kennedy Space Center] with the processing and testing requirements. SHARE was designed to fly along the port side longeron of the Space Shuttle Orbiter Payload Bay.

"Although it was installed horizontally in the OPF [Orbiter Processing Facility], once the Orbiter was rotated to vertical and moved to the launch pad, all 50 feet of the SHARE experiment needed to be hinged out to allow for installation of the IUS/TDRSS, and then rotated back to its stowed configuration for launch. It was a clever application, but made for just enough difference in ground processing to keep the Space Shuttle processing team on their toes."

The Chicken Embryo Development (CHIX)-in-Space experiment was an upgraded version of the middeck payload that had been lost on STS-51L. Mike Haddad, who was the Mechanical lead for this payload, remembered reading about the experiment from the 51L information, "So when it came to working that for STS-29, I jumped on it. There were actually two versions of the payload: the flight middeck version that would be loaded into *Discovery's* middeck, and a ground middeck version that stayed on Earth. These middeck containers were again made as identical as possible, even though only one would fly in space. The idea was to handle both the same, so the only real differences would be the time in space. Working with John Vellinger, who came up with the idea as an eighth grade student in Lafayette, Indiana, was awesome, and what better sponsor for chickens in space than Kentucky Fried Chicken [KFC]. I supported any offline work John had in installing the eggs into the flight middeck and ground middeck versions, the offline testing, and then preparations for it being taken out to the pad to be loaded into the Orbiter, I think within 24 hours of launch. Again, both would follow the same path up to the point of one being installed into *Discovery*. Installation and testing went fine, so we were able to get a photo of those of us that worked the late loading of STS-29 middecks."

"The plan for this mission was to land in California at Dryden Flight Research Center [DFRC]," Mike continued. "If we launched on time, we had to take the ground version to a local lab, located in Titusville just west of KSC, to be shaken using the planned ascent profile, again to try and duplicate as much as possible between the flight and ground version. We launched, so this first part went well. Then Kentucky Fried Chicken had their own corporate jet waiting at the Titusville airport, with engines running, so once the shaking was done we immediately transported it to the jet, loaded it onto one of the seats and strapped it down. Of course John was part of all this work. I remember to this day, him waving goodbye as they closed the cabin doors, and standing there to watch that jet take off and head west at the fastest possible speed to get them there ASAP."

Jeannie Ruiz was a Mechanical lead for all the STS-29 middeck experiments. "Following landing, we got the experiment off the Orbiter and took it up to Buckhorn [Shuttle special purpose station, California]. The road was full of pot-holes [raising concerns whether] the eggs were going to be alright. We were afraid they were going to break because they weren't hatched yet." Less than a week after *Discovery* landed, the first space-flown egg hatched. The 1.5-ounce (41.8g) male chick, named "Kentucky", broke through its shell at KFC's headquarters ("Colonel Sanders Technical Center") in Louisville. "The first space chicken,

Kentucky, was donated to the Louisville Zoo and put on display," said Gregg Reynolds, then Vice President for public affairs at KFC. [1]

THE PALLET-ONLY MISSIONS

During the next decade, several Shuttle missions with Level-IV involvement flew under specific payload titles in a series of flights. Though far fewer than originally planned because of the developing International Space Station (ISS) program and restricted budgets, these are summarized here under their series headers. There were other Shuttle missions, including DOD, satellite, observatory and planetary payloads, that had little to no Level-IV involvement, and will only be mentioned in this book if they contained an experiment worked by Level-IV. For continuation, we have split the Spacelab missions flown in the 1990s into two divisions: pallet and MPESS missions; and those which primarily featured the pressurized Long Module configuration. In this chapter, we focus on the work carried out by Level-IV on the missions which *did not* include the pressurized module. The first of these pallet-only flights flew at the start of the decade, featuring the much-delayed first flight under the Astro series.

ASTRO SERIES

This was originally designated the Space Telescopes for Astronomical Research Series (STARS), formerly listed under the Office of Space Science payloads which were later renamed the Astro Series. Commonly mistaken for an acronym, 'Astro' is just a shortened version of '<u>Astro</u>nomy'. The "Astro Observatory" was specially developed as a system of telescopes that could fly multiple times on the Space Shuttle. The telescopes were mounted on a Spacelab pallet in the Orbiter's Payload Bay. The Spacelab Instrument Pointing System (IPS), pallets, and avionics were utilized for attachment to the Shuttle and for control and data handling. Initially, there were plans for six Astro flights, but they were subsequently cut to three, and in the end only two missions were flown (STS-35 and STS-67).

STS-35, *Columbia*, launched December 2, 1990, 8+ days
Payload: Astro-1; 1 Igloo; 2 Pallets (Fwd Pallet F010, Aft Pallet F002); Instrument Pointing System (IPS).

The primary objectives were round-the-clock observations of the celestial sphere in ultraviolet and X-ray astronomy, with the Astro-1 observatory consisting of four telescopes: Hopkins Ultraviolet Telescope (HUT); Wisconsin Ultraviolet Photo-Polarimeter Experiment (WUPPE); Ultraviolet Imaging Telescope (UIT); and Broad Band X-Ray Telescope (BBXRT).

Mike Wright told the authors, "Most Level-IV engineers had become interested in the space program, not only through real-life space programs like Apollo, but also from television programs like *Star Trek*. One particular day of Spacelab Astro-1 testing was pre-planned as a *Star Trek* day, complete with costumes, control room sound effects, a cardboard viewing 'scanner' for the participating astronaut, and even 'tribbles'. Some engineers also dressed as Starfleet crewmembers. Due to the Level-IV team's dedication and professional expertise, this seemingly frivolous event had no impact on actual test operations. Management was not made aware of the plan in advance, but the next day, a couple of the test engineers were called to meet with the Director of Payloads; reportedly, he was not happy. The Director told the engineers not to do something like that again without permission."

Scott Vangen added that, "One time, the Astro payload crew [Sam Durrance, Ron Parise and Ken Nordsieck] got into *Star Trek* day, such as Ken looking into the 'Spock' viewer which was just a cardboard dome that was just bolted onto the Data Display System. Most people had fun with that, because many of us were influenced by science fiction."

Riley Duren told the authors that, "I was probably 19 or 20 years old and I was dressed like crewman Jones with a red shirt. What we didn't know was that the PBS [Public Broadcasting Service] show Nova was shooting a documentary on the Shuttle program, and they were trying to get in to see some action. The branch chief said, 'Oh sure, we're testing this thing called Astro. We've got a control room and they're testing today'. Well, nobody had let him in on *Star Trek* day. I remember he literally showed up at the door of the control room unannounced and he had a film crew in tow. He opened the door and the door went 'smoosh, smoosh', and I saluted him. And then you heard all the sound effects and he just stopped dead. And his face turned bright red. He didn't say anything, but he just turned on his heel and walked out and they [the film crew] didn't come in. That was one of my favorite memories. I can still vividly remember it today." *Star Trek* day was fun for those involved, not so for management, but everyone lived through it. Level-IV continued doing what it did best, preparing Astro (and other payloads) for launch.

Elias Victor explained that as a Co-op student, "I had a little bit of knowledge with AutoCAD. We had some of the checkout equipment that needed to have cables run under the raised floor, so I was doing AutoCAD and getting underneath the floor and pulling the cables as they were needed. I was also involved with PPCU coming online and there was a good opportunity to try out new equipment. We also had to design and build cables, [which were] taken down to the tech shop so that they could put the cables together and then brought back up to hook them back up. The first thing that I had to realize was that these technicians really were the knowledgeable people."

There were Mechanical as well as Electrical personnel working as Co-ops in Level-IV. Mike Hill recalled that, "When I came back, Luis Moctezuma taught me metrology and I fell in love with it, working with the theodolites and doing the math and all that stuff. It was great work. Luis was getting tired of doing it, so I jumped in with both feet and started doing it."

Crew members also worked problems that would have a direct effect on Level-IV operations, as Astro Payload Specialist (PS) Sam Durrance recalled, "For their CCD [Charge Coupled Device] detector, WUPPE designers had only been given an upper limit of what the given current should be. And then when you turned it off, it still had a residual current in it. So we were looking at the data, looking at the command chain, and there was the question, 'Is the current within tolerance?' Basically, there was no lower limit. The answer was yes, but I said, 'That's not right because there should be no current at all when it's not on'. It was just the problem was with the procedure not having a lower limit. I remember Ken Nordsieck was pretty upset about it because it had been that way for a while, but it turned out it hadn't been damaged."

Sam Durrance also recalled that, "I was essentially responsible for trying to figure out how to operate as a single observatory instead of three separate instruments, but not totally. My name was on it, but this was something that the three of us, Ron Parise, Ken Nordsieck and myself, and also the Level-IV folks at times, put together as a Joint Operation Procedure [JOP]. This was my first encounter with this kind of development of a procedure and how it worked. There was a lot of crosstalk with Level-IV on that one."

Mike Hill recalled, "For Astro-1, they wanted a particular alignment set, and then there were some cases where they just wanted to know what they could calibrate on orbit. Astro-1 was the one that comes to mind as the most difficult one, because those were big telescopes to do crane ops with and they had adjustable mounts. We had to put them on and off probably two or three times before we could get the alignment. So you would do a crane lift, put it on, do the alignment then figure out how to move the mount. You took it off, made an adjustment, put the instrument back on, and would shoot it again. That was a good two weeks or so to do all that right, with all the ops and everything."

Jim Sudermann added that, "One of the things that I got involved in, at some point, was that we started building the PPCU, the Partial Payload Checkout Unit. That was the project that we did with DE [Design Engineering]. Basically, Payloads gathered up 10 million bucks [to fund it]. DE had the reputation of taking your money and saying, 'We understand your requirements', and then going off for four or five years, spending all your money and showing up with something that had nothing to do with your requirements. So Polly Gardner and myself were going to watch them like a hawk. We were going to make sure we got what we wanted, so I spent several years working with DE."

Riley Duren's first assignment at Kennedy, working for Zoot (Jim Sudermann) and Polly Gardner was on the design and development of PPCU that was taking place during the Astro-1 flow. "They asked me to shadow the DE team that were working across the street on developing, testing and delivering the PPCU. I would spend a few days a week with them in the lab, where the electronics engineers were putting the wraps together and testing the boards, and software engineers were developing the user interface. I would say in particular I was hanging out with the software team." Riley had a little bit of training in C (programming language) in college, but he had to teach himself Unix and C, learn how to do some coding and actually immerse himself in the operating system. He also had to learn AutoCAD so he could do the layout of the control room. It was a lot of input, doing the drawings, then he worked with the DE facilities teams to set it up. The actual setup and installation happened while he was away during his last year at Auburn University.

While new people were being brought into Level-IV, the work for those already there was not slowing down. For the Astro-1 payload, the O&C (Operations and Checkout) operations were coming to a conclusion, and then it would be off to the next phase of payload processing, the transportation and installation into the Orbiter. But new problems seemed to crop up every day, including some that would follow the payload all the way through to pad operations.

Sam Durrance explained that, "On HUT, we couldn't open the aperture because of the cesium iodide coating and the gradings. I made the detectors, but we wanted to test, so we got a Celestron telescope and mounted it on the aluminum door in the front of the Orbiter, aimed at the primary mirror. We got a 35 millimeter camera and took some astronomical slides, and it mated exactly with the detector itself, but we could then look at the actual motion of the stars through the field as the telescope turned. So we nailed that with the test. Also, we had an issue with the HUT vacuum pump. The detector had to stay under vacuum at all times, through all the testing, moving operations, lifting operations. It was a total nightmare."

Luis Delgado added that to keep continuous power on the HUT Vacuum Ion Pump, there were certain operations for which connecting a hardline to the telescope was just not possible, so a battery unit was secured directly to HUT to keep the pump operating. When it got to the OPF, it became a priority operation, and ended up with a little cable going to the T-0 plate. During the rollover from the OPF, Doug Butler and Jim Nail dragged the battery cart behind the Orbiter as it was moved to the Vehicle Assembly Building (VAB). Once in the VAB, with the cable still connected, Shuttle personnel hooked up the sling and the Orbiter. Once the Orbiter was rotated, and just before it was hoisted to go over and onto the External Tank (ET), Level-IV technicians reached over and disconnected the cable from the T-0 plate. They then took the cart and ran into the elevator, up to the Mobile Launcher Platform (MLP) as the Shuttle was being lifted to mate to the ET. After mating, and with the cart positioned in the proper location, the cable was reconnected to the T-0 plate and power applied to HUT through the Orbiter T-0. Once the stack went to the pad, the power was controlled from the Launch Control Center (LCC). Luis Delgado added, "We didn't fly hot. We shut down at a negotiated time, maybe T-15 minutes before launch."

Roland Schlierf, recalled, "I was sitting in the Aft Flight Deck [of *Columbia*] and monitoring the parameters of the telescope while we were running IVTs [Integrated

Verification Tests] and monitoring the ion pump. If you lost the pump for two hours, the spectrum analyzer was degraded. We left the OPF to go to the VAB and then out to the pad. I went down there to check on my auto dialer in the MLP. I had my little auto dialer plugged into a phone on the wall, so if there was a problem it would call me up and let me go fix it. Then one of the pad techs decided to use the phone. He had the phone off the hook and was using it to call his wife or something. So I got really annoyed. It was probably the first time I blew my stack at work. And this was one of those older, senior techs, the kind that thought they owned the world, knew everything or whatever. I said, 'There's a sign right there that says, "Do not use this phone, what are you doing?"' And he started arguing with me about it. I ended up taking off the little cord between the receiver and the phone. That was my solution. I still have that cord to remind me of being seriously annoyed."

Mike Hill recalled that, "Astro-1 was a lot of fun, because we followed that out to the OPF and onto the pad. We took the 'remove before flight' baffle covers off at the pad because we didn't want contamination getting into the telescopes. We had streamers and little grabber things that you use to get things off high shelves. And then you couldn't just pull [the covers] off. We had to go around and get them from different sides and just slowly pull the things off. I think Tim Owens was a Co-op at that time. He went out with me and helped me with that."

Mike Hill removing baffle covers at the pad. [Mike Hill.]

Cheryl McPhillips explained that the BBXRT was manifested for the flight and was cooled by a supply of solid argon. There was no way to relieve the argon tank, and it was a flight safety requirement that the tank had to be monitored to make sure it did not get over-pressurized. Cheryl went out with the tech to install the hardware in the MLP, which ran up to the telescope and the Orbiter, and then came down from there and into the LCC before eventually running back to 1263 (the room in O&C where all the electrical patching took place) and up to the User Room (also in the O&C). Then BBXRT had a failure out at the pad, with the telemetry lost, meaning the tank pressure could not be monitored. They also had the Launch Readiness Review (LRR) that day, very close to launch. Brewster Shaw was a Program Manager at that time and asked what had happened. Cheryl McPhillips said, "We lost our telemetry," and Shaw replied, "You've got 24 hours to fix this or you're going to fly as ballast." The "B" word. Cheryl recalled, "I remember 'ballast'. We did not want to be ballast, so we had to go out to the pad and troubleshoot it. We wrote up the troubleshooting steps, finally got the paperwork approved, and went out to the pad. Then the QE was a woman who had a dress on. Well, you couldn't have a dress on at the pad, so we had to wait for her to go home and get changed and come back. Luckily, the engineers that were there from Goddard were right on it. The telemetry box failed and we could see that it wasn't putting out any telemetry. Luckily they had a spare up at Goddard, which they overnighted to us. We got it installed and then they flew. I worked 35 hours straight. I ran into one of the techs who said, 'You're a hero. People know how much you worked and how hard it was, and that you were stressed out'."

'Twas the Night Before… Launch

Sam Durrance recalled the special time with his family shortly before the launch: "This was a night launch, so the barbecue was at night. We got on the porch [of the Astronaut Beach House] with the Shuttle on the pad only about a mile away. It had the big xenon lights shining on it and it was really a cool time, some time with your family and all the rest of the crews' families. When the tanking started, it showed excess hydrogen gas in the engine compartment. They said, 'Go back to Houston'. We got really good at that, we did it five times altogether, the whole thing."

At this point, Astro-1 had become the most traveled payload on the ground at KSC. Due to numerous delays and issues with the Orbiter, it went through many facilities multiple times, where normally they would only go once before launch. As part of that unenviable "record" established, a special button was made that depicted this epic set of events.

STS-35 button, showing all the locations at KSC the payload travelled to before launch. [Mike Hill.]

With all the bugs finally worked out and the problems fixed, all that was left to do was launch. Roland Schlierf explained that he worked the C-1 Console for launch because of the ion pump, "I think Sue (Sitko) was [also on] console. Scott (Vangen) and some other folks, I think, were on console for the launch and then as soon as it went, we all got on a NASA-4 plane up to Huntsville, Alabama and worked the POCC [Payload Operations Control Center]. I just remember being there when the first science came. That's the other thing, to actually be able to see the fruits of your labor, to be there. When the first images came down, we were sitting around guys that had worked there for 10 years, and we could see them just erupt when that first science came in, like kids in a candy store. All those years of work for those guys paying off. That's the kind of inspiration you hold on to your whole career, and when you got mired down in the mud, that's the kind of thing you'd grab onto: 'Wow, I know what it's like to see somebody commit 10 years of their life to something'."

A wild ride

PS Sam Durrance took up the story from the unique position of sitting inside *Columbia* as it finally left the pad. "So finally we did launch, a wild ride. It's more just jerky motion, back and forth, up and down, right, left, as the solids [Solid Rocket Boosters, SRB] are burning. You get to two minutes in the launch profile and the SRBs are jettisoned and it sounds like someone hit at the tank. The thrust drops immediately from 2.8g back down to 1g; it sort of jerks you up like that. The main engines burn for six and a half more minutes and are really smooth. The thrust picks back up from 1g to 3g very, very smoothly, but it never stops getting acceleration. So it's pushing you back into the seat with the force of a couple of hundred pounds. You're breathing against the compression, then all of a sudden you go from that to nothing, no force at all. Your stomach comes up in your throat and you're going over a roller coaster. Your hands just float up in the air. You release all five harnesses on your seatbelt and just float up out of the chair. It's just a wild ride. Then you're there and you're instantly – it seems like instantly – in another world. So as soon as you get there you head upstairs to see, and the view just blows you away. But at that time you're pretty busy because you've got to do a lot of setting up. You've got to set up the Spacelab computers. I don't remember how long it took us go from orbit insertion to operating the telescopes, but it wasn't that long."

With Shuttle missions averaging just a week or two, every minute of the time on orbit was crucial for gathering as much data from the payload and experiments as possible. With the crew on orbit powering up the various Astro systems, the teams on the ground already had their systems all checked out, so when the crew was ready to go with on-orbit operations they could begin immediately, without wasting a moment before it was time to come home.

"[For] Astro-1, I was a member of the POCC team, and the WUPPE experiment had two experiment computers," recalled Scott Vangen. "They were totally redundant. They were called DEPS [Dedicated Experiment Processors], DEP A and DEP B. You could operate on DEP A or DEP B. They were just redundant. If A failed you switched over to B. Not a big deal, but you didn't want to do that. Well, WUPPE powered up and we were a day into the science. We were doing fine and things started getting squirrelly and then DEP A just went belly up and this was a big deal. I mean, you've lost science, you're dead, and the other instruments are going on. I noticed something on separate telemetry that the PIs weren't looking at and, sad to say, neither the POCC or JSC was looking at. I had built a kind of little secret private page with some PIDs [Pocket IDs] – basically telemetry of things that I wanted to look at and the health of the RAUs. I plotted the last six or eight hours of cycling, and the [RAU] heater was not on, else the heater would've stabilized it I told Art Code this, and this was all happening in minutes here, real-time; the pressure was on and I didn't want them to switch to DEP B. I said, 'There's something else that we don't know'. I talked to Art Code and I got in the comm loops per the protocol to talk to the POCC principal. I said, 'I believe JSC [who controlled these heaters, not Marshall] didn't turn the heaters on. RAU 5 and RAU 6 are freezing'. So he connected me the POD to talk to the Flight Director, and connected me with the Payloads Officer at JSC. We had a conversation. JSC said, 'I'm sure we turned them on'. I said, 'Well check, or just send the command again, there's nothing to lose. It's two commands'. Well they turned them on, and instantly they started warming up, and within 30 minutes they were exactly where they needed to be and it ran, it hummed the rest of the mission. I think that was one of the things that kind of punched my card with Art Code when they picked me as an Alternate Payload Specialist."

Sam Durrance recalled on STS-35 (Astro-1) that, "There were not many things that were too bad, but as we were going along in the first day or two, one of the Spacelab computers had an electrical fire in it. I figured out later it was a resistor that had shorted."

Scott Vangen explained that because Astro-1 had the longest timeline on the ground before flight of any Spacelab mission due to all the hydrogen leaks, the Data Display Systems (DDS) were operated much more often than envisioned. Every time it powered up, it started to collect little fine particulates, and the filters and the DDS got clogged. This stopped the cool air from flowing, so some of the electronics overloaded and burned themselves up. "So now we had the crew 'x' days into the mission and they had lost all of their communication and display means with Spacelab, Now the crew couldn't do anything." Sam Durrance added, "We had no way to command the instruments from the flight deck." (See Callout Section: *The Heart of What Made Astro-1 a Success on Orbit*.)

The Heart of What Made Astro-1 a Success on Orbit

The following is a continuation of the exchange that took place during a gathering of Level-IV personnel and crew members, recalling their experiences on Astro-1.

John-David Bartoe: "We had TV visibility of some of the cameras in some of the instruments on the ground, where they did not have it on orbit. However, on orbit, the manual controller for the IPS was still functional but all the screens were blank, because both DDS were dead. But the Orbiter cameras in the aft flight deck were still operational and we could see [one of the instruments]."

Sam Durrance: That was HUT, the primary acquisition telescope of the three. It had a 14-arc-minute field of view, a low-light-level camera looking at the field of view, reflecting off the aperture itself. So we had the target itself. We had this star tracker that was built at JPL. It was on the Goddard experiment [and] was one that was used for tracking the Goddard experiment [UIT]. The Star Tracker would have had a nice field of view, but they didn't provide it. It could have been provided on that instrument, but it was not. We were able to get from them a readout of what was essentially the motion of the telescope itself, so we had a number for each of those. One was right ascension and the other was declination. When the IPS was slewed between observations, it would not come up close enough [to reach] either of those numbers. It wasn't accurate enough, but they could see them on the ground. [The ground] started uplinking the numbers to us that would describe the right ascension and declination of HUT when it was in view of the target. Once it was in view of the target, we could see it. We could then tell the ground what we used to use on orbit, which was the numbers that we could actually drive the HUT to. But we had to reverse that role and [the ground] would drive the numbers, and we would just be calling them out because they couldn't see it [the target]. We'd say, 'Up, up, up, up, left, left, right, right', until it was in the field of view. Once it was in the field of view, then we could manually guide it."

John-David Bartoe: "I remember that we would sit there with a pencil watching it, and we would say, 'Up, right, left, up, down'. I tried to get to keeping it right in the right place, in the middle. If anybody were to ever listen to that tape, they'd wonder what was going on."

Scott Vangen: "I remember you [John-David] sketching out that control loop. You eventually sent down a draft to me and said, 'What do you think?' Well, it worked and many people looked at it and said, 'This looks like the ONLY way it can work'."

John-David Bartoe: "There was another big moment there in the POCC too that really helped, which was when it was realized that all the instruments would have to be commanded from the ground. Each of the instruments. Marshall said, 'Okay, the commands will be sent by the Command position'. There was a position called Command. We pushed back big time against it, because management was wanting to go that way, the Mission Manager was wanting to go that way. We said it would never work, because we'd need to send hundreds and hundreds of commands and there was no way one human person could do it. They kept objecting that the commands might hit each other on the way up, that they could be rejected. There was always this story about the simultaneous commanding causing rejections. We basically said, 'We'll just take the chance. If it's rejected, we'll send it again. It's not that big a deal'. So they finally gave up. It was unbelievable to watch all of those teams in the POCC run their instruments from the ground through all that time. This was really interesting."

Scott Vangen: "It became *modus operandi* going forward to enable the POCC team to take care of their instruments."

Sam Durrance: "Yes turn it, put it in the right configuration."

Donna Bartoe: "So we were doing all of the commanding [from the ground]."

Sam Durrance: "And it worked like clockwork. We had lots of issues on Astro-1 in addition to that. The difference between that and Astro-2 is just unbelievable. Astro-2 was just smooth as glass the entire time. We ended up with more than 50 percent, something like 70 percent I think, of the targets that we had anticipated. And we actually did have one day less. We originally were going to be a ten-day flight, but Vance [Brand, STS-35/Astro-1 Commander] was getting kind of antsy and there was adverse weather at the landing site. We couldn't land in Florida because the Shuttle Landing Facility [SLF] wasn't recertified yet. So we were scheduled to land in California, and there was weather approaching, so I guess he [Brand] decided to come back early. So it was nine days instead of ten, but it was an adventure."

Scott Vangen: "The fact is that the mission was saved and we got good science out of that. It's a great example of having the right people, being able to adapt real-time, and flexibility with the team."

As with most missions, the work did not end after landing. There was a great deal of postflight work involved, as Roland Schlierf recalled: "I remember the term they used; it was 'calibrations'. They wanted to have them before the mission, during the mission and after the mission, to refine the science that they got. So it was still important. We had to fire up the Hopkins Ultraviolet Telescope in the O&C postflight to calibrate the spectrometer. It was me and Sam Durrance who ran that out. For whatever reason, all the Hopkins engineers and whatever were off doing something else. So there I was, sitting in the O&C Building right after the mission, with Sam Durrance, the astronaut on the Orbiter during the mission, calibrating his telescope. That was so cool. All the coolest stuff happened the first time round. I knew I'd never beat that."

STS-67, *Endeavour*, Launched March 2, 1995, 16+ days, Payload: Astro-2; 1 Igloo; 2 Pallets (Fwd Pallet F010, Aft Pallet F002); Extended Duration Orbiter (EDO) kit.

Astro-2 was the second dedicated Spacelab mission to conduct astronomical observations in the ultraviolet spectral regions, but for this re-flight of the Astro set of telescopes, the Broad Band X-Ray Telescope (BBXRT) was omitted. As with most re-flights, the Astro-2 processing flow was fairly clean, with little as far as problem solving was concerned, so most of what is discussed for this mission was something unique, even for Level-IV.

Scott Vangen (see Callout Section: *Alternate Payload Specialist*) recalled that the Astro suite of telescopes, as a payload, was to going to fly three times. "It was manifested to fly a third time and I would have gone into a flight rotation. That was the understanding. Ron [Parise] and Sam [Durrance] flew again because Ken [Nordsieck] chose not to and I was okay with that. I knew we were not taking Ken's spot on Astro-2. I was going to be the new backup and then on Astro-3 we would fly again. After the very successful Astro-2, Astro-3 was manifested and I was going to go back into flight rotation. Astro-3 would fly refurbed instruments with whatever the latest technology was, then we would fly Astro-4, -5 and -6. But then suddenly, Astro-4, -5 and -6 disappeared and the intention was to have Astro-1, Astro-2, Astro-3, but it ended up being just Astro-1 and Astro-2. It was like the Atlas series; we were going to fly a boat load of those, [but it ended up] we flew two and a half."

At this point during the Shuttle program, the Space Station program development of the ISS was really gaining steam, and was having direct and significant influences on the Spacelab program. With only so much money to go around, NASA began scaling back – as well as cancelling – some Spacelab missions to fund ISS.

Alternate Payload Specialist

One of the main highlights to mention for Astro-2 is that a member of the KSC Level-IV team, Scott Vangen, was selected as an Alternate Payload Specialist (APS). He would train with the prime flight crew and could be called in at short notice to fill in for any Payload Specialist that was supposed to fly. Mike Haddad recalled that, "Being an APS was huge in my opinion. That somebody that I knew personally had been picked to be an APS. What was the process? How did it work and basically, who picked him?"

The Payload Specialist (PS), as the name sounds, was a specialist for a specific payload, so the possible flights for a PS were a lot less than an MS. The PS was actually selected by the science community, not by the standard NASA astronaut selection process.

Selection Process

Scott Vangen, recalled, "The PS had to be picked by the science community and my name was out there in the PS community. As an outsider, which I was being a guy from KSC, as an engineer not a scientist, and as a non-PhD, I broke three of the rules, because the PS typically was a PhD, a scientist, and from a specific institution. I was none of those when Astro-2 was manifested. Astro-1 had flown and it had three PS, Ron Parise, Sam Durrance and Ken Nordsieck. The idea was the mission was going to fly three times, Astro-1, -2 and -3, and they were going to go Round Robin. Each would PS fly two times. So after Astro-1, Ron and Sam flew Astro-2. Ken was going to rotate into a flight position [for Astro-3] and one of the other two would become the backup, but Ken decided not to take that flight position. I talked to Art Code, before I even put my name forward, to see whether I even had a chance. He said, 'Well Scott, you're kind of a known man. We like you. And you never know till you try'. So I applied and went to Goddard for an hour-long interview with Art and a few others. Afterward, I was getting ready to leave and

Art walked out of the meeting saying, 'Scott, that was a good interview. I think we're going to pick you. We learned a lot from you during Astro-1, and we learned that what we're doing on the Shuttle platform for astronomy is 90 percent operations and 10 percent science. When we came into this, we thought it'd be the other way around. So we'd be honored to have you because you come prepackaged with the 90. We'll teach you the 10'. And they did, and that's the first half of the story. Then it was off to training and all the things that we did. But that's how I got from the dream of being an APS."

Training for the Mission
"Steve Oswald was our Commander," Scott continued, "and 'Borneo' [Bill Gregory] was our Pilot. Tammy Jernigan was the Payload Commander and [the other Mission Specialist was] John Grunsfeld, his first flight. And then of course there were the other APSs, Sam Durrance and Ron Parise, and me. I started about 18 months before the flight. The training at JSC was focused around the Orbiter and the vehicle; the payload development and training and simulations were all run from Huntsville. We on the payload side spent a lot of our time hopping back and forth: JSC to Marshall, to JSC, to Marshall, to JSC. We had offices at Marshall as well. You learn interesting things: 'So, this is going to be your house for 14 days', which was the planned duration of [STS] 67, our flight. Ron and Sam and I already had the basics [Orbiter and Spacelab Systems]. I had been doing that with every payload since the start of my career, so we kind of kicked it up a level and we probably dived in deeper, I would probably say, than most if not any other Spacelab team. So you have your core systems, you have your safety contingency, and that includes survival training. They put you in a centrifuge, to model the asset profile from liftoff all the way through to eight and a half minutes after SSME [Space Shuttle Main Engine] cutoff. What was easy on the ground is now much more difficult. Then there was the 'Vomit Comet' and they brought us out to Ellington and my first orientation flight, We landed and the next group up, to my surprise, was Tom Hanks and Ron Howard, the team from the *Apollo 13* movie. They were there for their orientation flight, which would eventually be multiple, multiple flights, and that footage became the Apollo 13 movie microgravity shots. Probably about six months out from flight we started doing Joint Integrated Simulations [JIS, simulations of mission activities]. It started with just Marshall, and then we had the payload teams come in at the POCC, and we'd run those nominals and malfunctions and the timelines.

"Getting closer to launch and suddenly Astro-2 was number one on the runway, the Orbiter was rolling out of the OPF, getting stacked and heading out to the pad. And this was suddenly very, very real. Then came TCDT [Terminal Countdown Demonstration Test]. I think it was after TCDT that Oz called me the office. He said, 'Scott, just to let you know that we got the final flight medicals back and there may be a situation on one of the two Payload Specialists that you're backup to, and I only share this with you because your life could change really fast, really quick. Are you good to go?' I said, 'Of course I am. That's why I'm here'. I was trained, but that was all they could tell me. That particular medical issue got resolved, and I'm glad."

On launch day, Scott watched from the viewing area at KSC, then got on a charted plane at the SLF with the PI for HUT, Art Davidson. They flew to MSFC and supported the POCC for the mission.

"When Astro-2 landed, and I really have a biased view on this," Scott added, "of course we should have just landed, refueled and flown again, because it was a finely tuned machine. Just go up and do another two weeks of science, two weeks of hundreds of new targets, IPS work, great. The three instruments were fantastic. The team was all clicking, and we'd done that before [re-flown] on other Orbiters and other Spacelab missions. [We] should have just topped the tanks and done it one more time. Then I would've flown into a flight position of course, which would have been absolutely incredible. It would only have been a question of whether Ron or Sam swapped out, but they were fully supportive of that, of course."

The Other Side of Astro-2:

STS-67/Astro-2 Mission Specialist John Grunsfeld conveyed, "The interesting thing about Astro-2 as a Spacelab pallet mission is that we didn't do module training at the Johnson Space Center because there was no module. But we had Scott Vangen as our Alternate Payload Specialist, who of course was a world-class payloads expert from Kennedy Space Center, so that was a huge advantage for us. And as a result, we took trips to KSC a few times during the buildup of the Spacelab pallet to see the hardware and, especially Tammy Jernigan and myself, to practice the EVA interfaces. That was something that I was very interested in, although they were contingency only. What would we do if we had to hard-stow the Astro telescopes, or even worse, do some kind of a jettison? We got to do those activities in the O&C Building. And then, once the hardware was connected electrically, we participated in some of the electrical tests and such down at the Cape. That was our first opportunity to actually interact with the real hardware, versus simulations where we were typing in commands and switches. But stuff doesn't really happen [in simulations] the way it will for real, commanding through real GPCs to real hardware in Igloo and then to the end elements. The Astro telescopes weren't designed to work in 1g, so we weren't actually raising and lowering the telescope or pointing it, but at least we were getting a feel for what it was really like and seeing the real hardware for the first time. The point I was trying to make is that we trained it in Houston on the best available hardware. Some of the long sims we would actually do inside the Spacelab module at Johnson Space Center, but it bore no resemblance whatsoever to what we were going to fly on orbit. The only reason we used it was because it had a GRID computer interface that was the same as the Spacelab computers. So it was not realistic training at all, whereas when we went to the Cape, which was probably three or four times during the integration and tests, we actually got to see and work with the real hardware. I thought that was really valuable to be able to see, versus some cartoon out the back window."

"Actually, Astro had two encounters with Level-IV," explained Sam Durrance. "After the *Challenger* accident, we went through the whole season, the whole scenario again. Everything was as assembled and aligned, so we didn't have to do that, but we did have to do all the data rejects, adjusting and stuff, the testing. After Astro-1, there had been a significant improvement in coding and technology for the ultraviolet reflective optics. By recoding those, we could increase the sensitivity by a factor of four, almost. So it made sense for us to take it apart and redo it, which meant Level-IV support. We brought it back and we took it apart in the Apollo [Apollo Telescope Mount, ATM] Clean Room again. So I spent half of my life in there, and took the mirror out and had it shipped up to Goddard. They had it stripped of the iridium and coated with a silicon carbide coating, then sent it back down to Florida and back into Level-IV. We put it back into the Apollo Clean Room plenum, rebuilt it and measured the focus, did the alignment."

The second PS on the Astro missions was the late Ron Parise, a ham radio amateur who operated the SAREX (Shuttle amateur radio experiment). During one of the loss of signal (LOS) periods that happened during a mission, Scott Vangen was able to talk to Ron via a fancy ground communication setup unknown to most personnel in the POCC. Scott said, "Ron before we lose you, we need one last entry. We need an Item 4 space SIR". Ron Parise's call sign was W4SIR, so on hearing this request, Ron connected the dots quickly that Scott was going to contact him via the ham radio set on the Orbiter. Nobody else understood that; their own special code of what was to occur next during the LOS. Through some tricky configurations on the ground, Scott was able to talk to Ron via the ham radio while still on his headset at the POCC. They talked for bit until they were about to come out of LOS. Scott had the information for the upcoming target numbers to hand, which was filters and apertures and all kinds of numbers. He called this information up to Ron on the amateur radio, but nobody else knew Ron had it. When contact with the Shuttle was regained (the term used was "coming out of LOS"), Ron read back the numbers Scott had given him, as Scott recalled: "Everybody in the POCC, their eyes got big as saucers. 'How the hell does he know that?' For years, for years, no one knew."

One other aspect of these types of missions was the pointing requirements on orbit. They were based on when the Shuttle launched and whether it achieved the desired orbit, a major problem for the Spacelab-2 mission that had the Abort-to-Orbit incident. As Sam Durrance explained to the authors, "On Astro-2, after the experience with Astro-1, we didn't take any planning at all into orbit with us. Well, we had a few contingency targets and things like that, but for the entire mission, that's what we did. They set up a 12-hour back room, where they would plan the mission for the next 12 hours and then uplink it to the teletype. Then they would plan the next 12 hours and so on. So the entire mission was run by uplinking the payload Flight Data File, essentially."

"For the Astro-2 mission, I had started working in Level-IV," explained Alex Bengoa, "and I worked really closely with the payload developers for the telescopes. They were great people really, older gentlemen. The UIT Telescope used actual film to take pictures in the ultraviolet spectrum and the only company that made that film at the time was Kodak. They didn't have much business for that type of film, so the Goddard UIT people went out and bought all the stock of that film from Kodak and had it stashed in a refrigerator at Goddard, so the supply was pretty limited. As part of my assignment, I had to install the film before launch, while the Orbiter was at the pad in that vertical orientation. We had to go in to the PCR [Payload Changeout Room], with GSE set up all around the telescope, and climb up to the telescope and install the film."

Astro-2 was essentially a refight of Astro-1, except for the large BBXRT which had only been manifested for one flight. "We had operated the telescopes with our Ground Support Equipment and so I knew which button to push," said Sam Durrance. "And when we were setting it up in Level-IV, we were actually using the Spacelab computers, using the interface, and having all of the Level-IV folks right there with us issuing the commands and just having discussions about [what we were doing], or software issues and most of what we hoped could be fixed with software problems."

John Grunsfeld recalled, "We had very nominal mission compared to the others. We did have access to the PCMMU [Pulse Code Modulation Master Unit] data stream for the first time, so Sam Durrance and I did get a chance at the Cape to plug in and validate that it worked. Working with the coms folks at JSC, we had to have that capability on the Orbiter and then also in the test labs, to make sure that we could receive the data we were using to extract the telemetry from the Astro star trackers, in case the automatic star acquisition didn't work. On Astro-1 that Guide Star Acquisition failed repeatedly, and as a result, they only got a fraction, a very small fraction, of the data that they had intended. [On Astro-2] The automatic acquisition worked some of the time better than Astro-1 but not all the time, but because we had this other capability that we had developed and then tested at the Cape to make sure it worked right, we were able to achieve I think well over 100 percent of what was expected for the flight. So that was really useful and it was because we had the actual flight hardware, and the interfaces, and the ability to test it at the Cape, that we were able to make sure that it would work as a backup. One thing I should mention was that Scott Vangen was the Alternate Payload Specialist. During the flight, he also became our payload Capsule Communicator [Capcom], and so the relationship that we developed over a couple of years of training together and doing the test together at the Cape then transferred into having this close connection. Talking the same language and understanding each other as team members really helped execute the mission once we were in space. There were things that I could communicate to Scott without having to use extra words or to be verbose, which he understood because he had trained to do the same tasks. That was really valuable."

Some of the same Level-IV people that supported STS-35 in the POCC also supported STS-67, but others had moved on. Donna Bartoe (Dawson) left Level-IV and was actually in law school working as a law clerk at (NASA) Headquarters when Ted Stecher, for UIT, brought her back to the POCC for Astro-2. Basically she was a lawyer for Astro-2, as Donna Bartoe recalled: "The experiment people pushed [to have us at the POCC]. UIT just said, 'She's part of our team. We don't want to train anybody else'."

John David Bartoe and Sam Durrance stressed how important Level-IV personnel were to their training and support during the flight. John-David Bartoe said, "[There was great] value in having the KSC people in the POCC who had gotten deeply involved with the experiments that were going on. That was a really important part of their contribution." Sam Durrance added, "We had worked exclusively with the Level-IV people in the labs that were set up. I mean, we had our own little POCC for each instrument. We worked extensively, hundreds of hours I would guess, with people from Level-IV sitting there at the console. That was invaluable; you can't learn any better how to do it than by doing it."

Some of the data was being relayed to the ground, but some was stored on board. The question was how to ensure that data would be recovered properly to be analyzed later. Level-IV personnel worked hard to make that happen.

Alex Bengoa continued, "So, you've performed the mission, and now this film has come back and it's precious data,. For that mission, the original planning was to land at KSC, but at the last minute we were informed, due to circumstances, that they had to go on to land at Dryden. So we had to pack up all our GSE at the last minute and ship it overnight to Dryden. Then, myself and my team of technicians had to call last minute for flight tickets and fly out to California. I remember getting to California at 11 in the morning, driving to Dryden to verify that my GSE made it intact, doing all my preps and then going to the hotel, sleeping for a couple of hours, and showing up at 1 am, per the schedule, to remove the film. We had already moved [the Orbiter] over to the mate/demate area; it was already off the runway. We had to go through the middeck, through the airlock and ended up in the Payload Bay. They had put up planks along the payload for us to walk on. We had another team that had come up in advance and set up all the GSE so that we could climb on the top of the telescope. The UIT telescope was actually the one near the top, closest to the Payload Bay doors, if I remember correctly. The doors were still closed."

Scott Vangen added, "Probably one of the more interesting [experiments] to capture the significance of things, from a mission perspective, was HUT, the Hopkins Ultraviolet Telescope. One of Art Davidson's prime directives for this one-meter UV telescope was to just stare at certain objects for long periods of integration time. What they were searching for was something called the helium abundance curve. On Astro-1, they didn't have enough integration time and we had other problems. There was good science that was taken, but he couldn't get the long durations. On Astro-2 they had many targets to look at, but they kept coming back to this one too, to get hours and hours of integration time, like deep field. It needed lots of time.

"Each day as they got more time, you could see the corners of Art Davidson's mouth starting to crack a smile, because what he was seeing was that a peak was starting to form right at the anticipated frequency, kind of indicating that the model of helium abundance ratio was correct, the theory. Now he wasn't going to reveal the results − that needed post-processing − but by the end of the mission he was just a happy, happy camper. Art went from just kind of sweating bullets at the beginning of the mission to this; it was fantastic. At one of the American Astronomical Society [AAS] meetings, Art Davidson got up and presented [the findings] and as the chart came up and it showed the integration function, it validated the models to astrophysicists. This was really big news. It showed that all their decades of building these hypotheses was validated for the first time. That's a big thing. This is the data. This is it. This confirms our thinking of the early formation of our universe, or at least this parameter is valid. It was big news and it got a standing ovation for Art."

After just two out of the hoped for half-dozen missions, the Astro series was now complete. The telescopes would be removed from the cruciform and sent back to their respective offsite locations, and the rest of the payload complement and IPS would be removed from the pallet to prepare the pallet for its next flight. Both pallets would fly again six years later on the ISS assembly mission 7A, (STS-104) in 2001. This time, they carried not science instruments but major space station components: *Quest,* the Joint Airlock Module, and the high pressure gas assembly, four gas storage tanks (two gaseous oxygen and two gaseous nitrogen), on the Spacelab Logistics Double Pallet (SLDP) configuration.

The Astro Restoration Project

As of the writing of this book (2021), a team of people that worked the Astro series have now begun the Astro Restoration Project (ARP), to restore the elements of the Astro payload to be displayed for the general public. This began with restoring the basic cruciform structure that the telescopes were mounted to and will be followed by installation of each instrument so the entire telescope package will be back together again as it was over 25 years ago. For more information about this project, just do a search on "Astro Restoration Project[1]."

ATLAS SERIES

If we hope to preserve our fragile environment, we must first understand our planet's major components (the land, oceans and atmosphere), how they interact with one another, and how other forces such as the Sun and Earth's magnetic field interact with them. A series of NASA Space Shuttle missions assisted this effort through detailed studies of one part of the complex system that supports life on Earth: the atmosphere. The series, initially called Earth Observation Missions, was to study the long-term variability in the total energy radiated by the Sun and determine the variability in the solar spectrum between the Sun and the Earth's atmosphere. They would fly the Short Module and a package of experiments on pallets. The first two missions were later merged into one and renamed ATmospheric Laboratory for Applications and Science (ATLAS for short), but without the Short Module. The pallet-only ATLAS missions were now part of Phase I of NASA's *Mission to Planet Earth*, a large-scale, unified study of planet Earth as a single, dynamic system. Throughout the ATLAS series, scientists gathered new information to gain a better understanding of how the atmosphere reacts to natural and human-induced atmospheric changes. That knowledge has helped us identify measures to keep our planet suitable for life for future generations.

A total of 12 ATLAS flights were planned, flying once a year and aimed at studying solar inputs and atmospheric responses over an 11-year solar cycle, but after only three missions the rest were cancelled. Tracy Gill recalled that, "For those ATLAS missions especially, it seemed like we would have very diverse crowd. We had a friendly rivalries between the two countries participating [Germany and Italy]. I remember one time, during the [soccer] World Cup, there was a lot of animated conversation while we were testing about what was going on with the World Cup, especially between the Germans and Italians. It was during one of the ATLAS missions. I think Germany and Italy were playing in the World Cup and we knew that for a lot of our clients, their attention was going to be elsewhere during a certain, set, two hours

---

[1] Also see Mike Haddad's article *Lost and Found* in the British Interplanetary Society Space *Chronicle* magazine, January 2022, Vol 2-1, pp. 2–7.

during the mission. They were watching the [soccer] game because that's what they did. It was their World Series and Super Bowl, all wrapped up in one."

STS-45, *Atlantis*, launched March 24, 1992, 8+ days, Payload: ATLAS-1; 1 Igloo; 2 Pallets (Fwd Pallet F004, Aft Pallet F005)

ATLAS-1 consisted of 12 instruments and 13 experiments to study the Sun and Earth's atmosphere. Additional payloads on the Shuttle included the Shuttle Solar Backscatter Ultraviolet spectrometer (SSBUV, its fourth Shuttle flight), and several Get Away Specials (GAS). Science experiment payloads included: Space Tissue Loss (STL 1); an Investigations into Polymer Membrane Processing (IPMP) experiment; the Shuttle Amateur Radio Experiment II (SAREX II); Visual Function Tester 2 (VFT 2); Radiation Monitoring Equipment III (RME III); and the Cloud Logic to Optimize Use of Defense Systems (CLOUDS-1A) Experiment.

A number of integration problems were overcome by Level-IV personnel. Sharolee Huet explained that, "The experimenters had made the feet of the MAS [Millimeter-Wave Atmosphere Sounder] antennae detachable, since they had to be precisely aligned on the Orthogrid – the four detachable feet that is – so we couldn't leave the main structure sitting on the floor. We had to build up wooden blocks to set the structure on while installing the detachable feet. This, however, was not known until the day of the crane lift. Also, helicoils were sticking out of the Orthogrid, so we inspected each one [top side and bottom side of the Orthogrid], put tape by the bad ones, and then started replacing them. We ran out of helicoils several times, and also had to get new tools."

ATLAS-1 PS Byron Lichtenberg told the authors, "Clearly, by ATLAS-1, all the bugs had been worked out of the ground system software and all the interfaces, so all of our tests went very smoothly as I can recall. We were able to get through the different portions of the timeline that they were interested in, primarily looking at either power usage or data delivery and usage. We had experiments on there that really couldn't be turned on, so they had to give us some simulators in there to mimic the hardware behavior of the flight equipment. But in terms of the controls and displays and computer software and things, again, it gave you that familiarity. It gave you that confidence that, yeah, I could do this on orbit."

Sharolee Huet revealed how the ATLAS pallet was accessed underneath the Orthogrid without damaging any cables: "The first ATLAS mission had cables run on top of the Orthogrid, which were damaged during experiment integration, while some were run on the pallet which had been covered by cushions and then punctured the pallet. A strut was also damaged from side loads. You had to squeeze between the struts from the aft of the pallet while stepping over two side-mounted pallet boxes. So a strut was removed at the front of the pallet, two bridges were made to cover the cables on the pallet and provide a sturdy step, and handholds were mounted to the underside of the Orthogrid with variable length. The pads were changed so that the cables would not be crushed, while not damaging the pallet from your own weight."

Kevin Zari remembered that, "MSFC Teledyne Brown engineer [TBE] Bill Telesco crashed HITS [made the High-rate Input/output Test System fail] not once, but several times as we were getting close to end of day on the ATLAS payload." Troubleshooting began, and Kevin remembers it took something like three hours to come back up. It was discovered that the same TBE person ran the special comp [computer program] again as soon as the system was back up. "He wasn't aware of the differences between KSC and the MSFC system. The special comp was relying on system time, and that system at KSC didn't allow for it [KSC had a small subset of the capabilities of the systems at MSFC]. Needless to say, it was another few hours before everything came back up and they could do an orderly shutdown of the Spacelab."

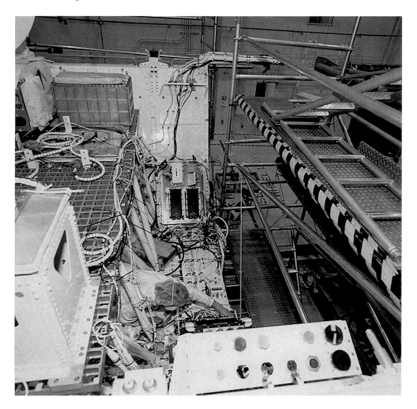

Tight access on ATLAS. [NASA/KSC]

While work on the flight hardware, flight software, and GSE continued to increase, Level-IV would also investigate ways of improving the processes.

Darren Beyer explained that the payload was being damaged during ground processing. "They had an experiment called the Millimeter-Wave Atmosphere

Sounder which was this big antenna. So of course it would get brushed into, it was going to get damaged and everything else. But unless you have somebody specifically looking at these things, you're in a scenario where nobody's going to be trying to fix it. But looking at the root causes of this, [improvements came about] through doing this TQM [Total Quality Management] process. We did everything from changing where people got on and off the thing to coming up with a new foam we could use to protect the payload during processing. And we carried this stuff on over to ATLAS-2, which was the next mission in the series to run. David Sollberger and I worked on this, and it was the first time that ground support hardware was actually designed into the flight hardware specifications." When the time spent repairing damage was examined between ATLAS-1 and ATLAS-2, it had been reduced by 98 percent. Darren still refers to that to this day as one of his crowning accomplishments. "It wasn't putting hardware together, doing the cool stuff at NASA. It was this behind the scenes thing, but it had a major impact on the next mission that flew. So I'm still pretty proud about that one. That was a fun one. And actually, Dave Sollberger and I then got interviewed for some NASA TV thing, and that team was entered into the national TQM contest and we got talks on it everywhere. So it was big. It was a big success."

Rey Diaz explained that, "ATLAS had three manifested missions, ATLAS-1, -2 and -3. I was with the experiment team from the beginning, starting in 1988. I was able to travel to several contractors in Europe to see the hardware development as they were going from Spacelab-1 to this new mission and the new ATLAS complement. For three days we tested at the Level-IV area. We tested the ATLAS payloads to their maximum level, with either data, power, cooling or any other conditions. We had all seven astronauts from that mission at the time, which included the Commander, Charles Bolden, who became our Administrator later on, and also Kathy Sullivan who was the Payload Commander." So the two very, very experienced and highly accomplished astronauts were part of that testing. They tested just to ensure that everything was fine before placing it in the Spacelab Carrier. After it was moved to the carrier, there was further discrete or single testing, for example with Space Experiments with Particle Accelerators (SEPAC), one of the US experiments.

Juan Calaro explained that, "Cindy Martin-Brennan had two experiments that she wanted to offload, and they said, 'Juan, why don't you take SOLSPEC? We still need to finish up writing the procedure to test'. I remember, I had to go through typing. Everyone was so busy and I just had to figure out how to write a procedure for this experiment that I'd just quickly read up on. It was finally SOLSPEC's turn to test out." That was his first experiment, and it was weeks into the ATP [Accelerated Training Program, a program that would condense the normal learning time by up to half]. It was also his first taste of working with international partners, which he found so awesome about his job. "You always have a handful of PRs and Devs [Problem Reports and Deviations] too when you write the procedure, but SOLSPEC was pretty straightforward, pretty easy."

With O&C operations complete, like all other Spacelabs the payload was moved to the OPF for installation and testing with the Orbiter.

"It was an End-to-End test for the STS-45 ATLAS mission," said Rey Diaz, " There I was on a Saturday and the End-to-End test required us to have the KU-Band antenna on the side of a Shuttle. They put a hat on the antenna [to protect personnel] and then the piped the signal into the network that took you all the way up to the TDRS. The two of them connected to the POCC and Mission Control in Houston, so the entire network was connected from the OPF. So I was doing this and something was not working. We were supposed to be [done] at eight hours, but here we were eight hours still doing all the testing, putting in all the things that we needed to. We asked for additional time and got four additional hours, so 12 hours. I remember we were delaying closing of the Payload Bay doors and I was receiving phone calls first from Joe Born. Then they went up to Tom Breakfield and then it went all the way up, the phone calls were getting higher and higher [up the chain of command]. I knew that I was bumping the Payload Bay operations and remember [John] Conway and others said, 'We need to stop or we'll delay any Shuttle operations'. We had accomplished as much as we could, but we didn't complete what we wanted because we were having some difficulty getting all the rest of the information. The high data rate wasn't really working for us, and after doing all the troubleshooting, pretty much we had to give up."

Sharolee Huet added that, "[We were] Trying to keep the MAS antennae from getting damaged after integration onto the Orthogrid, and during an OPF closeout we found what appeared to be a footprint on the antennae." No one could pinpoint when this mark had happened. To limit the possibilities, a curtain was made to go around two sides during pallet processing, until the pallet left for the OPF. "The poor antennae had also lost much of its edge tape during processing, which had to be replaced and repainted," Sharolee continued. "There was also some damage to the honeycomb inside the antennae near the edge. If that wasn't bad enough, it came back from space with a micrometeorite skip across the face of the antennae."

As mentioned, payload closeouts mostly happened in the OPF before being moved to the pad. At the pad, final closeouts were performed and the payload readied for launch. This launch was originally scheduled for March 23, but was delayed by one day because of higher than allowable concentrations of liquid hydrogen and liquid oxygen in the Orbiter's aft compartment during tanking operations. During troubleshooting, the leaks could not be reproduced, so STS-45 was launched on March 24 and the on-orbit operations began

Cindy Martin-Brennan recalled that the biggest memory that she had – and she actually got an award from it – was the Imaging Spectrometric Observatory (ISO). It reminded her of a stand-up piano, in that it would have different lids that would open up and expose the instruments within to the environment of open space. Data collected that way was imaging data, but whenever the Shuttle fired the Reaction Control System (RCS) engines to maneuver, the instruments needed to be locked out (the lids closed) because they were collecting space environment data, and that environment could be contaminated when the RCS fired. Cindy added, "You're sitting on the console during a Shuttle flight, listening to everything that's going on and listening up for your experiments. But there was an unplanned RCS firing and it was going to happen very quickly. I was the only person on the console who caught that this was going to happen. We had to get permission from Houston to upload this command to quickly shut down our instrument, because it was fully operating at that time. I guess the experimenter at the time − Dr. Marsha Torr and her husband Doug were the two PIs − basically recognized that if I hadn't caught that, the instrument would have been totally ruined. So I got an award from them saying about my contribution. I was just doing my job, but apparently they wanted to commend me for paying attention."

"For ATMOS, I was the test engineer," explained Tia Ferguson, "learning to really enjoy the electrical side and the science team. I got to travel up to Marshall and sit in the HOSC [Huntsville Operations Support Center that housed the POCC], with their payloads and while they were doing their mission. It was really fun to be up at Marshall and be a part of that operation. Sending commands while it was on-orbit was pretty cool experience."

Rey Diaz remembered, "We were going through some of the ATLAS-1 mission profile and a number of things that were happening were similar to what we had already experienced in Level-IV. I remember at the time we had either Sue Sitko or Cheryl McPhillips who worked on the payloads before launch, and the first thing that came to us was an explained condition. We had seen that and that [fact] was sufficient for the crew to continue on with whatever they were doing. That saved some mission time. So, we even contributed during the actual mission. We contributed with our expertise."

With the mission completed and the Orbiter returned to the O&C, the work was not yet done. De-integration would take place, and sometimes interesting things would be noticed following the return of the hardware.

Sharolee Huet explained that, "With one poor experiment, we had to cut a chunk out of its Charge Collection Device [a huge sphere that looked like Mickey Mouse ears in the Orbiter Bay] because it had been hit by a micrometeorite and MSFC wanted it examined. The SEPAC experiment did not return for the remaining ATLAS missions."

Atlas Follow-on Missions

As with most re-flights, the Level-IV processing flows went fairly well and ATLAS flew two more times, mostly carrying the same instruments, and without any major issues occurring. ATLAS-2 was flown on STS-56 in April 1993 and ATLAS-3 completed the series on STS-66 in November 1994.

> **STS-56, *Endeavour*, launched April 7, 1993, 9+ days**
> **Payload: ATLAS-2; 1 Igloo; 1 Pallet (F008).**

> **STS-66, *Atlantis*, launched November 3, 1994, 10+ days**
> **Payload: ATLAS-3; 1 Igloo; 1 Pallet (F008).**

TETHERED SATELLITE SYSTEM SERIES

The Tethered Satellite System (TSS) evolved from a 1984 Memorandum of Understanding (MOU) between NASA and the Agenzia Spaziale Italiana (ASI, the Italian Space Agency). The Italian agency would develop a reusable satellite, while NASA would develop the deploying system and the tether, and provide payload integration (Level-IV) and flight facilities on the Space Shuttle. There were also a number of proposed follow-on missions for the late 1990s [2]. The *Near Term Mission* was designated TSS-1R (Re-flight), an improved version of the science objectives that would be ready to fly within two years of the first mission. This proved very useful in scheduling the re-flight after the tether reel mechanism jammed on the first mission, flown in 1992. A *Medium Term Mission*, designated TSS Atmosphere (TSS-A), would use a 62-mile (100km) downward deployment of a non-conductive long tether in investigations of Earth's atmospheric and magnetic properties from an orbital altitude of 81 miles (130km). The *Long Term Mission* was called the Shuttle Tethered Aerothermodynamics Facility (STARFAC), in which the downward deployment would be extended to 81−87 miles (130−140km). At the time, it was linked to research at 56−62 miles (90−100km) for the National Aerospace Plane (NASP, subsequently cancelled). Apart from TSS-1R, none of these follow-on missions reached flight status.

> **STS-46, *Atlantis*, launched July 31, 1992, 7+ days**
> **Payload: Tethered Satellite System-1; EOIM III (on MPRESS F004); 1 Pallet (F003).**

The Tethered Satellite System-1 (TSS-1) satellite was to have been attached by a 12.5-mile (20km) cable to the Shuttle Payload Bay to explore the dynamics of electricity-generating systems. However, attempts by the crew to release the cable failed and the experiment was incomplete. The European Retrievable Carrier (EURECA), developed by the European Space Agency (ESA), was deployed. The satellite, carrying 15 experiments in materials science, life science, and space physics and upper atmosphere research, was retrieved by STS-

57 a year later. Other experiments included: Evaluation of Oxygen Interaction with Materials III (EOIM); Thermal Energy Management (TEMP 2A); IMAX camera production; materials processing experiments; Limited Duration Space Environment Candidate Materials Exposure (LDCE); Pituitary Growth Hormone Cell Function (PHCF); Air Force Maui Optical System (AMOS); and Ultraviolet Plume Experiment (UVPI).

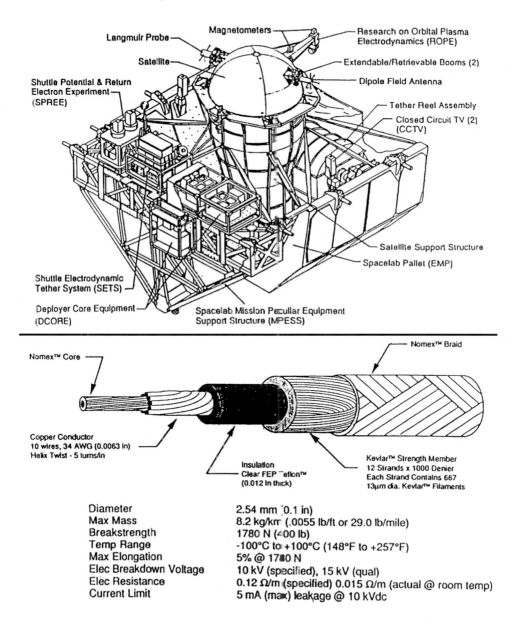

Diameter	2.54 mm (0.1 in)
Max Mass	8.2 kg/km (.0055 lb/ft or 29.0 lb/mile)
Breakstrength	1780 N (400 lb)
Temp Range	-100°C to +100°C (148°F to +257°F)
Max Elongation	5% @ 1780 N
Elec Breakdown Voltage	10 kV (specified), 15 kV (qual)
Elec Resistance	0.12 Ω/m (specified) 0.015 Ω/m (actual @ room temp)
Current Limit	5 mA (max) leakage @ 10 kVdc

TSS-1 and Tether composition. [Graphic: *The Dynamic Phenomena of a Tethered Satellite, NASA's First Tethered Satellite Mission (TSS-1)*; R. S. Ryan, D. K. Mowery and D. D. Tomlin, Page 9.]

Mike Haddad remembered, "In planning for the TSS mission, I got a trip to Colleferro, Italy. I met with the Italians to learn about the GSE that would be used to install the flight batteries into the satellite, and GN2 servicing GSE to pressurize the satellite's bottles with 137 lbs (62kg) of GN2 [GN2 jets were used to control the satellite on-orbit]. The trip was required so that I could write and perform a battery installation procedure and GN2 servicing operations for the TSS. I went there to understand what their plans were for these operations, determine whether they would work at KSC and, if not, figure a way to make changes or updates to the GSE or processes so they could make those changes before arriving at KSC. Pressurization would happen in the O&C, but battery installation would occur at the pad late in the flow, so everything had to work correctly. One of the updates I suggested was to have a second set of GSE for the battery operations. If a part failed and we were at the pad with no backup, we would have to scrub the launch."

Johnny Mathis said, "I do remember, when we were working getting TSS ready, we had problems with the paint that was on the tethered satellite and we had to repaint it. We had to pull all the panels off, repaint them and put them all back on. This was not planned. I remember in six weeks I earned 120 hours of comp time, because I worked 60 hours a week, six weeks in a row."

Because of the Magnetometer, a sweep was performed in the High Bay just to look at magnetic influences that were present. There was more concern about the influence on the Magnetometer, as opposed to what the Magnetometer might influence, because all the tools being used around it had to be degaussed. There were placards and procedures, to ensure doing certain things was avoided. Manette Smith remembered, "I was involved just as an integration engineer. They had pyros for wire cutters for the Tether, in case they had to release it if they couldn't get everything back into the Payload Bay and close a payload door. We had long telecoms about unit testing, all the testing. [There was] a lot of work when you had to put the NSIs [NASA Standard Initiators] in because you needed to keep them capped, and when you did the final connections there was a lot of discussion on that. And also because of access considerations, particularly when it was vertical and seeing how late could we do it. [That's on top of] concerns about RF [Radio Frequency] fields and all that kind of good stuff. So we had a little bit of everything on TSS."

Dan Kovalchik added, "The cables interfacing our roll-around rack with the test stand were always getting trampled. I designed a simple 'S'-shaped bracket and we'd hang the top half over the test stand rails and then drape the cables over the bottom half. The challenge, though, was getting the thing made in a timely fashion. Normal design and fabrication channels could take months, but my experience with the Employee Suggestion program provided the answer. Whoever managed this program always rushed to approve and implement suggestions made by technicians. I got our tech Vic to write up the suggestion and instructed him to put my name down as the person best qualified to judge its worth. The Suggestion team quickly routed the paperwork back to me. I gave Vic's input a glowing review, it was approved, and the brackets were delivered in record time."

There were a lot of issues with the US side when Level-IV received the TSS from Lockheed Martin. Johnny Mathis said, "I remember one of the first days I was going to work on the pallet, and we already had hundreds of PRs open on TSS as part of the receiving inspection. I remember I was trying to start working on it and I had been down there for an hour. I had accomplished nothing and I had picked up five PRs. I got frustrated and had to go take a break and come back. Eventually we got there. I used to tell people when they came to work for us that when they wrote a TAP [Test and Assembly Procedure] to just write a TAD [Test and Assembly Deviation] when their stack of deviations was thicker than the original TAP was."

Jim Sudermann also worked on the Tethered Satellite: "That one worked out pretty good for me, because in the early part of any mission you did a lot of groundwork. You went out to there where they were building the things and watched them test, and you'd try to understand where you were going to put in your test procedure and stuff. So I got a week trip to Turin, Italy and then I got two weeks later on in Munich, Germany. We wanted to go because it was where we could start writing our test procedures. We would be watching their test and working out what was important and what was not important; what parts of this we could pick out and put in our procedure. It was very important for us to get this early visibility." This sort of cooperation early on often depended on the PIs as well. Sometimes, the PIs would want Level-IV to be involved early on, and later, almost inevitably, they saw the value of the Level-IV involvement. Jim added, "I think it was very rare that we got any pushback."

Maynette Smith said that, "Right after I turned it [TSS] over, I moved on to a Section Chief [position] for the Partial Payload Checkout Unit folks and the folks that did the high rate data systems stuff. PPCU was brand new and Tethered Satellite was going through it the first time, so it worked out really well then. It was a challenge, and then the satellite also radiated so we hit an anechoic cap that we could radiate into to make sure that that interface worked properly. Johnny Mathis was the mechanic on it, but there was some change later on in the flow."

Dan Kovalchik added that, "The TSS experimenters didn't trust their system [with good reason, it turned out] and wanted a hex dump of every telemetry frame received by the PPCU after every day of testing. I reluctantly complied and had the techs deliver the resulting four-foot-tall pile of line printer paper to the experimenter's workstation." Dan learned that the experimenter was not really interested in *all* the telemetry, just one of the subframes, so he fine-tuned the playback parameters and subsequent paper dumps were only a foot tall. Ultimately, the experimenter shared his true concern with Dan; he suspected his telemetry processor was dropping a minor frame. This would show up as a skip in the minor frame counter, so they were essentially looking for the presence or absence of just one bit throughout the whole day's testing. Armed with new UNIX tools and a bit of knowledge about the TSS telemetry frame structure, Dan constructed a series of commands that searched a 10 Mb file of hex data and found that single bit. "After that," Dan recalled, "there were no more requests for printout. I thought this was a pretty cool feat and I was looking forward to the big cash reward I'd get from the environmentalists for saving all that paper, but I'm still waiting for the check."

Lockheed Martin, who built the US half, assembled it in Denver on a GSE pallet. The cable that would run everywhere was built as one piece, not multiple cables, and was placed on a ground pallet. So then the question was 'How do we get it off the ground pallet onto the flight pallet?' Johnny Mathis told the authors that, "I actually think that Sharolee [Kepler] Huet was the one who did this. They basically got a big cargo net and tied strings down from the cargo net to the harness assembly so they could pick up this harness as an assembly. It was essentially about 15 feet (4.5m) by 10 feet (3.05m), picked up all in one piece and moved over and put onto the flight pallet."

Sharolee Huet recalled, "[The problem was] How to move a 150-lb (68-kg) cable that conformed to a pallet shape. Martin Marietta had an EM [Engineering Model] pallet to do their integration on [a bad idea] and built their cable to run all over the pallet, but with no way of transferring it from the EM pallet to the flight pallet. After finding a little-used lifting beam by the altitude chambers, and a cargo net, we suspended the cable from the cargo net that was stretched across the beam. It was then moved to another stand so the pallets could be switched. But then the whole flight configuration was shifted to the aft a bit, I don't remember why, so it wasn't as easy as re-clamping in the same locations on the flight pallet. The lacing had to be reworked where possible. That was the stiffest cable I've ever handled."

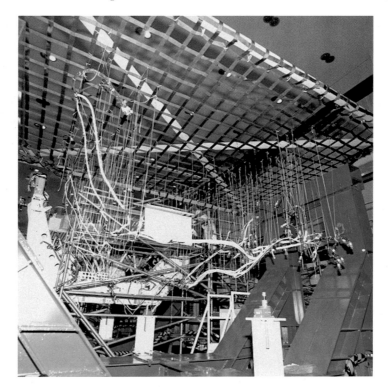

TSS cables hanging from webs in preparation for pallet integration. [NASA/KSC]

Mike Wright said, "I was the networks engineer for the payload, which turned out to be quite interesting due to the number of pyro circuits that were required to fulfill flight safety requirements. It also involved a rather unwieldy electrical harness that included several dozen connectors, which had to be integrated on a non-flight pallet, and then moved in one piece for reintegration onto the flight pallet."

"We were already going into CITE [Cargo Integration Test Equipment stand]," recalled Roland Schlierf. "We'd already finished all our Level-IV testing, Level-III/II. SPREE [Shuttle Potential and Return Electron Experiment] had this recurring isolation problem where they would get noise and it would shut them down. If it were to happen on orbit and was happening often, they would lose all their science." It was a big problem they were trying to fix, and they had tried two or three different fixes through the Level-IV and Level-III/II processes. They kept thinking they had it fixed and then it would happen again, and they were never able to emulate it back in their lab up in Massachusetts. They were basically just guessing. Everything worked fine in their lab and in Level-IV, but as soon as they got connected to the CITE simulator it wasn't good, so they were going to fly as ballast. There was one day left of CITE testing and they called to say they'd finally got to recreate the problem in the lab in Massachusetts and had found a solution. They said, 'We need you to get in there and change out an inductor,' a job for someone like [Jim] Pope or [Mike] Stetzer, one of those guys. I called, I think it was Jim Pope, at home at six, seven o'clock at night and said, 'If we don't take this thing out tonight, they're flying as ballast'. So they told me the steps and I had a Mechanical tech stay over and get inside the experiment. The Mechanical tech knew what he was doing, and he removed the component. I gave it to the payload developers and said, 'It's all yours. You got ten hours to go fix it and give it back to me'. I basically just slept in a chair, let them do their magic, and in the morning they said, 'Okay, wake up. Here you go'. They handed me an updated board, we installed it and it flew."

Talking of sleeping, Riley Duren recalled, "I never slept at Kennedy, though I know people who did, and here's why. Once, before I was going on shift, and this was probably for TSS testing, I came in around midnight and all the lights were off in the office suite. I opened the door and fumbled around with the light switch and when I turned the lights on, I saw all these bugs going in different directions. I mean, all over. It wasn't like the offices were dirty, but there were roaches everywhere."

With O&C operations complete and TSS installed in the Payload Canister, it was transported to the VAB where it was rotated to vertical. It then moved on to the VPF to pick up EURECA and finally on to the pad for installation into the Orbiter

Sharolee Huet commented, "I was also claustrophobic, so going into the mid-deck the first time at the pad almost got me. It was very disorienting and fast paced, You had to keep your wits and I was always scared of screwing up, but I lost the claustrophobic feeling."

Johnny Mathis recalled, "The satellite had batteries in, but not like today's lithium ion batteries that you see in your cell phone. These were big, almost like car batteries, that we had to put in a satellite and there were four of them. They had to put them in late, out at the pad, because once charged they lasted only a certain amount of time before they started to degrade. It was very difficult to get into the locations to put those in. At first evaluation, [I thought] we can't get there and be able to reach in with two hands and put this battery in. Then they came up with some slides to help put it in. I thought, 'It's still tough', but finally I said, 'Well, if you can get me to these locations, I can get them in'. I couldn't find a platform that would get me to those locations, but the pad team had these flip-down platforms that they got from Vandenberg, and they said those could get me into that location. So now, I had to do it. Actually, they put them in, because access was only available via pic-board off the platforms. They had to wear a harness, and the job was at vertical in the PCR. All the integration to this point had been horizontal integration. So a few days before launch, they were installed. They knew they fit because they did fit a check in the O&C."

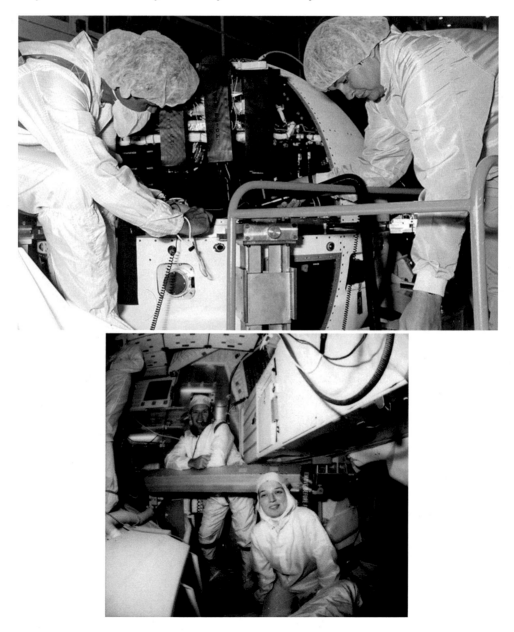

Top: Johnny Mathis performing battery fit check operations in the CITE stand, O&C. Bottom: Sharolee Huet and astronaut Brian Duffy performing middeck operations at the pad. [NASA/KSC]

"There were panels on the deployer and satellite that both had to be removed and reinstalled for each battery, "Johnny continued. "The panels were relatively easy, it was reaching in far enough for the GSE and battery installations that was the hard part. It was just a slide guide. You bolted the battery in and just turned the crank, and it cranked the battery in till the bolt holes aligned. Then you connected the cables and you did that for each of the four batteries."

Johnny Mathis also recalled another concern: "I was working the satellite but on the reel side. The Americans built the reel. After we had done all the testing, they were doing some analysis at Marshall and they determined that for certain landing scenarios, the feet for the reel, where it attached to the pallet, could potentially have a problem. So they came up with this system that would clamp around it to beef it up. You had to slide in underneath the reel and then bolt it on and you had to do that to all four feet. All the testing was done and that [fix] wasn't expected to affect the testing." When TSS got on orbit and the tether was reeled out, the force consistently increased on having to turn to turn reel and the crew could not figure out why. Johnny continued, "Well what was happening was that there was a tether guide, just like on a fishing pole, that went back and forth, and one of these feet stuck up a little too high. So when the guide came across, it hit this and the guide would quit moving. They made the right call in saying, 'Stop, come back down, and we'll figure out what the problem is', but literally it was a piece of metal sticking up into the mechanism for the reel, and it got added in late. We were in a CITE stand when it got added."

STS-75 *Columbia*, launched February 22, 1996, 15+ days
Payload: TSS-1R (Re-flight) and USMP-3; Pallet (F003); 2 MPESS (Fwd F006, Aft F002); EDO kit.

Johnny Mathis told the authors, "This one I put my heart and soul into and I can remember standing in the PCR after we did our closeouts, looking at it and thinking, 'It's perfect, I wouldn't change a thing', and walking away. Then they ended up losing the satellite when the tether broke during on-orbit operations. But the second thing that I liked most was because they had a big investigation afterwards into why we lost the satellite. There was a combination of five or six reasons why we lost the satellite [and] I remember after the audit was over the investigation team standing up and telling Mr. Breakfield that when they got through it, they couldn't find one thing KSC had done wrong. So I remember being very proud of that, even though it failed. I was very, very proud of TSS."

NASA Level-IV Mechanical Engineer Glenn Chin said, "The major unexplained anomaly with the TSS-1R tether snapping on orbit drove significant ground testing on mission return. I had the cool job of using a self-designed LN2 contraption to freeze the tether control/guide mechanism, in an attempt to recreate the on-orbit issue that could have contributed to or caused the tether to snap on-orbit."

With the satellite gone, no more missions were planned, but the one important lesson learned was not to perform late changes to hardware before launch. If late changes *were* required, it was essential to test after those changes to ensure things would still function properly on orbit.

UNITED STATES MICROGRAVITY PAYLOAD SERIES

The United States Microgravity Payload (USMP) program was originally mani-fested as a long series of missions, flying at least one a year through the 1990s. In the event, only four missions were flown. These missions were designed to per-form materials processing experiments in low-g conditions, building upon the experiences with the Material Science Laboratory (MSL) flown on STS-61C in January 1986 that used the MPESS structure to support three experiments. Initially, there were plans for up to 19 MSL MPESS (structure only) flights, but the pro-gram was significantly scaled back following the *Challenger* tragedy and emerged as the USMP series. Though based upon the design of its predecessor, the new improved carrier could accommodate twice as many investigations, with increased resources available for each experiment together with much improved data han-dling and command capabilities. The USMP-3 payload was also included with the TSS-1R mission in a combined payload configuration, and flown as STS-75.

STS-52, *Columbia,* launched October 22, 1992, 9+ days
Payload: USMP-1; 2 MPESS (Fwd F006, Aft F002).

USMP-1 contained several microgravity experimental packages. Among them were: the growth of cadmium telluride crystals from vapor phase; growth of protein/enzyme crystals; and a number of high school experiments, such as the clotting action of snake venom on blood plasma proteins, germination of Florida's official flower seeds, and microgravity effects on dry mustard seeds that were germinated after return. Also on-board were six rats that had been given anti-osteoporotic treatment with an experimental drug. One of the main experiments was the Lambda Point Instrument.

Riley Duren explained, "What made [the] Lambda Point Instrument[2] unique from an integration perspective at Kennedy was that it really had two parts. It had the cryogenics and the electronics that interfaced between the instrument. The payload avionics were built by Marshall, which in turn interacted with the Space Shuttle avionics that was all built by JPL [Jet Propulsion Laboratory]. It was a descendant from a previous version of this instrument called a Superfluid Helium Experiment that flew on Spacelab-2, but the new part of the instrument was built by Stanford University. My understanding is that those two things weren't completely, fully wrung out and tested before they were delivered to Kennedy because of schedule pressure." When the instrument got to Kennedy, a lot of time was spent not only testing and troubleshooting the interface between the JPL backend of the payload and the Orbiter and the Marshall electronics, but also debugging things between the

---

[2] The original investigation determined the way in which the heat capacity of bulk helium changes as the material makes the transition from normal to superfluid state. The name derived from the graph that was a result of plotting the specific heat capacity as a function of tempera-ture, and which resembled the Greek letter lambda

JPL and Stanford parts of the instrument. This should have been debugged long before it arrived at Kennedy, so it was more work than anticipated.

There were a lot of issues. The instrument was about to leave the O&C Building and go to the Pad when there was a GIDEP alert that had ramifications for that mission. GIDEP (Government-Industry Data Exchange Program) is a program that tracks flaws or failures in electronics. An alert arose that some of the components and the Lambda Point Instrument had a manufacturing failure, a flaw, that could lead to them failing in flight. That triggered a mad scramble to assess whether the payload had that issue, as Riley Duren explained: "After we'd finished all this testing in the O&C Building, we literally had to open up the instrument, and then the JPL team and the Stanford team had to come out, pull the electronics, and basically tear the instrument apart. I remember sitting in the living room of the beach house in Cocoa Beach on a Saturday morning with everybody on the phone, including Headquarters, to make a decision that we were going to rework the electronics. The payload would be sent to the pad and we would reassemble it and retest it at the launch pad. So we did that."

The USMP-1 MPESS was an open truss assembly, with fittings in the bowels of the truss. "After installing the fittings – we had three quarter inch fittings to torque and those were really hard to get to – one of the things we had to do at the end was lock wiring," explained Alex Bengoa. "Basically, we had to reach in. It was really uncomfortable for the technicians to do the lock wiring. I remember one time the techs couldn't get the torque wrenches the way they wanted them. At the time I was a little bit lighter weight, so the way we did that torque was that they grabbed me by my ankles and I was lowered upside down through the MPESS from the top. I was holding a wrench backing the nut, and the technicians were doing the torquing."

Aaron Allcorn recalled that, "During its trip from JPL to KSC, the valves inside the LPE [Lambda Point Experiment] cryostat froze. As a result, the payload had to be maintained at cryogenic temperatures using LHe for the duration of its processing at KSC, including integration onto the MPESS, testing, and installation at the launch pad. This necessitated the development of cryo transfer processes and procedures, and additional support from the Level-IV fluids team, including John Bruno, Glenn Chin, John Monkvic, Joe Delai, and myself. It was a challenging period, particularly during pad ops which typically were done during the overnight shift to avoid conflicts with other operations. Yet this was one of the most memorable and rewarding projects of my time in Level-IV. And yes, I still have the burned electrical plug that one of the techs blew."

Riley Duren said, "The PI and/or the Payload Manager asked me if I was interested in supporting the mission. Then I had to put a formal request through, I think in this case through Marshall and Kennedy management. In both cases, the reason why they asked was that those payloads were, shall we say, 'twitchy'. A lot of these instruments were not at all turnkey, they were not simple to operate. They were really kind of cutting edge technology and very finicky. There was an awful lot of troubleshooting on the ground before launch, and when you're leading a test team and troubleshooting, you get to know the ins and outs of the instruments. So they hired me and other experiment engineers to go in and basically work on the on-orbit operations. I remember Lambda Point was pretty uneventful in operations. It was not the same on the ground; it was a handful on the ground."

STS-62, *Columbia*, launched March 4, 1994, 13+ days
Payload: USMP-2; OAST-2; 2 MPESS (Fwd F006, Aft F002); plus EDO kit.

USMP-2 included five experiments investigating materials processing and crystal growth in microgravity, while Office of Aeronautics and Space Technology-2 (OAST-2) featured six experiments focusing on space technology and spaceflight. Both payloads were located in the Payload Bay, activated by the crew and operated by teams on the ground. USMP-2 experiments received emphasis at the beginning of the flight, but later in the mission *Columbia's* orbit was lowered about 20 nautical miles to facilitate OAST-2 experiments.

STS-75, *Columbia*, (see above)
Payload: USMP-3.

Flying on the Shuttle for the third time (along with TSS-1R), the USMP-3 package included US and international experiments, all of which had flown at least once before. These included: Advanced Automated Directional Solidification Furnace (AADSF), a crystal growth facility; Critical Fluid Light Scattering Experiment (Zeno), to study element xenon at its critical point; Isothermal Dendritic Growth Experiment (IDGE), to study the formation of dendrites, the tree-shaped crystals that dictate the final properties of the material in metals manufacturing; and Materials for the Study of Interesting Phenomena of Solidification on Earth and in Orbit (MEPHISTO), to study how metals solidify in microgravity using a furnace.

STS-87, *Columbia*, launched November 19, 1997, 15+ days
Payload: USMP-4; 2 MPESS (Fwd F006, Aft F002); plus EDO kit.

USMP-4 research was deemed to be highly successful. This fourth flight focused on materials science, combustion science and fundamental physics. Experiments included the Advanced Automated Directional Solidification Furnace (AADSF); Confined Helium Experiment (CHeX); Isothermal Dendritic Growth Experiment (IDGE); Materials for the Study of Interesting Phenomena of Solidification on Earth and in Orbit (MEPHISTO); Space Acceleration Measurement System (SAMS); and Orbital Acceleration Research Experiment (OARE). The Microgravity Glovebox Facility (MGBX) featured several experiments: the Enclosed Laminar Flames (ELF), Wetting Characteristics of Immiscibles (WCI) and Particle Engulfment and Pushing by a Solid/Liquid Interface (PEP). Highlights included the fastest dendritic growth rate ever measured and the highest level of supercooling ever obtained for pivalic acid, a transparent material used by researchers to model metals, in IDGE. With CHeX, the most precise temperature measurement ever made in space was achieved.

SPACE RADAR LABORATORY SERIES

Part of NASA's *Mission to Planet Earth* series (and originally designated OSTA-3/5 and 7), the objective of the Space Radar Laboratory (SRL) series was to obtain photographic and radar images of the Earth's land and ocean surfaces using the JPL-developed Spaceborne Imaging Radar (SIR) system. There were three missions planned, but only two were budgeted and flown. NASA's first large-scale radar observations were conducted from the automated Seasat satellite in 1978, which operated successfully for 105 days. Then, in November 1981, spare hardware from Seasat was flown on the Shuttle (STS-2) and designated Spaceborne Imaging Radar-A (SIR-A). This equipment was then rebuilt and flown a second time in 1984, as SIR-B on STS-41G. For the SRL missions, the improved SIR-C payload was combined with a German-Italian instrument called the X-Band Synthetic Aperture Radar, which used higher-frequency radar than the American instrument. This package flew twice, both in 1994, on STS-59 and six months later on STS-68. A planned third SRL mission was deleted from the manifest.

STS-59, *Endeavour,* launched April 9, 1994, 11+ days
Payload: SRL-1; 1 Pallet (F006); MPESS (F003).

SRL-1 was the first in a series of flights of this payload. It was designed to: acquire radar imagery of the Earth's surface for studies in geology, geography, hydrology, oceanography, agronomy, and botany; gather data for future spaceborne radar systems, including Earth Observing System (EOS); and provide measurements of the global distribution of carbon dioxide (CO_2) in the troposphere. Instruments on board included: the Shuttle Imaging Radar-C (SIR-C) with multi-frequency (C- and L-Bands), multi-polarization (HH, VV, HV, VH), and multi-incidence angle (15 to 55 degrees) capabilities, thus lending itself to a wide range of Earth surface applications; the X-band Synthetic Aperture Radar (X-SAR), an X-band, VV-polarized imaging radar system built by Dornier (Germany) and Alenia (Italy) for the German Space Agency (DARA) / German Aerospace Research Establishment (DLR) and the Italian Space Agency (ASI); and Mapping Air Pollution from Space (MAPS), for the study of global air pollution. Also on board the SRL was an ocean wave spectra processor, designed and built by Johns Hopkins Applied Physics Laboratory, which collected data on ocean surface wave length, direction, and height. Four 45-Mbps data channels were recorded on special high data rate tape recorders, and real-time data was transmitted to ground stations. About 50 hours each of SIR-C and X-SAR data were recorded during the mission. The combined SIR-C/X-SAR science team was made up of 49 members and three associates, representing 13 countries. SIR-C/X-SAR data collection was focused on several worldwide supersites and correlated with ground and aircraft measurements. Radar data was also calibrated to allow comparisons with other operating spaceborne radars.

Rich Jasnocha started work at KSC in April 1991 and was assigned to SRL-1, where Riley Duren was the lead engineer. Rich was the engineer for the antenna interfacing with the JPL PI, and went to JPL with Riley to learn about the antenna. For the SIR-C antenna, there were L-band, C-band and X-band components. The X-band antenna was the only part of the antenna that moved.

Riley Duren remembered meeting the JPL team for SIR-C when he was still a Co-op student, during a meeting at KSC while designing the PPCU. "I probably started working with the SIR-C team six years before they ever showed up at the Cape. I remember Rich Jasnocha and me went out to JPL for radar testing because we could never do this at Kennedy safely. They had rotated the entire radar structure including all the phased array panels vertically in this High Bay. They put a giant absorber wall in front of it and they could actually fire up the radar and transmit in the High Bay without worrying about frying people with microwaves. We participated in that testing and got familiar with all the people in the GSE, the procedures and the equipment on it, before it showed up [at KSC]."

Rich Jasnocha added, "One time, when we were rotating the X-band antenna, it started making a loud noise and no one knew what was happening. I don't know if that ever got resolved or anything, I don't recall it doing it more than once."

STS-68, *Endeavour*, launched September 30, 1994, 11+ days
Payload: SRL-2; 1 Pallet (F006); MPESS (F003).

SRL-2 was the second in a series of flights of this payload. As with the first, it was designed to: acquire radar imagery of the Earth's surface for studies in geology, geography, hydrology, oceanography, agronomy, and botany; gather data for future spaceborne radar systems, including Earth Observing System (EOS); and provide measurements of the global distribution of carbon dioxide (CO_2) in the troposphere.

To support the calibration and validation of the radar, engineers often used something called a corner reflector on the ground. Imagine taking a box and slicing it from corner to corner, so you have half a box, though it is technically called a tetrahedron, or a corner reflector. The idea is that when the electromagnetic wave comes in, regardless of the angle it comes from, it will yield a specular or a very direct return, as opposed to getting something scattered off at odd angles. Typically, with SRL, official calibration sites were set up in the Mojave desert, for example, and different places around the world. Riley Duren told the authors that, "Somehow, we got to talking during some long tests one night at Kennedy, and I said, 'When you look at the radar image of a corner reflector it appears very bright in the image, and rough surfaces appear dark. You could build an array of these corner reflectors and spell something out that could be seen from space'. Knowing ahead of time that the Kennedy Space Center region would be included in a regularly-scheduled radar pass, we gathered a group of about 30 or so volunteers who'd worked on SRL, one of them being Luis Moctezuma. I was the mission Electrical lead and he was a Mechanical lead, but he was also a very talented surveyor. I said to Luis, 'Let's lay out 30 corner reflectors'. We designed a set of radar reflectors from common construction materials and laid them in out in a dot-matrix pattern that would spell out the letters 'KSC' for Kennedy Space Center. You can see in the image, the letters 'KSC' right out front of our building.

Main: This image was acquired by the Spaceborne Imaging-C/X-Band Synthetic Aperture Radar when it flew aboard the Shuttle *Endeavour* on Oct. 4, 1994. Notice the letters "KSC" spelled out by corner reflectors. Insert: KSC SRL-2 Corner Reflector Team: (left to right) Mike Haddad, Ross Nordeen, Beverly Sudermann, Dave Olsen, Susan Hutchison, Ed Koshimoto, Sue Sitko, Mike Lombardo, Michele Smith, Maynette Smith, Rich Jasnocha, Dan Kovalchik, (Unknown), Riley Duren (sitting). Missing is Luis Moctezuma. [Riley Duren.]

OTHER PALLET-ONLY MISSIONS

Due to the ruggedness and flexibility of the pallet design, several other non-Spacelab flights utilized Pallet hardware to support other mission objectives. These included:

> **STS-64, *Discovery*, launched September 9, 1994, 10+ days**
> **Payload: LITE (Lidar In-Space Technology Experiment); 1 Pallet (F007).**

STS-64 marked the first flight of the Lidar In-space Technology Experiment (LITE) and the first untethered US Extra-Vehicular Activity (EVA) in 10 years. The LITE payload employed LIDAR, a type of optical radar which stands for LIght Detection And Ranging, using laser pulses instead of radio waves to study Earth's atmosphere. The LITE instrument operated for 53 hours, yielding more than 43 hours of high-rate data. Unprecedented views were obtained of cloud structures, storm systems, dust clouds, pollutants, forest burning and surface reflectance. Sites studied included the atmosphere above northern Europe, Indonesia and the south Pacific, Russia and Africa. Sixty-five groups from 20 countries made validation measurements with ground-based and aircraft instruments to verify LITE data.

LITE lift over Astro-2 and SRL-2. This photo shows how crowded it was getting in the Level-IV area, where the teams were working on three different missions at the same time. [NASA/KSC]

STS-61, *Endeavour*, launched December 2, 1993, 10+ days
Payload: Hubble Space Telescope Servicing Mission 1 (HST-SM1); Pallet
(F009).

The objective of this flight was to repair, replace, and/or update the instruments on the Hubble Space Telescope (HST). During several days of EVA, the crew installed corrective optics (COSTAR) in the light path after removing the High Speed Photometer (HSP) instrument; replaced the older Wide Field/Planetary Camera (WF/PC) with a newer version (WFPC 2); and replaced malfunctioning solar arrays.

Work on the first HST Servicing Mission by Level-IV began before the *Challenger* accident and then, after the HST was launched and the problem was discovered with the primary mirror, the hardware to be integrated on a Spacelab pallet changed to support correcting the mirror. What follows deals with some of the work that occurred before *Challenger* to support this first servicing mission.

Debbie Bitner explained that "HST, which wasn't Spacelab but used Spacelab hardware, that was my biggest assignment. The interesting thing was we used the Engineering Model pallets, so some of the integration procedures were a little bit different. I had to install hard points on the pallets but instead we did something called bolt stretching. The equipment that we used for bolt stretching was garbage because it was these little tiny sensors. You had to put this little drop of some kind of goop in the head and then you put the transducer on, and if you just wiggled it, it would change the reading. So at the same time that we were stretching the bolt and taking measurements, we were also torquing. I always felt like if I got an actual torque number, that made sense to me versus a transducer reading that might be a little bit screwy. Jay Smith worked on installing the keel latch and the pallet. It was a crane operation and we started at eight o'clock in the morning, and I remember we were still there trying to do that crane operation at midnight that night. We didn't have rules back then about not working more than 16 hours in a day or anything like that. We eventually got it done, but we were tired."

Between 1997 and 2009, four further Hubble Servicing Missions were flown (SM-2, February 1997, STS-82, *Discovery*; SM-3A, December 1999, STS-103, *Discovery*; SM-3B, March 2002, *Columbia*, and SM-4, May 2009, *Atlantis*). All resulted in extensive upgrades to the telescope. Each mission used the same pallet (F009) to support the equipment and hardware required on each flight [3].

THE FINAL SPACELAB PALLET MISSION

Flying in February 2000, the STS-99 mission was heralded not as the final "official" Spacelab pallet mission[3], but rather as the first Shuttle flight of the new millennium. The mission had its genesis in the SRL missions of STS-59 and 68, flown in April and October 1994, respectively. That December (1994), the idea was put forward of acquiring a global high-quality, high-resolution topographical data set by reusing the SIR-C and X-SAR instruments. During 1995, discussions were held with the Defense Mapping Agency (DMA), exploring the potential for such a mission. In 1996, the National Imaging and Mapping Agency (NIMA, which had replaced the DMA) signed an agreement with NASA to fly the SRTM mission during 1999–2000. That October, the SRTM project officially began.

STS-99, *Endeavour*, Launched February 11, 2000, 11+ days
Payload: Space Radar Topography Mission (SRTM); 1 Pallet (F006).

The National Imagery and Mapping Agency (NIMA) mission utilized the Shuttle Radar Topography Mission (SRTM) radars. The 30,000lb (13,600kg) SRTM instrument consisted of a pair of transmit/receive antennas below the Payload Bay, and a pair of receiving antennas at the end of a 196-ft (60m) "rigid" tower. The accuracy of the mapping was modest: in any "terrain" segment, the relative height accuracy was 32ft (10m) and the relative horizontal accuracy was 64ft (20m); for forests, the surface of reflection at both frequencies was the tree canopy, not the terrain. The tower was an assembly of stacked cubic frames made of steel, titanium and plastic, initially contained in a 9.8ft (3m) can that was pushed out by a motor and held in rigid shape by a thruster at the high end. It was erected at an angle of 45 degrees from the vertical. The Payload Bay antennas and the transmitters were the same as those used in the SIR-C/X-SAR radars that were flown on two Shuttle missions in 1994. SRTM included additional receiving antennas at the end of the mast for interferometry.

During one of those long shifts, the Test Conductor said they were running up against 12 hours and needed to shut down the payload to perform troubleshooting. He was on the floor running the test and Phil Mead wanted to try something, but

---

[3] STS-85, flown in November/December 1997, was another non-Spacelab mission which utilized Spacelab hardware in support of its mission objective. Located in *Columbia's* Payload Bay was MPESS F004, on which the Japanese (NASDA) Manipulator Flight Demonstrator (MFD) was mounted. This consisted of three separate experiments to demonstrate applications of a mechanical arm for possible use on the Japanese Experiment Module of the future International Space Station.

the steps would take too long because they had to power down and then power back up again. Juan Calaro said, "Phil disconnected a live power cable because he needed to know what was coming out of the connector. He pinned it out and we got the measurements and put it back on, and we got our data point. That didn't happen that often. There was just a situation where we knew it was safe to do and we could get away with it and get our data point."

Kevin Zari was the lead electrical engineer for the SRTM payload processing activities, and after the SRTM post-test debriefing, they had a splinter meeting with JPL. It seemed that a cable problem would cost them another day of testing. JPL was not convinced the problem was on their end, so Kevin needed to use the next day's planned test to troubleshoot the problem. Kevin Zari said, "We needed to prove to the team that the problem seen with the commands not being received by the payload was on the payload side, not a checkout system or mission-specific ground cable. On talking to a checkout systems expert, this would be a two-man job. It was 8:00 pm. He said, 'You've got to be kidding me, we just got home'. I replied, 'Well, we can't let an entire test team just stand around tomorrow and wait while we prove our cables are built correctly. Quickly eat some dinner, and meet me at the LPIS stand in an hour'. I could not believe we were going back in. We were going to break the 12-hour rule, but we could not waste another test day with everybody sitting around waiting. We spent the next 10 hours testing the ground mission-specific cables, pin to pin, demonstrating that the cables met the drawings – the entire copper path from the checkout system right up to where the payload connected to the partial payload pallet. We reported our findings to the Payload Test Director at the 8:00 am meeting. The resolution to the problem now lay in the hands of the SRTM Payload Developer."

Juan Calaro added that, "Something else that was late on SRTM. There was a communication issue too, so Phil and I went and met with astronaut [Don] McMonagle to propose this test in the Orbiter. I remember it was just the three of us and he said, 'Let's go ahead and do this test'. There were some data issues on the Orbiter, and connecting to SRTM. I actually went 24 hours straight at work, and I remember on SRTM we busted so many hours. I was testing so much, it wasn't funny at the time. I got a prostate infection and I remember going to the bathroom and it would hurt like a mother and it wouldn't go away. I finally found some time to go to the doctor and he actually asked me if I was a truck driver. He said that kind of stuff only happened to truck drivers who sat in their trucks so long they got prostate infections. I actually got one because I was just sitting down for so many hours doing all the testing. It's kind of embarrassing now to mention that."

Scott Vangen also told the authors "I was kind of a seventh crew member on SRTM. I wasn't trained to fly, but Kevin Kregel, the Commander, considered me the seventh crew member and I was brought into that because I was a Payload Specialist (PS). After Astro, I was working a few other Spacelab missions then went over to Station and worked the US lab in SSHIO [Space Station Hardware Integration Office], Tip Talone's organization. John Grunsfeld called me and said, 'Let me read you a letter from John Young in the crew office. The short and skinny of it is, we're in trouble. SRTM, STS-99, originally included a PS, but George Abbey has decided there's going to be no more PS, and now we have a rightly disgruntled JPL'. John's letter basically said, 'We don't have anyone who knows anything on this hardware. You [Scott] have got the background of what it's like to be a crew member and to get ready and to train. You've got the background in avionics and checklists and procedures. You know some people out of JPL and they respect you, like Riley Duren'. That was the big connection, and John knew that too. I knew Riley and I knew the SRTM team because SRL flew. I didn't work SRL, but I knew Riley and some of the connections and the astronaut. I was at JPL for over a year. I became kind of the seventh crew member of the interface and that's how I got to be so close to Janice Voss. JV was the Payload Commander, and Kevin Kregel is a friend of mine. We interviewed together in my first interview, back in 1993. And I said, 'Kevin, I'll work for you at the discretion of the Astronaut Office' because I could not be happier. I read John Young's letter and I thought, 'How are we going to do this?' So between JV and Kevin and [astronaut] Janet [Kavandi] – and the German and Japanese Mission Specialists came on board – it was full speed ahead. Riley was excellent. He immersed me into the JPL team and they embraced me. As soon as the crew visited the first time, made arrangements and everybody came together, instantly they clicked. And then Kevin enabled me. He said, 'Scott, what do we need for simulators? What can you set us up with?' So I got to know JPL avionics and their equipment for testing SRTM."

The rest of the flow went smoothly up to the day of launch. This being the last Spacelab mission, it would be the last time many of the tasks would take place as part of the Spacelab program. One such task was the crew leaving the O&C Building (called the Crew Walkout) to get in the Astro Van to go the pad for launch, as the last Spacelab crew. Mike Haddad recalled, "I had moved on to working Space Station by the time the last Spacelab mission flew, but the STS-99 crew walkout was very important to me because my good friend JV [Janice Voss] was flying that mission and it would be her fifth and last flight. JV got my cousin, Lisa Thomas, a pass to get onsite for the launch as a guest of the crew, so we went to the crew walkout together. I was almost directly across from the O&C's double doors when they came out, while Lisa had worked her way a little more to the right towards the front of the Astro Van. There were so many people there, like always, so if you were short and not right up at the front, it was a challenge to see the crew. My poor cousin Lisa is not that tall, but she was able to convince one of the photographers [who normally stood on a stepstool or ladder to get a good shot] to let her stand in front of his stepstool, not blocking his view but in a position where she could see the crew. She told him she knew one of the crew members, JV. His response, rolling his eyes, was 'Sure everyone says that, but you can still stand there.' The photographer tried to impress Lisa that he knew more than her and said, 'I know Janice Voss, she is married to James Voss, another astronaut'. Lisa said, 'No she is not, she is single'. Again the photographer rolled his eyes. I should note that JV had told us to wear bright Hawaiian shirts so she could see us in the crowd, and as the crew came out, I yelled to her and she waved. Lisa yelled her name and JV faced right to Lisa, waved and said hello. As the crew left, Lisa thanked the photographer for letting her stand there. He was shocked and said, 'You really did know one of the crew! Plus, I should thank you. When JV saw you and looked this way, I got one of my best photos ever'. As guests of the crew, Lisa and I got to do all kinds of fun stuff the day of launch. Attending all the family/friends gatherings, eating lots of great food, pictures, meeting new people, and finally taking our special positions at the viewing site. Finally… 10, 9, 8, 7, 6, 5, 4, 3, 2, 1, liftoff and JV was on her way. Successful ascent, and now on-orbit operations would begin."

Scott Vangen added, "The idea was that the Shuttle orbits the Earth around and around and around. They were building up a contour map which would be the equivalent of having surveyors go out about every 10 meters and make a measurement, but all over the Earth. We got the crew procedures, people clicked, [and it was a] super successful mission. In fact, it was 99.7 percent successful. It had an easy metric. We did 99.7 percent of the entire land mass planned, less the 0.3 percent, and that was because there was a burp in the system when the recorder didn't turn on at the right time. But where it happened was over the US, and we had all of that data from the US Geological Survey, so it was 100 percent. I was honored that Kevin and JV considered me as the seventh crew member, and my little fantasy secret was wondering, if something happened to one of the MSs, if I could have stood in. They didn't have a backup, but I felt I could have, because I knew the systems. It probably never would have happened, but that's how much Kevin trusted me with the system. And then at JSC for mission ops, I was again air-to-ground coordinator, and that was called the CIC position, Crew Interface Coordinators. Great mission, highly successful. They're still using that data. Think about it: every point on the planet with the precision down to centimeter resolution. And it all started with John Young's letter and John Grunsfeld saying, 'Scott, make a phone call'. So you never know."

Janice Voss, busy on the Aft Flight Deck of the Orbiter *Endeavour* during the STS-99 SRTM mission.

With the mission completed, the landing successful and the crew departed, work and fun dealing with the post-mission would continue.

Scott Vangen explained, "Post-mission they invited me as a seventh crew member on the foreign trips. We went to Japan for a week and Germany for a week. Gerhard [Thiele], JV and Mamoru Mohri. Mohri's face is a recognizable, famous face. I mean [in Japan], he's Yuri Gagarin, Neil Armstrong, Alan Shepard all wrapped into one. He's Japan's first astronaut, so everybody knows who he is, and he's about the nicest guy you'll ever meet. We walked into a nightclub and the guy who's at the door, his eyes [just widened]: 'Oh, come on in'. And the crew still are good buddies. Those are lifetime relationships. I would imagine if you flew with them, it's even a tighter bond."

SUMMARY

Now came the hard truth. It may have been a new millennium, but in Level-IV not many celebrated. It was over. No more Spacelab missions *per se*. The formal Spacelab era had ended. After STS-99, while some of Spacelab's flight components and GSE would be used on future missions, such as the pallets being used in support of the International Space Station (see Appendix 2), most of the other flight hardware and GSE would need to be removed to make room for something new. The O&C would be transformed to support ISS. It was a new era in spaceflight and payloads, but before calling time on the program, we review the series of successful Long Module missions in the next chapter.

References

1. http://www.collectspace.com/news/news-062917a-kfc-zinger-chix-space-mission.html
2. STS-46 Preflight Press Briefing, TSS-1 background, by Gianfranco Manarini, ASI TSS-S Program Manager, NASA JSC, June 11, 1992.
3. Shayler, David J, with Harland, David M., **The Hubble Space Telescope: From Concept to Success, and Enhancing Hubble's Vision: Service Missions That Expanded Our View of the Universe**, Springer-Praxis, 2016.

11

MODULE MISSIONS

"The resulting experiments [from Neurolab]
fully demonstrated the capabilities
of the Spacelab program."
Jay C. Buckley and Jerry L Homich
The Neurolab Spacelab Mission,
NASA SP 2003-535, 2003.

In this chapter, the authors review the background and activities of the Level-IV team in supporting the 13 Spacelab Long Module missions flown in the seven years between June 1991 and June 1998. We begin with the first of these under the Spacelab Life Sciences mission series and continue with each series in turn, ending with Neurolab, the final Spacelab Module mission.

PREPARING FOR STATION

The 1984 Presidential authorization to create Space Station *Freedom* initiated not only an extensive increase in building the hardware for the station and assembling the team to fly the missions, but also in planning the significant number of Shuttle flights required to loft the hardware, assemble and maintain the facility. After almost a decade on the drawing board, and with the plans for *Freedom* becoming too large, too complicated, and certainly too expensive, the program evolved into the International Space Station (ISS). The inclusion of Russia as the new partner in the space station program, to initiate assembly and to support rotational resident crewing and resupply using Soyuz and Progress spacecraft, was key to

© Springer Nature Switzerland AG 2022
M. E. Haddad, D. J. Shayler, *Spacelab Payloads*, Springer Praxis Books,
https://doi.org/10.1007/978-3-030-86775-1_11

commencing orbital operations. The Shuttle was utilized to complete the assembly and stock up the station with tons of supplies but there was another side to the development of the ISS, which was the eventual retirement of the Space Shuttle system and the demise of Spacelab-related missions

By the mid-1990s, the projections for actually getting ISS hardware into orbit looked more promising than in the decade before. Plans were being developed to provide American astronauts with the opportunity to fly long duration missions on the Russian Space Station Mir, *before* embarking on similar sojourns on ISS. This was sensible, because the Russian program (and the Soviet one before it) had gathered extensive experience in space station operations since 1971 with their series of Salyut space stations and the more recent Mir. Cosmonaut expeditions had lasted up to six months or more, with many flying international crewmembers. In contrast, the longest Shuttle flights were no more than 18 days, and the *only* American missions longer than that had been the three Skylab missions of 28, 59 and 84 days flown two decades earlier.

Another important factor to consider with the American program of the early 1990s was the significant lack of experience in rendezvous and docking with a large-mass space station. There had been a number of Shuttle missions which had *rendezvoused* with satellites or free-flying payloads, and crews had extensive experience in using the versatile Remote Manipulator System (RMS) to capture and deploy them. They had also begun to gain additional Extra-Vehicular Activity (EVA) experience in preparation for the upcoming extensive and complicated "wall of EVA" that would be key to the assembly of the ISS. However, no American astronaut had physically guided a spacecraft to dock with a second vehicle since July 1975 and the Apollo-Soyuz Test Project. The decade of experience built up in the Astronaut Office between 1965 and 1975 in the Gemini and Apollo era programs was long gone. By the time Shuttle missions were being planned to dock with Mir, American astronauts had not completed such an exercise for two decades.

To address all this, the focus of the Shuttle program changed, in part due to the consequences of *Challenger*, but also through a growth of interest in looking back at Earth from orbit. NASA also began to focus on lightening the manifest, with less commercial and military missions and more scientific and technology missions aimed at gaining experience that would help in the creation and operation of ISS. To assist in this, several missions using Spacelab hardware were devised and amended to support the growing emphasis in providing baseline data for subsequent ISS operations after assembly complete, when the space station would change from an assembly site to a fully-functional international research facility.

This was a boost for Spacelab, but the signs were also there to indicate an end to the program due to tighter budgets, fewer Shuttle missions and greater emphasis on getting ISS operational. Indeed, once the Shuttle retired the Spacelab hardware could not be flown in its existing configuration, and even ESA had progressed to developing its own integral research laboratory as part of ISS, called *Columbus*. The writing was on the wall. By the time a second Shuttle, *Columbia*, and her crew were lost in 2003, the Spacelab program had already been completed, with only a few flights of the pallet planned to support outfitting the ISS. The ISS program was about to take over. Before that happened, however, Spacelab module missions would provide both the experience and database to furnish the ISS science program with a firm platform to build upon.

THE LONG MODULE MISSIONS

Originally, there were to be missions utilizing both the Long and Short Module configurations, but only the Long Module version ever flew. The Long Module design afforded the opportunity to expand scientific research on Shuttle missions beyond the smaller experiments carried in the middeck or on unpressurized Payload Bay carriers. They also offered much needed experience in life and microgravity sciences, bridging the gap between Skylab and the ISS. Like most of the Shuttle system, the hardware was not afforded its full potential, and with the rise of Shuttle-Mir docking and Space Station assembly missions, coupled with the demise of adequate funding, the Spacelab Long Module missions were not utilized beyond 1998, barely 15 years after the first module flight had flown so successfully. Indeed, in 16 total flights of the program, the Spacelab Long Module hardware performed well, and many supported two-shift operations, doubling the scientific return from a one- or two-week mission. Clearly, the Spacelab Long Module missions were one of the successes of the Shuttle program from the point of view of extending the scientific return of human spaceflight activities.

SPACELAB LIFE SCIENCES

Originally designated the Space Biomedical Laboratories (SBL), these were first proposed in 1978, with an initial experiment selected in 1981 to investigate the effects of exposure to weightlessness using both human and animal specimens. It then became Spacelab-4, and was eventually flown as Spacelab Life Sciences (SLS) on STS-40 in 1991. When Spacelab-4 was divided into two missions in 1984, the second mission, originally a NASA Life Sciences mission called Spacelab-10, was intended to re-fly some of the Spacelab-4 experiments with additional ESA involvement. This eventually became SLS-2 with the Extended Duration Orbiter (EDO) pallet, retaining the plan to re-fly SLS-1 experiments to provide comparable data. A third mission, SLS-3, was proposed to further investigate the effects of acute weightlessness on living systems, and included a significant influx of French life sciences experiments. It was planned as a 16-day EDO mission in early 1996, and included musculoskeletal investigations as well as American and French primate experiments. Had it flown, this mission might well have become known as Spacelab F-1

(France). Early plans for a fourth SLS mission suggested a general life sciences laboratory mission, which was later renamed Neurolab in recognition of "The Decade of the Brain". There were also short-lived plans for a possible SLS-5 mission, though that idea barely got beyond the planning documents.

STS-40, *Columbia*, launched June 5, 1991, 9+ days
Payload: Spacelab Life Sciences-1 (SLS-1); Long Module #1 (CD MD001); Tunnel (MD001); Floor (FOP MD002); Rack (See Appendix 2).

This was the first Shuttle mission dedicated to life sciences. The goals of the Ames Research Center (ARC) were to prove the functionality of the improved Research Animal Holding Facility (RAHF) and ensure that the RAHF and General Purpose Work Station (GPWS) maintained particulate containment relative to the Spacelab environment during all operations. The crew itself became the focal point for the SLS-1 research to determine the causes of space sickness and note physical changes in microgravity conditions. The human crew was accompanied by 29 white rats and 2,500 jellyfish. The Orbiter also carried 12 Get Away Specials (GAS) and seven NASA Orbiter Experiments. It was the most concentrated life sciences research conducted in orbit since Skylab.

There were some unique experiences with SLS-1, because it was the first mission that the Johnson Space Center (JSC) had 'managed'. Most of the Spacelab-type missions were managed out of Marshall (Space Flight Center, MSFC) when the Spacelab Program Office was at that field center. The new integration between two centers (MSFC and JSC) led to new issues and new conflicts that Level-IV had never dealt with before. NASA and their contractors had to learn a raft of new processes and requirements that were not documented at JSC, as Bruce Morris explained: "That was a good learning experience for us, because there were times the JSC guys questioned [a process] and said, 'Why don't you do it this way, the way we do it for other kinds of Spacelab payloads?' We'd say 'That's a good idea. Let's try it that way'."

Damon Nelson added, "We had a lot of interesting dynamics. I don't think they initially understood the Level-IV responsibility for experiment integration." Level-IV had to work through some of those dynamics, and Damon was responsible for a large portion of the Mechanical integration of the racks for that mission. The biggest challenge was with the electrical routing, because the drawings for that were very open ended. They included a lot of pretty straightforward work when it came to trying to get the boxes in, but the rack rails could not be out of alignment or the box would bind during installation. Damon Nelson recalled, "It even had gas lines that had to be routed and the drawings were very nonspecific. They knew that we were going to have a mess on our hands and so they gave us a lot of leeway. Basically, we routed as best we could to try to get everything in."

Josie Burnett continued, "I did the Research Animal Holding Facility, integrating the mechanical structure to beef up those racks to be able to hold other experiments. One was for live animals and one was for the rodents that we flew, and it was the second time it was flown. The generation before me had flown this same rack on Spacelab-3." Kennedy Space Center (KSC) was empowered to make field engineering changes to modify hardware for fit or function. During integration, the drawings could be changed, as opposed to when MSFC was responsible and sent a team of their own engineers to monitor Level-IV and make such changes themselves.

By 1991, some of the new people coming into Level-IV were from outside KSC or the Cape Canaveral Air Force Station (CCAFS) and were new to the space program, while others transferred from elsewhere at KSC or CCAFS. Different groups had different requirements, processes, and approaches to how work was accomplished, which for some did not go with their mode of operation or thinking.

Kevin Zari said, "My first Co-op assignment in Level-IV was building and maintaining telemetry and command databases and display pages for the various Spacelab missions, starting with SLS-1. It was also difficult because I was a Co-op, and sometimes I couldn't make the whole flow of the payload because school semesters in those days were quarters but they ended when they ended. We'd find that a lot of issues with payloads in the early days were more about telemetry and command database errors. We'd have to transfer the data over the VAX system from Marshall to [Jim] Dumoulin's lab and then we'd have to burn it to these big old nine track tapes. I had to take those tapes and write them a certain way, and then bring them to the HITS [High-rate Input/output Test System] machine and load them onto the disk drive, and then finally, eventually, get the database loaded into the MicroVAX."

Joni Richards added that, "Bruce Morris was the lead Test Conductor and I had the LSLE [Life Sciences Laboratory Equipment] refrigerator freezers. They were garbage, but that was the best technology they had. We just kept them on during the test to make sure we could accept the loads with all the experiments we wanted to run in parallel. I needed a bio break, so I stepped out the door of the control room, and the guy that was from Johnson Space Center came into the hallway. He said, 'Hurry, get back on the NET'. I asked what was wrong and Bruce Morris stood up and said, 'Switch over to another channel'. I switched over to this channel and said, 'What's the problem?' and they said, 'The measurements on the refrigerator have gone all out of whack'. One of the Alternate Payload Specialists had opened the door, which wasn't in the procedure, and I remember asking why in the world they would do that. And then there was this pause, and you heard '… stupidity [come from the Payload Specialist]'. I didn't realize the Payload Specialists had also switched over to that channel. So I felt really bad."

Todd Corey worked as an Ops engineer for the SLS-1 and IML-1 missions in both the Experiment Integration and Spacelab arenas, and was responsible for coordinating and scheduling multi-discipline engineering tasks for assembly and testing of flight hardware. One day, for the SLS-1 mission, he was asked by Bruce Morris to fill in for him when he was out sick and to run an integrated experiment test in the control room. Todd Corey said, "As an Ops guy, you usually weren't broadcasting over the OIS [Operational Intercommunication System]. It was a bit intimidating coordinating the test over the OIS, especially with a bunch of seasoned test engineers, but I managed to survive."

While new people were brought in to work the flight hardware during the SLS-1 processing flow, new people also helped with other areas of the Level-IV world. Perhaps not as glamorous as other Level-IV assignments, they were still critical in helping personnel to do their jobs. Diana Calaro said, "When I hired on as a Co-op, one of the first things I had to do was help set up the first IBM 486 computers on the center. We had about 40 or 50 of them and they all went mostly to the Level-IV and Level-III/II folks."

With testing and operations complete in the O&C, the payload followed the familiar route to the OPF and then the pad, both locations part of the Shuttle's world as before. Level-IV needed to conform to their operations and schedules to make everything work and, as with other missions, this was a demanding compromise

According to Bruce Morris, "We were recycling people from their daytime job to come work things out on the pad with us. Or we would go in at midnight to go do third shift operations with the Shuttle because payloads was a top priority. So for hours you would sit around waiting to get called to maybe do three- or four-step operations. We were in their world, and their work had priority. How do you plan resources? How do you plan people? And how do you do your replanning quickly on the fly when you realize you've got somebody on your team that's been waiting to go out to do work for almost 24 hours, and they're about at the breaking point and you've got to get them home to get some sleep. As Test Conductor, if we couldn't get someone to come in we would become the test engineer. We'd have to go find somebody to sit in for us while the payload was powered up, and sometimes it was your boss."

Darren Beyer also admitted, "I slept in my office. I wouldn't say a ton of times, but enough times. I remember working an 18-hour day; I cheated. I was not gonna drive 45 minutes to get home and then be up for however long, then sleep for three hours, and then get up and shower and drive 45 minutes to get back in again. I took my bunny suit bag, and that was my pillow, and I just crawled up under my desk and went to sleep."

Finally, with ground operations complete, the mission launched and Level-IV personnel were called upon to support the flight, as they had all that ground time experience with the hardware. "SLS-1 was the first time I got invited by the Mission Manager to go to the Payload Operation Center in Huntsville," said Bruce Morris. Bruce would sit there during the mission and help them work through problems that they had seen on the ground, because JSC would say they were seeing a particular activity occur during the mission and ask whether the Level-IV people remembered something similar happening on the ground. If it was something they had seen before, then they could call up the problem report from KSC. Bruce Morris recalled that, "This was before you had the Internet and all the other things, so it wasn't like we went to a database and pulled the thing up. We literally had to try to remember some key words, call the Quality office where all the Quality records were kept, tell them about when it might have happened in the flow and give them some key words so they could find it, based on the title. They'd go back through a filing cabinet, literally going through pieces of paper. When you're young like that your memory recalls pretty good, so a lot of times you just remembered right off the top of your head. On that mission, I really learned a lot about how the mission ops worked. I'd done tests on the ground, but I'd never been involved in the hierarchy of how things were changed on orbit, reacting to problems and re-planning and things like that. When you were troubleshooting, you could just get on the headset and say, 'Okay, I've got the troubleshooting guide laid out, I'll give you a summary of it', and then go off and start running. The thing is, you had to send new instructions up to the astronauts, and it could get complicated."

"We were also in the landing convoy afterwards," explained Darren Beyer, "and, just like we were the second to last people to go in before launch, we were second truck in the landing convoy to go in, not just to pull the middeck experiments out, but the critical life science and other critical science that was in Spacelab. We'd have to send people out to Edwards [Air Force Base, EAFB] to be there in case there was an abort, and we would go out three days after launch, because if it had to do an abort before three days there wouldn't be any useful science anyway, so we wouldn't need to be there for that. Edwards is a massively huge facility with a huge main landing strip and then another smaller one that wasn't connected. The only way to get there was over roads and that was going to take too long, so the experimenters chartered a small Learjet that flew into Edwards to wait near the Orbiter [landing zone]. We were literally rushed in, grabbed the science out and then I escorted it, put it onto the Learjet and we flew about three or four miles to the other facility so we could take it off and give it to the science guys as quick as possible after the flight."

Josie Burnett added, "I was part of the landing team as well on that mission, so I got to see the whole Shuttle on operations from cradle to grave, from launch to landing. It was because there were live animals. If something were to happen on orbit and they needed to do an emergency landing, Level-IV would have to remove the animals as soon as it landed."

Alex Bengoa said, "I would say my first real Level-IV job was working with the SLS-1 module after it returned, and the action that I was given was to investigate why the water pump had unexpectedly shut down on orbit. I was not able to recreate the problem. We had a little side joke that it might have been the crew who, for some reason, may have gone to sleep on the SLS module and may have inadvertently kicked off the pump switch. At the time that was just an internal joke we had within the team because we couldn't recreate it. The systems back then were manually controlled, all activated via manual switches, so any inadvertent bump of this switch may have set it off."

STS-58, *Columbia*, launched October 18, 1993, 14+ days.
Payload: Spacelab Life Sciences-2 (SLS-2); Long Module #2 (FOP MD002); Floor (CD MD001); Tunnel (MD001); Rack (See Appendix 2); plus EDO kit.

The primary payload for this mission was the second flight of the Spacelab Life Sciences (SLS-2) Payload Bay cargo. In addition, seven experiments provided further data for on-going medical studies supporting the Extended Duration Orbiter (EDO) Medical Project. The EDO Medical Project was designed to assess the impact of long duration spaceflight (over ten days) on astronaut health, identify any operational medical concerns, and test countermeasures for the adverse effects of weightlessness on human physiology. Only three of the EDO experiments took place in-flight; the other four occurred prior and/or subsequent to the mission.

Like most re-flights, the bugs tended to have been worked out on the first flight, so SLS-2 was mainly uneventful.

Alex Bengoa became involved with the Module Vertical Access Kit (MVAK) while he was still part of the fluids group. "We were allowed to diversify a little bit within the organization, because they were always looking for eager engineers. So I got involved with MVAK because there were not that many people trained for MVAK at the time. My role was as White Room task leader, passing on the cages to the technicians inside the Orbiter, which were then lowered down into the SLS-2 module for installation."

This was another mission containing animals, which was always a challenge, but this time a problem occurred that again tested the ingenuity of the Level-IV team. Juan Calaro said that, "Mimi Shoa of Bionetics in Hanger L helped us a lot with getting the animals ready, but as I was getting the hardware ready for the rats − they had lights to simulate daylight and nighttime − the lights broke in the middle of testing. These lights were really hard to get to, so we had to find a way. That was first my first experience with a bore scope, and we had to go in there with the bore scope to find out where these lights were and how we could get access to them. The running joke was 'How many engineers does it take to change out a light bulb in an express rack?' This concept of Level-IV was ingenious because we had access and we could do these kinds of things and test them out before we integrated into the Spacelab [module]. With that configuration, you could get access to the back of the rack, so finally we got that fixed."

INTERNATIONAL MICROGRAVITY LABORATORY

The International Microgravity Laboratory (IML) was a series of cooperative missions devoted to materials and life sciences studies, including teams from NASA, the United States, ESA, Canada and Japan. The idea was for NASA to offer free launches and payload integration in return for the use of any equipment flown onboard the Shuttle. The launches were planned to occur 18 months apart, in order to fully evaluate the results of one flight before embarking on the next. There were at least three IML missions planned, but only two were flown. An IML-4 mission was listed in NASA documentation for a short time, "for planning purposes".

STS-42, *Discovery*, launched January 22, 1992, 8 days
Payload: International Microgravity Laboratory-1 (IML-1); Long Module #2 (FOP MD002); Floor (CD MD001); Tunnel (MD001); Rack (See Appendix 2).

STS-42 was the 45th Shuttle flight and the 15th flight of *Discovery*. The main objective of the flight was to carry out the International Microgravity Laboratory-1 (IML-1) mission, a collection of life sciences and microgravity experiments developed by more than 200 scientists from 16 countries. In addition to the IML-1 Module, STS-42 also carried 12 GAS containers housing experiments ranging from materials processing work to investigations into the development of animal life in weightlessness.

"I got the IML assignment, and that was probably the pinnacle of my career with Level-IV," admitted Damon Nelson. "We had a Japanese experiment, we had experiments that had flown on Spacelab-1 that were out of the former German Republic, we even had a PI from England. It was just an amazing point in time. We worked a lot with the European Space Agency, so I really got a flavor of it being an international thing, because up until that time all of my work had just been mostly with NASA centers." Damon had the opportunity to travel to Europe and went to the Noordwijk facility in the Netherlands, ESA's largest. He also went to a place called Oberpfaffenhofen, which at the time was the new mission control center in Germany. That was the German Space Operations Center (GSOC), comparable to NASA's Payload Operations Control Center (POCC), There were challenges with the schedule, again because Spacelab missions were still being re-examined in the aftermath of *Challenger* and the manifest changes. At one point, IML-1 actually changed Orbiters, which was very controversial. It was initially planned to fly on *Columbia*, which had a two-week on-orbit capability, but that would have meant delaying the mission an additional six months. Instead, it was switched to *Discovery*, which meant it would only be on orbit for nine days but would get to fly earlier.

Alex Bengoa noted, "Some of the payloads required a vacuum source, so we had vacuum pumps to simulate the vacuum of space. How hard a vacuum was needed depended on the mission. With some of these experiments, one of the things that they liked to do was vent gas, which was really damaging to a turbo pump that was spinning over 100,000 rpm. So we always asked them to let us know when they were venting so we could isolate the turbo pump. Of course, they never told us, which would end up completely shutting down the turbo pump and we'd have to go and reset. So we would yell back at the control team, 'You need to tell me when you're venting so I can be prepared'."

On Aaron Allcorn's very first day in Level-IV, he got a tour under the floor of the Spacelab module to see some of the fluid lines. Upon coming out from under the module, he stood up too quickly and hit his head on a metal plate that was sticking out, which could not be seen from under the module. Aaron remembers being worried that he would get fired for doing something that stupid on his first day: "My supervisor, George Veaudry, showed up and said [with a smile] 'Well, Aaron, I guess we'll just have to give you a desk job!' I left Level-IV in an ambulance and got stitches at a local hospital. Day 2, and every day after that, was much better. Now you know why from that day forward that plate was covered with foam padding."

George Veaudry said, "I was concerned that he knocked himself really hard in the head. He got a good cut there, but we pretty well contained all the blood. He did bleed on certain parts, but we cleaned up things like that. I was just trying to make light of a situation and I could tell he was really nervous. First day on the job and here he was bleeding all over the space hardware. We started putting cushions around sharp edges after that, so in a way he actually did a good thing." Alex Bengoa added, "I have no idea how many people busted their heads on that because it was low, I would say about four and a half feet. So you had to squat down and it was one of those where you had to squat so much that you couldn't look up. So you were hoping that you'd cleared the piece of GSE [Ground Support Equipment]. We put pipe foam with black and yellow stripe tape on to protect people. Bumping your head on that became a Rite of Passage."

Tia Ferguson, recalled, "I came down on February 5, 1990, and the first job I had was under Josie Burnett. She was installing the Fluids Experiments System [FES] and Vapor Crystal Growth System [VCGS] and integrating those racks into IML-1. I remember working closely with Marshall Space Flight Center's Todd Mcleod, and Byron Bonds from Teledyne Brown. Byron Bonds kind of taught me engineering, because in school you learned a lot of theory, so I was extremely green and had to be taught by the folks in the office like Josie and Carrie."

Tia Ferguson in front of the IML-1 she had integrated. [NASA/KSC.]

As described earlier, most refights of similar hardware did not have a lot of problems since most of the assembly and integration tasks had been completed previously. Tia Ferguson described how, during IML integration, there was one drawing that was causing some stress problems. "So they wanted us to open some holes after it had been installed in the module and install bigger bolts. We opened the wrong hole and this was in the module, with shavings coming out and everything. There was a big to-do because the holes were now open, they were too big and you couldn't un-machine something. Fortunately, David Johnson [a leak check engineer at Marshall] had invented an epoxy filling machine that you could insert into the hole, create a no-gap and fill the hole with the epoxy. That was strong enough to hold the smaller bolts. He got his only patent in NASA because of the errors that I made."

Mike Kienlen was a manager at the time, when Level-IV was trying to install some plumbing in the IML floor. It was pre-bent, quarter-inch-diameter (6.35 mm) stainless steel tube that was supposed to connect to different spots under the floor. A Level-IV engineering tech and Quality went in to install it, but there was no way it was going to fit. It had to go through hardware and was just completely the wrong design. Mike Kienlen said, "So the engineer called Marshall up and told them it didn't fit. He got told we weren't installing it right and that it did fit because they'd installed it and it fitted in their simulator at Marshall. We said, 'It doesn't fit on the flight floor'. So he got on a plane and flew down, went out on the floor, couldn't install it either and realized it was wrong. So he spent two days measuring the flight floor. All our guys kept saying to him, 'Why don't you just design it down here and install it down here? Then you won't have to build a simulator at Marshall'. Another example was [an item where] two threads had to stick out past the nut, and when they installed it [on the experiment] per the EO [Engineering Order], it was one thread *short* of sticking out. It was three threads short [in total], so I needed a longer bolt. I got on to Marshall and told them I needed to put a longer bolt in it, but they insisted it was the right bolt. I sent them a picture but it took a week to convince them that it was put together right and that we needed a longer bolt."

George Veaudry said, "Alex Bengoa did a lot of the rack flow balancing for us, as far as the air cooling was concerned. He came up with a really good system for the flow balancing loop, so that you could just make some adjustment and you could see all the flow that was coming out of each one of the stubs for the cooling of each one of the different experiments. He was able to look at all the stub flows all the way, all the Delta-P across each one of the stubs, plus the Delta-P across the rack itself. So he came up with a way of doing that, instead of each one individually then stopping every time to power up."

Alex Bengoa admitted, "That was really tedious work. When I came in, all my team members who had been doing it explained me how to do it. Coming straight out of college, I figured there had to be a better way to do this. It was tedious, it was time consuming. That was the early 1990s and computers, Data Acquisition Systems, were getting more into the scene. So I said to myself, 'Let me tackle this'. I talked to George [Veaudry] and we got some funds, and I start ordering piece parts out of catalog pages. I got Delta-P pressure sensors, and then I got what they called a Lunchbox Computer back then, not even a laptop. I loaded it with LabVIEW, which was off-the-shelf, and then I worked with the techs and we built some cables, then had a Data Acquisition Unit, and I hooked everything together. We were able to read the pressure from there, from the new sensors, and we tied the sensor tube to the pressure sensors all at the same time. I set up the Data Acquisition System and the LabVIEW program, then basically hit the button. Every time we got the right flow rate it would record a flow rate, it would record the pressure. When you were done, you told it to execute and it would automatically spit out the flow versus pressure curve parameters. It would give you that equation. So I could see all the data in real-time for all the flow rates, going into the total flow rate, going into the rack, and the individual flow rate that was going to each individual experiment. So in real-time you could go in and adjust the butterflies until you achieved the final requirements. Then I could use the same set up again to do that next level of flow balancing integration. It worked so well, I got a Silver Snoopy award for that."

Tia Ferguson said, "I remember dropping a washer. You had to write a PRACA [Problem Reporting and Correct Action], the problem report, for everything you did, because you did not want a loose piece of metal floating around in space. We looked for that darn washer for weeks. We never found it; it just completely disappeared."

O&C (Operations and Checkout Building) operations continued, testing was completed, and then it was off to the Orbiter Processing Facility (OPF) for final closeout of the payload before heading to the Vehicle Assembly Building (VAB) for stacking operations. The hardware had been through many checks to ensure all was in order, but everyone once in a while something could slip through. Hopefully, it would get caught on the ground before launch and not in space.

An ID plate that was labeled incorrectly was discovered during final closeouts. This was the last chance to touch the hardware before launch.

Label fix (top) and PR for label corrected by Tracy Gill. [Tracy Gill.]

Tracy Gill explained that, "I was doing close outs in the OPF and at the last minute, I found this label that closed the rack airflow valve that was backwards. Each side of the racks had a different label, but this had the same sticker on both sides even though the valves were oriented differently. So we found this, and then I had to get approval over OIS. I got all the concurrences and then Jim Pope, who was back in the O&C, had to go off and get all the signatures, I think eight signatures. I had the easy job just finishing the final switch list and configuration check while waiting for him."

"I will never in my life forget [IML-1]," Damon Nelson recalled. "We confirmed all the preflight switch configurations for this module because there was no plan to go into the module at the pad to load late samples. So once we left the module, it was closed out for flight. We were about 90 percent through our checklist and one of the final steps was to remove the protective film cover from the glove box, this pristine glove box that had a window that the astronauts would view through as they had their hands inside the glove box and were manipulating samples. We had just removed the protective film cover when a Spacelab tech – not my tech, he was a tech that was doing the Spacelab – came over. He took a bottle of MEK [Methyl Ethyl Ketone], which was a solvent, and he sat it on the glove box viewing cover because it was one of the only horizontal surfaces in the Spacelab. And I looked at him, and he looked at me, and I said, 'I don't think you should have done that. Please remove that immediately'."

As fate would have it, a little bit of that chemical had bled down the outside of the MEK bottle and when the tech removed the bottle, it had etched about a 300-degree ring on the cover. Of course, the tech started to try to clean it, but Damon Nelson said, "Immediately, I told him, 'Don't touch that! Don't do anything else'. And it was one of the saddest calls I ever had to make to the Mission Manager and the PI, and they ended up using [the glove box] as is and flew like that. They flew at least once, if not twice. I had an opportunity to go to Germany later in my career and the European Space Agency had created an unbelievable display of the rack on the floor that had flown many missions by this time. One of the experiments that they were displaying was the rack called Bio-rack that included this glove box. I looked at that glove box and that 300 mark was still there. I saw a co-worker and said, 'You see that ring? I know the story'. I know that as much as we protected that glovebox, I was just sure they were going to make us de-integrate it but, bless his heart, the PI could live with it."

Damon Nelson also got to go to Marshall to support the mission at the Payload Operations Center. "That was my one opportunity to actually go and participate in the actual mission, and they requested that I went because of my broad understanding of the whole experiment complement and how we had put it together. The mission was a very clean mission. We did have a launch delay and that set the timeline down a couple of hours, but they were able to recover that timeline and it was a very successful mission."

Tia Ferguson recalled that, "I was out there for IML [landing operations], and we actually went in the module and got the stuff off the module. There was this 80 pound (36kg) storage container that contained the crystals that they grew and we had to carry that thing out of the module, me and two technicians. It was right after it landed, to get that science. It had been really complicated to get it locked down for the launch and for the returns, and I remember training the crew over and over how to lock it down. Well, when the rack came back, it wasn't locked down and the interior of that storage was pretty beat up and mangled because they didn't lock it down properly, even though we'd trained them a lot on the ground."

"I also had the opportunity to fly to Edwards Air Force Base for the landing," recalled Damon Nelson. "This mission landed in California, and the reason they did that was that they wanted to have an assured landing site so they could have the team out there with all the equipment to retrieve the samples. It was very critical that they got the samples as soon as possible. So, the Mission Manager made a decision, working with the Shuttle program, to go ahead and just plan the landing even though we were starting to transition landings to KSC. He was able to sell the criticality of baselining Edwards as the prime landing site. So, again, getting to see this mission literally almost from the beginning to the end was really a highlight. When we went to Edwards we had to get in the module, kind of like what we'd done at the OPF but postflight. There were a lot of certifications to do so it just made sense to send just the techs in. We didn't need to have extra bodies inside the module. We were just going in there to retrieve stuff and get out. Jeannie Ruiz was the lead for those sample removals. I was more there just to run the trap line, working with the team and trying to keep her job as simple as possible. But I do remember we spent a sleepless night for sure. Getting everything set up and ready for the early morning landing, and seeing the crew right after landing was just really something else. That was cool."

Just over two years later, the second IML-mission was ready to fly.

STS-65, *Columbia*, launched July 8, 1994, 14+ days
Payload: International Microgravity Laboratory-2 (IML-2); Long Module #1 (CD MD001); Floor (FOP MD002); Tunnel (MD001); Rack (See Appendix 2); plus EDO kit.

Once again an international group of scientists collaborated on the mission. This time, scientists from ESA, Canada, France, Germany and Japan worked with NASA to provide the worldwide science community with a variety of complementary facilities and experiments. Research on IML-2 was dedicated to microgravity and life sciences. The life sciences experiments and facilities on IML-2 included: Aquatic Animal Experiment Unit (AAEU), Biorack (BR), Biostack (BSK), Extended Duration Orbiter Medical Program (EDOMP) and Spinal Changes in Microgravity (SCM) in the center aisle;

Lower Body Negative Pressure Device (LBNPD), Microbial Air Sampler (MAS), and Performance Assessment Workstation (PAWS) in the middeck; Slow Rotating Centrifuge Microscope (NIZEMI); Real Time Radiation Monitoring Device (RRMD); and the Thermoelectric Incubator (TEI). Microgravity experiments and facilities on IML-2 included: Applied Research on Separation Methods (RAMSES); Bubble, Drop and Particle Unit (BDPU); Critical Point Facility (CPF); Electromagnetic Containerless Processing Facility (TEMPUS); Free Flow Electrophoresis Unit (FFEU); Large Isothermal Furnace (LIF); Quasi Steady Acceleration Measurement (QSAM); Space Acceleration Measurement System (SAMS) in the center aisle; and Vibration Isolation Box Experiment System (VIBES).

Joni Richards was the lead for IML-2: "I oversaw the mechanical work, electrical work, and basically the whole integration and got to say go for launch. How cool is that? On IML-1, Damon Nelson had that role, so when the European partners, many of which knew him and worked with him on IML-1, were in the audience at the O&C and it was kind of the kickoff meeting, Damon walked in and everybody was all smiles: 'Yay, Damon's here. We know him, this is going to be great.' And then Damon proceeded to introduce me as his replacement and you could totally see the look on all the different faces, the look of, 'Oh my God, for the love of God. Damon, please do not leave us, please'. So I jokingly said to them, 'Wow, I can see that you all have a look of concern on your face.' They laughed that nervous laugh when you find there's truth in what someone just said, even if it's unnerving. But it turned out to be just fine."

As with other re-flights and preparations for a subsequent mission, some processing flows went smoothly and some did not, while some had only a few problems that the team figured out how to solve.

Tracy Gill admitted that they had some problems with the high rate data lines from one of the experiments during processing. "It was an intermittent failure. Sometimes the signal was there and sometimes it wasn't. Because it was an intermittent problem it was difficult to find, and we spent a bunch of time troubleshooting. They finally found the problem, which was from a bad solder joint on the pigtail for one of the Kapton wires, by wiggling the data cable coming from the experiment and going to the High Rate Multiplexer. The failure was intermittent because the gap in the broken solder joint was so small that just a change in temperature or someone walking in the lab would cause the pigtail to connect or disconnect. This type of failure was pretty common because the Kapton wires were very tricky to solder. After identifying the problem, the networks folks repaired the cable and they didn't have any further problems during the flow."

Joni Richards continued, "All the travel I had to do was to help cut down the number of problem reports we had. IML-2 was the most amount of European experiments to date and we had the least amount of PRs [Problem Reports]. So it was a good return on their investment to send me to these locations. I got to go to France and Germany and Italy twice, and I was in Japan almost three weeks because we went to three different cities, Tokyo, Kobe, and then Tsukuba where their launch facility is at.

"But one of my most interesting, important memories was in getting rid of one of our Mission Sequence Tests [MST]. There was a gentleman called Mo Lavoie who had charged down this path to try to get rid of it. Mo was a sharp guy. It was the one in Level-III/II that I had gotten rid of, because the original requirement for having that second MST was more to test out the subsystems, not for training. Even though I had a battle, we made our date that we'd set 18 months in advance. I actually presented to former astronaut [Robert L.] Bob Crippen. He was our Center Director at the time and did a very conservative calculation with our budget folks to figure out that what we'd saved was over $100,000, though they said it was probably double that."

With the IML-2 ground processing completed, the payload went through the launch preparation cycle, leading to a successful launch in July 1994. With the Shuttle safely in orbit, support for payload operations turned over once again to Marshall Space Flight Center.

Tracy Gill told the authors that, "ESA actually requested me to be there for support, and they funded my travel well and they had to pay a crazy tax on it too. I don't remember what it was; 25 percent or something, So that felt like quite an honor because I got this letter from ESA that said they wanted me there to support the mission and that I was critical to our success. Well, I remember there was one incident during a Sim [simulation] where they modeled some problem, I think for one of the experiments, CPF the Critical Point Facility. They recorded data on HRM and their data was basically snapshots of how the fluid was reacting, and video images of what to do if they didn't have that. This was the old days and we only had the one camera, one video feed at a time, so most of the time they couldn't get it. I said, 'Just every hour or so, film five seconds of video on analog video, and then you're going to have the same data'. And everyone seemed really surprised, like 'That's a good idea!' But to me it seemed really obvious because I'd seen it all and it was just different forms of data. Everyone was happy because it was easy to solve and they went off and modified the procedures and we carried on in the Sim. I think it helped me establish my credibility there at the HOSC [Huntsville Operation Support Center] – 'That guy knows what he's talking about because it's his stuff'. I knew how something worked that they weren't used to because we powered up every day and tested things and we were just used to it. So we had the benefit of that experience."

Tracy Gill did an IFM (In-Flight Maintenance) on IML-2 because the Japanese did not get their data right away on orbit. "I called them from Atlanta airport on my way up there, to see how it was going. I heard their data wasn't working and

I knew we had repaired a cable on that path and there was a solder sleeve in there. I said, 'That's the most suspect area; we repaired it'. We used to troubleshoot the problem by wiggling the cable bundle and I said, 'Just go out and wiggle the bundle'. I got to the HOSC about 20 hours after launch, and they still hadn't done this because JSC didn't want to go to the crew unless we had a full troubleshooting plan. The troubleshooting we had done at KSC and recommended was that they had the astronauts wiggle the data cable. They [JSC] responded with something like, 'We're not going to tell them to wiggle the cable. We need to have a detailed troubleshooting procedure'. They evidently didn't think a simple instruction like wiggling the cable was up to their standards. So I called John Lekki and let him know what had happened. I asked John to go in to work − it was on a weekend − find the paperwork from the troubleshooting we had done at KSC and fax it to the POCC. I said that I needed the procedure to convince them that they should wiggle the cable. Evidently, having this in the KSC procedure made it an appropriate troubleshooting step and so they included it in the procedure for the on-orbit troubleshooting. After 41 hours, and having a hugely ridiculous cable bypass plan written and approved because it involved building a bypass cable with the IFM tool kit, the crew were getting ready to execute our complete troubleshooting plan when the JAXA customers called on the Huntsville loop and said they just started receiving data. I asked the CIC [Crew Interface Coordinator, who talked to the crew on orbit] to ask the crew if they had started working the procedure early.

"The crew reported they were getting ready for the work and that they'd opened the panel and one of them had grabbed the cable to read the label. That fixed it, and it wouldn't fail again in microgravity unless someone else grabbed it, which wasn't going to happen since it was in the sub-floor. We said 'Don't touch it, it's fine. Close it up and walk away'. We asked for no further work to be done. They stopped troubleshooting right there and left it alone for the rest of the mission. They didn't have another on-orbit problem with that experiment and they got all their data. So the Japanese thanked us for fixing the problem before the formal troubleshooting plan even started. We were not happy that the problem had happened again, but it was nice to see how helpful we could be for on-orbit troubleshooting because of the experience that we had with the hardware. Postflight, we found the problem with the data cable by CB50 which caused the In-Flight Anomaly at the beginning of the IML-2 mission. As we suspected, it was a bad solder sleeve connection at the splice from the core cabling to the 22-gauge wire entering the connector right at CB50. Unfortunately, it was the same wire that we repaired at KSC in Level III/II, and the heat shrink around the solder sleeve did not appear to be fully heated during the repair, which allowed it to come loose during launch."

Early End of IML

The IML series ended after only the two missions, but Level-IV had learned more about the flight hardware and international cooperation which could then be used for future payloads, such as those described below.

UNITED STATES MICROGRAVITY LABORATORY

The United States Microgravity Laboratory (USML) program was initiated in 1987, in support of the development of Space Station *Freedom* and the subsequent ISS program, to fly in orbit for extended periods using the Spacelab module and EDO pallet system. This provided greater opportunities for domestic American research in materials science, fluid dynamics, biotechnology (crystal growth), and combustion science. Although at least four missions were included in NASA's long range planning documents, only two were flown.

> **STS-50, *Columbia*, launched June 25, 1992, 13+ days,**
> **Payload: United States Microgravity Laboratory-1 (USML-1); Long Module #1 (CD MD001); Floor (FOP II MD003); Tunnel (MD001); Rack (see Appendix 2); plus EDO kit.**

USML-1 was a Spacelab Long Module with an EDO pallet in the aft Payload Bay. It carried 31 investigations from five basic areas: The fluid dynamics experiments examined basic fluid phenomena, from movement caused by heating to the dynamics of individual liquid drops; the crystal growth experiments grew a variety of inorganic and organic crystals; the combustion science experiments examined the differences in the shapes of flames and how these spread in microgravity versus gravity; the biological experiments examined the production of various products and monitored the changes to human physiology as a result of extended exposure to microgravity; and the technology demonstrations tested experimental concepts and facilities for use on future missions. Four new experiment facilities were flown on the USML-1: the Crystal Growth Furnace, which was used to grow high-quality semiconductor and infrared detector crystals using both directional solidification and vapor growth techniques; the Glovebox, which was used for direct manipulation of experiments while keeping the crew isolated from the materials involved; the Surface Tension Driven Convection Experiment apparatus, which conducted studies of fluid mechanics and heat transfer; and the Drop Physics Module, which used acoustic force (sound waves) to position and manipulate liquid drops so that their physical and chemical properties could be examined.

Josie Burnett explained that, "There was a point in my career within that Level-IV timeframe where I thought it had become more that I was counting bolts. How could have this been such a glamorous job four years ago? The beauty of Level-IV was that I was able to walk across the hall. Maynette Smith had just been promoted to a supervisor position, and that's how I ended up as an experiment project manager and did the STS-50 USML. I walked across the hall and said to Herb Rice, 'I'd like to do that job and I think I could do it'. I didn't know any better. He looked at me and said, 'I can't promote you' and I said 'What do you mean promote me? That has nothing to do with it, I don't want to do it for a promotion'. I was thinking more 'Don't tell them you're giving them an impossible task, and see how they do'. Seriously, I think that's the key to developing new hires; you give them an impossible task, but you don't tell them it's impossible and just see what they do."

"[The] Crystal Growth Furnace [CGF] was an interesting payload," recalled Kevin Zari. "That was the first full payload assignment once I was a backup engineer again. Kim Jenkins, who at the time was the lead of a HITS, was the prime on that. I actually worked on writing the procedure for the system itself, the Ground Support Equipment that we used. I was tired of reading 50 pages of 'You go do this, go do that'. I needed something that a Co-op that was brand new off the street could literally just come in and use. So I came up with the idea of the book that I made, and it was called 'HITS for Dumb S**ts'. Of course I had to asterisk out some of the letters so that we wouldn't get in trouble. I'd walked through the activation procedure and it literally was nothing more than a bunch of screenshots, either highlighted or circled or whatever. But I drew attention to the fact that 'You need to press this and then you're going to see this and then this', so basically it was a series of pictorial images of what a user had to do. And so I tested it. We had a Co-op who came over as part of their coming onboard, and NASA had to do these detailed rotations. So I took this person that was completely cold, who had no idea, and I said, 'Can you execute with what you see in this book? See if you can figure it out'. And they were able to successfully bring up HITS."

Tia Ferguson remembered that, "When we knew when the Spacelab module was going into the Shuttle [Payload Bay], we had to go take the astronauts for their final walkthrough. With USML-1, it was with Bonnie Dunbar and the other astronauts, And Bonnie was talking to one of the other astronauts and she said, 'This a smart short' and she went and unplugged a cable. I was supposed to be there to prevent them from doing that, and I thought, 'Wow, you can't do that. You don't understand. Now we've got to go back and test that thing and run a whole bunch of tests because you just unplugged that cable'. So we had to go back and do more testing to reverify that it was connected properly."

Alex Bengoa started diversifying from doing fluids work and testing. He also became an experiment engineer for middecks and other major payloads: "For middecks, I worked with Crystal Growth experiments," he told the authors. "I picked up the payload from the developer at the lab and transported it to the pad. Then, once we arrived in the middeck, we would pass a payload to the Orbiter techs and they would do the installation into the middeck. Once that happened, we were part of the testing with the Launch Control Center [LCC]. We got on the headset, and then we'd do a power transfer from battery power to Orbiter power. Then I would get back in, the Orbiter techs would leave or get into the middeck and then I'd do all the switch throws for the transfer of power. But that was here locally at KSC. Of course, the Orbiter could land at KSC or Dryden, so depending on the mission I could be posted at the landing site or remain here at KSC to support the removal of the payload. Then we would take it back and give it back to the payload developers."

Tia Ferguson added that, "For USML, I integrated the Drop Physics Module, and for the Glove Box I was actually a test engineer, so I got to flip switches. I worked with Tracy Gill – I think I was his alternative – so I would go down into the module and flip the switches while he was running the test procedure. So I learned about testing and from that I really loved the electrical part and that inspired me to go back and get an Electrical master's degree once I moved up to Marshall Space Flight Center, because I really enjoyed the testing and doing that. I also did the Protein Crystal Growth. I spend a lot of time over in Hanger L working with a scientist, getting their proteins ready and then transporting those out to the Shuttle and installing them in time. We were the last ones on the pad before they started tanking [loading the External Tank with cryogenic fuel and oxidizer]."

Tracy Gill, added, "We'd launch and then go up to Marshall POCC right away, a couple of times to work normal console support, like STS-50 for the USML Glovebox that sat on console. And I ran a shift. I helped by talking to the ear of the CIC. We were talking to the crew about what we were doing with the Glovebox. They had to request our support and then somebody had to agree to pay travel money. They wanted us there because we put it all together. We'd seen all the anomalous behavior. We knew where the bugs were. After the first mission, they pretty much assigned me to be a floater and a general troubleshooter so that when any trouble came up I was put on a team that would help come up with work-arounds or solutions. That's basically what I did. Every other mission, I went to Huntsville."

"Then we would get on a plane as soon as the launch happened and head out to Dryden," Tia Ferguson added. "Back then it was Dryden, and every morning we would go at four in the morning. And of course the mission was going great, so they said, 'Well we don't need you today', and then we'd have all day to figure out something to do until 4:30 the next morning to see if they were going to come down the next day. We spent a lot of time out driving, playing volleyball and driving to the ski area, or driving to Sequoia National Park, because you could get there and back by 4:30 in the morning the next day."

STS-73, *Columbia,* **launched October 20, 1995, 15+ days.**
Payload: United States Microgravity Laboratory-2 (USML-2); Long Module #1 (CD MD001); Floor (FOP II MD003), Tunnel (MD001); Rack (See Appendix 2); plus EDO kit.

This mission carried out microgravity experiments related to fluid physics, material science, protein crystals, and combustion science. The USML-2 Spacelab mission was the prime payload on STS-73. The 16-day flight continued a cooperative effort of the US government, universities and industry to push back the frontiers of science and technology in "microgravity", the near-weightless environment of space. Some of the experiments carried on the USML-2 payload were suggested by the results of the first USML mission three years earlier. The USML-2 mission provided new insights into theoretical models of fluid physics, the role of gravity in combustion and flame spreading, and how gravity affects the formation of semiconductor crystals.

"Another area was training the astronauts" explained Tia Ferguson, "and I remember training Bonnie Dunbar on USML-2. The Drop Physics Module had a camera cartridge that was really finicky to be able to get in and out. The crew had a mock-up at JSC that they trained on, but the hardware acts differently than the mock-up. For the flight hardware you had to kind of jiggle it in to get it in right, otherwise it wouldn't seat and record properly. [I wrote] steps to open the Drop Physics Module and to take out and put back in the camera cartridge, and explained to Bonnie, 'Here's the finesse that you need to use'.

"The Spacelab racks were designed as if they were exactly like the drawing. I mean, they were designed to be square," Tia Ferguson added. After launch, the racks would get twisted and turned, and nothing ever fitted on orbit because the racks were distorted. Tia remembered that the Fluid Experiment System (FES) rack had four brackets in the back and required completely new drill holes. The holes were off from where the drawing said they were supposed to be, misaligning by about an inch (2.54 cm). Re-drilling holes, and sometimes re-machining brackets that didn't fit and shimming where required, happened quite often. Tia Ferguson continued, "The Drop Physics Module didn't fit. It was a double-width experiment that went across both sides. They had to take out the center rail and it didn't fit. So I had two technicians on either side basically pulling the rack apart. They put their hands inside the rail above and below the Drop Physics Module and pulled on either side, so that we could slide it in and get it through the opening. Things like that just help you learn."

Juan Calaro said, "I was a backup experiment engineer on CGF, Crystal Growth Furnace, and Kim Jenkins was the experiment engineer for that, and I think we had a communication issue with the payload. I remember Scott Vangen helping. It was a pretty challenging problem and no one could figure it out. I remember going into meeting rooms and talking about the problem, and Scott came into to help us try to troubleshoot the problem. These were typically 12−14 hour days. Kim was starting a family and I remember the stress that was on her going through that, so I tried to help her as much as I could. I remember her struggling with family and the work-life balance. Level-IV was really meant for someone who didn't have a family and was young, because the hours we put in were just [horrendous, but] you wanted to be there."

Darren Beyer added that, "For USML-2, I did astronaut training in the OPF, and then the final consumables loading in the Spacelab during CEIT [Crew Equipment Interface Test], where we got the astronauts into bunny suits and into the Orbiter in the OPF. With what had happened before on the last mission, I guess it had gotten kind of ugly between the crew side and our side, a little bit of a testy situation. So I was in a tough spot because I wanted to be accommodating to the crew, but then you couldn't let them do everything because they would invalidate a bunch of stuff. So then, one of the astronauts who was going to be doing experiments had gloves he would have to put on. He said, 'I want to put on the gloves'. Now these were things that got stowed long before the Orbiter was in the OPF, and if he opened up one of those things and put the gloves on we would have had to go through a whole process and procedure to replace the glove he'd used. I couldn't get a hold of anybody to authorize it, so I just made a call and said, 'All right, go ahead and do it'. The guy put the gloves on and they didn't fit and they ripped. I remember catching a bunch of hell on Monday: 'Why did you let him do this? Now we've gotta go back in'. I said, 'Well, wait a minute, the gloves ripped. What would've happened if this had happened on orbit? We can't go in and replace gloves on orbit'. We ended up, by virtue of the fact that there's gotta be a little bit of [camaraderie and harmony], that Cady Coleman made us cookies because she also knew that there were some issues going on."

Final switch list verification was normally done in the OPF just before the Spacelab module was closed up for flight. So how could something that was set correctly on the ground be different when it got into space? Tracy Gill explained: "We looked over everything and then we put the switches in the final switch configuration as well, so when they powered up [on-orbit], the things came on that needed to come on. [But with] the USML-2 Glovebox, one of the switches was in the wrong place. And there was this grumpy German guy who was the PI that I'd worked with for a long time, and he said, 'You were supposed to put these switches in the right place as I did'. I said, 'I did. I [took] pictures. I know I did. I don't know what happened. Something happened'. There was a ground version of the Glovebox downstairs in the HOSC, and at one point in the mission we were looking at something. The Glovebox had a cover with Velcro on it, to cover the window on top. You took it off when you were ready to use it so you could see inside. Well, the German guy was right in front of me and he lifted off the cover, and as soon as he did that the switch flipped, the one that was misconfigured. And he just looked at me and he didn't say anything. So probably when the crew pulled that thing off to get ready to use it, the switch flipped. So it wasn't the wrong configuration, but that is what they reported."

Tia Ferguson explained that, "There was a cargo area in the middle of the rack. It had a whole bunch of stuff inside and Teledyne got the idea – it got approval somehow – for us to put a flag on the foam at the top. So we all signed the back of this flag and we installed it with Velcro to the top of the foam before closing the storage container. And when they got on orbit they didn't know that; there were no instructions. Apparently the Ops folks didn't know it was there, and so neither did the crew. But underneath the flag were all of the lists of what was in the storage. They weren't able to find the lists until they figured out that they needed to remove the flag."

PRECUSOR TO KIBŌ

Japan had a long-held dream of flying its own astronaut in space to operate Japanese experiments and, having recognized that developing Japanese human spaceflight potential was a distant prospect, opted for a cooperative mission with its long-standing economic and scientific partner, the United States. The Space Shuttle offered better prospects for a citizen of Japan to conduct meaningful scientific research in orbit in the interests of Japanese science. Japan was an early partner in Spacelab flights, with one experiment flying on the very first Spacelab mission. In 1979, the National Space Development Agency of Japan (NASDA, the forerunner of the current Japanese Space Agency, JAXA) began developing a payload called the First Materials Processing Test. In 1984, Japan made contact with NASA about the possibility of flying their experiments on a dedicated Spacelab mission, to be operated by Japanese astronauts. In the event, the suggested Japanese experiments were not sufficient to fill a complete Spacelab module, so NASA developed and manifested experiments to complete the payload and complement the theme of what became Spacelab-J. When Japan became a partner in the US Space Station *Freedom* program in 1985, they proposed the development of their own Japanese Experiment Module (JEM) that went under the name of Kibō ("Hope"). The Spacelab-J mission became an important step in Japanese human spaceflight development, and in extended cooperation in the ISS program.

> **STS-47, *Endeavour*, launched September 12, 1992, 7+ days,**
> **Payload: Spacelab-J, Long Module #2 (FOP MD002); Floor (FOP MD002); Tunnel (MD002); Rack [See Appendix 2).**

Primary payload Spacelab-J (SL-J), or "Fuwatto [weightlessness] '92", utilized the pressurized Spacelab module. Jointly sponsored by NASA and the National Space Development Agency of Japan (NASDA), SL-J included 24 materials science and 19 life sciences experiments, of which 34 were sponsored by NASDA, seven by NASA, and two were collaborative efforts. The mission was extended one day to further science objectives. The materials science investigations covered such fields as biotechnology, electronic materials, fluid dynamics and transport phenomena, glasses and ceramics, metals and alloys, and acceleration measurements. The life sciences investigations covered human health, cell separation and biology, development biology, animal and human physiology and behavior, space radiation, and biological rhythms. Test subjects included the crew, Japanese koi fish, cultured animal and plant cells, chicken embryos, fruit flies, fungi and plant seeds, and frogs and frog eggs.

Cindy Martin-Brennan recalled, "I was in charge of the First Materials Processing Test [FMPT]. It was the materials rack, I think it was actually a couple of racks with multiple furnaces, a whole suite of different types of furnaces to do materials science payloads. I spent a month in Japan and worked with Mitsubishi Heavy Industries [MHI] who built it all outside Tachikawa, Tokyo. I went over there and then Bruce Morris joined me. The Japanese staff were fabulous, I don't remember any problems. Of course there was a little bit of a language problem, though it wasn't really that much, but I think there was a little bit of a problem culturally with me being a female engineer. When I went over to Japan, at the time all of their females were making coffee and running coffee breaks. I never received any disrespect, I think it was just a little bit of an adjustment for them."

Tia Ferguson recalled that she did the sub-floor for Spacelab-J. "I did the fluids routing and things like that. Working with the Japanese, we would go in in the mornings and you would pull up at 7:30 in the parking lot and they would all be behind the building in rows. They'd have maybe five rows of eight or nine people in each row. They were reaching their arms down and then touching their toes and doing all these stretches and everything as a whole organization. I remember seeing them do that. Then [there was] how they made group decisions, not individual ones. You would ask them a question and their immediate answer was 'Yes'. After a while, you learned that 'yes', didn't mean 'yes'. It meant 'I understand and I need to go congregate with my group and then we'll come back to you with the decision'. I loved working with the Japanese."

"Spacelab-J was my last mission in Level-IV," Bruce Morris told the authors, "and then I moved over to work Space Station. Like many of the Level-IV team, I had an experience of working with a very large group of international engineers and scientists, and they had their cultural way of approaching activities like this. I had to learn more about how the Japanese tended to operate, so I could negotiate and I could work with them in a way that was effective so we worked that way together. That was just so much fun." One of the interesting things concerned a whole list of documents that were required to show up with an experiment when it came to

KSC. It was called the data pack. There were things in the data pack that Level-IV actually never needed at KSC, but the Japanese culture is very much about completeness, and so every single thing KSC asked for in that document package, they delivered. Bruce Morris remembered, "Well, we decided that [some of that] wasn't really required anymore, but [then they came back with] 'No, it says in the guidelines that these are the rules and we are supposed to turn that [the data pack] over to you and you are expecting it'. We found these things out. It was really fun."

Alex Bengoa was also on the Spacelab-J mission, as part of the MVAK team. "We had some live animals, some fish. I did get to go and train at the old Hanger-L where the Japanese were processing the fish and putting them in the flight hardware tanks. One of the things I had to worry about was to make sure there was enough water in the tanks or in the flight hardware with the fish when the support equipment was disconnected. So they gave me this little hex Allen wrench, and I told them, 'Okay, stick it here. If it's to the red line, then we have to add more water'. So I had this 'calibrated Allen Wrench' with little marks on it. There was an accumulator inside so that if it wasn't within the mark in the wrench, I would have to go in with a syringe full of water that I could use to inject more water. The Japanese told me that when I did this I had to go really slow, because a High Delta change in pressure could really damage the fish's ear drums. So I was going really slow and of course I'm hearing people in my ear on the OIS saying, 'Alex, you're going too slow', but I was doing what I was told. Once complete, I had to go into the Orbiter middeck and transfer them to the techs, who would then be lowered down [via MVAK] to install them in the module for flight. I was not part of the team that that went into the module after landing and the story I heard was that there was one fish that was still alive when it came down on the Spacelab. We were expecting all the fish to be dead upon return and there was this big fish still breathing, still alive. People were scratching their heads figuring what to do with this fish."

"We did a lot of training," added Tia Ferguson. "MVAK for Spacelab-J and then out to the pad at the LCC to run that operation. Daniel Shultz was the prime and I was his backup. I got to ride in the bosun's chair and saw how the technicians had to install the fish and frogs."

Following the mission, Level-IV personnel were onsite to handle the animals. Sharolee Huet said, "Felix [Joe] and I were in a newspaper article about the SL-J landing, but we were noted as Lockheed techs. I always wore my hair braided and the teal shift with the Shuttle on the back, so I could find myself [in pictures and videos of landing operations] in the sea of folks at landing."

NASDA[1], who had paid for the mission, were disappointed at the time with the decision to fly just one Japanese Payload Specialist with four American career astronauts on the payload team. The issue was partly amended by NASA on later Spacelabs which flew a number of international crewmembers.

---

[1] Japan's Institute of Space and Astronautical Sciences (ISAS), the National Aerospace Laboratory of Japan (NAL) and the National Space Development Agency of Japan (NASDA) merged on October 1, 2003, to form JAXA.

A SECOND GERMAN MISSION

The second German-managed Spacelab mission was designed to focus upon a variety of studies under microgravity conditions. Some of the experiments flown on the first German Spacelab would be flown again, supplemented by other internationally-supplied investigations. The D2 mission, as it was commonly called, augmented the German microgravity research program started by the D1 mission. The German Aerospace Research Establishment (DLR) had been tasked by the German Space Agency (DARA) to conduct the second mission. DLR, NASA, ESA, and agencies in France and Japan contributed to D2's scientific program. Eleven nations participated in the experiments. Of the 88 experiments conducted on the D2 mission, four were sponsored by NASA

> **STS-55, *Columbia*, launched April 26, 1993, 9+ days,**
> **Payload: Spacelab D2, Long Module #1 (CD MD001); Floor (EM MD001); Tunnel (MD001); Rack (See Appendix 2); plus Unique Support Structure (USS).**

On SLD2, the crew worked in two shifts around-the-clock to complete investigations into the areas of fluid physics, materials sciences, life sciences, biological sciences, technology, Earth observations, atmospheric physics, and astronomy. Many of the experiments advanced the research of the D1 mission by conducting similar tests, using upgraded processing hardware, or implementing methods that took full advantage of technical advancements since 1985. The D2 mission also contained several new experiments not previously flown on the D1 mission, and conducted the first tele-robotic capture of a free-floating object by flight controllers in Germany. The crew also conducted the first intravenous saline solution injection in space, as part of an experiment to study the human body's response to direct fluid replacement as a countermeasure for amounts lost during spaceflight. They also successfully completed an in-flight maintenance procedure for collecting Orbiter waste water, which allowed the mission to continue.

Juan Calaro told the authors, "I had a robotics experiment on D2 and it was around the time, I think, of the transition with East and West Germany and the challenges that Germany was dealing with. I remember talking about those challenges with the German PIs. I had the opportunity to go to the Netherlands and to Germany to support one of the payloads they were developing."

This was a mission that was managed by the German Space Federation, and there were a lot of interesting cultural challenges with the Germans. The Experiment Integration role and Level-IV team were directly funded by DLR, as opposed to previous experiences where Level-IV was just part of the NASA agency, so in a sense they were working for the German Space Agency. That made for an interesting dynamic. "We got to know a lot of the engineers on a first name basis," said Damon Nelson. "One of my last trips to Germany was to actually participate in what they called the Mission Sequence Test, which was a test that I would then be responsible for running once the rack and floor were put into the Spacelab module. That was a great experience, getting to watch their team perform this Mission Sequence Test on just the racks and the floor. I saw the dynamic of the flight crew, which was quite revealing." The crew included a very hard-driving astronaut, Jerry Ross, who was the lead Mission Specialist. He was quite demanding of that whole team. There were also two German Mission Specialists. Once the hardware got to Florida, the racks and the floor were integrated with the Spacelab module. The KSC flow was not as long as it had been for other Spacelabs at KSC, because a lot of the Level-IV experiment integration part had already been done in Germany.

Sue Sitko (standing) and Theresa Martinez (sitting) working with the Modular Optoelectronic Multispectral Scanner (MOMS-02) Team in the User Room, during Spacelab D2 Mission Sequence Testing, November 1992. [NASA/KSC.]

"We had the Shuttle accident and also the bigger delay," said Hermann Kurscheid, "so over the years the payload for D2 developed more and more. It was so densely packed that we sometimes thought, 'How can we all handle that?' But finally we made it for the D1 mission. Also for the D2 mission, we had a flight floor there in Germany and we already mounted the payload on that. We could also then do the Level-IV testing in Germany and then we transported the floor and the bridge in specific containers directly to KSC. And then they could put it down right on the floor and stand. Also, we had a small park in Oberpfaffenhofen in Germany and for the D1 and D2 missions, the new park was built there in Oberpfaffenhofen, and the flight operations people on the ground were then located there in the POCC in Oberpfaffenhofen."

"The hardware came to Florida," said Damon Nelson, "and I remember they had so many issues. I had actually made a proposal that we should do a three-week Level-IV in the US, just so that we didn't have so much open work when we got to Florida, or when we got ready to put the racks and the floor into the Spacelab. I had sold that concept with our managers at KSC, but we could never sell it to the Germans. I think part of it was just political; they didn't want to look like they had shipped when they had not completed their work. I can understand why that was a bit sensitive, but regardless, we went into the Spacelab module with the rack and floor train. We had individual NASA system engineers for each of the German experiments and our engineers wrote the paper with their inputs. Any Problem Reports that we took, obviously we involved them in the problem resolution. It was an interesting dynamic, because you didn't necessarily have that when we were dealing with US-managed missions. This mission also did not involve any late stowage. We had the Long Module and we had an MPESS with some experiments on it and we went out and we integrated just like we had done other missions in the OPF. For the most part, everything went pretty well."

Hermann Kurscheid added, "There was a Mission Sequence Test and that was actually taken from the timeline. We selected the slices to put the most load on the system to see whether it could handle it or not, or problems might occur. I think this was a ground testing, but beneficial for the flight."

This mission also did not involve putting payloads like animals and plant samples into the module late. There was a refrigerator and a freezer in the middeck, and these were critical to preserving the samples from when the Spacelab was shut down on orbit until the samples could be retrieved from the middeck postflight. The interface tests of these freezers and refrigerator with the Shuttle systems were performed, and failed. Now, just weeks before launch, this critical piece of hardware was not working on the middeck area. There were backups, and the mission was going to rely on these backups, but there would be an elaborate changeout procedure on the day of launch. Damon Nelson recalled, "If one of the things went belly up... I never will forget being on the C-1 Console [the Payload console in the Firing Room] on the day of launch, knowing that if they made the call that the thing had failed when the Payload Specialist [PS] went into the middeck [to be strapped into his seat for launch], we were going to have to do a swap out two and a half hours prior to launch. The C-1 was going to be the console where we had to run the swap out procedure, and I remember being so happy to hear that call from the APS that the refrigerator and freezer were still working. So, we were all ready to launch STS-55, and we had a main engine shutdown [on the pad]. Needless to say it freaked everybody out, like they had always done in the past. I think we had

had three or four other main engine shutdowns and it required about four weeks to swap all the Shuttle engines out."

Alex Bengoa added, "On the second German mission, the D2 mission, we had an on-pad abort, so the decision was made to leave the Orbiter on the pad and do an engine R&R [remove and replace]. Of course, that took a lot of time, so the mission managers decided that there were multiple pieces of hardware inside the module that would have to be removed and replaced because of limited life, requiring MVAK operations." Level-IV had never trained to do MVAK on the D2 mission, so the existing team from the Spacelab-J mission had to turn around and do a fast-paced training to do the MVAK on D2. Again, Alex was called up to be middeck Task Leader with the team, who successfully removed the hardware from the module and replaced it with fresh hardware. Alex Bengoa recalled, "That was one of those quick [turnarounds]. Our team was able to completely come together again, we were all experienced already, and we were able to support the mission at a really fast pace. So that was really good. I remember working with Dan Schultz, who was one of the main guys. We would sit down over at the LCC and do all the control or the operation from the LCC. Everybody was on headsets. So I had my 15 minutes of fame."

Spacelab D2 MVAK. [NASA/KSC.]

Roland Schlierf remembered, "All of a sudden, Damon Nelson was our integration engineer and he said, 'We're going to do an MVAK'. We never planned for it, we never scheduled it, we never wrote any procedures for it. We were violating all of these requirements that they never gave us. So Dan Schultz and I stayed, and Damon and all of us and Jim Pope wrote all the requirements, with the Germans, for MVAK. Dan Schultz and I start talking about the long nights, night after night, writing the MVAK procedure. And we did it all in a matter of a few weeks; from nothing to executing an MVAK in a few weeks."

Hermann Kurscheid, "I really want to mention for Spacelab D2 the extraordinary work and jobs the Level-IV team did for the refurbishment after the launch scrub in March. And to get the payload again ready for flight and the actual samples on board and all that, and they went into the module while in the vertical position. Samples, batteries and all that, they got them again to flight status. It was such a tricky job to get it all done, and for the success of the mission it had to be perfectly done."

Damon Nelson, continued, "In our case, the Germans said, 'Not only do we want to swap out samples, but we want to do a couple of maintenance runs on some of the hardware that has pumps that need to be cycled'. So it was a remarkable operation, and the Spacelab guys were again able to get all the platforms installed first time. We actually didn't tether the technicians on lines. We elected to just put this platform in, and that way they would be more free to do these maintenance operations. It just gave a lot more flexibility. I will never forget this other thing that occurred. We had an experiment called WERKSTOFFLABOR. It was an experiment that had flown on Spacelab-1, and it involved some type of furnace or material processing hardware that utilized space. A vacuums experiment, actually, that was tied to a vent line that would vent to space on orbit. Our technicians went into the module and they reported that when they cranked this experiment up there was a leak in the vent system. I was working with Luis Delgado and we had the Spacelab activated all day. This took a lot of coordination because we were in the Orbiter and the Orbiter had to be powered for us to have power. After a couple hours, Luis reported back and said, 'Well, we've troubleshot two of the interface joints on the experiment and they are both good. We've got one more GSE interface to test and Damon, if this interface isn't leaking, we've got a leak on the flight side'. Sure enough, two more hours passed and he called back and said, 'All of our Ground Support Equipment joints are tight. Now we've got a leak on the flight side'. We found out that the vacuum valve on the flight side was leaking. We did some research, and we found out that during one of the crew training exercises called CEIT [Crew Equipment Interface Test], a crew member had cycled the valve and apparently had not put it back into the proper closed position. Bottom line, if we hadn't had the main engine shutdown and we hadn't have gone into the module, we would very likely have had the crew report that we had evacuated the Spacelab after we got on orbit. The whole module would have been in a vacuum. So, lots of lessons learned for that experience, but we were so lucky."

German astronaut Hans Schlegel floats "upside down" inside the Spacelab D2 science module during Shuttle Mission STS-55. In the background, astronaut Bernard Harris monitors one of the experiments. [NASA/KSC.]

Finally, after all the unplanned work was completed and all the problems solved, the mission launched and flew successfully. But *Columbia* was not going to land where planned, so the next problem was how to support landing operations.

The prime landing crew was in Florida, as was all the GSE and capabilities, but about four hours prior to landing, Mission Control decided that the weather was not going to be good in Florida and switched the landing to California. As Damon Nelson explained, "I was here in Florida, so I immediately got hold of our launch site support manager, Bob Sturm, and said, 'Well, Bob guess what, they're going to land in California and all of our equipment is in Florida. All of our personnel are in Florida. We've got to get this to California and we've got to do it ASAP'. So Bob got hold of the Mission Manager and I contacted my counterparts. They were pretty stunned, but the wheels got rolling and before we knew it, NASA had lined up a charter aircraft, so everybody and all their equipment and everything got loaded onto this charter aircraft. The charter aircraft was at the skid strip and when they cranked up the engines, smoke just went everywhere and the Germans were freaked out. The Germans refused to get on the aircraft unless Bob got on it. They had to get the samples off within about eight hours post-landing, so they had a little bit of time. I know they didn't get out there before it landed, but they got out there in time to get all their samples retrieved."

Roland Schlierf explained, "My mom escaped from communist Germany as a teenager. She left just a few months before the [Berlin] Wall went up. So the cool thing about D2 was that we got given all these flags and stuff for postflight because we did a good job. I got to give a flag to my mom and dad. So my mom's got the German Flag that flew on D2."

Hermann Kurscheid observed, "Spacelab-1 was a really excellent training for me on the job of D2, and on D1 I learned quite a lot. I think I also got an understanding of KSC, how KSC worked, what the procedures were, responsibility, access requirements and how things were handled there, transportation and all that. I was then also responsible for the D2, for ground operations activities at KSC. That was the best training I could get to do my job on the D2 mission."

In the NASA Payload Flight Assignments issued between 1987 and 1992, Spacelab D3 was listed as flying a Long Module with a Unique Support Structure and was scheduled for launch around 1994, but was later removed from the manifest. There were other, unrealized plans to fly a third German Spacelab mission, designated D4, which was envisaged as a pallet-only mission called the German Infrared Radiation Laboratory (GIRL). Had this mission flown, one plan indicated that it would not have used the Long Module, but utilized the Igloo and up to four pallets.

SPACELAB-MIR

The *first and only* Spacelab space station docking mission

Between the authorization of Space Station *Freedom* in 1984 and the early 1990s, the American Space Station program came under increasing pressure from budget restrictions and increasing complexity in the designs of the proposed station. At the same time, the decline and eventual fall of the Soviet Union offered the opportunity to bring Russia into the Space Station plans as a full partner. Ironically, there had been plans for a Space Shuttle to dock with a Salyut space station in the 1970s but this never progressed further than the drawing board and joint meetings.

Combining the Shuttle program, with its experience in space rendezvous, EVA and handling large payloads using the RMS, together with the Russians and their experience and expertise of multiple long-duration spaceflights, was ideal and logical. There would be huge difficulties, disagreements and doubts, but 30 years later and with the benefit of hindsight, it has proven to have been the right choice, with over 20 years of continual operations on the ISS using both American and Russian spacecraft to ferry crews to and from the station. The ISS program had been divided into three phases. Phase 1 featured plans for seven (with provisions for up to ten) American Shuttle missions to dock with the Russian Mir space station, as revealed in December 1994.[2] This would provide the Americans with much needed experience in space station operations and docking with large structures in space, something NASA had not attempted since Skylab in 1973.

The very first Shuttle docking mission would carry the Spacelab Long Module so that the first data in over 20 years could be gathered from an American astronaut towards the end of an extended stay in space. Unfortunately, this was the only Spacelab Long Module mission to dock with a space station – American or Russian [1]

STS-71, *Atlantis*, launched June 27, 1995, 9+ days
Payload: Spacelab-Mir (SL-M); Long Module #2 (FOP MD002); Floor
(CD MD001); Tunnel (MD002); Rack (see Appendix 2).

STS-71 became the first US Space Shuttle-Russian Space Station Mir docking mission, with joint on-orbit operations. In doing so, it created largest spacecraft combination to date and featured the first on-orbit changeout of a Shuttle crew. About 100 hours in total of joint US-Russian operations were conducted, including biomedical investigations and transfer of equipment to and from Mir. Fifteen separate biomedical and scientific investigations were conducted, using the Spacelab module installed in the aft portion of *Atlantis's* Payload Bay. They covered seven different disciplines: cardiovascular and pulmonary functions; human metabolism; neuroscience; hygiene, sanitation and radiation; behavioral performance and biology; fundamental biology; and microgravity research.

Juan Calaro said that the Spacelab docking with the Mir was pretty exciting. "We did have an ergometer and Lower Body Negative Pressure device, called LBNP, for Level-IV testing that was on the easier side. It wasn't like a big Spacelab mission with tons of experiments, this mission only had a handful of experiments. The big thing was we were docking with Mir. There was a problem with ergometer or something when they were already integrated into the Orbiter, and I remember suiting up and doing some checkout in the OPF. Joel Lacquer suggested there could be a problem, and I remember arguing with him that it wasn't a problem. It turned out he was right so I had to eat some dirt on that one."

There were an eventual nine dockings of the American Space Shuttle with the Russian Mir space station under Phase 1 of the ISS program. The second mission, STS-74, delivered the Russian-built Docking Module that, when attached to the Kristall Module, gave better clearance for future Shuttle dockings and therefore

---

[2] Phases 2 and 3 focused upon the assembly of the ISS and are not part of this current work.

did not require additional Spacelab tunnel equipment. However, each of the remaining seven docking missions (STS-76 S/MM-03; STS-79 S/MM-04; STS-81 S/MM-05; STS-84 S/MM-06; STS-86 S/MM-07; STS-89 S/MM-08 and STS-91 S/MM-09) carried either a single or double Spacehab module, which STS-74 could not carry in addition to the Russian Docking Module. Those remaining Shuttle docking missions therefore required a Spacelab tunnel configuration to gain access to the Spacehab module located in the Payload Bay, or the Mir station. As a result, all seven docking missions after STS-74 carried the Spacelab MD002 tunnel assembly previously flown on Spacelab-J. [2]

THE LONGEST SPACELAB MISSION

The Life and Microgravity Spacelab mission was created to perform research through experiments in a stable low-gravity environment, with emphasis on life and microgravity sciences utilizing the pressurized Long Module. The mission combined and continued previous investigations conducted on the Spacelab Life Sciences and USML missions.

STS-78, *Columbia*, Launched June 20, 1996, 16+ days, Payload: Life and Microgravity Spacelab (LMS); Long Module #2 (FOP MD002); Floor (FOP MD002); Tunnel (MD001); Rack (See Appendix 2); plus EDO kit.

The LMS mission featured a Long Module with 41 microgravity experiments, involving fish embryos, rats, Bonsai plants, fluid dynamics, metallurgy, protein crystal growth, etc. Five space agencies (NASA/USA; ESA/Europe; French Space Agency/France; Canadian Space Agency/Canada; and Italian Space Agency/Italy) and research scientists from ten countries worked together on the primary payload of STS-78. More than 40 experiments flown were grouped into two areas: life sciences, which included human physiology and space biology; and microgravity science, which included basic fluid physics investigations, advanced semiconductor and metal alloy materials processing, and medical research in protein crystal growth. LMS investigations were conducted via the most extensive telescience to date. Investigators were situated at four remote European and four remote US locations, similar to what would happen with the ISS. The mission also made extensive use of video imaging to help crew members perform inflight maintenance procedures on the experiment hardware. The most extensive studies ever were conducted on bone and muscle loss in space. STS-78 marked the first time researchers collected muscle tissue biopsy samples both before and after flight. Crew members also were scheduled to undergo other life sciences investigations, including the first ever comprehensive study of sleep cycles, 24-hour circadian rhythms and task performance in microgravity. The microgravity science investigations included the Advanced Gradient Heating Facility, in which samples of pure aluminum containing zirconia particles were solidified. The Advanced Protein Crystallization Facility was the first ever designed to use three methods for growing protein crystals. Electrohydrodynamics of Liquid Bridges focused on changes that occur in a fluid bridge suspended between two electrodes. This research could find applications in industrial processes where control of a liquid column or spray is used, including in ink-jet printing. The crew performed in-flight fixes of problem hardware on the Bubble, Drop and Particle Unit (BDPU), designed to study fluid physics.

Kevin Zari explained that he was the prime on this mission's Microgravity Measurement Assembly (MMA). "This was really what I would consider the height of my first real experiment. I was prime on it. I had backups that were working for me. I remember the first time I met my Payload Developer, coming into the offline labs. I was all excited. I set my alarm and everything and showed up, and I was waiting in the lab feeling like a parent expecting a baby. They came in and the Developer's name was Andreas Schütte. He was an amazing, amazing person.

"Another benefit was that I got to travel to Bremen, Germany a couple of times," Kevin continued. "I was still only 25 years old or something at this point, still wet behind the ears. One of the payload customers, from ESA, was expecting their first child. I decided there had to be a way to have him participate from Europe. I didn't care about IT Security, and none of these things mattered in those days, so I actually took an Sgi O2 computer and hooked it up in such a way − and had some software loaded on it – so that when the customer went to my web server and signed in with his credentials, it actually took our OIS channel and streamed it to him. So for the one person who I knew couldn't attend the final test, I pushed the boundaries. He was actually able to remotely listen in but couldn't talk. Masimo Bendeka was the name of the PI there. That's also when I got my first visit from IT security.

"MMA had one of the experiments, one of the accelerometers on it, called the quasi steady accelerometer, and it was really looking for almost DC-level components. Really ultra-slow oscillation, right? It was designed for one hertz or less. One cool thing was that as the Shuttle was orbiting, as it passed the side of the Earth where the Sun had heated up the atmosphere and that atmosphere was less dense than the other side, there was the thought that the Shuttle actually experienced an acceleration because there was less resistance. There were less molecules and that it would go faster. It actually proved the theory and it was beautiful."

The crew also asked the payload people for Kevin to support their team for 24 hours in a row. They were doing tests growing protein crystals and wanted to understand what the environment was, such as if a crew member bumped a rack or similar that caused an acceleration while making the samples, and might force them to redo the experiment. Kevin Zari added, "The MMA experiment was actually more like a utility that other experimenters used. Watching the console, looking at the telemetry, interpreting the data, working together with other payload teams to explain what we saw happening on board."

Sharolee Huet said, "I got to meet an astronaut that was from the area near my home town. Our fathers had worked together but I had never met him. After his first mission, thanks to a VITT [Vehicle Integration Test Team] friend of mine, I got to meet him. That was the start of a good friendship. He worked a couple Spacelab missions later and I got to work with him. During the LMS mission, the crew came to KSC to train on the hardware, since the mission was on a fast track. It was my last mission and definitely the highlight. There was no way to top that for me. I got to see him on landings on the runway. To me, those were the hardest launches and landings to watch."

MICROGRAVITY SCIENCE LABORATORY

According to the STS-83 Press Kit, the Microgravity Science Laboratory (MSL) mission was aimed at "NASA's continuous effort to understand the subtle and complex phenomena associated with the influence of gravity in many aspects of our daily lives." The mission was also to serve as a bridge to "America's future in space", spanning the relatively short missions of Mercury, Gemini and Apollo and the experience created by Skylab, Space Shuttle and Spacelab to the planned ongoing research on ISS. Indeed, MSL would test some of the hardware, facilities and procedures intended to be used on ISS.

Two for One

Only one MSL module mission was planned, but the initial flight of the MSL was cut short due to concerns about one of the three fuel cells (#2), marking only the third time in the Shuttle program's history that a mission had ended early. The STS-83 MSL mission landed after just 3 days 23 hours. It was therefore decided to re-fly the same payload, crew and Orbiter again quickly, to maximize the investment in the mission. The next available mission number was STS-94, which flew just three months after STS-83.

> **STS-83, *Columbia*, launched April 4, 1997, 3+ days.**
> **Payload: Microgravity Science Laboratory (MSL)-1; Long Module #1 (CD MD001); Floor (FOP II MD003); Tunnel (MD001); Rack (See Appendix 2); plus EDO kit.**

MSL was a collection of microgravity experiments housed inside Spacelab. Despite the early return, the crew was able to conduct some science in the MSL-1 Spacelab module. Work was performed in the German electromagnetic levitation furnace facility (TEMPUS) on an experiment called Thermophysical Properties of Undercooled Metallic Melts. This experiment studied the amount of undercooling that could be achieved before solidification occurred. Another experiment performed was the Liquid-Phase Sintering II experiment in the Large Isothermal Furnace. This investigation used heat and pressure to test theories about how the liquefied component bonded with the solid particles of a mixture without reaching the melting point of the new alloy combination. Also conducted were two fire-related experiments. The Laminar Soot Processes experiment allowed scientists to observe the concentration and structure of soot from a fire burning in microgravity for the first time. The Structure of Flame Balls at Low Lewis-number experiment completed two runs. This experiment was designed to determine under what conditions a stable flame ball could exist, and if heat loss was responsible in some way for the stabilization of the flame ball during burning.

Tracy Gill commented that, "The payload included a significant quantity of GSE cables used for KSC test purposes, and a design concept on STS-83 was to use T-0 capability for payload maintenance activity that precluded a labor- and resource-intensive Spacelab power up at the pad. More commonly worked issues were problem resolutions for problems uncovered during payload testing at KSC."

**STS-94, *Columbia*, Launched July 1, 1997, 15+ days,
Payload: Microgravity Science Laboratory-1 Re-flight (MSL-1R); Long Module #1 (CD MD001 – *Final Flight*); Floor (FOP II MD003); Tunnel (MD001); Rack (See Appendix 2); plus EDO kit.**

STS-94 was the re-flight of the primary payload of STS-83, the MSL collection of microgravity experiments housed inside a European Spacelab. It built upon the cooperative and scientific foundation of the International Microgravity Laboratory missions IML-1 and IML-2, the US Microgravity Laboratory missions USML-1 and USML-2, the Japanese Spacelab mission, Spacelab-J, the Spacelab Life and Microgravity Science Mission, LMS, and the German Spacelab missions, D1 and D2. MSL featured 19 materials science investigations in four major facilities. These facilities were the Large Isothermal Furnace, the EXpedite the PRocessing of Experiments to the Space Station (EXPRESS) Rack, the Electromagnetic Containerless Processing Facility (TEMPUS) and the Coarsening in Solid-Liquid Mixtures (CSLM) facility. Additional technology experiments were also performed in the Middeck Glovebox (MGBX) developed by MSFC, while the High-Packed Digital Television (HI-PAC DTV) system was used to provide multichannel real-time analog science video.

Kevin Zari recalled that, "STS-94, the re-flight of MSL-1, [resulted in] the quickest TAP [Test and Assembly Procedure] ever. I think it was literally less than five minutes. We had the payload activated, we had the technicians on the ground doing the checkout and then we met all of the OMRS [Operations and Maintenance Requirements Specifications] requirements. We issued some commands and then we turned off the payload. It was amazing. That's how streamlined and how efficient we ran as a payload team together, the Payload Developers, the KSC personnel and the telemetry folks and everything."

Sharolee Huet added, "After I left Level-IV, I did get asked to de-integrate an MSL rack since it was one designed for Space Station. I gladly went back to de-integrate the Express and the SMIDEX racks."

THE LAST MODULE MISSION

Originally designated Spacelab Life Sciences-4 (SLS-4), this was remanifested as Neurolab to mark the "Decade of the Brain" and became the final mission of the Spacelab Long Module configuration. The program began with the formal proclamation issued by US President George H. W. Bush in July 1990, which proposed to dedicate the decade of the 1990s as the "Decade of the Brain." In 1991, NASA proposed a Neurolab mission as its contribution to this mandate. What emerged was a single mission of 16 days, flown by a seven-person crew and using the Spacelab Long Module for what turned out to be the final time. In the 2003 NASA book on the Neurolab mission, it stated that the "resulting experiments fully demonstrated the capabilities of the Spacelab Program." [3] Unfortunately, despite the glowing recommendation, Spacelab had already been grounded never to fly again, its promising capabilities abandoned.

> **STS-90, *Columbia*, Launched April 17, 1998, 16 days,**
> **Payload: Neurolab; Long Module #2 (FOP MD002 – *Final flight*); Floor**
> **(FOP MD002); Tunnel (MD001); Rack (See Appendix 2); plus EDO kit.**

The primary mission of STS-90 was to conduct a comprehensive list of neurobiological experiments and observations on a number of species: seven humans, 18 pregnant mice, 152 rats (including 12 females with prenatal litters of eight each, and two with litters of seven each), 229 swordtail fish, 60 snails, 75 snail spawn packs, 824 crickets, and 680 cricket eggs. According to a PI (for rat research), "the findings from the microgravity experiments may help gain some more insight into the best way to treat neurological patients with Parkinson's disease, and balance disorders."

"I was preparing a message to begin recording what happened during the last Spacelab Level-IV Mission Sequence Test of all time, but a funny thing happened," recalled Tracy Gill, "When I heard that the payload was accidentally powered off, we got the whole complement of payloads into their maximum power configuration and then took our measurements. As the test leads for measuring voltage were being rearranged for another measurement on our power control box by the payload, the main 28V DC power was accidentally interrupted for about two seconds by an inadvertent switch throw. One of the astronauts said over his headset, 'Oops, we lost power. Hey, now it's back'. Realizing that all the computers were probably already in the middle of rebooting, I knew we had lost a lot of setup time, muttered a few inaudible curse words to the gods of spaceflight hardware, and then began polling the test team to assess what we needed to do to recover. All the RAUs skipped all channels and several pieces of experiment hardware reset. We took out a Problem Report to document this. I polled the test team to see if we had any anomalous conditions that needed to be addressed right away. We decided not to cut the activities of the [test] slice short since we had met our main objective, and after recovering from the problem, we didn't lose much time in the test day. No further anomalies resulted from this event and I had already dispositioned this PR as an explained condition due to human error.

"Around noon, all the analog values from the Pump Package went to zero. First we troubleshot at the FLMCP I/F [Fluid Loop Monitor and Control Panel Interface] to the RAU and then we also checked those analogs at the FLMCP to Pump Package I/F with the help of Spacelab DPA [Data Processing Assembly] engineers. Both I/Fs showed zeroes, which exonerated the RAU since that was the input into the RAU. At this point we were almost ready to make the decision to pull the Flight Pump Package from Rack 4, until we decided that it would be wise to measure the voltages coming into the Pump Controller and the FLMCP, since it would be easier to remove and replace the FLMCP if that had been the problem. Fortunately, Alex Bengoa had suggested that these were the same symptoms he'd seen when the FLMCP was off, so after asking management for an extra hour beyond 8:30, and with the help of the Networks group, we measured AC voltages at the Pump Package I/F and they were all zero. We went further back in the loop and measured the AC voltages at the EPSP [Experiment Power Switching Panel] to FLMCP I/F, and again we got all zeroes. At this time, we suspected a GSE power problem and sure enough, the 400 HZ GSE power supply was off. Apparently, a safety door in the supply was partially opened, which automatically shut down the power supply. [A technician had opened the door of the AC power supply because it was loud, which killed all AC power on the lab.] The problem was corrected and we verified the Fluid Loop powered up nominally, both pumps. I have to thank Robert Wark for actively participating with that entire troubleshooting effort. He had a long week fighting Fluid Loop fires, so to speak!"

Darren Beyer explained that, "My last one was Neurolab STS-90, and this one was sad for me for a number of reasons. It was the last time I was on board any

Orbiter. I had already left Level-IV when I got on board the Orbiter. I was working with Jim Dumoulin, and we did panoramas from inside the Space Shuttle. We did one in the middeck and one in the crew deck or flight deck. We had one in the tunnel, we did one in the Changeout Room right outside the hatch, and we did one in the module. It was really cool. But what was really sad about it was that it was on *Columbia*. That was the last Orbiter I was on, and it hit me a lot harder when *Columbia* didn't make it home [STS-107 in February 2003]."

Roland Schlierf, added, "This was back when they made the announcement that KSC was going to go from 2,400 people, civil servants, to 1,200 in a year or two. Everybody laughed. Well, they wanted us all to leave, that's for sure. These were 'progress projections'. I remember going on a trip to Houston because I was thinking that if I was going to keep working for NASA, having looked at every center and what they were forecasting and projecting, then if I was going to go anywhere, the only place I'd want to go would be Houston. I flew out there and thought about it and decided, 'Nope, I'm not leaving Florida. There's no way'. So thank God I never had to. I think the lowest [staffing] we got to was 1,700,"

END OF AN ERA

And that was it, the last Spacelab Module mission. It was a very sad day for many people at KSC, around the US and across the world. Many would ask 'What will happen now?' It was, once again, an uncertain time for many workers at KSC, none more so than the experienced and dedicated Level-IV team.

References

1. Shayler, David J., **Linking the Space Shuttle and Space Stations. Early Docking Technologies from Concept to Implementation**, Springer-Praxis, 2017.
2. For further mission detail on each Shuttle Mir docking mission, see Furniss, Tim and Shayler, David J., with Shayler, Michael D., **Praxis Manned Spaceflight Log 1961–2006**, Springer-Praxis, 2007.
3. Buckley Jr., Jay C., MD, and Homick, Jerry L. PhD. (Eds), **The Neurolab Spacelab Mission: Neuroscience Research in Space**, NASA SP-2003-535, 2003.

12

Spacelab Says Goodbye

"I [worked] on the last Spacelab
laboratory mission, which was Neurolab.
I stayed till the end to lock up the doors…
it was kind of bittersweet. You know it's going away."
Tracy Gill, Level-IV Electrical Engineer

Looking back over 20 years, it seemed to be an untimely end to the Spacelab program in order to make way for Space Station. There remained untapped potential in the Spacelab system, its hardware and capabilities, but there was already competition in the introduction of the commercially developed Spacehab middeck augmentation module, closely followed by a significant increase in the drive towards a larger, more permanent international (rather than national) space station program. Despite this the Spacelab series survived for a while, but the writing was on the wall as many of the planned Spacelab pallet and module missions were remanifested or cancelled.

LOST OPPORTUNITIES

After the last flight of a Spacelab Long Module on STS-99 in 1998, just 15 years after the first had flown on STS-9, the percentage of 'Spacelab' missions – both module and pallet-only – in the Space Shuttle program was 22 out of the 100 missions (over one fifth) flown by October of 2000. By then, assembly of the International Space Station (ISS) had taken over most of the remaining Shuttle missions. In fact, of the 35 missions flown from December 2000 to July 2011, only

© Springer Nature Switzerland AG 2022
M. E. Haddad, D. J. Shayler, *Spacelab Payloads*, Springer Praxis Books,
https://doi.org/10.1007/978-3-030-86775-1_12

three were not directly related to ISS assembly and resupply. In the new millennium, the final pallet flights may have supported the ISS assembly era, but Spacelab had long since given way to the Space Station.

In addition to the missions explored in the previous chapters, there were also numerous proposed, planned and potentially brand new Spacelab missions that never flew. That was a shame, as Level-IV would have ensured their efforts would have made those Spacelab missions as successful as those which did fly. Had they flown, the additional opportunities to provide greater understanding and experience would have helped ISS as it was assembled, and afterwards as it became a ground-breaking research facility.

Some of these unflown Spacelab missions were:

- *Space Plasma Laboratory (SPL)*. SPL was initially identified as Spacelab-6 and would have explored the ionized atmosphere of Earth. At least two missions were planned in early manifests, both without the pressurized module. The first would have carried the Igloo and one pallet, while the second would have utilized the Igloo and two pallets. These missions would have re-flown experiments from earlier Spacelabs, together with a new experiment using a pair of extremely long whip antennas to transmit very low frequency radio waves into the magnetosphere.
- *Shuttle High Energy Astrophysics Laboratory (SHEAL)*. Three SHEAL missions were planned to study astronomical objects, obtaining images, spectra and timing data on celestial X-ray radiation sources using at least four instruments.
- *Shuttle Infrared Telescope Facility (SIRTF)*. SIRTF was a large facility to support experiments to increase our understanding of the formation and evolution of stars, planets, galaxies, and unusual galactic objects. The one-meter telescope's cryogenically cooled instrument was identified in 1979 as a leading astrophysical observatory for the Shuttle-Spacelab pallet system. Several annual flights were planned for the 1990s, but experiences from Spacelab-2 revealed that the Shuttle environment was not suited to onboard infrared telescopes due to the "dirty" outgassing from the Orbiter. Plans were changed to make the facility a free-flyer, but the *Challenger* accident meant that the Centaur upper stage required to place the facility in a high orbit was eliminated from Shuttle use. That meant launch had to be on an Expendable Launch Vehicle (ELV) and it was designated the Space Infra-Red Telescope Facility (to retain the SIRTF acronym). The only one of NASA's four Great Observatories not to be launched on the Space Shuttle, it was delayed by redesigns and budget constraints during the 1990s, before finally being launched in 2003 on an ELV as the Spitzer Space Telescope. It subsequently operated successfully until its retirement in 2020.
- *Solar Optical Telescope (SOT)*. SOT was intended to fly a one-meter telescope on two separate missions utilizing the Instrument Pointing System (IPS), the Igloo and two pallets to investigate a range of solar phenomena.

This research would have focused upon solar plasma heating and the redistribution of energy. It was also planned to obtain very high spatial resolution observations of the Sun. These missions were cancelled and redesignated SunLab.

- *SunLab*. A series of at least three missions with the Igloo and one pallet was designed to study small-scale structures on the Sun's surface and to measure the coronial helium abundance. The Spacelab-2 instruments would have been flown on the first mission, and potential Payload Specialists (PS) were chosen from the back up PS on Spacelab-2. The loss of *Challenger* and the recovery from that tragedy contributed to the cancellation of these missions in favor of other objectives.
- *Dark Sky*. Dark Sky would have conducted a sky survey for extended infrared sources, X-ray imaging of galaxy clusters, and cosmic ray measurements, using the Igloo and two pallets.
- *Star Lab:* Star Lab was envisaged as a seven-day classified Spacelab Long Module and pallet mission, but was cancelled in the wake of *Challenger*. Originally planned for the early 1990s, its principal objectives were to demonstrate acquisition, tracking and pointing technologies relevant to the Space Defense Initiative (SDI), using a variety of onboard sensors and lasers to monitor ground-launched missiles and other targets. The program would also have included a variety of defense-related experiments, such as corrective optical aberrations, atmospheric effects on lasers, assessing submarine laser communications, and observing weather phenomena.

These were just some of the lost missions of Spacelab which Level-IV could have been involved with, mostly during the 1990s. But could Level-IV have handled such an extensive manifest had many of these payloads been retained? Those in Level-IV would have said "Yes." They would do what was required to make the launch, with the understanding that if outside influences changed (hardware arriving late, landing in California instead of Florida, etc.), it could have easily caused a launch slip. Having said that, Mike Haddad added that had the Spacelab program continued, the mindset in Level-IV was "We would have done what was necessary and safe to try and make the launch date." Unfortunately, the team was never afforded the opportunity to demonstrate this.

SPACELAB'S DEMISE

"It worked for a long time and I'm just really sorry to see it go," Bill Jewel reflected. As Spacelab was coming to an end, most of the Level-IV personnel had taken on dual roles, of finishing the remaining Spacelab missions or flights using Spacelab hardware, and beginning Space Station planning and operations. Mike Haddad explained the early transition towards ISS: "The first Space Station meeting I remember going to was led by Bill Haynes, in late 1985 or so, and at that time it was named Space Station *Freedom*. He was trying to take some of the lessons we had learned from Spacelab and incorporate them into Space Station planning, one of which was no 'Ship-n-Shoot'. Billy Haynes did not like the 'Ship-n-Shoot' concept at all and was actually preaching the need for integrated testing of the Space Station elements early on, something I and many from Level-IV totally agreed with. Years later [March 1991], we had begun construction of the Space Station Processing Facility [SSPF] and I was in a Space Station meeting when a person from one of the other space centers, not KSC [Kennedy Space Center], got up and was adamantly opposed to it, and literally laughed at us. 'Why are you building the SSPF? We won't need it, it's a waste of time and resources. The Space Station elements will come to KSC and go straight to the launch pad'. I could not help myself and I asked him why he was saying that. He stated, 'Everything is being built under NASA direction, so it will all work together'. There were some very talented NASA people working Space Station, but that was really no guarantee the hardware and software would work together. Well, I needed to really hold back and not tell that person what I really thought about his comment, but my response was, 'We are building the SSPF to ensure the Space Station elements will work together before launch and best ensure there are no problems on-orbit'. SSPF opened in June 1994, and Billy's concept of integrated testing would eventually become a reality when we performed the first ISS Multi-Element Integrated Test beginning in January 1999."

Long Module #1 being prepared by Level-IV personnel before being sent to the Smithsonian Air and Space Museum in Washington, D.C., c. 1998. [Jim Dumoulin.]

While NASA was preparing for humans to live in space for extended periods of time around the Earth, and then eventually deep space travel, Spacelab was probably the real beginning, the stepping stone for Space Station, as Mark Ruether described. "It was those experiments that we did during Spacelab that allowed us to better understand what we could and should do with the latest platform, the International Space Station, so that we could do things even longer term than that. That was a huge step."

"I helped them develop some design guidelines at that program, and they pretty much accepted them," said Tony Ornelas. "I think a lot of the things that I had success with at the program was a direct reflection of the fact that I gained that experience through Level-IV. The scheduling of the tasks, the testing, making sure it works right before you go too far into the procedure. The 'Ship-n-Shoot' philosophy was something that the program just pretty much wanted to do, but a lot of people back at KSC and everyone else said, 'You guys got to have some limited testing. You can't just throw it all up there'. Some of those things were derivatives of Level-IV. That's not necessarily from my perspective, but from a problematic standpoint."

As the SSPF was being set up and the Ground Support Equipment (GSE) started coming in, teams were established to perform certain tasks. One of them was the Activation/Validation Team. Bruce Birch recalled, "They would do activation and validation of the facility equipment, like the vacuum systems, the heating and cooling, making sure it was all working. A lot of the compressed air, the compressed gases, and a lot of the electrical for the control rooms and stuff to make sure of all the cables. They were working that while we were finishing the last few Spacelab missions."

Large pieces of Space Station were being assembled at many locations around the US, and as time went on some of the upper management of the Space Station program felt it was not progressing as fast as they would like. It was decided that most of that hardware would be shipped to KSC before it was fully completed, thus moving the schedule forward. This had not been planned beforehand and KSC had already committed the new SSFF to other Space Station elements, mostly the pressurized modules. The issue became where these items being shipped would be processed and prepared. With the Spacelab program winding down, the O&C (Operations and Checkout Building) seemed to be the perfect place. Bruce Birch continued: "They called a 'travel work'. A lot of the truss work that was being done by McDonnell Douglas out in California, they brought it to KSC, so we ended up putting the trusses up in some of the O&C stands. We modified the stands to put the trusses up and ended up putting a fence up inside [the O&C] High Bay. This was the demarcation line. 'This [area] is Station', but it was considered as factory work so they didn't have to follow all the KSC rules. They used the rules that the company had. So we had Spacelab on one side [of the fence] finishing up its cleaning, and the other side you had Space Station being worked before it was turned over to the government. This was not the SSPF, we split the O&C into two parts to do this work. The SSPF was not leased out yet, but there were commitments from commercial organizations to use it. Two sets of rules, two different companies. The forklifts and other items, going back and forth. So as a senior manager, I had managers in both groups. I was trying to keep them straight and keep them from fighting with each other. All that knowledge and experience transferred right over. It was part of the PGOC [Payload Ground Operations Contract]."

Many of the systems used on Spacelab would now have a counterpart on Space Station: structures, mechanisms, fluids, power, data and communications. With Level-IV having the experience with both the flight and ground side of these systems, this seemed to be the right group to start working Space Station.

"All that ground comm work that I did with Craig [Jacobson]," said Roland Schlierf, "all that technical expertise transferred over to the C&T [Communications and Tracking] system. They had high rate frame mucks [Multiplexer] and they had very similar kinds of comm systems and lines running throughout the building. Craig and I walked down the SSPF before the walls went up; we were walking on the girders and the interior walls were up. I remember walking across from one part to the other. There were these big six-inch gaps in between the parts of the building and in all the cable trays, and Craig and I basically checked all the comm lines that were going in."

Bob Ruiz explained that during the PDR (Preliminary Design Review) phase of Space Station and development of the station racks, there were several issues related to the interfaces with the experiments; some design flaws with the ISS racks. An expert on Spacelab racks, Bob commented, "I tried to simplify it for the payloads. In other words, make the interface simpler for the payloads and also have a capability to take parts of the rack apart. I think there was a center section so that if you went to a double rack, it was easier to do that configuration. I used some of the lessons learned from Spacelab for the Space Station racks. We also submitted comments to improve the rack removal and installation. There were some issues with our captured load, getting racks in and out of ISS, so we implemented some designs so that you wouldn't have to have a captured load when you were trying to lift a rack out [Captured load is where a device has to be removed before the load can be moved or manipulated]. So we put in some capability to improve the rack replacement. That was a very little known fact, one of those things that nobody ever knows about it. So there were lessons learned that were put into ISS. Definitely.

"The Orbital Replacement Units [ORU] were components like batteries that would need to be changed out at times over the life of ISS. The issue was how to build them so they could be replaced on-orbit. This was where a Level-IV person could help, someone who had spent their whole career working installation and removal of similar flight hardware components. I did look at the requirements and basically fell back on my Level-IV experience of putting stuff together and taking it apart, and the requirements that we should put into the design requirements for station."

When Space Station was still named *Freedom*, a number of Level-IV personnel helped out with planning, special tasks, etc. Still working Spacelab missions full time, they could only support *Freedom* between missions.

SUPPORTING ISS

Following the final Spacelab-type missions, five Shuttle Assembly Missions carried Spacelab pallets to support hardware to be installed on the ISS [1]. Though these were non-Spacelab missions, the experiments were worked on by Level-IV/ Experiment Integration, or the missions used Spacelab hardware. These missions and their payloads were:

- STS-92: October 2000, *Discovery*, 1 x Pallet (F005), PMA-3.
- STS-100: April 2001, *Endeavour*, 1 x Pallet (F004), Canadarm 2.
- STS-104: July 2001, *Atlantis*, 2 x Pallets (F010 Forward; F002, Aft), Four High Pressure Gas Assembly containers.
- STS-108: December 2001, *Endeavour*, Lightweight Mission Peculiar Support Structure Carrier (LMC#2) and MPESS Hitchhiker bridge structure containing various experiments and hardware.
- STS-123: March 2008, *Endeavour*, 1 x Pallet (ID unknown), Canadian robotic arm Dextre.

Bruce Morris explained, "We got calls to say, 'We've got this task going on where we need to figure out how to plan to do these kinds of things, would you review this for us?' If we had time, we'd do that kind of stuff. It helped you to get known among some of the more senior leadership at Kennedy, because we were very low down in the organization. All of a sudden, these gray beards from NASA were calling up asking us to review stuff. Telling them about our experiences all helped our careers grow, because we got exposed to the things that we didn't expect to ever see in the planning stages of a major program like Space Station. Those experiences in Level-IV all prepped me for moving over to Space Station. It was just very cool. We went through the redesign at the international level, and actually got to help out on some of the negotiations. We got called by the people in Washington to send data and things like that, to help out with their negotiations and provide information to the people that were trying to figure out how to make the International Space Station come alive. One of the big things about trying to get the Space Station to operate was that we were going to have to cut costs, because there was so much overrun on it. It was not only the redesign of the vehicle itself, but also redesigning the way we were going to process the vehicle, because the overhead of all those people working on it was just not going to work [on costs]. We'd got a lot of experiences from Level-IV that we could bring to that and say, 'Okay, well, the way we do this is like this', because we naturally were very lean because we were a young organization. We didn't have three-shift operations and things like that, so we had to develop approaches to doing stuff that were much leaner than the classic engineering way of doing things because we just didn't have the resources to do it. That was great, because we got to come in and help out and try to plan for how Space Station was going to work."

"We [Level-IV] were part of the initial design of the space station in 1984, during 1985, and yet again in 1986," recalled Rey Diaz. "There was a big push to work with the different work packages that were involved with the design and I was part of work package number five with the electrical group, which back then was at Lewis [Research Center]. Here we were, working all the other missions and also working on design for Space Station. That took some time, to review all those drawings and documents, and then [there was the] telecoms and traveling."

Level-IV personnel were getting exposure to a worldwide view of things in the space program, and the science that would be performed on Space Station was more of a direct follow-on from the science performed on Spacelab. Many of the international countries that worked Spacelab would now be working Space Station, marking an extension of the international partnerships that had been formed early in the Space Shuttle program via Spacelab.

Roland Schlierf recalled, "I was going to get to go work with all these different countries. All bringing the whole world together. To be able to be involved in something where you were bringing all these countries together for something really meaningful… it was powerful for me."

It was well known within NASA that the Space Station was going to be very expensive, and the agency began looking for things to cannibalize. Spacelab was a big target because it was basically already doing what Space Station was going to be doing. NASA could not afford to do both at the same time, so the agency began to do studies on how to save money in other areas of the space program to pay for Space Station.

Jim Sudermann explained to the authors that he got put on a study with several other engineers, who were supposed to figure out how to save money. They were told, as part of their brief, to find how to cut ten percent out of Spacelab operations. "I said, 'We could probably do better than that', but that was not the message that they were trying to give us. As it turns out, the message they were trying to give us was *not* to find more than ten percent savings, because they wanted to *cancel* Spacelab. [But] We didn't know that then."

Many people in Level-IV saw the writing on the wall. Spacelab was going to be cancelled so the money could be used for Space Station, and it was simply out with old, in with the new. At this time, PGCC was also starting to take over some of the Level-IV work, as Angel Otero recalled: "We were not hiring any more people and [yet] there was more work coming. If they started giving work to contractors, they were going to end up giving the whole job to the contractors, and I figured I would quit while I was ahead. That's why I left [in June 1989]."

Damon Nelson said, "My Level-IV career ended with Spacelab D2 and it ended pretty abruptly. My lessons learned were that there was definite commonality between the module for Spacelab and the modules that we used with Space Station, and I worked the US Lab and the Airlock. It was cool that I became responsible for the integration of the modules that were going to fly on Shuttle for Space Station. There were other components – truss assemblies and solar arrays – but my responsibility was the modules. But I have to tell you, we had a lot of integration between the payload and the Orbiter that sometimes people didn't acknowledge. And we had a lot of problems. I think that the Space Station guys that we initially worked with did not have any appreciation for that integration function between these two vehicles. Their focus was on developing a Space Station element and they didn't really think about [the fact that] it had to be able to communicate with the Shuttle. When it was initially on orbit, especially for the early modules, the power, the data, all came from the Orbiter, so I think it really helped a lot to have that appreciation of that aspect. I'd had international experience, which proved helpful when I worked with the Germans and Japanese later on in Space Station. So a lot of it was that working relationship, more than anything, that I think I took into Space Station."

Spacelab, like Space Station, had involved many international partners, but the one new partner that had had no involvement in Spacelab was Russia. With Space Station now named the International Space Station (in 1993), the Russians were a major player. A nation that was once the biggest rival was now going to be an ally.

Roland Schlierf told the authors, "I only worked for the Russians once. It was really interesting and it was for Station. I just remember being in a couple of week-long TIMs [Technical Interchange Meetings] with the Russians and Russian translators and all that stuff. It was quite the adventure. I didn't get ever go to Russia though."

Almost a Trip to Russia

ISS needed a lifeboat, some way to get the crew off the Space Station in an emergency; an Assured Crew Rescue Vehicle (ACRV). The Shuttle would only be docked for a couple of weeks at a time, so something permanently attached to Station, 24/7, was required. There were many different concepts of the ACRV, but one idea suggested by the Russians was to use their Soyuz vehicle. The initial physical review determined that two Soyuz crew ferry vehicles could fit in the Payload Bay of the Shuttle. In one Shuttle mission, two ACRVs could be taken to the Station to support the maximum crew of six planned for full-up Space Station operations.

NASA needed to determine how to process the vehicle through the Shuttle system at KSC. Many people were suggested, some from other centers, but it was decided to select a KSC person. Mike Haddad recalled, "I was selected to perform the special assignment as payloads processing expert, to study the feasibility of processing a Soyuz vehicle through KSC to be used as a Space Station Assured Crew Rescue Vehicle. I was told my Level-IV background of working flight hardware on Spacelab and how payloads flowed at KSC were the main reasons I was selected. I was to be the sole NASA representative to observe and participate in Soyuz TM-18 vehicle ground processing and launch operations during a month-long tour in Russia. To do this I needed to learn how to speak, read and write Russian, so I took a crash course in Russian, had many meetings with Russian personnel, learned about the Soyuz vehicle, and learned about the Baikonur Cosmodrome spaceport located in an area of southern Kazakhstan and leased to Russia. I also had to set up a housing in an apartment just outside Baikonur, which I was worried about because, talking to the Russians stationed in Moscow, when they traveled to Baikonur and had to sleep there they would not stay there any longer than they needed to. I guess it was not the best kind of lodging, but that was what was required to do the job. We had everything planned ready to go, and two weeks before I was supposed to leave, the whole concept was cancelled."

Maynette Smith was in Level-IV Electrical and then moved over to Space Station Utilization in the 1995/1996 timeframe when a reorganization and merger of the Level-IV group with the Utilization Group occurred. "I had both for a couple of years until 2000, and then I was out by then. We processed the last of the Level-IV things. So I was in different capacities for a number of years."

ISS Hardware Integration Office

Level-IV was gradually being broken up, but where would all those people be relocated and what would they be working on? While some in Level-IV were moving to the Utilization Office to work experiments, others would be selected into a new organization, the Space Station Hardware Integration Office (SSHIO). This was created to help prepare the large ISS flight hardware elements to fly into space. In March 1996, John J. "Tip" Talone Jr. became director of the SSHIO, which became the ISS/Payloads Processing Directorate in May 2000. SSHIO was created to bring the flight hardware processing experts from KSC together, go out into the field to help the contractors process the flight hardware in preparation for arrival at KSC, and then be responsible for the hardware after it arrived at KSC and was turned over to NASA for final phase of assembly, integration, test and closeout prior to launch.

Mike Haddad explained, "In March 1996, I was honored to be selected to be a member of the SSHIO as a Technical Integration Engineer [TIE], Systems Integration Engineer, Structures/Mechanical/Fluids Engineer, and KSC POC [Point-of-Contact] and Lead Engineer for EVA [Extravehicular Activity] and Crew Systems [EV&CS]. As the EV&CS Lead, I reviewed flight procedures, performed hands-on operations and technical integration, and was team lead for EVA/IVA [Intravehicular Activity] fit checks and On-Orbit Constraints Tests on flight Structures/Mechanisms/Fluids hardware associated with International Space Station missions. This included reviewing and updating EVA checklists and flight procedures that supported ISS mission operations, and direct contributions to Johnson Space Center Mission Control Center [JSC MCC] for real-time problem solving. My other duties included Specification & Standard reviews, requirements generation, prototype/development testing, structures/mechanisms installation, technical integration, and requirements generation/verification for Multi-Element Integrated Testing [MEIT]. There was also Human Thermal Vacuum Testing on EVA flight tools and mechanisms, research testing at NASA's Plumbrook Facility, Interim Control Module [ICM] design at the Naval Research Laboratory, Habitation Module design as well as Gamma fitting, fluid GSE/flight tool design, fit checks and flight procedure development.'

The Station Program Office and Boeing had decided early in the ISS program to go with no integrated testing on the ground, with everything simply packed into the Shuttle like cargo and launched (back to the 'Ship-n-Shoot' concept). It needed people from Level-IV, now working in SSHIO, to fight for why it made sense to put time back in the schedule to do integrated testing. Bruce Morris recalled, "It was us, the Level-IV people, who were the ones that came with the data that said, 'Look, somewhere in the flow, if you have the opportunity to integrate testing, it's going to save you big time. And here's why...' The things we found that you never discovered when individual-level piece part testing, because you never had a chance to actually try it out on the ground first.

"The Level-IV guys were the ones that came in and saved the Station program because of those kinds of experiences. That it is not exaggeration, it was a huge deal. For many of us that were in Tip's [Talone] group, our job was to go out and live with the contractor, embed ourselves in the contractor team and take that same kind of experience to help them, because they were in essence being a Level-IV group back at their factory. [We] helped them to get through integration and test faster, because we had to cut the timeline on assembling Space Station down because we just didn't have the money to be able to afford a long duration, integration and assembly operation as long as it was planned out the way it currently was.

Those kinds of schedules always suck up resources and dollars. Even if the team's waiting, you've still got to pay salaries and things like that. So we had to find ways to take time out of the schedule, in a safe way, to be able to afford the program the way it would need to plan out. At the same time, other people that worked in our same organization back in Level-IV were doing the right thing by trying to find out ways to put more work in the schedule, because it was the right kind of work to do; taking the non-value stuff out and putting some value stuff in there. So it was fun, and I do mean this sincerely. I think there were a lot of people from Level-IV that made a huge difference in the success of Space Station, because they knew from their training experiences the reality of what was going to happen [if there was no testing] and what we were going to have to do to make this work on Space Station. It was a mixed feeling. You were coming to the end of something, but you were starting something new."

Mike Haddad added that, "As part of SSHIO, I had to travel to those locations around the US that were preparing the ISS elements. That included the US Lab, the Nodes and Airlock at MSFC [Marshall Space Flight Center] in Huntsville, Alabama, and the Truss elements in Canoga Park and Huntington Beach in California. One thing I learned real quick was that the contractors did not like us, meaning NASA, being there. In fact, they hated the fact that NASA was now in their world. We were there to help them, but they did not see it that way at first. I remember they would not invite me to important meetings. I would hear about the meetings second hand and when I showed up in the meeting room, they would end the meeting and everyone would leave. It took a while for them to warm up to the fact that we did help them. We were not going anywhere, so we might as well all get along and do the job we need to do to make ISS a success."

NASA had a very rigid structure and once the ISS hardware/software was in KSC's hands they were accountable for it, so they were going to document everything that happened. Sometimes that worked against KSC, and ultimately it led to the demise of one of the checkout systems for Space Station, as Kevin Zari, explained: "We were too costly to the program. The program made a risk-based decision and said, ' PTCS, the Payload Test and Checkout System, was the best fidelity you could ever ask for, but we're paying for it'. They were willing to take a chance and use a lesser-fidelity PRCU [Payload Rack Checkout Unit] eventually."[1]

---

[1] The Payload Rack Checkout Unit (PRCU) could test experiments as well as ISS racks. It had some electronic boxes in the system, but much of the ISS simulation capability would be done with software instead of hardware. The ISS program wanted a distributed test system where they could test the flight hardware and software as they were being developed, before being shipped down to KSC. The ISS program built several of those at centers because they were a lot cheaper than the PTCS.

HELLO SPACEHAB

Spacehab was a commercial consortium established in 1983 to satisfy an expected expansion of commercial opportunities in space. Stowage is always at a premium on space missions and while the Shuttle Orbiter had the capacity for 42 lockers on the middeck, only seven or eight were available for scientific equipment. The remainder were filled with crew equipment, food, clothing, camera equipment and other miscellaneous items used by the crew on a nominal mission, so there was a need to expand locker space for longer missions and additional small scientific and commercial experiments. The Spacehab Company developed a pressurized compartment which effectively quadrupled the workspace volume available for a crew, adding up to 61 lockers of stowage space. The first flight of a Spacehab module occurred on STS-57 in 1993 and it proved a very useful addition to the Shuttle hardware inventory [2].

Spacehab Hardware for Space Shuttle Missions

The development and operations of the Spacehab module are beyond the scope of this book, but for information, Spacehab hardware consisted of:

- Integrated Cargo Carrier (ICC), unpressurized.
- External Stowage Platform (ESP-2 and ESP-3), an ICC variant.
- Logistics Single Module (LSM) and Logistics Double Module (LDM).
- Single Module (SM) and Research Double Module (RDM), pressurized.

The Spacehab hardware was specifically designed to fit inside the Payload Bay of the Space Shuttle and flew on a total of 22 Shuttle missions, including seven to the Russian space station Mir and eight to the ISS.

Spacelab was winding down and Space Station was ramping up, and in the middle of that, Spacehab was being processed and flown. Even though Spacehab was a commercial venture, some saw it as competition to Spacelab, though some saw it as totally different. There were some similarities between Spacelab and Spacehab, even though the hardware itself was very different. Level-IV personnel had some interaction with Spacehab processing.

"Actually, the first Spacehab module was one of the missions I was working on as the LSSM [Launch Site Support Manager]," recalled Mark Ruether. "So Spacehab was now this commercial module owned and operated by a commercial entity, but it was flying in the NASA Space Shuttle. So you had all these people at the center, through the Experiment Integration group [who were] used to the model of Spacelab, a NASA old model and module, and they worked hand in hand with experimenters to go make sure all that [integration] happened. Then you flip that and say that the module is now owned by a commercial entity, and it's the commercial entity that is making decisions versus NASA on what goes in there, how it's processed, who touches it and who's responsible for it. So you've had a lot of these things that a team of people have been used to with Spacelab and you just followed the process to go make all that happen. Now it was different. And that was probably the biggest challenge of Spacehab for me.

"We used to do it because you'd fly some of the same experiments. The same experiments that flew on Spacelab or the middeck were now flying on this module. But it was Spacehab's call on how those things were processed, how they got input, so it was just a whole different model. We had to change the culture, which is one of the hardest things that there is to do. We had to change the culture almost overnight, or make sure that people could flip the switch, because you could be working on the Spacelab module one day and then tomorrow you got to work on a Spacehab module. Your role and responsibilities were different even though the experiment was the same." So there were some significant challenges, and as the LSSM, Mark was busy translating the nuances of those differences back to the team. Sometimes the Spacehab people contacted him because the team was trying to do what they had normally done for a Spacelab module, but it was different for them. As Mark said, "That was a challenge. That was interesting. Challenging, but interesting."

There was emotional attachment to the work that Level-IV used to do, and that work was now being turned over to some other group. From a programmatic perspective, it was clear that there were huge operating costs to sustain a workforce and flight hardware for the Spacelab program; huge infrastructure costs, huge operating costs and overheads. By buying the module from a commercial entity, all they were buying was the mission it was needed it for, without all the overheads. Mark Ruether observed, "I'm sure, from a programmatic perspective, there was a huge cost saving by having a commercial entity do that, even though you had to pay them some profit. So it depends on which level you were at. If you were at a very high level, you understood you were no longer paying for the whole cow. You were just getting the milk that you needed. But if that was the work you used to do, and that work was going away and you liked that work, what could you do? You were upset because someone stole your cheese."

Cheryl McPhillips thought Spacehab was cool, as it was more of a commercial venture. "I think we had to take our money from Spacelab and put it in the Space Station because there is only so much NASA money every year, [so] I don't feel like it was the demise of Spacelab. I thought it was a good complement. I always thought it was pretty cool. I jumped out when we started to get rid of everything, when we shut down the Spacelab program. I jumped ship over to Space Station and it was just really a great experience that I had from working in software, power, data, trying to put Space Station together and test it. It was a real good fit for what they needed at that time over in the Space Station."

Tracy Gill agreed that, "Spacehab *was* commercial, but it was much less capable and we were helping them get it off the ground. We loaned them a bunch of equipment and we talked to them. We loaned them all our batteries and battery support equipment for middeck and went over and looked at it when they were first doing integration for STS-57, which was the first Spacehab mission. It was difficult to do that, because the writing was on the wall that they were going to get rid of Spacelab in favor of this, so that was tough."

Astronauts that flew Spacelab were now going to be flying on Spacehab, so they also had to adapt to the different type of hardware, the processing, rules, and everything else. Astronaut John Grunsfeld added a different perspective: "My second mission was Shuttle-Mir Docking Mission 5 on STS-81. Its primary [objective] was to go and pick up John Blaha and drop off Jerry Linenger, along with 2000 pounds (907 kg) of cargo. But in the Spacehab we had the Biorack facility, which was an international facility to do biology experiments in microgravity. We had some simulations at Johnson Space Center, but Jeff Wisoff and myself as the prime operators came down to the Cape two or three times, to Hangar L across on the Canaveral side, which NASA owned or operated to do life science work. Inside of Hanger L, there was a smaller building, temperature-controlled, and all of that had the Spacehab operations and experiment areas. And there, Jeff and I would go through and do real operations on the real hardware in preparation for flight.

"It was a parallel to the kind of Level-IV activity that went on, but it had a higher emphasis on crew training. And so our prime crew training, in many cases, was to go there and work with the actual hardware, relocating hundreds and hundreds of these tiny little experiments that were running, along with a freezer that kept the samples pristine. That training over in Hanger L was as flight-like as we could get and that really trained us on the actual operations since it was life sciences. Most of the action was occurring inside of sample vials and we were just the operators to take things out of little centrifuges, put them into the freezer, take stuff out of storage, put it into the centrifuge or into the incubation chambers, and hundreds and hundreds of these relocations. While we were on orbit, our feet were underneath a restraint and we were standing up relative to the rack just like as if we were on the Earth. So the training in Hanger-L was almost exactly like what we had on orbit, both procedurally and the motor functions, teaching the muscle memory and such. It was just like being on orbit, although being on orbit was still more fun. There was some criticality but it wasn't make or break. The other part of that though, is that while we were doing them on orbit and communicating to the ground what we were doing, at times there was a team on the ground with the training hardware that we had trained on in Hanger L doing exactly the same experiments as a ground control, maybe delayed by a minute. I know they were doing the same experiments around the clock with us, so that they could then compare a sample of Arabidopsis seeds sprouting in space and on the ground done at the same time. All of the operations occurred just the same way they did on orbit.

"The Spacehab training was like the Spacelab training, in that we got to crawl around and experience all those interfaces prior to it being in the Orbiter. Plus, once it was in the Orbiter, we got to crawl around on it too, and in buckets, and occasionally knock it down and actually walk out of platforms. But that was a little bit different. But I think the experience, that whole team flight crew, even flight controllers in some cases but in particular the technicians and the engineers working on Spacelab missions, knew this folded right into what we were doing with International Space Station assembly and integration at this Space Station Processing Facility. And I think without having that Spacelab experience, it probably would've been very difficult to spool up that kind of knowledge that we needed to assemble the Space Station. I made numerous trips for electrical and functional mating and checkout of Space Station elements, but also the external EVA interfaces that were so critical in assembling the Space Station."

NASA was now talking about long-term living in space, for months on end, but Space Shuttle, Spacelab and now Spacehab could only stay on orbit for 10–20 days at most before having to return. Whatever experiments were included on the mission, that was the duration they had to work with. With ISS being on orbit for much longer, it would be a platform to do so much more over an extended period of time. Mark Ruether said, "I wouldn't think of that [ISS] as a competitor to either Spacelab or Spacehab just because of the duration of time that it would be up there. I think it was the right next step for the agency to ask what was the type of testing it needed to do to further the ability to have humans in space for a longer period of time. You've just got to extend the duration and that's what Space Station provides. So, I don't think people really saw it as a competitor. I think that, for those that really enjoyed Spacelab and then that went away, they looked at [the fact that they'd got] another opportunity that was very similar to go on to when Space Station came along. I don't know what the gap was between the end of the Spacelab and the beginning of the Space Station, but there was a big gap and so it's probable a number of those people moved on to other things. But it became that next opportunity for some young people to go learn on some flight hardware."

MPLM, MELFI AND MEIT

The transition to Station included processing some Spacelab payloads over in the SSPF. The O&C was being reconfigured for ISS elements, so many would now find their working days filled by processing flight hardware in the SSPF.

The Ninja Turtles

Alex Bengoa recalled, "It was 1996, and I was assigned to the MPLM, the Multipurpose Logistics Modules. We had three modules, provided by the Italians to support the station, and it was pretty funny. The Ninja Turtles, we used to call them, *Leonardo, Donatello* and *Raffaello*. I also always joked with the Italians about what happened to *Michelangelo*, and they kept saying, 'No, he was not an engineer [referring to the sculptor Michelangelo] so it doesn't count'." When Alex transitioned to Station, he was assigned to the MPLMs as the Environmental Control and Life Support Systems [ECLSS] engineer. He was responsible for several components, including a fan for air circulation inside the module, some positive pressure relief valves, some negative pressure relief valves and some isolation valves. As part of the MPLM design, there was an active water loop. At the time, the objective was for that system to be active during the mission on orbit before connecting to the station. The Station program was going to fly some active refrigerator freezers inside the MPLM, so the water loop would help transfer the thermal energy between the freezer and MPLM, out to the Orbiter, and then get the cold water back in to help cool the refrigerators, such as the Minus Eighty-Degree Laboratory Freezer (MELFI). "We never flew that module," Alex recalled. "*Donatello* never flew. We only flew *Leonardo* and *Raffaello* and they were not in that configuration."

Bruce Morris explained that, "The excitement of what we did in Level-IV kept us there and was invaluable to Space Station, especially going through some of the craziness with the station elements. It was the direct experience of Level-IV that put the Mission Sequence Tests in at Kennedy for all the elements, because Boeing had taken the position we didn't need to do that. We took the position that if the first time the flight hardware was going to see each other was up in space, that was a bad idea. So it was invaluable in keeping problems from occurring on Space Station on orbit. We would find and fix them during assembly and checkout at KSC. When we started working on Space Station, NASA started planning for Space Station activities the same way we planned Shuttle activities, where we were going to work out a timeline for the crew, for those months at the time, right down to the minute. But that was one of the things we learned from the Russians. They said, 'Your astronauts are going to revolt. They won't do that and it's not going to be worth your planning because things are going to change so much'. So we learned from them that instead of doing that kind of *mission* timeline planning, we should do *timeline* planning that gave the crew a general time to do something, gave them the goals and objectives, and gave them more of an interface between the flight and the ground systems experts, so that the crew could get directions from the ground more quickly than routing it through all these people."

Before each critical component of the ISS was launched from KSC and assembled in orbit, it was tested and integrated with other components here on Earth. The responsibility of testing the early components of the ISS rested in the hands of hundreds of men and women, led by NASA's Cheryl McPhillips.

Multi-Element Integration Testing

SPACEPORT NEWS June 22, 2001, Cheryl McPhillips MEIT Article.

Cheryl McPhillips was the winner of the 2001 National Space (NSC) Club Eagle Award, an award she said that she accepted "on behalf of the Multi-Element Integration Test [MEIT] team as the project lead of MEIT-1." MEIT-1 was a joint effort between NASA and The Boeing Co., and consisted of ISS personnel who saw to it that all of the Space Station's components were in perfect working order and able to communicate with each other as they were prepared in the SSPF for launch. MEIT-1 took three years of intense planning and is credited with the successful testing of some of the Station's primary components, including the US Laboratory *Destiny*, the Canadian robotic arm and the large Z-1 truss. Since its inception, MEIT has detected and resolved thousands of anomalies. The National Space Club reserves the annual Eagle Award for those who have had the most significant influence over the successful and safe completion of a human spaceflight mission. Rick Hauck, former astronaut and Chairman of the NSC Awards Committee, said, "Those who have led the MEIT project are perfect examples of individuals who work on the front line towards a successful and safe space program." Boeing's Christian Hardcastle and NASA's Kenneth Todd from Johnson Space Center in Houston were also named in the award in recognition of their leadership during MEIT-1.

Mike Haddad added, "Many were against MEIT, but once it was successful those same people jumped on board about their involvement and what they did to make it happen. There were hundreds of people that made MEIT a success, but I'll still say to this day that Cheryl was the one who spearheaded the effort that made MEIT happen when others were trying to take credit for the work she did, at least for MEIT-1. I remember sitting next to her for years; all the phone calls she did, all the coordination, all the meetings she led and the travel she did to pull off MEIT-1."

Cheryl McPhillips added her own recollections: "Billy Haynes was the one who really got it approved. It was his child. He wanted it to happen, and Tip set him up to get it approved by the program. It was great fun, of course, now looking back at it. It seemed like it was great fun. I know Mike Haddad would leave me little chocolates when I'd had a tough day. It was great because we got to lead it out of Kennedy and so we got all the engineers from JSC, the subsystem engineers supporting us. And then we had our engineers that were smart in those subsystems too. So we got to go and work on what the requirements were.

"We just got to create the whole test and we got to put four modules together. They weren't going to test it on the ground they were waiting till it got in orbit, kind of like the 'Ship-n-Shoot'. This was a 'Ship to orbit and see if it works'. Luckily, we got to test everything out and that's where we found [the problem with] the jumpstart, or I'll call it the Auxiliary Power Unit [APU] from the Shuttle that was supposed to start up the first Solar Array module [P6]. Every time we went to start P6 up, it would short and blow a circuit and shut down. That was the first thing we encountered right before Christmas, when everyone was getting sick in the User Rooms and the control rooms. Flu went through the whole group, it was terrible. We were trying to work and finally they said, 'Wait, you guys go take a Christmas break and we're going to check it out in California', out at the SPEL lab, the Space Station Power Electronics lab run by Rocketdyne. So they got it to repeat and damned if they didn't figure it out while we were on Christmas break and got to come back. It was a change they'd made late to the APCU and then they didn't go back through qualification, and that's when they found out they had a short in the printed circuit board that was inside the APCU. That was a big one. That was the biggest one. It would've caused a re-flight of the P6 and we found it on the ground."

Mark Ruether added, "[For] the initial integration with MEIT, the big module pieces came in here and you did some level of integrated testing. I know there were some people that worked on Level-IV that ended up working MEIT too. Rather than just integration of experiments in a rack, you're taking huge modules and plugging them together."

"MEIT for Space Station," recalled Maynette Smith, "…the problems that were found and corrected. Connecting all those elements on orbit without doing that [MEIT], you wouldn't have a working Station. So this whole concept that the first thing that needs to go is testing down at the launch site… they needed to go back and look at all these programs and the problems we found and corrected. I mean, something as seemingly simple as a grounding problem. We chased grounding problems all over the place. It was a big bugaboo because [they couldn't figure] 'Why is it doing this? I don't understand why this is doing that', and we would finally chase it down to a grounding problem. The more complex we get, software and all of that [the more likely that is to happen], especially if you're dealing with pieces that have never seen each other."

Mike Haddad worked System engineering in the control room for MEIT-1 testing, basically trying to take one step back from each of the subsystems to get more of a big picture, integrated aspect of the testing, and verify tracking of the requirements. "The KSC subsystem experts were the real heroes of MEIT. Their background on the systems they worked in the past, the understanding of the ISS subsystems, the time to review the requirements, review the flight hardware/software drawings, generate the procedures, create GSE drawings and then build the GSE to support MEIT; to do all of that and then implement the series of tests was just extraordinary, especially when many were sick for days at a time because the flu was going around KSC."

Kevin Zari told the authors, "If something failed on orbit, they [ISS program] could come back and say, 'What happened when you were testing at KSC? Show us your test procedures'. And I could say, 'Oh yeah, we saw that same thing', or 'Wait a minute, we did that at KSC and we never saw this issue, what caused that?' We had a payload that failed on orbit. That was the CGBA [Commercial Generic Bioprocessing Apparatus] and it failed because after testing the payload on orbit, they transferred the file to Huntsville, which we had done at KSC, and the systems that it passed through at Huntsville sent it FTP [File Transfer Protocol]. FTP had two modes, ASCII or binary. If you sent a text file, you typically wanted to send it as ASCII, which meant it took chunks of seven bits and transferred seven bits at a time. If you did it in binary, it took chunks of eight bits and transferred at eight bits at a time. Normally it didn't matter, because in the end it reconstructed the file, but this was a configuration file that they were only going to be able to create once it was on orbit. They knew certain things, they loaded it into the payload and then they would power on the rest of the payload and life would be good. Well they transferred it in a way that worked at KSC, but when they transferred it through the systems at the HOSC, it went from a PC, from a Windows® system, to a Unix® system and then maybe back to a PC. It flip-flopped between systems and in doing that it added the end-of-line marker, whatever it was. There are different ways to represent carriage return, and if you transfer in ASCII, it preserves the way that you've done it. If you do it in binary, it doesn't. So what had happened was that in transferring the file to the HOSC, they corrupted their own file by adding these escape characters that weren't able to be parsed by the payload. [They] uploaded the config file and lost the payload; it was dead. It came back down, we examined it, with lessons learned if we found out exactly what happened. That was another beautiful thing. There was a payload failure investigation board, and guess who got asked to be on the payload? The Level-IV guy that was working the payload. The root cause was the same person who worked in a laboratory environment and was used to fixing things, who was used to finding a problem and then fixing it right away. They found a problem and tried to fix it right away, but didn't have that same engineering discipline that we would have had. Again, it was the discipline learned in Level-IV, that doing things the right way pays it off, even if it costs money doing it that way."

Tracy Gill said Level-IV were consistently called on to troubleshoot hardware and software problems on Spacelab and then ISS payload hardware. This classically involved isolating problems to test systems or flight hardware, and then often between payloads, flight hardware and the payload carrier, such as EXPRESS. For example, on the first two EXPRESS racks on the STS-100/6A mission, Tracy worked over 120 Problem Reports (PR), including procedurally-induced conditions, cable problems, software faults, and failures requiring replacement of box-level components. It showed what an empowered team could accomplish. Only in hindsight did he realize how much authority he had to engineer and implement solutions, which required much more program coordination in the ISS era. "I was disappointed that many in the ISS program regarded Spacelab as some sort of anomaly that was a slipshod and happy-go-lucky effort, when it was far from that. There were dedicated and brilliant teammates who did what needed to be done for the missions to succeed, using guile, smarts, and will power."

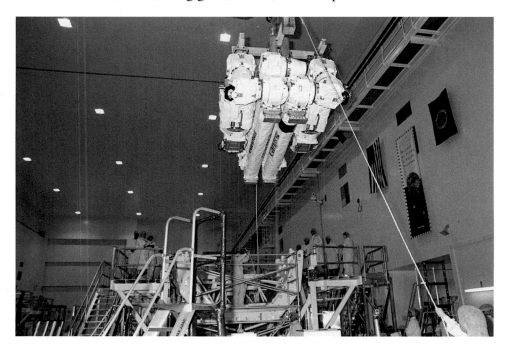

The Canadarm2 robotic arm, also known as the Space Station Remote Manipulator System (SSRMS), being installed onto a Spacelab pallet in the Space Station Processing Facility (SSPF) at Kennedy Space Center, Florida, September 28, 2005. This hardware would fly on STS-100. [NASA/KSC.]

One of the other major elements Level-IV personnel were assigned to was the robotics elements of Space Station, the Canadian elements. Josie Burnett was assigned to be a technical liaison with Spar Aerospace, which was developing the Canadian elements for Mission 6A (STS-100). She observed their systems integration testing to prepare them to come to KSC where they did the MEIT, hooking up the US lab with the Canadian arm. Even back then, the Canadian partner was on the critical path for the ISS assembly sequence, so it was a big deal. The Canadian Space Agency provided the arm as an equal partner to NASA. Josie Burnett recalled, "I taught a lot − respect the politics of the situation − but really it was good work too. By then the Spacelab program had ended, but we used the legacy Spacelab pallets for transporting the Canadian arm to ISS on orbit."

THE BITTERSWEET FEELING

Tracy Gill added that, "I tried to go work Station, but in 1998 I [worked] on the last Spacelab laboratory mission, which was Neurolab. I stayed till the end to lock up the doors, which I'm happy with in retrospect because I could've been shuffling my feet for a while on Station and there was plenty to do when I came over here. But I got to close it [Spacelab] out with a bang and have fun on that. Okay, well no, it was kind of bittersweet. You know it's going away."

Saving Spacelab Hardware

Bruce Burch stayed through all the Spacelab flights, and actually helped close down the program as they got rid of items that were Spacelab unique and transferred what they would need for Space Station. "Some stuff went over, like the strongback [used to lift a completed Spacelab payload complement] and some of the rack holders. We didn't take any of the Freon® units but the rack conditioning units are still in use now. They will be using them for Gateway to cool the modules, some 30−40 years later."

There was a zero-based review in 1995 that created some difficulties, because the agency was flying to get into a faster, cheaper, better mode. KSC was still adding new engineers to the workforce but if there was a reduction in personnel, as suggested by the review, then anyone that came new could be candidates to be removed. People were scared so they left, with some of them having been with KSC less than two years. Rey Diaz explained that "the group of people that stayed not only had the experience from early on, but also carried that on to getting the Station ready, closing out the Shuttle missions and all that. That was really commendable of all those people who went through a lot of changes, including the end of the Spacelab program."

"On the outskirts of that," recalled Elias Victor, "I remember the tunnels of equipment underneath the O&C Low Bay and High Bay where we had all the extra gear, the spares. I remember all of that being torn up, taken out and removed. It was very sad to see that happen, and it was also very sad to see people who had put their heart and soul into that witness it. That happened also when a lot of that gear got sent to Ransom Road. Ransom Road was basically a junkyard where items were sent to be sold, pretty much auctioned off for pennies on the dollar. I know that some of our folks, Scott Vangen being one of them, made a bid to try to recover some of that equipment. I helped Scott go to Ransom Road and just get whatever he was able to buy off and store. I used my little truck and we loaded it up and then drove multiple times back and forth. We tried to get the gear out and save some of that gear. Some of it went into museums for people to see instead of being scrap metal. Other generations, future generations, won't forget the things that were done. I'm glad that we were able to save some of that stuff for the new kids and that we'll be able to talk to them later on and say, 'This is what this was about. This is what happened during these years'. It was good to save that. We found houses around Merritt Island and we talked to people who could lend us a garage or a room, some space to shove this stuff in so that it wouldn't get lost. So with my little Mazda B-2200 we drove back and forth to Ransom Road, loaded up and took it to rooms, dropped it off and then did it again."

"Actually, I think Spacelab was pretty quickly closed out," recalled Frank Valdes. "One day there was just a flatbed truck and the module was put on top of it and strapped down, taped over and shipped out. There was really no transition from Spacelab. There were really not too many lessons learned that were transferred over, because either the wrong people were working on it or they didn't know any better. And then they had to institute for the module, for the ISS, for the Node. They had the same growing pains that we had for the pallet; things were being crushed, people were stepping on cables and fluid lines. They were doing the same mistakes, that [could have been avoided] if they had worked Spacelab or if they had listened or asked the right people. Basically I think the attitude was, 'We know what we're doing. We don't have to do it your way'."

As with Spacelab, then Spacehab and now Space Station, Level-IV personnel and the astronauts would work together in preparing a new ISS component for spaceflight. Many of those same relationships established during Spacelab flourished with ISS and those relationships extended to the Payload Developers (PD) and Principal Investigators (PI).

Former astronaut John Grunsfeld said, "I recall the time I flew to the Cape with a signal strength meter and a network analyzer, and Scott Vangen and I carried it into the US Laboratory before it was fully outfitted and did a survey of the RF environment to show how the wireless network would work. If I'd dropped that network analyzer, we'd never have had an International Space Station. We were hardly supervised, but that's another story."

NEW HORIZONS

Kevin Zari explained that "After Level-IV, I went on to International Space Station payload processing, working with many of the same PIs and PDs that were working on Spacelab. The Payload Test Engineer [PTE] role for Spacelab payloads naturally rolled into PTE assignments for ISS experiments. In parallel, I developed some systems to deliver test data to payload developers [Spacelab Payloads Operations Online – SPOO], and was project engineer for the follow-on equipment named Settebello, which delivered audio and video to Payload Developers remotely, and the Espresso [an EXPRESS rack payload simulator]. These assignments led to exposure to IT systems and technology infusion. I finally got to laugh at those paper strip charts in Hangar AE. This then led to my Chief Technology Officer role for ISS at KSC. I am now [2019] the ISS Integration Manager in the Strategic Implementation Branch for Exploration Research and Technology Programs, ensuring that processing of ISS Orbital Replacement Units [ORUs] and associated hardware goes smoothly, and passing lessons learned on to Gateway [and perhaps the new lunar program]."

Working on the floor and doing hands-on gave Level-IV personnel a different perspective on operations. Even as of the writing of this book, working on the Space Launch System (SLS) part of the Artemis program and working very closely with the operations contract with Jacobs, the Test and Operations Support Contractor (TOSC), the Level-IV experience helps to understand their perspective when they have questions and when they want something, where they are going to go, and right down to the basics of nonconformances. Alex Bengoa said, "I understand where they come from with the questions. I understand their pain and I'll always try to be an advocate for what they need, and be at the forefront so they can do their job. I make sure that they have the data that they need for them to do their job. I try to work that out with my counterparts at Marshall and JSC, just make sure that we have the right data so we can do the work here at KSC. So, being on the forefront, being hands on, I got to learn all those things that are helping me out now to do my job. Of course, sometimes people look at me a little bit sideways, so I tell them where I come from, my backstory, and they understand."

What is very unique is that the reputation of the Level-IV people went well beyond KSC, as stated many times before. People who worked Level-IV moved to other centers to perform whatever their hearts desired. Those same people became leaders at their centers and would go out looking for their own new talent. The PIs and PDs that worked the Spacelab program, both foreign and domestic, are also now part of the ISS team.

Angel Otero explained that, "Bill [Sheredy] was at Marshall when I interviewed him. He was there because NASA had offered to send some Level-IV guys to help Boeing with the processing of the US Lab for ISS. But then Boeing didn't want the help. Bill was at Marshall supporting that, and when I was doing interviews he came in and said he was originally from Michigan, so he wanted to get back closer to home. Then he told me he was from this group [down at KSC] called Level-IV. I said, 'You're hired'. He kind of looked at me, and I said 'I'm one of the old Level-IV guys, so if you're from Level-IV, you're coming with me to Cleveland. So he ended up working for me."

Kevin Zari noted how his reputation expanded. "[Payload Developers would say] 'We work with this really decent guy and he's fun' and would actually recommend me to their colleagues: 'If you're going to KSC, ask for this guy if he's working.' It was really cool. I have letters from Payload Developers, even through to ISS payload processing. Payload customers wrote thank you letters to the Center Director saying in particular that if they came back again they looked forward to working with Kevin Zari. I mean it was pretty much guaranteeing me. A funny story is that even when I stopped being a payload test engineer and I started to do more. it worked to help with some gap that existed in my directorate. They were continuing to ask for me. I was fully in my job, about a year into it, and I was still flying over to Italy to support a payload that was flying on station. Again, that relationship you build is really great."

Most of the discussion up to this point has been from the engineering perspective, but there were also Level-IV quality control people that assisted with the transition to Space Station and the ground operation of ISS elements. They would travel all over the country to the different hardware sites because it was considered necessary with all the hardware coming to KSC. KSC had a reputation for being the strictest of all of the centers when it came to Quality Assurance (QA) and adhering to the engineering. That was out of necessity, because KSC was where the buck stopped. Everything was finally integrated there, as Bob Raymond explained. "All the piece parts had to fit and work together. You'd run into these problems when these things started to show up at KSC, when you were starting the final assembly of these things. I would have liked to have seen more travel actually, during the Level-IV days. I think we would have had fewer problems if we had [done that], because we could go work with the QA and the engineering departments at the design centers in the manufacturing centers, to be sure that those sorts of problems didn't find their way to KSC. We were on a short leash when we got the hardware showing up at KSC. We had six months to get all of this stuff put together and get it ready to go fly. Maybe a little bit more than six months, but sometimes as little as six months."

Bob Raymond enjoyed his time with the technical team and Tip Talone. "Tip Talone is how I ended up over in Space Station, because he remembered me from the technical team and the *Challenger* investigation and asked, 'Maybe you want to come over here?' And I said, 'Okay'. Actually, when I went to Space Station they hired me as a data systems manager. I was doing administrative data systems and it was good. I had written a couple of application software packages that were used at KSC, but eventually I migrated back towards the flight [side] and my first step was as a project manager on several projects, one of which was the engineering. I'm talking Space Station now, the things that I learned in Spacelab that would carry over into Space Station. It was a carry-over, because of the control room environment, data communication, bringing in the customers' home sites. That was part of what I did, and so I was chosen to do a couple of projects while I worked in Level-IV. I was asked to be the data systems manager for the Engineering Support Room, ESR, and the job with the ESR was to get the data down from the Space Station and be able to display it up in our control room on the third floor of the SSPF, so that we could avoid having to ship engineers to Houston for every flight with arms full of drawings and documentation and the like. Instead, they could stay at our site and if help was needed, then they were right there instantly instead of having to travel. So we used Houston's resources as much as possible, used their system to get it down from the Space Station to Houston, and Houston would transmit it to us and then we would distribute it to the other Engineering Support Rooms. So that was a fun project."

Mike Haddad explained, "All the fit checks of EVA and IVA hardware, cable connectors, fluid lines, etc. that we performed at KSC prepared the crew for successful on-orbit EVA and IVA operations. Then I supported the ESR for crew on-orbit operations, EVA and IVA. It worked great. We were on-console, like being in Mission Control in Houston, but at the ESR. When the operation was complete, you went back downstairs to work in the SSPF on the flight hardware you'd been processing. During one of the earlier missions, the crew was having a problem with some blanket material. When two pieces of Space Station come together there's an interface there and they had to install this blanket material, in essence to make it like a nice smooth tunnel. The attachments weren't working, so they couldn't get the blanket material to physically attach and hold in place. Initially they tried to describe the problem to us, but then they went ahead and created a video and downlinked it to us here on the ground. I was sitting on console at the time and saw the video of what they were doing, and because I had worked on that piece of hardware here at Kennedy Space Center, I knew pretty much from the video what the problem was. Again, describing it with words was very tough, but as soon as we saw the video, we knew exactly what the problem was they were encountering. We had a similar piece of hardware in the Space Station Processing Facility High Bay, so we went ahead and validated what we thought would be the solution to the problem, and then related it up to the crew, which then of course fixed the problem that they had on orbit.

"It made it all successful, in that it was a combination of seeing the video and us having the hardware at Kennedy Space Center to be able to come up with a real-time solution to the problem. It did not take long, in reviewing the video. It was pretty instant when we saw that. So within an hour or two, we knew what the problem was. We got the hardware here set up, and then went through a trial run before we sent the procedure up to the crew. It took a number of hours [because] again, we wanted to make sure it was correct before we sent it up to the crew. This specific problem was inside, so there was no real time rush. They weren't outside during the spacewalk, it was inside, so we had some time to come up with a solution. It took about 12 hours from the time of the initial problem till we had a solution and they had fixed it.'

"Space Station was basically a Level-IV in the extreme," explained Herb Rice, "because we were preparing much more significant payloads that were much more active and were going to stay there much longer, and that required a lot more testing, a lot more operations. Some of them were coming from different countries, just like Level-IV. So it made all kinds of sense to have it done the way it was done. It was done right and many of the folks in it, of course, were former Level-IV payload folks, and they all performed quite well in Space Station I thought. It was fun. It was the most ideal career that an idict from eastern Kentucky could have."

Eli Naffah, stated, "I think Spacelab was the precursor to Space Station. A lot of the experiments that were conducted on Spacelab gave us the impetus for what we were going to do on the Space Station. In fact, a lot of the microgravity research that we do on the Space Station has been done at the Glenn Research Center [GRC] and there were a few Level-IV people that came up. Angel Otero and Bill Sheredy are two of the folks that I know came up to Glenn and worked microgravity, where they actually designed the experiments that were integrated. I think the significance of Spacelab was to begin doing science, more long duration science, in the microgravity environment."

"It was really a great experience," said Kevin Zari. "It was the experience and the opportunity that Level-IV gave you as an engineer, fresh out of college or not. I mean, you could have been there as a seasoned engineer, but it really was a great experience. You literally traveled the world. You traveled with the payload and you gained competency in areas. I was an electrical engineer but I didn't pay attention to acceleration other than gravity constant and understanding that. But then I started to understand how triaxle sensors worked and stuff like that and found that kind of interesting. Then you start to develop your understanding that everything was related. So when some Spacelab microgravity payloads came to fly on Station, they naturally came to me because I had established that relationship at the microgravity measurement group conference. So all of a sudden I had a close relationship with the folks at Glenn Research Center. It was a telescience center that was up there and it had a special name, but the folks that ran that knew me because I was going up there. And when accelerometers came in, I got assigned them because I was the microgravity guy. So the thing that was beautiful about Level-IV was that it really was the foundation for many things from a technical standpoint. It was foundation for you to develop skills. It was the ability for you to build a name for yourself within your community of folks, but also within the Payload Developer community and their payloads. You had Payload Developers that were busy building things and you had Principal Investigators that were the scientists and you started to get a name for yourself among both of those communities. You might be coauthoring a paper on the success of a payload on mission. You also were known and recognized by the person doing the science that was grabbing the information. So it was really cool, and naturally they are the same people that would develop similar types of experiments for Space Station."

Space Station utilization was a transition from Level-IV to Station. Utilization consisted of small payloads that could be launched in any vehicle, but once the Space Shuttle was retired these payloads were much smaller than those launched on the Shuttle.

Tracy Gill said, "Once Station started, they asked me to support the first two EXPRESS racks coming up because they knew I was experienced, they were comfortable with me, and we had those relationships built from Spacelab. Once they'd done those first two racks, they decided they were used to that now, which was fine because things change. But it was nice that we had built a credibility, '*Instant credibility*', and that they still wanted that comfort level with our team, to have them there to just in case something went wrong. We had a lot of ground experience with them and because of the comfort level and experience, they wanted me to go up there for that. I was Spacelab and we carried right over to Station. I think, from the ground processing experience, when I was working Spacelab it seemed like the paperwork system was overwhelming to me. It seemed like it took a lot of agreements to do anything. Then when I got to Space Station, it made it look like my life was easy on Spacelab, because it was another magnitude more complex."

Station Takes the Lead

So Space Station took the lead for the Level-IV personnel still at KSC working payload operations. It was a very different type of work to when Level-IV started in the early 1980s. There was a big change following the *Challenger* accident, still lots of Spacelab payloads to work through the 1990s, and then Spacelab came to an end by 2000. Over all those years, all those payloads that flew, and all the experiments on those payloads that flew, so much science was achieved.

There were lots of lessons learned that then carried into Space Station, but also throughout other areas of NASA, private industry and wherever else Level-IV personnel went once leaving KSC Level-IV/Experiment Integration. "Don't reinvent the wheel, try and learn from the past, take that knowledge and apply it to the new programs, hardware, software, assembly, integration and testing philosophy." Those who worked Level-IV did that, and in the process made many NASA projects and missions very successful. To this day, they continue to spread the wealth domestically and internally throughout the space communities across the entire planet.

References

1. For further information on Shuttle ISS assembly missions see: Shayler, David J., **Assembling and Supplying the ISS: The Space Shuttle fulfills its mission**, Springer-Praxis, 2017.
2. Shayler, David J., **Linking the Space Shuttle and Space Stations: Early docking technologies from concept to implementation**, Springer-Praxis 2017, pp. 172–175.

13

A Place in History

> *"The thing that I learned from Level-IV*
> *was that no matter what the problem is,*
> *don't jump to a conclusion.*
> *Observe and try to decipher what's going on*
> *before you open your mouth."*
> Juan Rivera, NASA Level-IV Software.

As Mike Haddad explained, "Everyone who worked Level-IV came away having learned their own lessons, but also with a larger picture of lessons learned. I know I did. So much of what I am today, not only professionally but also personally, can be traced back to those Level-IV days. Someone may say, 'How can work affect you personally?' Well, the excellence that was demanded by the job we did carried over into everyday life. 'Do it the best you can or don't do it at all'."

LESSONS LEARNED OR OVERLOOKED FROM SPACELAB

The lessons learned from the Spacelab program have been documented via a number of different publications. While the operational side of Spacelab provided significant experience gains in processing and operating scientific payloads in orbit – which had huge benefits for the International Space Station (ISS) program – there were other issues which were not so useful or positive.

The cooperation between NASA and the European Space Agency (ESA) meant Spacelab was the US space agency's largest international cooperative program to date, and the first for the Space Shuttle program. By agreeing to "no exchange of

© Springer Nature Switzerland AG 2022

M. E. Haddad, D. J. Shayler, *Spacelab Payloads*, Springer Praxis Books,
https://doi.org/10.1007/978-3-030-86775-1_13

funds", the use of common hardware was encouraged as much as possible. High expectations were placed on both the Shuttle and Spacelab to deliver, and despite many hurdles and setbacks, both did indeed deliver on that promise of carrying a pressurized research laboratory into space many times.

For the Europeans, the agreement meant selecting an industrial prime contractor to lead the coordination between several companies from nine European nations. Without previous experience in human spaceflight, it necessitated employing US industrial consultants to advise and prepare both the hardware and the paperwork. In addition, all levels of governmental, industrial and academic institutions were called upon behind the scenes to become both mediators and valuable sources of information. All of this was helpful in developing the European role in the ISS, and helped NASA and the United States forge deeper and closer relationships with its international partners two decades before embarking on operations at the Space Station.

Early application of state-of-the-art technology helped reduce the risks and helped keep the Spacelab program on schedule and within costs. The design of the whole system afforded the opportunity for repeated flights of the same hardware, with relatively short turnaround times between missions. There was easy access to spare parts initially, but over a period of nine years of development and more than 15 years of operations, it was found that access to electrical and electronic equipment became a problem. Over the 24 years the program was in operation, this did lead to problems in the supply of parts for Spacelab.

While the production and delivery of just one prototype Spacelab (EM-1) to Marshall Space Flight Center (MSFC) was the result of cost saving measures, it was detrimental to developing the system further in Europe without an engineering model to work on. It became clear that following the completion of the infrastructural element to the program, NASA, ESA or the European countries did not take up the opportunities to develop and utilize the system to the advantage that the Europeans would have expected and preferred. Surprisingly, there were no agreed organizational systems in place for evaluating the results, their importance in science and technology, or their application for on-going studies. It was clear that the direction to create the Space Station was chosen too soon, without considering the dual application of Spacelab flights or hardware to supplement research on the station. The limited mission time of the Shuttle and the delays in bringing the ISS to fruition created a void of almost a decade between the final Spacelab module mission and fully operational science on the ISS laboratory modules.

Despite this, valuable experience was gained in both the US and Europe in preparing, launching, monitoring and operating scientific instrumentation on Spacelab, which had direct application on ISS. The slowdown of Spacelab operations saw the Europeans look toward the Soviet Union, and latterly Russia, for extended duration experience on their Salyut and Mir space stations, which they could then apply to preparing for subsequent flights of European astronauts on the ISS.

Level-IV Lessons Learned

Throughout this book, many references to lessons learned by Level-IV personnel are attributed in their own words. Most of what this section will entail are the lessons learned as suggested by those who lived it. While most of these will be from a working level point of view, some do touch on more of a Spacelab programmatic, Kennedy Space Center (KSC), and NASA overall level. Of course at the time, most of the personnel were at a similar point in their lives, with the same ambitions. They were young, driven to perfection, and few had family ties, offering their full attention to the job at hand. Such lessons learned included:

- Hands-on experience with the hardware and software (flight and GSE)
- Responsibility
- Accountability
- Excellence
- Team work, leading, working together, learning as a group, covering each other
- Attention to detail
- Freedom to work and think out of the box
- Empowerment
- "Can-do" attitude
- Desire to exceed
- Understanding other cultures
- Importance of the work, keeping crew safe, making things work, mission success
- A way of thinking and learning about complex systems (end-to-end, big picture perspective, and how things come together)
- *Esprit de corps*, camaraderie and friendships, how close everybody was
- Social aspects (play and work together), social network.
- Meaningful work
- Institutional knowledge and experience
- Learning engineering and different engineering disciplines
- Training ground for young engineers
- NASA understanding contractor work and role
- Decision making, how to make sometimes hundreds of decisions a day
- Problem solving, (techniques, not create other problems with fix, etc.)
- How to work under pressure
- Bonding
- Dealing with success and failure
- Doing the work yourself, the whole process from cradle to grave
- Dealing with different personalities
- Training ground for everything else that happened
- Working with the different cultures and the different nationalities

- Learning to use logic
- Learning how to listen
- Don't jump to conclusions
- Importance of documenting things, configuration management
- Reducing the number of signatures based on the need
- Discovering limitations
- Communications, clear and articulate
- Infusion of new technology
- The experiments, new and unique every time
- Doing science (working with best scientists on the planet)
- Handling million-dollar hardware; very direct, hands-on exposure to the science with all the experiments and the experiment teams that came through Level-IV
- Imparting skillsets and mindsets that were pretty unique
- Relationships
- Assigning challenging and meaningful work to a young person
- Intricate choreographed workflow
- Managers that were there and supported us and defended us; very little intervention into the day to day operations by our management
- Overcoming the intimidation and the overall awe of the place
- Being flexible and knowing when it was okay to push back on unnecessary processes
- Avoiding late changes
- Don't break configuration
- Understanding how things go together.
- We worked hard and we played hard.
- If something needed to get done, we would do it.
- If you give a group of young, semi-talented engineers the responsibility and trust to go out and do the job, they will get it done
- Trusting your people
- Interaction with the crew (astronauts)

VOICES FROM LEVEL-IV

What follows is a collection of first-hand comments from some Level-IV alumni, looking back from the perspective of two decades after the Spacelab program ended and providing a representative cross-section of their experiences, reflections and memories of their time spent on the team.

Aaron Allcorn: "I believe the hands-on experience that I was able to gain while in Level-IV was formative for my career at NASA. My supervisor in Level-IV was George Veaudry and I remember him saying that one day we would look back at our time in Level-IV as one of the best times of our career. How right he was!"

Alex Bergoa: "It's hard for me to remember every nuance, but working with the people, all those good friends that I made over those years, it was really, really good. And not just work. After work we always ended up either playing volleyball or going down to the Keys together. It was a lot of fun. It was good."

Angel Otero: "I actually believe that Level-IV saved the Spacelab program, because we used to laugh at the old 'Ship-n-Shoot' concept that stuff would go to KSC and just get shot up into space and nothing would have to be done at the Cape. But just look at Spacelab-1, Spacelab-2, D1, Astro, all those missions. The amount of work, the amount of 'repair work', that the Level-IV crews had to do to get those experiments in working order, and the opportunities that the crew had to train with us on what they needed to do. One of my favorite people, one of my mentors that I had not seen in 20 years was Dean (Hunter) and I cried the day he passed away. I learned a lot from all those guys. I was a supervisor for many years and though I don't think I ever came close to what those guys were, I tried to emulate the way they treated us, the way they expected us to perform. They supported us and they demanded that we did things right. Remember the pump that was supposed to pump us out, but nobody wanted to leave? [Explained below]."

The pump Angel is referring to was the concept that Level-IV was a like a pump. You have something (new engineers) coming into the pump, the pump does its job (Level-IV/Experiment Integration, training ground for new engineers) and then the pump pushes out the result to do the next thing (now-trained engineers go out into the world to use that experience to do wonderful things). The analogy was that the Level-IV pump stopped working because everyone was having so much fun and nobody wanted to leave.

Bob Raymond: "I think the big lessons learned [were key]. People came out of Level-IV with a better understanding that we had to pay attention to that detail. We made mistakes, everybody makes mistakes, but [the question was] how we could do a better job of making sure the design engineering mistakes didn't make it to KSC, because we were working with a limited timeline by the time it got to us; people wanted to go fly. Another thing I hope we learned from Level-IV is that you need to hire people with relevant capabilities. It was a training ground. That's how you taught the next generation of space engineers, so you knew there were going to be challenges. [We] Couldn't let egos get in the way. Some did, but overall that was the big value of Level-IV, taking those skilled technicians and experienced QA and engineering staff. Some were experienced and some weren't, but as long as they worked together the taxpayers got a real 'bang-for-their buck'."

Bob Ruiz "The valuable lesson was the attention to detail. The details really made the difference to make or break you. Missing small details would create problems if you didn't get it right. You really had to pay attention, make sure you got everything straight. All your facts straight, your entire data straight, everything recorded properly, because all it took was one mistake and off you went. It could be drastic."

Bruce Burch: "I can't pick out any specific lessons, there are tons. There was caution and warning, or 'Why is this?' It was nice to be able to go back and tell them and give them the story behind why this was there. And then if that drove it home, it was one thing that you'd updated in the procedure. You're trying to fix this, but it can't put the history of why it was. But you can tell them and show them why it's still relevant. All the 'holes in the cheese aligning' and stuff. You wouldn't think something could happen, but it did all the time. You could say, 'Well for this to happen, you have this, this, this, and this happened to line up, and that's impossible'. I've seen it happen so many times. You just shake your head, 'No, it couldn't happen'. But it does. And it's trying to teach the new engineers and the people that, based on experience, it can happen and it does. And it's the simple things that get you."

Bruce Morris: "One of the greatest things about Level-IV was you got to learn engineering on things you never thought, on stuff you would never see in a particular field. I was in aerospace, but I never saw these kind of problems in anything we talked about in class. And so you learned all this very, very practical experience, across a bunch of different engineering disciplines. And all of us got to learn this and then turn it around, and when we got the chance, we could help the next mission along with lessons learned about, 'We've seen this problem before'. We still identified with that first job, and hands-on and everything. Nobody wanted to leave. One of the great things about having such a young group was we were able to respond so well because we didn't have an outside life yet. We were engineering nerds working for NASA. The team that you just spent nine months, a year with or whatever, working together and becoming a cohesive operating unit."

Byron Lichtenberg: "We've made some really good friends for life. Donna Bartoe married John-David Bartoe. She was one of the engineers down there before she went to law school. And the whole bunch of people, it was a great experience, very, very bonding. I think, in the NASA framework of teamwork and everybody pulling together, you had different jobs and different responsibilities, but it just showed the dedication, the sincerity of folks. I mean, to be there until three or four in the morning running these tests was really a sight to see. I was just honored to be part of it."

Cheryl McPhillips: "It's complicated because you had different cultures and so we set the tone for that. I think being young, like we were, we got a lot of people that came out of Level-IV and ended up in leadership positions now around the agency. A lot. The thing that I appreciated and loved most about Level-IV was the opportunity to be responsible for the success or failure of very expensive flight hardware experiments. It was a time of accelerated learning. We had classes, some that we taught each other, and OJT [On the Job Training] that taught us about the Space Shuttle systems that supported the experiments with communications, power, cooling, pointing, and structure. I think because of this responsibility levied on us, many graduates from Level-IV went on to significant leadership positions."

Cindy Martin-Brennan: "From the 'big pictures' perspective, I think most of us benefited from having that hands-on experience. At NASA, I think a lot of us moved on to positions where we could be better employees or better managers, even outside in the space community. I think that having that hands-on experience just afforded us the opportunity to be better civil servants or better employees. I think the lesson learned for me is that we do not have enough of those experiences for NASA engineers these days. It's hard to tell somebody what to do when you haven't done it yourself. Just having the experience and knowing what I'm talking about and being able to talk it technically. I think that is missing in NASA now. I don't know where people get that anymore, that type of hands-on experience. So if I didn't have that [experience] I don't think I'd be doing [my job now] and have the credibility I have today."

Craig Jacobson: "I consider Level-IV and the first part of my career to be the most rewarding, because they were doing science, the consequences of the results of those payloads. Riley [Duren] told a story about one instrument that started collecting some kind of data, climate data, and that just turned into mission after mission after mission. And some people in Level-IV, like Mike Wright, Bill Sheredy, Riley and some others, went off to do development experiments at other centers. With the technicians, really it was phenomenal. These guys and gals were there to help us and to do work, but they taught me just everything about removing and inserting pins in a connector, building cables, all the soldering stuff. The only thing would've been to have some sort of place to go after Level-IV that was as rewarding. I considered it important because those experiments went on to define whole areas of science, space science, Earth science and astronomy and everything, so [it would've been good] to continue to do that. When Space Station came along we were working on infrastructure, we weren't working on science, so it just wasn't the same. There were some people that came to Level-IV that didn't stay. For some reason, they didn't synchronize with Level-IV. They would go do other stuff. There's nothing wrong with them, they were great people and everything, but it really had this self-filtering function."

Damon Nelson: "What a great opportunity. I had started my career with Shuttle and I just got this early feeling in my career that I wanted to do something on the payload side, because it would be much more unique and every mission would be a little bit different. And so looking back, I'm really, really happy. I definitely felt like I checked the box from a lot of perspectives, because it was very taxing on the family, super-long hours. We were having our third child while I was working Spacelab D2. But yeah, it was all good."

Darren Beyer: "What I found immensely enjoyable was getting to work directly with the techs and QA guys. And I still think to this day that we had some of the best technicians that you could have at the space center. I mean, there were a few exceptions, but for the most part they would all work with you if you worked with them. They would work with you to get the job done. They were a union, like all technicians in a space center, but they wouldn't be strictly union on a lot of things, just to be able to help you get the job done. And I really appreciated that with those guys. Also, whenever you mention [you work for] NASA, people's eyes light up, and whenever you mention that you were a Space Shuttle Experiment Engineer at Kennedy Space Center, then people's eyes *really* light up. It gives instant credibility to you because NASA is a pretty prodigious place to work."

Debbie Bitner: "I think for me it was doing [things] myself, like writing the procedures, reviewing the drawings, things like that. Because then I went to Spacelab oversight, looking at and approving procedures that other people wrote. I don't think I would have been as good at reviewing those procedures and being able to read drawings if I hadn't had to do it myself. So I think it made me a better engineer for the next step that I went on to."

Diana Calaro: "I think one of the things that helped me, although I did some of the design stuff that was pretty focused, was you still had to see it as a whole. There wasn't a whole lot of that discipline called system engineering. So I think what Level-IV helped me do was develop that. You had to understand how the interfaces fit together and a certain level of testing. We had to learn how the whole system worked as a unit, not just one particular piece. And I think inherently that learning that from Level-IV helped me out when I went into Station."

Don Dolby: "Working at Boeing, working in Level-IV, that was definitely the best working experience. Plus the people too. I think we had a really good group and everybody seemed to get along really well, both engineers and technicians. It was a great job while it lasted but unfortunately it didn't last very long. I think it prepared me for everything because I think the kind of work I'm doing now is like Level-IV. The whole idea was for companies to ship their equipment in and then they would have them put together on site. And what I'm doing is almost same way. Level-IV work is the way we worked for different companies."

Eli Naffah: "I think the aspect of leading teams, and leading diverse teams, to achieve a goal. I remember more than the technical problems 30 plus years later. Just leading people and understanding how to get a team to all pull together. Those were really important things that I learned in Level-IV, even more so than the technical things that we typically fixed as engineers. That's what we did. There really wasn't anything technically that we couldn't fix. But you weren't going to be successful if you couldn't lead the team. There was a great *esprit de corps* and there was definitely that NASA 'can-do' attitude, and that was fostered by the original five people that started Level-IV and everyone who came to join. It was contagious. I imagine the stories are countless of the impacts of the Level-IV people. That's the legacy. There were so many people in Level-IV that went on and had tremendous careers and Level-IV was that foundation that really gave us the core that we needed to go out and be successful at NASA. NASA people in many instances no longer have the technical expertise to do what they need to do, and that's why programs fail. That's why they cost way too much, because NASA is just monitoring the contract without really any knowledge or insight into the how. Our second family was Level-IV and there's a bond still today between all of us. That's a strong bond that we built, and frankly, I've had many different jobs at NASA, they're all very good jobs and very interesting jobs and I love it all, but I've never had that bond again."

Elias Victor: "The best times that I've had were with the Level-IV team. I've been through four or five different organizations at the center, doing different kinds of jobs, but I have never found what we had back then. And I'm afraid I'm going to retire and never, ever find it again. Hopefully we can inspire those new kids coming in from school the same way that Level-IV did. I had such a great technical experience with Level-IV that I've spent my entire career with NASA looking for that same thing, looking for that same feeling. That same camaraderie that I felt with a group of people, that same opportunity, the same technical hands-on capabilities that were offered to a brand new kid out of a school. That's the environment I've tried to build at work for the new kids that come see me. That's what I was taught, to try to make their lives fun, their technical work fun, and try to use their strengths so that they can produce a better product. That's what Level-IV was."

Frank Valdes: "Once you do the work, you know all the hooks and you know the loopholes and you know the excuses. So it was a lot harder after that for anybody to say they couldn't do this or that, because you'd been down path and what they were telling you was BS. 'I've done that. It *can* be done'. You had that perspective. I think Level-IV, as the concept of the pump, was good but it was also poorly implemented, because there was no plan on how to leave. The Level-IV effort, the hands-on to actually try to get the engineers to have some real experience was good. It was a good program and probably also the thing that made it work was the people. I think if we had different people it might not have worked as good as it did because everybody was pretty much at the same time, at the same place, in their lives, and they had the same ambition or at least the same mentality of getting the work done."

Gary Bitner: "That's what I liked about that group. We all got along really good. There was a little tit-for-tat once in a while, but when it came down to it, we all put our heads together and got something done. So that's what I liked about it. Plus all the training we had, torquing and stretching and that kind of stuff. Running torque that helped me later on rebuilding chillers and things like that. You got used to those procedures so that when you tore something apart and put together, you wanted to make sure you did it right. So that definitely helped me."

Gary Powers: "That was a plus of the whole thing, watching people come together, including the contractors in the mission itself. The overall mission was the driving force, not just to Level-IV, although the Level-IV [personnel] should have got a lot of awards because it accomplished exactly what it was set up to do, getting the NASA engineering involved and motivated. We got a lot of lessons learned and it was important to document it. I wonder sometimes if anybody ever reads it, because we'll make the same mistakes over and over again. But if you document lessons learned and put it into the database, the artificial intelligence would have spit it out on the computer for you. One of the highlights of Level-IV was how the astronauts came in and did their laptop routine like they would do on orbit, and played with us. It helped the entire team understand the importance of what they were doing and it drove them closer to the mission, so when these guys flew they knew them by face and they knew what they were doing. As a matter of fact, if the flight crew ran into a problem, guess who they'd call? Someone from Level-IV to get on the headset and talk them through it. Experts from the specific Level-IV areas. And that was a real plus about the whole thing."

George Veaudry: "I left Level-IV as the Branch Chief there, I went to Expendable Launch Vehicles and I was the Branch Chief there, doing the work there and looking at all the different spacecraft and launch vehicles and how they fit the overall mission for NASA. Looking at all the tests and checkout of that equipment, but those were obviously done by the engineers that were there. My past experience from what I learned in Level-IV, my hands-on experience from the OSTA-3 and MSL, that went with me as far as a supervisor goes. That was one of the best things I think NASA did, at least at Kennedy Space Center, was to provide those people with that kind of opportunity and work experience for sure."

Gerry Rivera: "I think in general there were technical aspects that I kept applying throughout my career of 38 years total. But the basics of teamwork and the basics of working with people from all cross-sections, ethnic, national, men, women, everybody, was sharing the same respect for each other. All were very, very knowledgeable and very capable people. I think that was the best thing that I took with me from Level-IV, because when it boils down to the basics, it's all about how you personally treat other people that are either at your level, working for you, or above your level, from management or supervisors. You treat them with respect. You empower them and you trust them and always leave that communication line open. I mean, nobody's perfect. Everybody is gonna make errors, but you have to be open about being able to sit down and discuss them, fix what's wrong and press on. So I think that my main lesson is you can have all the assets in the world, in terms of material, equipment, budget and everything, but the best, the top asset is the people. Those are the ones that you really, really have to care for."

Herb Rice: "The division chiefs and the directors that were there had the same longstanding view of, 'Why should we let somebody else have all these good people?' and that carried all the way up through the director. The problem was that

Tom Walton, Willie Williams and Bill Jewel and all the folks at the top really liked what they were doing and didn't want to change. And every director outside of that organization wanted some of those people. So you wound up having all the other directors opposed to what Level-IV were doing [going] in front of the Center Director. But the Center Director supported it for a long time."

Hermann Kurscheid: "It was really a great endeavor and I am proud to be a member of that and I think quite a lot about what we have achieved there. Lessons learned. I think for me it was necessary to have right in the beginning the common understanding what the procedures should look like and which inputs had to be provided, and what procedures had to be reviewed. So definitely, I think in this respect it is important right from the beginning to have the same structures and procedures. [That makes it] easier to fill in the procedure and also do changes and review the changes, the revisions."

Jay Smith: "We were all young kids. We knew we all had our experiences, but really when you look back on the stuff we were doing, then it was unbelievable for us. It's really surprising, really it is. We pulled it off. And to this day, working as an auditor for American Airlines, wherever they work on our aircraft, we go and make sure that they're doing it correctly. I mean, we work with these other groups and we want them to succeed. Attention to detail!"

Jim Dumoulin: "One of the things that made a big impression on me was when I did the first couple of tests. I realized how dependent a test was on the assembly process, of the whole picture you have as an experiment engineer. When you were laying out how you were going to test something, you'd get a mental picture of exactly how everything was cabled. One of the cool things about Level-IV was that we were taught in the training auditorium [at KSC] and we could go right there from our control room. They brought professors to Kennedy Space Center because they wanted Level-IV engineers to learn new skills. Essentially, NASA offered training, graduate degrees, while you were working for NASA and they paid for it. Living out there on 16-hour days, it wasn't too big a deal to take a break during a test, go do a two-hour lecture in the training auditorium, and then go back to your console. A number of us did that. There were just all sorts of things that were beneficial for when you started dealing with payloads. You could say, 'I know something about that because I took a course in it in college', so it was a big help. Also, we had the control to be able to fix problems and the visibility to understand the problems because every day the guys were there. The right people in a kind of a 'can-do' attitude, people who knew enough about everything to solve problems. You didn't have to go back into a dozen meetings and figure out how you were gonna do something. People just did it."

John Conway: "[There were] Two things in my judgment. Regardless of what your disciples say, it was only for two things. One, it was an engineering development program. Because you didn't want to hire bright young engineers who would never build anything, who would never design anything, who would never have to make something work, or manage people who were doing it, whether they were contractors or civil servants. Career wise, management wise – to a person in my opinion – NASA were successful with a strong technical discipline. Didn't matter what it was, they learned that disciplined engineering process and became expert in it. The second thing is it was a way for NASA to assist the experiment developers. They would work with a PI who was the expert, who was the knowledgeable guy, and help him figure out how best to integrate the experiment into the Spacelab system. They had a chance to work with very bright people, and that rubs off on people when you do that. So it was a way to assist the scientific community and at the same time it was a way to develop engineers and talent for the future."

John Grunsfeld: "Having cut my teeth on Spacelab, that taught me a lot of things that I should be looking for in subsequent missions, as far as in what kind of testing was going to be done, and what was successful. Our Astro-2 Spacelab mission was very successful, with everything operating the way it was supposed to on the hardware. So I looked for those same kinds of things on the Spacehab mission and then also on the Hubble mission. There's no question that the integration and test work that went on at the Cape, with the Space Shuttle, was one of the large reasons that we had so many successful Hubble missions. For me, flying three of them, just imagine if the crew trained in Houston and never had the opportunity be up close and personal with the hardware on the floor, to walk around it, look at it, do EVA interface checks. If we just had the simulations in Houston and then went to orbit, [think of] all the things we wouldn't know, having never actually operated the real hardware, or seen displays of the real hardware working. It would definitely have been a different experience."

Johnny Mathis: "I think payloads used to pride themselves on the fact that they hadn't ever held up a Shuttle [launch]. We can't say we've never held up a launch anymore, but we certainly could through Spacelab, which to me was a matter of pride with all the folks that worked there. We were always ready on time. Time being relative, the schedule always slipped, but we didn't slip as much as Shuttle did. A lot of the lessons from Spacelab were that things got simplified, interfaces got standardized. To me, one of the best things about Spacelab was it was always something new. Always someone came in and they had spent probably 10–15 years of their life developing this experiment or this test. They wanted to do this research, they wanted to perform it, and you were there to help, in my case as a Mechanical, to get their experiment integrated into a rack, get it ready to fly. The excitement from them was always infectious. You got to meet people from all over the world. I dealt with experimenters from Canada, from Japan, from Italy, from Germany, you name it."

Joni Richards: "[With] The Level-IV core team, everybody was wanting to scratch each other's back. There was no backstabbing. [If] Somebody had to go sit in for somebody so they could grab a quick bite, they would do whatever they needed, or go get you lunch. [If] You came in late, no worries. But I think it was just as much that the stars align the personalities, the right mix of characters. You can't recreate that and we are still learning from what we did from Spacelab that has folded into the ISS and so on. The benefit of having been in Level-IV was to see that 'cradle-to-grave' when things got started. Now I knew the whole story, the whole lifecycle process. Reiterating on the whole Level-IV mentality, when we had a problem, well it wasn't just Scott [Vangen], but Scott often comes to mind, saying, 'Go look in the green log book', that typical, ugly green government style notebook. It was like a hardbound, almost burlap-looking book that Cindy and Scott and a lot of those folks used that had been there years before I even showed up on the scene. Cindy and Sue Sitko and Jim Dumoulin would make sure to log all the different problems that would occur. So you could look at that tab for that particular subsystem and say, 'I found it, I found the error message and this is what we did last time. Let's see if it'll work this time'. It was just working with a diverse group of engineers and international partners and everybody just did what needed to be done to make it happen. We were that kind of mentality of just making things work, and as sappy as it sounds, I get goose bumps because that was really us and was just so cool."

Detail of the Green Logbooks. [Jim Dumoulin.]

Josie Burnett: "Level-IV definitely was the best training ground for everything else that happened in my career. I mean, I can almost tie every lesson learned to that experience. Just something that I either observed and did right, or screwed up, during Level-IV. Working with the different cultures and the different nationalities has also been just a blessing. Being able to understand how, say, a Japanese community designed and how they solved problems differently than we did. A different mindset, a different culture and just respecting that they still came up with an engineering outcome, but there were different ways of doing that. I've had the chance to work with Europeans and a little bit with the Russians. They do a better job of passing on, doing succession for their successors. And it's hard for an average person who might be struggling financially to say, 'Yes, I'm willing to spend money on in the space program'. It reminds me of Level-IV when I run into the SpaceX teams. I indirectly know of the Blue Origin teams because friends and my husband have gone there. I hear their stories and it's like Level-IV."

Juan Calaro: "We worked issues all the time and that's why, I think, Diana and I talk a lot. I don't get too worked up about things and try to keep a level head. I got that from working Level-IV, because issues just came up all the time and you had to just stay calm and work through them and not get too excited. I think that was a lot of the Level-IV training and working under pressure. We made lifelong friends, the friends that I still have today. Your career moves on, you meet new people and you have different interactions, but I think my closest friends are still the ones from Level-IV. You had your good techs and your bad techs, and when you got a job schedule then depending on who you got, you knew it would make the job pleasant or the job difficult. Same with Quality, that was part of the game. And you learned how to deal with people. There was so much that we got out of Level-IV at such a young age; how to work with others, how to work as a team, how to work with people who could be difficult. Those were some of the things that maybe I didn't appreciate much at the time, but I think I'm using those skills today and it's what's gotten me through. It's the experience that I had opening my own paper, writing troubleshooting plans, and the way of thinking of how to resolve issues. That's what I'm doing these days. So it's kind of what I learned in Level-IV. It's carrying me through my career right now. I wish I knew back then how fortunate I was to start in Level-IV. But now is when you appreciate it. I was super lucky that I landed in Level-IV."

Juan Rivera: "I think the best thing that I learned in Level-IV, and I'm still very grateful to be part of that, was to use logic; 'Why this is not working'. I learned why testing is important. The discipline is also important. Another thing is you had to learn how to listen, then talk. I learned from Level-IV that no matter what the problem is, don't jump to a conclusion. Observe and try to decipher what's going on before you open your mouth. That's how it was. Also, I was lucky to work with great people."

Kevin Zari: "The experience really helped me grow good engineering discipline. When you're in school, you're a student and you'll follow the exercise, but this taught me the consequences of not following through and the importance of documenting things. One of the most unique things about Level-IV was that you really got your hands into it, and it gave you a real-life meaning for what it was you had just read in the textbook. I thought that was a great experience. Everyone felt accountability. Everybody felt responsibility. But the amount of signatures for the payload processing was nuts. We could have improved on that; that was just tedious and a nightmare. That was one thing I would have suggested, to reduce the number of signatures based on the need. That was one of the things we took with us out of Level-IV and a bit out of the beginning of Space Station processing, to streamline things to make it more efficient, because while there's value in having things very well documented and very rigid, you pay a price both in time and money, and certain processes were just too much attention to detail. Level-IV definitely made me the engineer that I am, there's no doubt about it. If I could be a Center Director or an Agency Administrator for a day, I would bring back Level-IV. I would bring back the opportunities it gave to a fresh-out-of-school young person to actually contribute to science, to critical space flight hardware. There is nothing like it. There has been nothing like it since either."

Liz Schlierf (wife of Roland Schlief, did not work Level-IV): "I remember about that whole time just how smart, how funny and how close everybody was. I had heard about this from Mike Wright for years, and he told me about the parties before I ever went. I just thought it was a really cool group."

Lori Wilson: "Nothing could ever reach the level of satisfaction provided by working with this group on this type of mission. Level-IV Integration was a unique experience because of the attitude of all the co-workers, most of whom were also new to engineering, young [unlike me], intrigued by the challenge of the work, and given the freedom to go 'all out' and think creatively about how to get the job done and to be successful at it. Yes, we had time constraints and safety rules, but very few 'company policies and rules' to interfere with the job. This resulted in 'can-do' attitudes and an extreme desire to succeed. The camaraderie and good-natured jokes made the 16-hour days a pleasure."

Luis Delgado: "You had to be on your toes, but I loved the hands-on part. I loved dealing with the technicians and the work being done. They really taught me how to be an engineer; how to torque, what was running torque, or breakaway torque. I started as a Fluids guy and ended up being a Mechanical guy, and I did crane operations. So I was blessed with the opportunities that gave me unique work."

Mark Ruether: "Experiment Integration was unique in that you had all of these young, relatively fresh out of school engineers come together and it was a good lesson of assigning challenging and meaningful work to a young person that has a lot of energy and a desire to go make something happen. The Experiment Integration experience was good because then you could ask the questions of the payload customer that would get to the heart of what the Experiment Integration person needed to know. The better you understand that flow from a working level, the better you can ask the right questions to drive through the information as needed."

Maurice Lavoie: "Well, I guess one of the big things that I learned was to accept other cultures. Spacelab taught you to work with a wide variety of people. You had to work with the Germans, you have to forget the history. You had to work with the Japanese, forget the history. You got to learn and know these people and learned to accept where they were from, how they reacted and how they dealt with things. We found that they were no different than us. They were no different than my wife or our daughter and all of our friends. There were perceptions, there were differences, perceived things, but actually they were like any other person. And I think my experience from Level-IV and working with all of these Europeans was just getting to the point of accepting people are different, but they're the same, and you can have great times with them."

Maynette Smith: "Level-IV was very special. If we could ever replicate it again, that was just an amazing place with amazing people working together to make a really fantastic things happen. And I can't imagine a better place to start my career than there because I learned so much from so many people. And not only the people in Level-IV, but everything that came through, every team that came through, and they knew it was special too. Even the ones that came in kicking and screaming because they didn't want to test, at the end were usually appreciative of what you were doing. So it was just phenomenal. And the managers that were there supported us and defended us, and I think it was recognized within the agency what an asset this kind of environment was. And out of it came a ton of people that went to a ton of different places and took that expertise that they learned everywhere."

Mike Hill: "I overcame the intimidation and the overall awe of the place and got to where I was giving [everything to] the real work. I didn't want to go home. I wanted to stay and work. I did not look forward to the weekends because I would be away from doing the work. It was so new and so fun and such interesting work that it still shocks me to this day that I had that attitude. Most of what I learned carried onto later years. How to deal with the technicians or the QA or QEs. The whole spectrum of human behavior that I don't even remember coming across going through college. I can probably think of a few a-holes, but to the extent of trying to get work down in what we were trying to do, if they had somebody that was obstructive [you learned] how to deal with that. Some people you never could come to terms with, but you had to figure out how to get around them, the politics of all that. I have always felt the skills I learned writing procedures have been a foundation that I have built on the rest of my career. Level-IV work was invaluable, and much appreciated. I didn't realize how much it meant to me until many years later, looking back at how far I had come."

Mike Kienlen: "I go way back after 33 years of working for NASA. The Level-IV experience was the training ground for everything from then on. I mean, everything. All the decisions that I was involved in the rest of my career all fall back to lessons learned during Level-IV. Everybody who came to work a mission, who showed up at Kennedy, realized how awesome Level-IV was. We were committed to the success of the mission and did whatever it took. We worked a 16-hour day, broke all the rules, but we didn't put 16 hours on the time card. We put 11.5 because that's all management said we were allowed to work, but we weren't going to leave because job wasn't done. The Spacelab missions would not have been successful if it wasn't for the Level-IV group. We were responsible for building hardware and testing it and flying it. You learn powerful lessons from that. One other thing was don't make last minute changes without testing the change. I can't tell you how many times I sat in a Launch Readiness Review where engineering had an issue and wanted to make a design change, and argued it was a simple fix and we didn't need to run a complete End-to-End retest. I would be one of the adamant people to say, 'Nope, no way'."

Mike Wright: "KSC's Level-IV was a good example of how a culture traditionally based on a government-contractor counterpart structure could address, incorporate, facilitate and develop long-term institutional knowledge by attracting and developing young engineers to do more than just 'push paper'. If a 'Level-IV' were ever to be established in the future, one lesson would be to institute an objective, well-defined method for assigning work to engineers. More philosophically, organizations at all levels of the agency must develop and then maintain their institutional knowledge and experience. No book on KSC's Level-IV would be complete without presenting the camaraderie that developed within the organization. This was facilitated by multiple factors, such as age [most of the engineers were in their twenties], and the size of the organization, with each discipline branch [i.e., Electrical, Mechanical, Integration, etc.] having only a couple dozen engineers. While every engineer felt a part of the whole, the relationships among the engineers went beyond just working as a team; most were friends, both during work and after hours. A long day of testing didn't see most of the team go home for a good night's sleep, but rather at the local pizza parlor. Many weekends and other time off was spent with each other, everything from beach parties to ski trips."

Rey Diaz: "We had really a pretty good team. It was very hard to stop this team. This team always wanted to get maximum out of everything."

Rich Jasnocha: "I liked seeing the results, seeing the science data when we had a successful mission, so seeing the end result. That's something that gives you motivation to go forward, that you can take with you. You feel a sense of pride, accomplishment. Everything was done with paper and it seemed like it took things so long to get done, and we always wanted to make the process more efficient. It was so unique and it was really awesome, while I was doing it. I enjoyed it."

Riley Duren: "One of the things that I really took away from Level-IV was the very direct hands-on exposure to the science, with all the experiments and the experiment teams that came through Level-IV. That was the most exciting thing for me, being able to get to work with some of the leading scientists and researchers in their fields. Level-IV laid critical groundwork for future advances both in the technology of the instruments on the payloads and the scientific discovery in the Earth and space sciences. Some of those future advances really were transformational. Also, there's a balance in doing complex engineering efforts, between doing the right kind of processing and being flexible and knowing when it's okay to push back on unnecessary processes. We figured out how to innovate and to cut out unnecessary processes and do it in a way that was safe. What I got out of Level-IV was this end-to-end mindset that when you work on a system, you're not just working at this interface between what bolts to the Shuttle, but you need to understand why we were flying and what problem we were trying to solve. What science questions are we trying to answer? How does the hardware work? How does the software work? How does the thing work as a system and how does it operate, including normal and off-nominal scenarios? And so Level-IV at that time was probably one of the only places in the world where you could go get that kind of end-to-end experience, from cradle to grave."

Roland Schlierf: "I'm definitely a 'work hard, play hard' kind of person, but a lot of it just seems like it's the quote of the day, or just a cliché or whatever. But it seems like we really lived it back then, like it was natural, just the way things were. A lesson I carried throughout my career was to be really careful about any late changes, because that's what bit Tethered Satellite. Sometimes someone wants to make a change after it was tested; it still happens all the time. 'No, you're not changing the hardware the day of the test, don't you understand?' So you've got that career progression that you get someone that cuts their teeth in Level-IV testing payloads. We had a diverse team before it was cliché. It seemed like we were out in front of that, to have leaders in influential positions that were not white males. I don't know if you have anything in there about team diversity, but we had very strong diverse teams. Also, if we didn't understand what was going on, we'd open an IPR [Interim Problem Report], right out of the gate, and then we'd figure it out. We were going to document and then we were not going to stop pulling on that thread until we understood what it was. And then if we never understood what it was, then it was a UA [Unexplained Anomaly] and it was going to CoFR [Certificate of Flight Readiness] and we were going to say that to you and that we don't understand what it is. I think it comes back down to a strong lead, like Scott saying, 'This is what we do. This is how it's done. We're not going down any other path'. And I still have [that mentality]: 'No. We're getting ready to go do some operation that's going to cost a whole lot of time and money with a whole lot of people. Often hazardous. It's an IPR; open it and work it. Period'."

Sam Durrance:, "Well it's a shame we're not still doing Level-IV. The Shuttle was an incredible vehicle and a little riskier than we had anticipated, but I think it settled down by the 130th flight. It was very expensive, but it was an incredible machine. And as far as staying in touch with the people I worked with in Level-IV, not as much as I'd like. Scott Vangen, Craig Jacobson, Mike Haddad and a few others are some of my best friends!"

Scott Vangen: "All fun experiences and lifelong friends and relationships with all these people. I don't see some of them as often as I'd like to, but John [Grunsfeld] of course, is an extremely close friend. And Ron [Parise] sadly passed away several years ago, succumbed to cancer. Sam [Durrance] of course is local and a friend, and will be for the rest of our lives. I have a little different take on that. When you have a culture and a group of people that work well together, the lessons learned are infused already in the environment. So from Spacelab-1, we tended not to make the same mistakes on Spacelab-3 and we didn't make them on Spacelab-2 and Astro. I mean the sequence of things. I had a lot of conversations with Tom Breakfield who was our director at the time. We took paper on everything we did – IPRs and PRs – and what bothered me and others, not so much flight but especially ground systems, is that you opened up the paper, saw what you were doing, resolved it, and you put it away. But then it doesn't link. So myself and Cindy, Jim and Craig and some others, put together an argument that we should be able to map these observations to the IPR so we could reference them. And that's where the idea of the green books came from."

Sharolee Huet: "The people I remember best were the Germans. Having a famous German last name [maiden name of Kepler] came in handy, but they thought I could speak German. I wished I had spoken German, it would have been fun. They always brought me chocolates and told me wonderful stories of Germany and Europe. Working a series of missions [ATLAS for example] you got to see the experimenters several times which was rewarding. The same was true of the TBE [Teledyne Brown Engineering] folks. They tried to take our inputs on board to make the next mission easier for processing, so hopefully less PRs [and therefore time]. I didn't like working JSC missions. They did not listen to our lessons learned and cost us processing time. Middeck experimenters were fun."

Tia Ferguson: "I spent a lot of time in the machine shop getting things re-done, using epoxy and gluing things back together. It was a great learning environment. Integration and testing was just such a great learning environment to see what not to do; how to design things, how to not design things. I think it was critical, not just the competence but the mechanical skills, to know exactly how things go together and to have been able to torque a bolt myself, to know that you can torque the head off a zero or a number two bolt. Those kinds of feel things we got is a lot of what engineers don't get enough of today I think. There's so much analysis now instead of testing, and now we lean so much on the analysis that you really don't get that hands-on feel. Mechanical engineering that you can understand physically. And I'm a feel person, I like to touch things. So you just really got a good understanding of the mechanics and the engineering from the bottom up, of the technicians and the machine shop. Those are the people who really understand design. They understand how things go together. What will work, what won't work. To be able to work side by side with them and do the same things they were doing was just an incredible opportunity. But one of the other things about that was the quick turnaround. You couldn't analyze something for months. You had to get that thing together. It was such an urgency, and learning how to make quick decisions and not sit back. We didn't sit back. We had to make spur-of-the-moment decisions, especially when doing the testing stuff. We had to dig in and understand how the electronics worked so that we knew what wasn't working when we had to troubleshoot. And we had to do it now! That's the other thing that you really got from that Level-IV KSC integrations."

Tony Ornelas: "If something needed to get done, we would do it. We weren't gonna sit back and wait for somebody to authorize this or authorize that. We were willing to take the risk and do the job. Now with that came a lot of responsibility too, so you had to be careful that you didn't just jump in at everything, and did what you thought was the right thing to do. So we did not hesitate to do what we thought was the right thing to do and I don't know of any incidents where we didn't do the right thing. That was one of the nice things about the Level-IV guys. They were willing to take responsibility and take corrective action as required. That's just the way we were. We did what had to be done whenever it had to be done and I think one of the biggest lessons learned from that was that if you put a group of young semi-talented engineers together and give them latitude and leverage to go out and do the job and you entrust them to do that, they will get the job done. You just have to give them some guidance and help them out. You support them and give them the resources they will need, they'll get the job done. And I think that's something that's true for any organization. You trust your people and you guide them and work with them then you'll ultimately have success. I think that was what Level-IV was really all about. I also think the one biggest piece that is probably going to be missing from your book is because of the passing of Dean [Hunter]. We will never get his version of it. We've really got to pay homage, to look at what he did and what he did for us. Without Dean, we couldn't have done this, I don't think, because he was always going to put himself in front of anybody and everybody to defend his people. Shame he's not around to see this book come about."

Tracy Gill: "We worked hard and we played hard and I know that's the kind of thing that made our team work well, because we had to. We had those bonding experiences. So we had parties during Level-IV and then everyone knew each other really well. With our PIs, we'd go to KARS [Kennedy Athletic, Recreational and Social], or we go to somebody's house, or they'd have a beer social. We'd all go to Cocoa Beach pier or to the PI's hotel, and we all knew each other really well at the beginning, so that all those relationships were built into the rest of the mission. That's when the team would really work at its best and people would go to the extreme to make sure everybody was successful. I remember going to the bathroom at some restaurant, and seeing a D2 sticker because all the Germans were everywhere. They were pervasive. Everyone was pervasive. Then we would have those Engineers Skip Days [ESD]. Every once in a while, we'd just say, okay it's ESD, because it was electric. We didn't encourage people to just skip work. 'Let's call in sick, take the day off and Friday we're going to have a barbecue and we're going to do this or whatever'. We did that every once in a while. We established great relationships around the world and those things carry over even to today, because I can go to a meeting and talk to ESA people about Spacelab, and when they know that you were involved in Spacelab they look at you a little bit differently, because it was such a momentous effort that carried a lot of weight around the world. Same with the Japanese. I worked with the Japanese a lot back then and that was an *instant credibility*. Same thing with other people at Marshall. They say, 'We know Tracy back from Spacelab', and it immediately gives you a level of credibility that you might not have had. We had great times socializing with all of the payload customers from other NASA centers, from around the US, and from around the world."

Beginning with OSTA-1 in November of 1981 and ending with Neurolab in March of 1998, 36 of NASA's Space Shuttle missions were considered Spacelab missions because they carried one or more of the Spacelab components, including the Spacelab module, the pallet, the Instrument Pointing System (IPS), or the Mission Peculiar Experiment Support Structure (MPESS). The experiments carried out during these flights included astrophysics, solar physics, plasma physics, atmospheric science, Earth observations, and a wide range of microgravity experiments in life sciences, biotechnology, materials science, and fluid physics, which included combustion and critical point phenomena. In all, investigators from the United States, Europe, Russia and Japan conducted over 760 separate experiments. These experiments resulted in several thousand papers published in refereed journals, and thousands more in conference proceedings, as chapters in books, and in other publications. A number of these investigations are considered landmark experiments, in that they produced results that set the tone for new vistas to be explored, and subsequently added greatly to our body of knowledge of the universe, the planet on which we live, how our bodies and other biological systems function, and the science involved in materials processing.

THE LEGACY

Starting near the beginning of the Shuttle program, Spacelab grew as the Shuttle program grew. The Shuttle was a "remarkable flying machine'; Spacelab was a "remarkable science platform". They worked together to stretch human experience in space, to discover things that help people on Earth and will help people leave Earth for the Moon, Mars and beyond. Spacelab was a huge part of the human experience and will go down in history as one of the great stepping stones that hopefully one day will see people head to the stars.

Spacelab also brought people together from across the globe to work on very complex, but short duration human spaceflight missions. There were many issues to overcome: different measuring systems (American measurements and the European Metric system) to agree; language barriers (including French, German, Italian, Japanese and Russian); and understanding different cultures and approaches to work ethics. But it worked. It took time, a lot of discussions, and give and take on both sides, but it worked. There were great successes and disappointments, notably the demise of the series and the lack of early flight opportunities for ESA countries that had built the hardware.

All of these were valuable legacies going forward into the ISS program, where national identity and a 'go-it-alone' approach was shown to be expensive and fraught with delays. It could be argued that the genesis of ISS operations lay in part in the Spacelab series of missions.

14

Closing Comments

*"Level-IV was the coolest job
at the coolest location for the
coolest agency in the world."*
Darren Beyer
Level-IV Aerospace Engineer.

In closing the book, both authors give their own thoughts on the Level-IV era at KSC.

Mike Haddad:
The lessons I learned from Level-IV taught me how to be a practical, everyday engineer. I learned a lot of theory in school, but Level-IV taught me some basic engineering functions, like the proper way to torque bolts, leak checking of fluid systems, the proper way to lift hardware using cranes. My gosh, I could go on and on. It taught me how to work with a variety of very skilled and smart people from all over the world. That ended up creating friendships that will last a lifetime. It taught me how to work through any differences to achieve a goal, problem solving, and the need to always find an answer or solution to a problem; the 'can-do' attitude, to never give up. It taught me attention to detail, and how the smallest item you overlooked could really come back and bite you. It taught me how to multitask and still achieve excellence, about the need for play to keep work from making you lose your mind. I think that part of the passion we had was because we loved the science end of it. We loved the science that was being done by Spacelab, the experiments, the PIs. We loved that these guys were going and discovering stuff that nobody else had done. That science was a real motivation for us. We were doing things to help us learn more.

© Springer Nature Switzerland AG 2022
M. E. Haddad, D. J. Shayler, *Spacelab Payloads*, Springer Praxis Books,
https://doi.org/10.1007/978-3-030-86775-1_14

David Shayler:

The visible element of each Shuttle mission was the flight itself, but for me as a researcher there was also the background story. The development of the missions, the experiments flown, the crew training and assignments, the flight control team, and the hardware used and prepared. On top of this were the postflight activities, the reports and results, lessons learned and applied. It is like putting together a huge jigsaw puzzle, with information derived from many sources that, at times, frustratingly offer conflicting facts.

This attention to detail makes the research even more challenging, yet rewarding and above all, fun. Finding that elusive item of data to complete a bigger picture is part of securing space history in print before it is forgotten or lost forever. This is what attracted me to the cooperative telling of the Level-IV story.

As a European, I have always been fascinated by the Spacelab program and saddened by its early demise. One of the challenges in recording each Shuttle flight is to detail the allocation of hardware and date the process of ground operations in between the missions. For me, the real story of the Space Shuttle is not just that bit between the launch pad and the landing site, but also the time from wheel stop to the next lift-off. This might not be as attractive to the headline makers, but without the pre-mission and post-mission processing on the ground by teams of dedicated and often 'invisible' workers, there would not have been a mission to fly. Detailing the physical movement of Shuttle hardware, and talking with those who actually put the elements together, is a largely overlooked story in the space book library, one which this book goes a little way to address, at least for Spacelab.

When Mikey Haddad welcomed my help to tell his and his colleagues' story of Level-IV at the Cape, it was an opportunity not to be missed. Hearing the accounts of members of the Level-IV team, giving a personal connection to the nuts and bolts, enriched the story of how multiple elements came together in order to fly safely, successfully and mostly on time as designed. And this is just the Spacelab side of the story. Future accounts from the non-Spacelab missions and the processing of the ISS assembly missions beckon.

Just as the long bibliography of astronaut accounts have recently been expanded by stories and recollections from NASA's Mission Control, so it is hoped similar accounts will emerge from the vast facilities and directories of the Kennedy Space Center, to fill in more gaps in the Shuttle story. This is but one of those stories.

Mike Haddad:

As we end this book about Level-IV, I've realized how hard it is to put all the work that was performed and all the accomplishments into one volume. The work we did could fill a book for each mission individually. But I think a statement made by my good friend Darren Beyer says it all:

"I think doing hands-on work and climbing around spaceships is pretty darn cool. When I got into Level-IV, it dawned on me that I was twenty-something, climbing around real spaceships, and handling multi-million-dollar hardware. That has real-world implications. So I think if you go back up the chain, I think you could say that Level-IV was the coolest job at the coolest location for the coolest agency in the world."

Afterword

The original objectives of Level-IV – to meet the experiment integration needs of the Spacelab program while nurturing a new generation of NASA engineers with hands-on technical assignments and mentoring – are easy to grasp. As are the most obvious outcomes. Those science investigations were almost universally successful, and NASA benefited from a new crop of experienced engineers who went on to do other great things at KSC, across NASA, and beyond. This all comes across clearly in the chapters of this book. My assignment here is to close with some thoughts about additional impacts, particularly science and societal benefits, systems thinking, and human dimensions. I'll start with painting a picture of what it was like for some of us back in the day when we first arrived in Level-IV, and how it influenced our careers. I hope this provides some context for how things evolved over the ensuing decades – including outcomes we couldn't foresee at the time.

Beginnings

When Mike Haddad shared his plans to write a book about Level-IV, my initial response was excitement that someone was going to tell our story. That was followed by: "Wait, has it really been 30 (+) years already?" Like many of us, I suffer from time dilation. Some events or entire spans of years are a blur to me, while others feel like they happened yesterday. I'm lucky that some of my clearest memories today include some extraordinary moments I shared with my friends and colleagues at the Kennedy Space Center.

My first day at KSC was September 12, 1988, a few weeks before the Shuttle program's Return-to-Flight following the *Challenger* disaster. Like many at NASA, I got my start as a "Co-op" student. I'd just finished my second year in Electrical Engineering at Auburn University and managed to land a job at NASA through a combination of dumb luck, an encouraging director of Auburn's cooperative education program, and a first boss (Enoch Mosier) willing to hire a kid with minimal qualifications but tons of enthusiasm. I'd been enamored with space

© Springer Nature Switzerland AG 2022
M. E. Haddad, D. J. Shayler, *Spacelab Payloads*, Springer Praxis Books,
https://doi.org/10.1007/978-3-030-86775-1

since my grandmother introduced me to *Star Trek* and after watching every *Cosmos* episode with my dad.

I never admitted this to any of my early supervisors, but I really struggled to keep my head above water in those initial weeks and months at KSC. I'm sure they could tell by my trademark blank stare, as I wobbled back to the cubical "pen" I shared with my new roommate Elias Victor and the other Co-ops after a training session. My technical retention at the time was maybe five percent on a good day. I recall recopying my notes every evening and then asking my saintly mentors the next day to please explain things again to their dim-witted understudy. One week at KSC was harder than any course in college, but also exhilarating. The first time Jim Sudermann (Zoot) gave me a tour of the High Bay and control rooms of the Operations and Checkout (O&C) Building. I was instantly embarrassed by the naiveté of my prior vision and how much I had to learn. Imagine structures, machinery, computers, cables, and pipes receding into the distance in all directions, with people quietly but busily rushing about. It was like being in the engine room or on the bridge of some enormous ship. The complexity and activity in the O&C Building was both overwhelming and reassuring. I told myself "Wow, this is where I need to be" One caveat: I know everyone in Level-IV had their own unique initial experience, and certainly not everyone was as wet behind the ears as me. Most folks arriving in Level-IV were already college graduates and some transferred in as experienced engineers from other directorates at KSC, or other NASA centers or industry. I recall one evening that first fall, when Dave Sollberger took a group of us new-hires on an impromptu tour of Cape Canaveral Air Force Station. We all gawked at the old Mercury and Gemini launch pads while looking back towards the Shuttle launch complexes, soaking up some of the richest portions of space history while recognizing we were about to help write a new chapter.

Over the next eight years, including graduating from college in 1991, I was proud to be part of the Level-IV team and family. I suppose I was part of the "second wave" in Level-IV. I was there for nearly 50 Shuttle launches and played at least a small role in about a third of them. Like most of us in the Electrical branch, I had two main responsibilities: leading experiment testing through the launch integration flow, and operating and maintaining the test/checkout equipment (hardware simulators and software). While I initially worked as a High Rate Input/Output Test Set (HITS) operator for several Spacelab missions, my own focus on this front was supporting Zoot and Polly Gardiner as they led the development of the Partial Payload Checkout Unit (PPCU) systems. I still laugh at the term "partial payloads", as it implied some half-finished gizmo used primarily for ballast. In fact, some of our partial payloads involved some of the largest and most ambitious science investigations ever performed in Earth orbit. I understand why "Partial Payloads" was a more convenient term than "Non-Spacelab Science Payloads that Are (mostly) Not Deployed from the Shuttle", but it still makes me chuckle. Supporting a payload test team as an operator of PPCU, HITS, or other checkout equipment was always interesting, and invaluable in developing skills like multi-tasking, communications, and troubleshooting that would prove important for other assignments. However, the real passion Level-IV awakened for me was for the *science* itself.

Before arriving at KSC, my exposure to science was too many sci-fi novels and popular science books, followed by lower-level undergraduate engineering courses. That changed when I started working alongside the many experiment teams coming through KSC to help prepare their instruments for launch. In many cases, we spent several years working closely with the Principal Investigator (PI) teams from some of the top research institutions around the world. Those teams often included the PI themselves, project managers, chief engineers, technicians, staff scientists, post-doctoral fellows and graduate students. I quickly learned that if I was sufficiently curious and persistent, it was possible to get a crash course in a science discipline. Some of these were admittedly introductory level, but others involved deeper dives that would lead me way down the rabbit hole of science.

Two missions I worked on at KSC in particular put me onto even longer-term paths that I don't think would have otherwise materialized. One was Astro-2 – in particular the University of Wisconsin team led by Art Code and Karen Bjorkman – that ignited a passion for astronomy and nearly resulted in me shifting away from engineering and probably to a dubious sojourn through graduate school. The other was the Space Radar Lab (SRL) that included a team of radar engineers and scientists from NASA's Jet Propulsion Laboratory (JPL) and the German space agency. It was a combination of those teams that ultimately led me to reluctantly leave KSC for the Jet Propulsion Laboratory (JPL) in early 1996. Today I continue to work for JPL, as well as a new, joint affiliation at the University of Arizona. Ironically, the move to JPL resulted in me returning to KSC as a PI team member on two separate occasions (so far). In 2000, we flew the Shuttle Radar Topography Mission (SRTM) on the Shuttle *Endeavour*. SRTM involved a rather audacious transformation of the original SIR-C/X-SAR payloads from the SRL series into an *interferometer* with a 60-meter deployable boom and outrigger antenna. SRTM was my last Shuttle mission, but it was great to experience Level-IV "from the other side". A big plus was when my Level-IV colleague and friend Scott Vangen joined the SRTM team at JPL and JSC, where we benefited enormously from his experience in crew training and mission operations. In 2009, I returned again to launch the Kepler mission, this time on a Delta-II rocket from the Cape. The experience on Astro-2 had resulted in a passion for finding Earth-like planets around other stars and led to me serving as Kepler chief engineer. After SRTM and Kepler, I felt I'd finally come full circle to my KSC roots. It was good to come home.

Science and Societal Benefits

Much of the scientific and technological impacts of Spacelab and the related Shuttle science programs ("partial payloads") have been summarized before. In particular, the Spacelab Science Results Study (Naumann *et al.*, 2009) offers a comprehensive synthesis of the primary research domains, including: astrophysics, solar physics, Earth science, and microgravity science (both physical and life sciences). That report cited over 760 experiments, including the results of purely exploratory research as well as some key technology spinoffs. It also acknowledged some limitations – chiefly the reliance on available public journal publications and results that were not yet available at the time of that study. I won't attempt here to offer an update to that assessment, nor will I venture a guess about the ultimate number of scientific publications, breakthrough discoveries, or number of students and scientists whose careers directly benefited from these missions. Instead, here is one person's perspective on broader outcomes, including successor programs and societal benefits that might have taken longer to materialize (or never happened) without the Shuttle science programs and the key enabling role of Level-IV.

One overarching point I'd like to make about the Shuttle science program is how it fitted into the broader space and Earth science enterprise. The Shuttle program didn't offer the first Earth or astronomical observations from space, but it played a key role in the expansion of other programs – in some cases expanding the development of previously flown (on free-flying satellites) key instrument technologies and observational techniques, and in others providing the first orbital testbed. Importantly, the "co-development" offered by the Shuttle platform also extended to people, scientists and engineers who would go on to design, operate and analyze data from many follow-on missions. One example is NASA's series of Great Observatories for astrophysics that included the Hubble Space Telescope, Compton Gamma Ray Observatory, Chandra X-ray Observatory, and the Spitzer infrared Space Telescope. The Shuttle obviously played a direct role in launching the first three of those observatories (and multiple HST servicing missions), and many of the KSC people involved were former Level-IV personnel. Additionally, Spitzer, while ultimately launched on a Delta-II rocket, benefited from lessons learned on prior Spacelab missions – Spacelab-2 in particular.

The societal benefits of Earth observations represent some of the most direct and far-reaching impacts of Level-IV and Shuttle science programs. There were many Earth science experiments on the Shuttle, including the tightly coupled discipline of solar physics, that fitted into a much larger and longer-term program. For example, the Active Cavity Radiometer Irradiance Monitor (ACRIM) experiment that flew on Spacelab-1 and the three ATLAS missions on the Shuttle also operated on the Solar Max, UARS and ACRIMSat missions, contributing a multi-decadal time series of solar irradiance that is important for accurate climate change assessments. Here, I'll focus on two central disciplines that in my view exemplify the broader societal impacts of Shuttle Earth science experiments: atmospheric science and radar science.

Anyone with an interest in breathing cares about the composition and chemistry of Earth's atmosphere. Some of the most iconic and humbling images taken from space are views of the Earth's limb – a thin, tenuous blanket that separates humanity and the rest of the biosphere from hard vacuum. In addition to immediate life support, the atmosphere is a dominant factor in the Earth's energy budget, where small changes in trace gas concentrations can have profound impacts. Starting in the Shuttle era, scientists began conducting experiments that probed the composition of the lower atmosphere from space, particularly the troposphere, with direct consequences on human health and climate change. For example, the Measurements of Air Pollution (MAPS) experiment that flew on the SRL missions made the first space-based measurements of carbon monoxide (CO) in the lower atmosphere, which were subsequently expanded with multiple NASA and European Space Agency (ESA) satellites. Space-based observations of CO, nitrogen dioxide and other pollutants are becoming increasingly important in populated areas that lack surface measurements of these quantities. The Atmospheric Trace Molecules Observed by Spectroscopy (ATMOS) experiment flew five times on the Shuttle (OSTA-3, Spacelab-3, and all three ATLAS missions). ATMOS provided a benchmark inventory of the chemical composition of the Earth's atmosphere for comparison and trending by follow-on satellite missions. ATMOS also pioneered radiative transfer algorithms which continue to be exploited by multiple instruments on NASA and international satellite missions that track greenhouse gases, and the three-dimensional structure of temperature and water vapor in the atmosphere. These are critical inputs to weather forecasts and climate models.

Space imaging radar is perhaps one of the biggest science success stories from the Shuttle era. The Spaceborne Imaging Radar (SIR) series began with OSTA-1/ SIR-A – the first Shuttle science payload in 1981. It was followed by SIR-B, SIR-C/X-SAR (two flights) and ultimately SRTM in 2000. Space-based Synthetic Aperture Radar (SAR), where the motion of the platform is used to reconstruct high resolution images independent of cloud cover, time of day and some surface coverings, has become a critical tool for remote-sensing, both for Earth observations and other solar system bodies. Interferometric SAR (InSAR) in particular provides critical three-dimensional information and change-detection. Shuttle imaging radar even contributed to the new field of space archeology, including discoveries of ancient settlements obscured by desert sands. Today, space-based SAR and InSAR satellites routinely generate data used by a huge community of decision makers, for everything ranging from earthquake hazard assessments, forest management, and commercial aviation, to understanding the impacts of ice sheet collapse and solid-earth deformation on sea level rise. The legacy of radar technology, algorithms and phenomenological understanding pioneered by Shuttle imaging radar has resulted in no less than 21 follow-on satellite missions, with 11 more currently in development (Freeman *et al.*, 2019).

Systems Thinking (and learning)

We received on-the-job training in *how* to learn about complex systems, most of which didn't come with a manual or text book. At the time, there were fewer formal training programs in space systems than there are today. Many undergraduate university programs now give students the opportunity to build and launch cube-sats. That was still pretty rare in the 1980s and 1990s, so we had to learn systems thinking on-the-job in Level-IV. While local universities taught space science and aerospace engineering *theory*, real systems engineering training occurred at KSC through our mission assignments. There are different ways to learn a complex topic. Many of us are visually oriented and benefit from immersion and repetition, so we'd cover the walls of my office with block diagrams, flow charts and error budgets for a given system so that we could step back and look at it throughout the day. I recall taking apart large binders to copy key pages for my own "wall of learning". Eventually we got to where we could carry most of it around in our heads, which became invaluable during testing, troubleshooting and mission operations. Some of our colleagues were ultimately so successful that they became regarded by a PI team as the expert on how to operate one of the more finicky instruments. In addition to all the mission-unique experience, we benefited from local discipline experts like Jim Dumoulin, Cindy Martin-Brennan, Craig Jacobson, Luis Moctezuma and too many others to list here – people with an ency-clopedic understanding of how things worked. The need for systems learning is not a fault of inadequate documentation, but rather an aspect of complex systems: there's no substitute for hands-on experience in exposing real behaviors.

Why is it important that so many people went through the Level-IV training program on systems thinking? Because increasing complexity is a feature of both NASA programs and many other technological endeavors confronting humanity, and we need people with that training. That's saying a lot, because the Shuttle is widely regarded as the most complex machine ever built, with over 2.5 million individual parts and a bewildering array of switches, displays and nested software. However, we are already seeing future space missions that involve increasing complexity, where success demands hardware that is much larger (or much smaller), systems that are spatially or organizationally dispersed, and/or equipment and software that are ever more capable.

Human dimensions

Last but not least, a core element of the Level-IV experience has to be the human dimension. Attention to human factors has long been recognized as essential in any complex endeavor, and was even more so in the Shuttle program. A friend of mine once remarked "rocket science is easy, it's the *socio-thermodynamics* that's hard". At the time we laughed, but over the years I've come to recognize it as true in almost every project involving a large group of people. The fact is that every Spacelab mission or equivalent Shuttle science payload involved a cast of hun-dreds if not thousands. It was not unusual for the KSC payload team alone to deal with three NASA centers and over a dozen PI institutions, including international partners. These missions involved a confluence of social and engineering cultures and often conflicting protocols and norms. While there were technical interface standards, requirements and operating procedures, there was no real guidebook or training about how to manage the human dimensions with so many actors. To be sure, by the time a PI team arrived at KSC they'd been through multiple Ground Operations Working Groups and had been somewhat indoctrinated in the NASA bureaucracy, which provided some much-needed structure. However, I always found that to be necessary but not sufficient. None of us are robots, and making things work on the ground involves human *relationships*, which can be tricky under the best of conditions.

At KSC, we had a standard government chain of command (line management). Additionally, every payload and mission was managed through one or more program organizations within NASA, frequently distributed across Headquarters and the centers. This was mirrored by ESA and other international partner agencies, as well as the PI's home institution. Now must of us in Level-IV were working stiffs, with no real formal clout other than the project engineer assigned to every payload who did have formal reporting paths. In cases when a key decision needed to be made – for example involving risky decisions like rework vs use-as-is for an experiment experiencing difficulties – the cognizant Level-IV engineer's ability to influence the decision depended on a) their relationships with those in the decision chain and b) their technical credibility. And there were definitely cases when the Level-IV person *needed* to weigh in, particularly in unusual cases that defied a simple decision. I watched some of my senior colleagues "work the system" successfully, learned from it and applied it myself a couple of times. That ability to manage through influence has served many of us well over the years that followed. I myself have mostly avoided being in a position of holding the gold because I hate the attendant paperwork, a common trait among many scientists and engineers. But as a result, I've used my Level-IV "manage-through-influence" training to good effect on every project since then.

In reflecting over the years, I'm impressed by what I can only describe as a team-building culture. Now this mostly preceded the subsequent formal management fads like rope courses and paintball (although I recall a number of friendly rubber-band fights breaking out spontaneously in the office suite on Friday evenings). What I recall most was senior members of our division making a persistent effort to bring new arrivals into the fold. I was particularly struck by a tendency of some colleagues to take visiting PI teams out to dinner or invite them into their homes. I know that many of those visiting teams spent weeks or months at a time at KSC, often away from their families, and I think gestures like these from Level-IV people probably went a long way to making them feel welcome. It's not a surprise then that many of these visiting PI teams struck up friendships with their KSC colleagues, which resulted in relationships that have spanned the decades and the globe. And of course there was the infamous Level-IV Halloween Party, where many members of the division would turn out most evenings starting in late September to construct an automated haunted house. The fact that people volunteered their free time to work on electronics, robotics and props was a testament to that culture of belonging. They say friends are the family you choose and for many of us Level-IV was and remains our second family.

This is probably a good note on which to conclude. The people in Level-IV played a unique role in advancing Earth and space science, with major societal impacts and programs that continue to expand today. More importantly, Level-IV taught us how to think about complex systems, how to manage by influence and how to build teams – all of which proved invaluable as we moved on to other challenges. I know I speak for all Level-IV alumni in expressing gratitude to Bill Jewel, John Conway, Tom Breakfield, and many others for the foresight and leadership in establishing an organization that would ultimately do so much for NASA, its partners, the public and so many of us personally. Thanks to Mike Haddad for the herculean effort in completing this book and chasing us all down for interviews! Most of all, to my friends and former colleagues in Level-IV: I deeply respect your dedication and tremendous accomplishments. It was an honor to share that special place and time with you. Best regards.

Riley Duren. [Riley Duren.]

Riley Duren
Research Scientist, University of Arizona, Tucson, AZ
Engineering Fellow, Jet Propulsion Laboratory, Pasadena, CA

APPENDIX 1

Level-IV Personnel Biographies

Listed here is a sample of some of the Level-IV alumni, in alphabetical order, and a brief resumé of their careers before, during and after Level-IV.

© Springer Nature Switzerland AG 2022
M. E. Haddad, D. J. Shayler, *Spacelab Payloads*, Springer Praxis Books,
https://doi.org/10.1007/978-3-030-86775-1

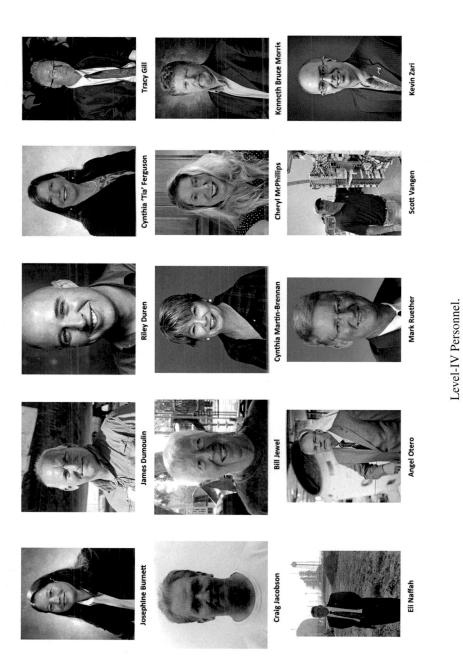

Level-IV Personnel.

482   **Appendix 1**

Josephine Burnett

Josephine Burnett graduated from the University of Florida in 1987 with a BS in Aerospace Engineering and earned an MS from the Florida Institute of Technology in Space System Operations in 1991. She retired in 2020 as Director of Exploration Research and Technology Programs at NASA's John F. Kennedy Space Center (KSC) in Florida, Josephine had been responsible for strategic leadership, and program and project management, for KSC's support to the Exploration mission, including the Space Life and Physical Sciences (SLPS), International Space Station (ISS), Advanced Exploration Systems (AES), and Space Technology (ST) programs. Josephine Burnett began her career with NASA in 1987 at KSC as an aerospace/mechanical engineer, integrating experiments onto Spacelab racks and pallets. In 1995, she served as the Center Director Management Intern, and performed in this capacity until 1996 when she joined the Space Station Hardware Integration Office. For three years, she supported the test and checkout of the Canadian Space Station Remote Manipulator System (SSRMS) in Brampton, Ontario, and the Canadian portion of the Multi-Element Integrated Test (MEIT), after which she was selected to lead the Element 6A Office as the acting chief. In 2000, Burnett joined the ISS/Payload Processing Directorate as the chief of the Future Missions and International Partner Division. In this position, she led the office responsible for the advanced planning of ISS international elements, Shuttle payloads, and concepts for improved ways to test payloads in the future. In 2003, she became deputy for Program Management for the ISS and Payload Processing Directorate at KSC. In this capacity she supported the director in the development of organizational roles and responsibilities, as well the development of long range organizational direction for KSC and ISS program needs. Burnett was selected as a member of the Senior Executive Services Career Development Program (SESCDP) class of 2004 and completed the program in March 2006. She returned to KSC as the chief of the Systems Engineering and Integration Division for the Design Engineering Directorate. In this capacity, she led a team of highly experienced systems engineers in support of the Constellation Ground Operations Project. Burnett then became the deputy director of the Space Transportation Planning Office, providing strategic guidance and leadership for the program management, systems engineering and integration of the Constellation Block 1 Orion/Ares configuration to low-Earth orbit. Mostly recently, she was the director of the ISS Ground Processing and Research Project Office, responsible for all ground processing of space station elements from around the world. These elements are operating in orbit and supporting the largest, most complex space station in human history.

James Dumoulin

By the 1990s, computer networks had become more essential to daily operations, and Jim could no longer run more than 2000 individual computers and remain active in Spacelab testing. He stopped taking an active role as a Spacelab experiments engineer and concentrated more on improving KSC's Information System technologies. In this role, he operated KSC's Internet Service Provider (ISP) dialup system, ran KSC's first Internet News Server (INN), and NASA's first web server (1992), first search engine (1993), first Voice Over IP (VoIP) system, and first streaming video system. All those systems now run under institutional contracts. Jim Dumoulin is currently the oldest non-retired original Level-IV engineer still working at KSC. He serves as Chief Engineer of the NASA Telescience and Internet Systems Lab, where he is responsible for building the Space Launch System (SLS) Telemetry System that will feed radio frequency downlinked telemetry and rocket camera video to the Launch Control Center after T-0.

Riley Duren

Riley Duren is a Research Scientist at the University of Arizona's Institutes for Resilience, and an Engineering Fellow at NASA's Jet Propulsion Laboratory (JPL). His 30+ years of complex program development experience include nine space missions (with the *Shuttle Radar Topography Mission* and *Kepler* being personal favorites). From 2008 to 2019, he served as Chief Systems Engineer for JPL's Earth Science Directorate, which is responsible for planning and implementing a broad portfolio of NASA satellite and airborne missions, research, applied science, and technology development projects. In 2008, he also began to extend and apply the discipline of science systems engineering to climate change decision support. His primary research today involves multi-scale carbon monitoring systems to support mitigation efforts, including the coordinated application of Earth observations from the surface, air and space. Riley is currently Principal Investigator (PI) for five research projects that are developing prototype monitoring systems for methane and carbon dioxide emissions, with public and private sector stakeholder participation. He is also leading the formulation of a new program to launch a constellation of small satellites to support decarbonization and conservation efforts. He is the recipient of two NASA Exceptional Achievement medals and the agency's System Engineering Excellence award.

Cynthia "Tia" Ferguson

Ms. Ferguson earned a BS in Mechanical Engineering from Tulane University in New Orleans, Louisiana and an MS in Electrical Engineering, with minors in Optics and Micro-electronics, from the University of Alabama in Huntsville (UAH). Ms. Ferguson is the Deputy Director of the Space Systems Department in the Engineering Directorate at Marshall Space Flight Center (MSFC). She was appointed in February 2017. The Space Systems Department is responsible for designing, developing, assembling, integrating, testing, and delivering flight, ground, prototype, and development products for human spaceflight programs, science investigations, and exploration initiatives. Reporting to the Director, Space

Systems Department, Ms. Ferguson assists in the oversight of an annual budget of over $110 million and the management of a diverse, highly-technical workforce. Throughout her 27 years of experience working for NASA, Tia Ferguson has served in multiple technical leadership positions. She began her career in 1990 at KSC as a mechanical systems engineer, where she performed integration and testing of Space Shuttle experiments. In 1994, she was the KSC lead project engineer for STS-66 payloads. Tia transferred to MSFC in 1995, where she was the Multi-Purpose Logistics Module (MPLM) Cargo Element Integration Project Manager. From 1999 to 2006, Ms. Ferguson was a mechanical design engineer in the Space Systems Department, and from 2006 to 2011 she was the Branch Chief of the Structural and Mechanical Design Branch, also in the Space Systems Department. From 2011 to 2012, she served as the Engineering and Science Services and Skills Augmentation (ESSSA) Source Evaluation Board Chair. From 2012 to 2013, she served in a detail as the Assistant Manager for the Science and Technology Office, after which she became the Project Manager for SERVIR, an Earth Science project that helps developing countries use Earth-observing satellites and geospatial technologies to manage climate risks and land use. In 2015, Tia Ferguson was selected as one of 24 participants in the Senior Executive Service (SES) Candidate Development Program (CDP). As part of this program, she completed several details: at Glenn Research Center (GRC) as a Special Assistant to the Office of the Center Director, where she created strategic messaging for internal and external center communications; at MSFC as the Acting Center Chief Technologist, where she was responsible for ensuring technology activities and the technology portfolio were managed strategically for the center; and at NASA Headquarters, as the Special Assistant to the Deputy Chief Financial Officer for NASA, where she led strategic planning and coordination for the Agency Strategic Implementation Planning (ASIP) meeting, and participated in budget formulation and implementation efforts, Strategic Workforce Planning (SWP) and acquisition strategic planning efforts. Tia Ferguson successfully completed the SES CDP program in October of 2016. She has completed multiple executive leadership development programs, including graduating from Harvard's Senior Executive Fellows program in February 2016. She is a Professional Engineer and holds a US Patent on a Micro-Electro-Mechanical Systems (MEMS) micro-translation device. She is married to Jim Ferguson.

Tracy Gill

Tracy Gill works for NASA at KSC. He has 30 years of experience on Space Shuttle payloads and on Space Station elements and experiment payloads, gaining valuable experience doing "hands-on" work on flight and ground support hardware and working with people from all over the United States and the rest of the world. He currently supports the NASA Artemis program, including the Gateway and NextSTEP (Next Space Technologies for Exploration Partnerships) commercial partnering efforts for lunar exploration. He also manages the X-Hab Academic Innovation Challenge for NASA. Tracy holds a BS in Electrical Engineering and an MS in Aerospace and Mechanical Systems from the University of Florida, an MS in Space Systems from Florida Tech, and is a graduate of the International Space University Space Studies Program in 2006. He is also an adjunct professor for the International Space University.

Craig Jacobson

Craig Jacobson has supported several NASA human spaceflight programs, including the Space Shuttle, the International Space Station, and future launch vehicle programs that will return humans to deep space, including to the Moon with the Space Launch System (SLS). His early career roles on the Space Shuttle program included spacecraft electrical test engineer for Spacelab flight experiments, and communications engineer for numerous Shuttle missions. In this role, he authored and executed flight experiment test and verification procedures, troubleshot flight problems and anomalies, provided high rate and fiber optic systems engineering expertise, and planned and configured test systems and data links to support major tests. Craig was also critical in the technology infusion into Shuttle payload operations, where he helped automate many labor intensive operations such as scheduling, technical documentation management, and logistics. In this role, he helped bring the latest computer hardware and software applications online, thus substantially reducing costs, eliminating human errors, and reducing process time. Craig has served in several engineering leadership and management roles, including computer systems administration, test and checkout systems operations, and sustaining engineering and development. This leadership experience crossed many different NASA programs, including Space Shuttle payload checkout, International Space Station system integration and operations, and International Space Station payload checkout. He also provided systems engineering expertise for the SLS test, integration and launch system. In this role, he ensured interfaces between subsystems interoperated and were compatible, requirements were traceable down to system implementation and up through verification, and software met integration requirements. Craig Jacobson is recognized for his expertise and leadership across the agency and has served many additional cross-agency and local roles. For example, he has been the subject matter expert supporting various space industry symposiums, future technology forums and trade shows, organizational change and development, and engineering support for major acquisitions. He has served in a wide range of aeronautic disciplines across NASA's most important human spaceflight programs.

Bill Jewel

Bill Jewel signed on to NASA in December 1959, his senior year at University of Tennessee, and started work in June 1960 at Langley Research Center, VA. He earned a BS degree in Engineering Physics and worked mainly in re-entry and meteor physics on solid rockets at Langley and Wallops Island, VA, until 1964. He transferred to KSC in 1964, and was assigned to assembly, test and launch operations for Saturn, Saturn 1B, and Saturn V. In the 1970s, he was asked to develop an operation plan and origination (and helped to sell the concept to NASA management) for KSC to assume the role of assembly, test, integration of experiments and processing of proposed Spacelab missions, which at the time were projected to involve about 60 percent of Space Shuttle missions. This was accomplished in summer of 1980. Towards the end of 1981, the Level-IV Division assumed the added responsibility of all Spacelab and experiment processing at KSC, becoming the Spacelab and Experiments Integration Division. Some called it "Hands-On". In May of 1984, Bill Jewel was assigned to the Space Station Skunk Works[1] in Houston to develop the Space Station requirements as input to the Source Evaluation Board (SEB) for the development of a NASA Station contract. Bill finally retired from NASA in April 1985, did some aerospace consulting work until 1995, then retired a second and last time.

Cynthia Martin-Brennan

Ms. Martin-Brennan began her career with NASA by providing technical and program support to scientists and astronauts as they prepared their science experiments for spaceflight on Space Shuttle Spacelab missions. In the early to mid-1990s, she managed a $300 million life science research program at NASA Headquarters. In the latter 1990s, she managed the Life Sciences Space Flight Support Laboratory located at KSC, including the operations contract. She was also the NASA manager for the Collaborative Ukrainian Experiment, which flew the first Ukrainian astronaut in space (STS-87 in 1997) as part of an incentive for Ukraine to join the Treaty on the Non-Proliferation of Nuclear Weapons, and for which she received the NASA Exceptional Service Medal. In 1998, Cynthia left NASA and worked as a consultant for industry, government and academic institutions. As a consultant, her jobs included serving as the executive secretary for the NASA International Advanced Life Support Working Group, business development analysis and support, and providing strategic planning and implementation for new programs. She was Director of Space

---

[1] A Skunk Works is a small project team and loosely structured group of people brought together to research and develop a new innovative project.

Programs at CSS, a small business located in Fairfax, VA, that included federal, state and commercial customers, with an annual revenue average of $10 million. She also served as the Executive Director of the American Society for Gravitational and Space Research, a non-profit organization 501(c)(6), whose mission is to provide a forum to foster research, education and professional development in the multidisciplinary fields of gravitational research. During her nine-year tenure as Executive Director, Ms. Martin-Brennan successfully executed a strategic vision to integrate the external research community (e.g., physical, biological) into one gravitational research community, in an effort to widen the pipeline leading to discovery and innovation. Currently, Cynthia is the Director of Stakeholder Management for the Center for the Advancement of Science in Space, a non-profit 501(c)(3) organization that manages the International Space Station National Laboratory. She works closely with the White House, Congress, National Academies of Sciences, and many others, to educate and inform about research and development activities on board the International Space Station National Laboratory. Ms. Martin-Brennan provides information pertinent to evolving US civil space national policy. She has over 30 years' experience with the US and International Space Community. She holds a BS in Computer Science from the University of Central Florida, and an MS in Space Technology from the Florida Institute of Technology.

Cheryl McPhillips
Ms. McPhillips graduated from the University of South Florida in 1984 with a BS and MS in Electrical Engineering. In 1989, she earned an MS in Space Technology from Florida Institute of Technology. Currently (2021), Ms. McPhillips is the Certification Manager for the SpaceX crewed Dragon and Falcon 9, ensuring SpaceX satisfies all NASA crew safety requirements. She has a distributed team of Program, Engineering, Safety and Mission Assurance, Flight Operations Directorate and Health and Human Performance that support her in verifying and approving Commercial Crew Program (CCP) and ISS Crew Safety requirements. Additionally, Cheryl has led the execution of SpaceX payment milestones leading up to certification, including: Critical Design Reviews, Integrated System Reviews, Certification Checkpoint and Flight Test Certification Review. Prior to becoming SpaceX Certification Manager, Ms. McPhillips was the Partner Manager for Sierra Nevada Corporation (SNC) in the CCP, serving as a liaison between NASA and SNC. Ms. McPhillips became Chief of the Space Station Engineering Division, leading 100 engineers in five branches and chairing the Materials Engineering Review Board. In 2003, she became Utilization Division Chief at the project office for integrating International Space Station research hardware into the Shuttle. In 2007, Cheryl became the Orion Ground Operations

Planning Manager for the ISS and Spacecraft Directorate. In 2009, Ms. McPhillips joined the Commercial Crew Planning Office, where she focused on acquisition planning, and spacecraft design and development. In 2010, she served as the Technical Chair on the Commercial Crew Development (CCDev) 2 Participant Evaluation Panel (PEP). Lastly, Ms. McPhillips served as the Chair for the Commercial Crew integrated Capability Space Act Agreement awards. Ms. McPhillips has received three NASA Leadership Medals, the Eagle Manned Mission Success Award and a Silver Snoopy. She is an Instrument Flight Rules (IFR) and multi-engine-rated private pilot.

Kenneth Bruce Morris
Bruce Morris joined Level-IV at KSC in 1983 and worked as an assembly and test engineer before becoming a Test Director for Shuttle payloads. He served as the Integration Manager for the ISS Laboratory Module (*Destiny*) before leaving KSC to join MSFC. During his 28 years with NASA, Bruce served in senior positions at NASA-MSFC, NASA-Headquarters, NASA-KSC, and Air Force Space Command, and has been responsible for a variety of functions and duties including certifying the readiness of teams and hardware for human spaceflight. Since leaving NASA for industry, he has been a Senior Associate, Director and General Manager, often holding multiple roles at once, while leading teams and organizations and delivering multi-million-dollar wins in commercial and government sales in Engineering Services and Flight Hardware Design Development Test and Evaluation (DDT&E).

Eli Naffah
Now the Chief of the Space Communications and Spectrum Management Office at NASA-GRC in Cleveland, Ohio, Eli started his career with NASA at KSC in Florida, where he integrated Spacelab experiments for launch on the Space Shuttle. He then moved back to Cleveland to work on the Space Station Power System, with responsibility for integration and processing of the Power System at the launch site. Eli later joined the NASA Glenn Office of Chief Counsel, serving as lead counsel for a number of significant matters, including NASA's Space Act Agreement Process, Export Control, Nuclear Reactor Decommissioning, Expansion of Cleveland Hopkins Airport, the Glennan Microsystems Initiative, Ultra-Efficient Engine Technologies, and the Exploration Technology Development Program. He then moved to the Space Flight Systems Directorate to lead the Orion Project Office at GRC, with responsibility for development of the Orion Service Module, Orion Requirements and Orion Space Environmental Testing. He went to NASA Headquarters to serve as Deputy Chief Engineer for the Science Mission

Directorate (SMD), where he was responsible for assessing SMD programs and projects, and a number of technical excellence initiatives involving space weather, flight software complexity, laser technologies, radio isotope power systems and orbital debris. He returned to GRC to serve as the Deputy Chief for the Avionics and Electrical Systems Division, responsible for avionics, power systems, diagnostics/data systems, electrical and electromagnetic systems, flight communications and flight software engineering for a diverse division portfolio, including CoNNeCT, SCaN CTS, Shuttle, ISS EPS, Orion, Ares I, Ares I-X, Ares V, Altair, EVA ETDP, NEXT Ion Thruster and the Radioisotope Power System Program. In 2011, Eli Naffah moved to his current position to establish the Glenn Space Communications and Spectrum Management Office. The Office currently serves the NASA Space Communications and Navigation (SCaN) program, providing leadership in the areas of spectrum management, architecture, standards and technology development. In addition to Agency Level Spectrum Management, his responsibilities include: operation of the SCaN testbed on the ISS; next generation architecture development and modelling; cognitive communications; integrated radio and optical communications; high data rate architectures; quantum communications; and atmospheric propagation.

Angel Otero

Angel Otero started his career with NASA at KSC as an engineer, installing and testing payloads flying on the Space Shuttle. He then moved to the Lewis Research Center (now GRC) where, among other duties, he managed the NASA Combustion Research program and oversaw the Shuttle Return-to-Flight work. In 2008, Angel moved to Washington, D.C., to serve as Program Executive for the acquisition of four new Tracking and Data Relay Satellite System (TDRSS) satellites. This was followed by serving as the Program Control Manager for the Space Communications Network Services Division, where he was responsible for managing the $650 million annual budget. In 2012 he moved to the Space and Life Sciences Research and Applications Division (SLPSRA) at HQ as Program Executive, before being selected as the Deputy Division Director for SLPSRA. In that position, he was responsible for overseeing the selection of research and project implementation for space biology and physical sciences experiments, mainly targeted for flight on ISS. After his retirement from NASA in 2018, Angel went to work for NATECH Corp., providing project management support to the National Air and Space Museum (NASM) Transformation Program. After two years at NASM, Angel returned to NASA as a contractor for Technology Project Managers (TPM), serving as the Deputy Study Manager for the LISA project. He also serves as TPM's Director of Aerospace Services.

Mark Ruether

Mark Ruether holds a BS in Mechanical Engineering from the University of Minnesota and an MS in Engineering Management from the University of Central Florida. He is the Chief of the Integration Office at NASA-KSC in Florida. The KSC Integration Office serves as the center's focal point for policy, institutional and program accountability, and issues and concerns that require senior management's attention. The primary functions of the office include: planning and implementation of strategic management; management and oversight of KSC governance; management of KSC's Business Management and Configuration Management systems; conducting independent assessments of agency programs and projects managed at KSC; identifying, guiding and resolving the center's cross-organizational issues; and coordinating actions assigned by the Center Director. Mark began his career at KSC in 1984 as a systems engineer responsible for ground systems in support of the processing and launch of the Space Shuttle. For the next ten years, he served in various engineering, test and systems integration positions on Spacelab systems and experiments, and other payloads. He was also a Launch Site Support Manager (LSSM) on several missions, including the first servicing mission to the Hubble Space Telescope and the first commercial module to fly on the Space Shuttle (Spacehab-1). In 1994, Mark Ruether was selected to serve on a one-year detail assignment at NASA Headquarters in Washington, D.C., where he supported the Life and Microgravity Sciences and Space Sciences organizations and served as the assistant program manager of the Chandra Observatory. He returned to KSC in 1995 and served as lead LSSM and Chief of the Launch Site Support Branch. In 2000, he became the Chief Knowledge Manager of the Launch Services Program (LSP) and then LSP's first Strategic Planning Manager. In 2009, he became the Chief of KSC's Integration Office. In the years since, the office roles and responsibilities have continued to grow into the functions assigned to the office today. Mark Ruether has received numerous achievement, performance and certificate awards, including KSC's Strategic Leadership Award and the NASA Exceptional Achievement Medal.

Scott Vangen

Scott Vangen worked on numerous Spacelab missions and experiments, beginning with STS-9 Spacelab-1 and running through till the end of the Space Shuttle payloads era. In 1987, Scott was a member of Dr. Sally Ride's post-*Challenger* accident task force team (*NASA Leadership and America's Future in Space*), and in 1988 he was awarded a Silver Snoopy Space Flight Awareness Award. Scott Vangen was the first KSC employee to be interviewed as a Mission Specialist astronaut candidate in 1992 (and again in 1995, 1997 and 1999). At the completion of the successful Astro-2 ultraviolet astronomy mission for which he served as Alternate Payload Specialist, Scott went on to support ISS as the avionics lead for the US Laboratory element in the Space Station Hardware Integration Office from 1996 to 1998. He was the Lead Technical Integration Engineer for MEIT-II, and then served as the Division Chief of the Future Missions & International Partner Interfaces Office. From 2003 to 2008, Scott Vangen was the Chief Operating Officer for the KSC Space Life Science Laboratory (SLSL), a world-class multi-disciplinary science laboratory formed by a joint partnership between NASA and the State of Florida. In 2008, Scott was selected as the Deputy Project Manager for the Lunar Surface Systems Office at the Lyndon B. Johnson Space Center (JSC) in Houston, Texas, supporting the Constellation lunar program. In 2011, he began serving on various future Lunar and Mars advanced study teams, including support on the multi-center Human Spaceflight Architecture Team (HAT). In addition, Scott served as a senior engineer supporting the International Space Exploration Coordination Group (ISECG) as the NASA lead for the Technology Working Group. Scott Vangen retired after 37 years of NASA civil service in 2019.

Kevin Zari

Kevin Zari serves as the ISS Integration Manager in the Exploration Research and Technology (ER&T) Programs Directorate at NASA-KSC in Florida. In that role, he continuously monitors space station processing at KSC to identify areas of risk and opportunities for improvement, and provides processing expertise and lessons learned with emerging programs such as Gateway. Kevin is currently serving as the acting chief of the Strategic Implementation Office for ER&T. Prior to his current role, he held the chief technology officer position for the Space Station at KSC and was responsible for technology infusion, information technology systems security, and innovation in both processes and technology. Kevin Zari also served as the special assistant to the director for the International Space Station and Spacecraft Processing Directorate, where he was responsible for technical integration of activities and data products used for agency, program and center actions. In 2011, Kevin supported the associate administrator for NASA's Office of Education at NASA Headquarters in Washington, D.C., where he led the strategic planning and development of the tools necessary for monitoring the performance metrics for the agency's education portfolio. As the Spacelab Program ended, he began leading various facility class payloads, and EXpedite the PRocessing of Experiments to the Space Station (EXPRESS) sub-rack payloads, including: Anomalous Long Term Effects in Astronauts (ALTEA), ELaboratore Immagini TElevisive 2nd generation (ELITE-S2), Fluids and Combustion Facility (FCF), Mice Drawer System (MDS), Microgravity Science Glovebox (MSG) and Portable AStroculture Chamber (PASC).

APPENDIX 2

Spacelab Hardware Assignments

This table lists the majority, but not all, items of Spacelab hardware flown.
Information courtesy Debbie Bitner/NASA-KSC/BD-C, dated August 5, 1997.

PRESSURIZED MODULE ASSIGNMENTS

LONG MODULE #1 (CD MD001)	LONG MODULE #2 (FOP MD002)
SL-1 (STS-9)	SL-D1 (STS-61A)
SL-3 (STS-51B)	IML-1 (STS-42)
SLS-1 (STS-40)	SL-J (STS-47)
USML-1 (STS-50)	SLS-2 (STS-58)
SL-D2 (STS-55)	SL-M (STS-71)
IML-2 (STS-65)	LMS (STS-78)
USML-2 (STS-73)	NEUROLAB (STS-90)
MSL-1 (STS-83)	
MSL-1R (STS-94)	

NOTE: Earth Observation Mission (EOM) 1/2, an unflown mission in its original configuration, would have utilized the LM#1 Core Segment in the Short Module design. The payload was subsequently flown in 1992 as the pallet-only STS-45 ATLAS-1 mission.

FLOOR ASSIGNMENTS

CD (MD001)	FOP (MD002)	FOP II (MD003)	EM (EM-MD001)
SL-1	SL-D1	USML-1	SL-D2
SL-3	SLS-1	USML-2	
IML-1	SL-J	MSL-1	
SLS-2	IML-2	MSL-1R	
SL-M	LMS		
	NEUROLAB		

NOTE: Earth Observation Mission (EOM) 1/2 would have utilized Floor Assignment EM (EM-MD001) in the Short Module design.

© Springer Nature Switzerland AG 2022
M. E. Haddad, D. J. Shayler, *Spacelab Payloads*, Springer Praxis Books,
https://doi.org/10.1007/978-3-030-86775-1

TUNNEL ASSIGNMENTS

MD001   SL-1, SL-3, SL-D1, SLS-1, IML-1, USML-1, SL-D2, SLS-2, IML-2, USML-2, LMS-1, MSL-1, MSL-1R, NEUROLAB

MD002   SL-J, SL-M (STS-71), S/MM-03 (STS-76), S/MM-04 (STS-79), S/MM-05 (STS-81), S/MM-06 (STS-84), S/MM-07 (STS-86), S/MM-08 (STS-89), S/MM-09 (STS-91)

NOTE: S/MM-02 (STS-74) carried the Russian-built Docking Module and therefore did not require a transfer tunnel system to Mir. All Shuttle-Mir docking missions from STS-76 carried a Spacehab module, requiring the use of a Spacelab tunnel system.

MPESS ASSIGNMENTS

F001	OSTA-2
F002 (MD-001)	LFC/ORS, USMP-1(AFT), USMP-2 (AFT), USMP-3 (AFT), USMP-4 (AFT), MSP-1 (AFT possibly assigned to unflown USMP-5 mission)
F003 (003)	SL-3, EOM-1/2, EOIM-III, SRL-1, SRL-2
F004 (004)	OAST-1, EASE/ACCESS, TSS-1, TSS-1R, MFD
F006 (001)	MSL-2, USMP-1 (FWD), USMP-2 (FWD), USMP-3 (FWD), USMP-4 (FWD), MSP-1 (FWD possibly assigned to unflown USMP-5 mission)

DOUBLE RaCK ASSIGNMENTS

DR-C003-	SL-1 (4) - SYSTEM, LIFE SCIENCE - CAMERA
(MD002)	SL-D1 (4) - SYSTEM, BOTEX/STATEX
	USML-1 (4) - SYSTEM (4 VCR)
	USML-2 (4) - SYSTEM (4 VCR)
	MSL-1 (4) - SYSTEM (4 VCR)
	MSL-1R (4) - SYSTEM (4 VCR)

DR-C005 -	SL-3 (10) - FES
(MD006)	IML-1 (10) - FES
	LMS (10) - SIR

DR-C006 -	SL-1 (7) - VITR
(MD004)	SL-3 (3) - VIDEO
	EOM-1/2 (3) [Unflown]
	USML-1 (8) - DPM
	USML-2 (8) - DPM

DR-C007 -	SL-1 (10) - CENTRIFUGE
(MD005)	SLS-1 (8) - LIFE SCIENCE
	SLS-2 (8) - LIFE SCIENCE
	MSL-1 (8) - CM-1
	MSL-1R (8) - CM-1

DR-C008 -
(MD001)

SL-1 (9) - VWF - CAMERA
SL-D1 (7) - RSC
IML-1 (8) - STOWAGE
SLS-2 (9) - REFRIGERATOR - FREEZER
SL-M (9) - REFRIGERATOR - FREEZER
NEUROLAB (9) - REFRIGERATOR - FREEZER

DR-C009 -
(MD003)

SL-1 (3) - OAS - ANALYZER
SLS-1 (7) - SMIDEX
USML-1 (10) - SMIDEX
LMS (7) - SMIDEX

DR-F013 -
(MD007)

SL-3 (4) - SYSTEM
SLS-1 (4) - SYSTEM (2 VCR)
SL-J (4) - SYSTEM (2 VCR)
SLS-2 (4) - SYSTEM (2 VCR)
SL-M (4) - SYSTEM (2 VCR), NO FLUID PUMP
NEUROLAB (4) - SYSTEM (2 VCR)

DR-F014 -
(MD008)

SL-3 (9) - STOWAGE - ACTIVE
EOM-1/2 (4) [Unflown]
SL-J (7) - NASDA - LIFE SCIENCE
IML-2 (8) - BDPU
LMS (8) - BDPU

DR-F016 -
(MD009)

SLS-1 (10) - GPWS
SL-J (3) - GPWS
SLS-2 (10) - GPWS
SL-M (3) - LSLE
NEUROLAB (8) - GPWS

DR-F018 -
(MD010)

SL-J (8) - NASDA - MATERIAL SCIENCE
IML-2 (7) - NIZEMI
MSL-1 (9) - LIF
MSL-1R (9) - LIF

DR-F019 -
(MD011)

SLS-1 (9) - REFRIGERATOR - FREEZER
SL-J (9) - SMIDEX
USML-2 (10) - SMIDEX
MSL-1 (10) - SMIDEX
MSL-1R (10) - SMIDEX

DR-F020 -
(MD012)

SL-3 (7) - RAHF
SLS-1 (3) - RAHF
SLS-2 (3) - RAHF
SL-M (7) - STOWAGE

DR-F021 -   SL-3 (8) - DDM
(MD013)   IML-1 (9) - CPF-FRE
           IML-2 (9) -BIO/FRE
           LMS (9) - LSLE
           NEUROLAB (10) - VFEU

DR-F023 -   SL-J (10) - NASDA - MATERIAL SCIENCE
(MD014)   IML-2 (10) - TEMPUS
           MSL-1 (3) - TEMPUS
           MSL-1R (3) - TEMPUS

DR-F026 -   IML-1 (4) - SYSTEM (4 VCR)
(MD015)   IML-2 (4) - SYSTEM (4 VCR)
           LMS (4) - SYSTEM (4 VCR)

DR-F027 -   USML-1 (3) - STDCE
(MD016)   USML-2 (3) - STDCE

DR-F028 -   IML-1 (3) - STOWAGE
(MD017)   IML-2 (3) - FFEU/AAEU
           LMS (3) - AGHF
           NEUROLAB (7) - RAHF

DR-F029 -   IML-1 (7) - GPPF
(MD018)   SLS-2 (7) - RAHF
           SL-M (10) - BATTERY
           NEUROLAB (3) - RAHF

DR-M025 -   USML-1 (7) - CGF
(MD019)   USML-2 (7) - CGF

DR-EM5 -   SL-D2 (10) - STOWAGE
(MD001)

SINGLE RACK ASSIGNMENTS

SR-C001 -   SL-1 (5) - STOWAGE
(MD003)   SLS-1 (12) - CENTRIFUGE
           SLS-2 (12) - CENTRIFUGE
           LMS (12) - CENTRIFUGE
           NEUROLAB (11) - STOWAGE

SR-C002 -   SL-1 (6) - LIFE SCIENCE
(MD004)   SLS-1 (6) - LIFE SCIENCE - ECHO
           SLS-2 (6) - LIFE SCIENCE - ECHO
           NEUROLAB (6) - BAG-IN-BOX

SR-C003 - SL-1 (11) - MATERIAL SCIENCE
(MD001)   SLS-1 (11) - LIFE SCIENCE
          SLS-2 (11) - STOWAGE - ACTIVE
          SL-M (11) - BAROREFLEX

SR-C004 - SL-1 (12) - STOWAGE
(MD002)   SL-D1 (12) - STOWAGE
          USML-1 (6) - DPM
          USML-2 (6) - DPM

SR-D009 - USML-1 (12) - GBX
(MD011)   USML-2 (12) - GBX

SR-D010 - IML-1 (5) - BIORACK
          IML-2 (5) - BIORACK (NOT A SPACELAB-OWNED RACK)

SR-EM1 - IML-1 (6) - STOWAGE
(MD002)   IML-2 (12) - STOWAGE
          LMS (6) - STOWAGE

SR-EM2 - SL-J (12) - STOWAGE
(MD003)   IML-2 (11) - STOWAGE
          LMS (11) - STOWAGE

SR-EM3 - USML-1 (11) - STOWAGE
(MD001)   USML-2 (11) - STOWAGE
          MSL-1 (5) - STOWAGE
          MSL-1R (5) - STOWAGE

SR-EM4 - USML-1 (5) - STOWAGE
(MD004)   LMS (5) - STOWAGE
          MSL-1 (11) - STOWAGE
          MSL-1R (11) – STOWAGE

SR-F005 - SL-3 (12) - VCGS
(MD005)   IML-1 (12) - VCGS

SR-F006 - SL-3 (6) - STOWAGE
(MD006)   EOM-1/2 (6) - [Unflown]
          SL-D2 (12) - BAROREFLEX

SR-F007 - SL-3 (11) - GFFC - MICG
(MD007)   IML-1 (11) - IMAX - MICG
          SLS-2 (5) - RAHF
          SL-M (12) - STOWAGE
          NEUROLAB (12) - STOWAGE

SR-F008 -	SL-3 (5) - PRIMATE - RAHF
(MD008)	SLS-1 (5) - SMIDEX
	SL-J (6) - STOWAGE
	MSL-1 (6) - CM-1
	MSL-1R (6) - CM-1

SR-F012 -	EOM-1/2 (5) - [Unflown]
(MD009)	SL-J (11) - REFRIGERATOR - FREEZER
	IML-2 (6) - RAMSES
	MSL-1 (12) - MGBX
	MSL-1R (12) - MGBX

SR-F013 -	SL-J (5) - FEE
(MD010)	USML-2 (5) - GFFC
	NEUROLAB (5) - LSLE

PALLET ASSIGNMENTS

F001 (MD001) -	SL-1, EOM-1/2, HST SM-03
F002 (MD009) -	ASTRO-1 (AFT), ASTRO-2 (AFT), 7A (AFT)
F003 (MD004) -	SL-2 (3), TSS-1, TSS-1R, UF-1
F004 (MD002) -	SL-2 (1), ATLAS-1 (FWD), 6A
F005 (MD003) -	SL-2 (2), ATLAS-1 (AFT), 3A
F006 (MD005) -	OSTA-3, SRL-1, SRL-2, SRTM
F007 (MD006) -	SRM (WESTAR), LITE-1
F008 (MD007) -	SRM (PALAPA), ANT, ATLAS-2, ATLAS-3, 1J/A
F009 (MD010) -	HST SM-01, HST SM-02, HST SM-03, HST SM-04
F010 (MD008) -	ASTRO-1 (FWD), ASTRO-2 (FWD), 7A (FWD)
E001 (E001) -	
E002 (EM-MD001) -	OSTA-1
E003 (E003) -	OSS-1
E004 (E004) -	
E005 (E005) -	TSS-1 (MOCKUP), KSC Robotics Lab

PALLET CONFIGURATIONS

EMP	MDM	IGLOO	IPS
TSS-1 (HDRS)	OSS-1	SL-2	SL-2
TSS-1R (HDRS)	OSTA-1	ASTRO-1	ASTRO-1
LITE-1	OSTA-3	ATLAS-1	ASTRO-2
	SRL-1	ATLAS-2	
	SRL-2	ATLAS-3	
	ASTRO-2		

APPENDIX 3

The Spacelab Missions

Shuttle Mission Orbiter	Launch Date Launch Pad	Landing Date Landing Location	Duration (dd:hh:mm:ss)	Mission	Spacelab Elements (Serial #)*
STS-2 *Columbia*	1981 Nov 12 LC-39A	1981 Nov 14 Edwards Runway 23	02:06:13:13	OSTA-1	Pallet (E002)
STS-3 *Columbia*	1982 Mar 22 LC-39A	1982 Mar 30 White Sands Runway 17	08:00:04:45	OSS-1	Pallet (E003)
STS-7 *Challenger*	1983 Jun 18 LC-39A	1983 Jun 24 Edwards Runway 15	06:02:23:59	OSTA-2	MPESS (F001)
STS-9 *Columbia*	1983 Nov 28 LC-39A	1983 Dec 8 Edwards Runway 17	10:07:47:23	Spacelab-1	Long Module #1 CD (MD001) Pallet (F001)
STS-41D** *Discovery*	1984 Aug 30 LC-39A	1984 Sep 5 Edwards Runway 17	06:00:56:04	OAST-1	MPESS (F004)
STS-41G *Challenger*	1984 Oct 5 LC-39A	1984 Oct 13 Kennedy Runway 33	08:05:23:38	OSTA-3	Pallet (F006)
STS-51A** *Discovery*	1984 Nov 8 LC-39A	1984 Nov 16 Kennedy Runway 15	07:23:44:56	SRM	Pallet (F007) Pallet (F008)
STS-51B *Challenger*	1985 Apr 29 LC-39A	1985 May 6 Edwards Runway 17	07:00:08:46	Spacelab-3	Long Module #1 CD (MD001) MPESS (F003)
STS-51G *Discovery*	1985 Jun 17 LC-39A	1985 Jun 24 Edwards Runway 23	07:01:38:52	SPARTAN	MPESS
STS-51F *Challenger*	1985 July 29 LC-39A	1985 Aug 6 Edwards Runway 23	07:22:45:26	Spacelab-2	Igloo #1 Pallet (F004) #2 Pallet (F005) #3 Pallet (F003) IPS
STS-61A *Challenger*	1985 Oct 30 LC-39A	1985 Nov 6 Edwards Runway 17	07:00:44:53	Spacelab-D1	Long Module #2 FOP (MD002) German Unique Support Structure

© Springer Nature Switzerland AG 2022

M. E. Haddad, D. J. Shayler, *Spacelab Payloads*, Springer Praxis Books,

https://doi.org/10.1007/978-3-030-86775-1

Shuttle Mission Orbiter	Launch Date Launch Pad	Landing Date Landing Location	Duration (dd:hh:mm:ss)	Mission	Spacelab Elements (Serial #)*
STS-61B** *Atlantis*	1985 Nov 26 LC-39A	1985 Dec 3 Edwards Runway 22	06:21:04:49	EASE/ ACCESS	MPESS (F004)
STS-61C** *Columbia*	1986 Jan 12 LC-39A	1986 Jan 18 Edwards Runway 22	06:02:03:51	MSL-2	MPESS (F006)
STS-29** *Discovery*	1989 Mar 13 LC-39B	1989 Mar 18 Edwards Runway 22	04 23:38:50	SHARE	Special Structure
STS-35 *Columbia*	1990 Dec 2 LC-39B	1990 Dec 11 Edwards Runway 22	08 23:05:08	Astro-1	Igloo Fwd Pallet (F010) Aft Pallet (F002) IPS
STS-40 *Columbia*	1991 Jun 5 LC-39B	1991 Jun 14 Edwards Runway 22	09:02:14:20	SLS-1	Long Module #1 CD (MD001)
STS-42 *Discovery*	1992 Jan 22 LC-39A	1992 Jan 30 Edwards Runway 22	08:01:14:44	IML-1	Long Module #2 FOP (MD002)
STS-45 *Atlantis*	1992 Mar 24 LC-39A	1992 Apr 2 Kennedy Runway 33	08:22:09:28	ATLAS-1	Igloo Fwd Pallet (F004) Aft Pallet (F005)
STS-50 *Columbia*	1992 Jun 25 LC-39A	1992 Jul 9 Kennedy Runway 33	13:19:30:04	USML-1	Long Module #1 CD (MD001)
STS-46 *Atlantis*	1992 Jul 31 LC-39B	1992 Aug 8 Kennedy Runway 33	07:23:15:03	TSS-1	Pallet (F003)
STS-47 *Endeavour*	1992 Sep 12 LC-39B	1992 Sep 20 Kennedy Runway 33	07:22:30:23	Spacelab-J	Long Module #2 FOP (MD002)
STS-52 *Columbia*	1992 Oct 22 LC-39B	1992 Nov 1 Kennedy Runway 33	09:20:56:13	USMP-1	Fwd MPESS (F006) Aft MPESS (F002)
STS-56 *Discovery*	1993 Apr 8 LC-39B	1993 Apr 17 Kennedy Runway 33	09:06:08:24	ATLAS-2	Igloo Pallet (F008)
STS-55 *Columbia*	1993 Apr 26 LC-39A	1993 May 6 Edwards Runway 22	09:23:39:59	Spacelab-D2	Long Module #1 CD (MD001) Unique Support Structure (USS)
STS-58 *Columbia*	1993 Oct 18 LC-39B	1993 Nov 1 Edwards Runway 22	14:00:12:32	SLS-2	Module (FOP)
STS-61** *Endeavour*	1993 Dec 2 LC-39B	1993 Dec 13 Kennedy Runway 33	10:19:58:37	HST SM-01	Pallet (F009)
STS-62 *Columbia*	1994 Mar 4 LC-39B	1994 Mar 18 Kennedy Runway 33	13:23:16:41	USMP-2	Fwd MPESS (F006) Aft MPESS (F002)
STS-59 *Endeavour*	1994 Apr 9 LC-38A	1994 Apr 20 Edwards Runway 22	11:05:49:30	SRL-1	Pallet (F006) MPESS (F003)
STS-65 *Columbia*	1994 Jul 8 LC-39A	1994 Jul 23 Kennedy Runway 33	14:17:55:00	IML-2	Long Module #1 CD (MD001)
STS-64 *Discovery*	1994 Sep 9 LC-39B	1994 Sep 20 Edwards Runway 04	10:22:49:57	LITE	Pallet (F007)
STS-68 *Endeavour*	1994 Sep 30 LC-39A	1994 Oct 11 Edwards Runway 22	11:05:46:08	SRL-2	Pallet (F006)

Shuttle Mission Orbiter	Launch Date Launch Pad	Landing Date Landing Location	Duration (dd:hh:mm:ss)	Mission	Spacelab Elements (Serial #)*
STS-66 *Atlantis*	1994 Nov 3 LC-39B	1994 Nov 14 Edwards Runway 22	10:22:34:02	ATLAS-3	Igloo Pallet (F008)
STS-67 *Endeavour*	1995 Mar 2 LC-39A	1995 Mar 18 Edwards Runway 22	16:15:08:48	Astro-2	Igloo Fwd Pallet (F010) Aft Pallet (F002) IPS
STS-71 *Atlantis*	1995 Jun 27 LC-39A	1995 Jul 5 Kennedy Runway 15	09:19:22:17	Spacelab-Mir	Long Module #2 FOP (MD002)
STS-73 *Columbia*	1995 Oct 20 LC-39B	1995 Nov 5 Kennedy Runway 33	15:21:52:28	USML-2	Long Module #1 CD (MD001)
STS-75 *Columbia*	1996 Feb 22 LC-39B	1996 Mar 9 Kennedy Runway 33	15:17:40:21	TSS-1R USMP-3	Pallet (F003) Fwd MPESS (F006) Aft MPESS (F002)
STS-78 *Columbia*	1996 Jun 20 LC-39B	1996 Jul 7 Kennedy Runway 33	16:21:47:45	LMS	Long Module #2 FOP (MD002)
STS-82** *Discovery*	1997 Feb 21 LC-39A	1996 Feb 21 Kennedy Runway 15	09:23:37:09	HST SM-02	Pallet (F009)
STS-83 *Columbia*	1997 Apr 4 LC-39A	1997 Apr 8 Kennedy Runway 33	03:23:12:39	MSL-1	Long Module #1 CD (MD001)
STS-94 *Columbia*	1997 Jul 1 LC-39A	1997 Jul 17 Kennedy Runway 33	15:16:45:29	MSL-1R	Long Module #1 CD (MD001)
STS-85** *Discovery*	1997 Aug 7 LC-39A	1997 Aug 19 Kennedy Runway 33	11:20:26:59	MFD	MPESS (F004)
STS-87 *Columbia*	1997 Nov 19 LC-39B	1997 Dec 5 Kennedy Runway 33	15:16:35:01	USMP-4	Fwd MPESS (F006) Aft MPESS (F002)
STS-90 *Columbia*	1998 Apr 17 LC-39B	1998 May 3 Kennedy Runway 33	15:21:49:59	Neurolab	Long Module #2 FOP (MD002)
STS-103** *Discovery*	1999 Dec 19 LC-39B	1999 Dec 27 Kennedy Runway 33	07:23:11:34	HST SM-03A	Pallet (F009)
STS-99 *Endeavour*	2000 Feb 11 LC-39A	2000 Feb 22 Kennedy Runway 33	11:05:39:41	SRTM	Pallet (F006) [Recommissioned?]
STS-92** *Discovery*	2000 Oct 11 LC-39A	2000 Oct 24 Edwards Runway 22	12:21:43:47	ISS-3A	Pallet (F005)
STS-100** *Endeavour*	2001 Apr 19 LC-39A	2001 May 1 Edwards Runway 22	11:21:31:14	ISS-6A	Pallet (F004)
STS-104 *Atlantis*	2001 Jul 12 LC-39B	2001 Jul 24 Kennedy Runway 15	12:18:36:39	ISS -7A	Fwd Pallet (F010) Aft Pallet (F002)
STS-108** *Endeavour*	2001 Dec 5 LC-39B	2001 Dec 17 Kennedy Runway 15	11:19:36:45	UF-1	MPESS
STS-109** *Columbia*	2002 Mar 2 LC-39A	2002 Mar 12 Kennedy Runway 33	10:22:11:09	HST-03B	Pallet (F009)
STS-123** *Endeavour*	2008 Mar 11 LC-39A	2008 Mar 26 Kennedy Runway 15	15:18:10:54	1 J/A	Deployable 1 (SLP-D1) Pallet (F004)
STS-125** *Atlantis*	2009 May 11 LC-39A	2009 May 24 Edwards Runway 22	12:21:38:09	HSM-04	Pallet (F009)

Shuttle Mission Orbiter	Launch Date Launch Pad	Landing Date Landing Location	Duration (dd:hh:mm:ss)	Mission	Spacelab Elements (Serial #)*
STS-127**	2009 Jul 15	2009 Jul 31	15:16:44:57	2 J/A	Deployable 2
Endeavour	LC-39A	Kennedy Runway 15			(SLP-D2)
					Pallet (unknown #)

KEY

* Spacelab Hardware Assignments, August 5, 1997, Debbie Bitner/NASA-KSC/BD-C
** Non-Spacelab mission but experiments worked by Level-IV/Experiment Integration and/or use of Spacelab hardware

ATLAS:	Atmospheric Laboratory for Applications and Science
Astro:	Not an acronym; abbreviation for "Astronomy"
CHROMEX:	Chromosome and Plant Cell Division in Space
HST SM:	Hubble Space Telescope Servicing Mission
IML:	International Microgravity Laboratory
IPS:	Instrument Pointing System
LC:	Launch Complex
LITE:	Lidar In-space Technology Experiment
LM1:	First Long Module
LM2:	Second Long Module
LMS:	Life and Microgravity Sciences
MFD:	Manipulator Flight Demonstration
MPESS:	Mission Peculiar Equipment Support Structure
MSL:	Materials Science Laboratory
OSS:	Office of Space Science and Applications
OSTA:	Office of Space and Terrestrial Applications
PMA:	Pressurized Mating Adapter
SHARE:	Space Station Heat Pipe Advanced Radiator Element
SLS:	Spacelab Life Sciences
SRL:	Space Radar Laboratory
SRM:	Satellite Retrieval Mission
SRTM:	Shuttle Radar Topography Mission
TSS:	Tethered Satellite System
USML:	US Microgravity Laboratory
USMP:	US Microgravity Payload

APPENDIX 4

Current Locations of Major Spacelab Hardware

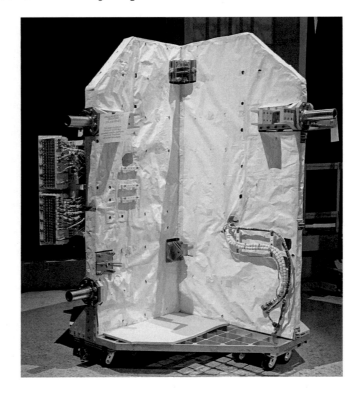

Astro-1 and Astro-2 Cruciform on display in the Atrium at the US Space and Rocket Center, Huntsville, Alabama. [Mike Haddad.]

A total of 32 shuttle missions flew Spacelab components, with the habitable modules flown on 16 missions. But what happened to all that hardware after the close of the program?

© Springer Nature Switzerland AG 2022
M. E. Haddad, D. J. Shayler, *Spacelab Payloads*, Springer Praxis Books,
https://doi.org/10.1007/978-3-030-86775-1

Following the retirement of the majority of the Spacelab system in 1998, some elements of the hardware were put on public display, while other elements were scattered throughout the world. Some items were salvaged but the remainder was mostly destroyed, and it is known that one flight floor was melted down for scrap. The locations of most of the ten space-flown pallets are also vague.

The known displays are [as of 2021]:

- Long Module #1 (CD MD001) is on display behind Space Shuttle OV-103 *Discovery*, at the Udvar-Hazy Center at the Smithsonian Air and Space Museum in Washington, D.C.
- From 2000 through 2010, Long Module #2 (FPO MD002) was displayed in the *Bremenhalle* exhibition at Bremen Airport, Bremen, Germany. From 2010, this unit was relocated to Building 4C at the nearby Airbus Space and Defense plant and can only be viewed during guided tours.
- Tethered Satellite System (TSS) hardware was, at one point, at the Florida Institute of Technology (FIT) in Melbourne, Florida.
- The Astro cruciform (see image) is in Huntsville, Alabama.
- A Spacelab-1, Astro-1 and Astro-2 Instrument Pointing System (IPS) is at the Smithsonian in Washington, D.C.
- The pallet used during the 1992 STS-46 mission (F003) that carried the joint NASA/Italian Space Agency Tethered Satellite System (TSS-1) was transferred to the Swiss Museum of Transport for permanent display on March 5, 2010. One of the crew of STS-46 was Swiss-born ESA astronaut Claude Nicollier. The mission also deployed ESA's European Retrievable Carrier (Eureca).
- The Spacelab pallet (F004) used to transport both the Space Station Remote Manipulator System (SSRMS), known as Canadarm2 (STS-100 in 2001), and the Special Purpose Dexterous Manipulator (SPDM), known affectionately as *Dextre* (STS-123 in 2008), to the ISS was loaned by NASA to the Canadian Space Agency for display at the Canada Aviation and Space Museum.
- Another unidentified pallet is on display at the U.S. National Air and Space Museum in Washington, D.C.
- An unidentified Spacelab Igloo is on display at the James S. McDonnell Space Hanger, Steven F. Udvar-Hazy Center, USA.

The authors would welcome any further information on the whereabouts of former Spacelab hardware components to update their database.

Bibliography

This book is a collective recollection of a particular function at NASA's Kennedy Space Center in Florida. Specifically, the development, structure, operations and personalities of the Level-IV function, which was responsible for the pre-flight preparation and mission processing of science experiments, payloads and hardware under the generic 'Spacelab' label for the Space Shuttle program, between 1981 and 2011.

The Interviews

The majority of the content of this volume has been gained from interviews and personal recollections from many of those who worked in Level-IV during this time frame. The main interviews, conducted by co-author and former Level-IV engineer Michael E. Haddad, included:

ALLCORN. Aaron, NASA Level-IV Mechanical Engineer, [Questionnaire] June 23, 2020.

BARTOE. Donna, NASA Level-IV Software Engineer, May 21, 2020

BARTOE. John-David, NASA Spacelab-2 Payload Specialist, May 21, 2020

BENGOA. Alex, NASA Level-IV Mechanical Engineer, August 14, 2019.

BEYER. Darren, NASA Level-IV Aerospace Engineer, August 29, 2019.

BITNER. Debbie, NASA Level-IV Mechanical Engineer, June 3, 2019.

BITNER. Gary, SEIS Level-IV Technician, June 3, 2019.

BURCH. Bruce, SEIS Level-IV Technician, July 12, 2019.

BURNETT. Josephine "Josie", NASA Level-IV Mechanical Engineer, August 20, 2019.

CALARO. Diana, NASA Level-IV Electrical Engineer, December 15, 2019.

CALARO. Juan, NASA Level-IV Electrical Engineer, December 15, 2019.

CHIN. Glenn, NASA Level-IV Mechanical Engineer, [Questionnaire] April 30, 2020.

CONWAY. John, NASA Payloads Director, February 18, 2019

COREY. Todd, NASA Level-IV Operations, [Questionnaire] December 12, 2019.

DELGADO. Luis. NASA Level-IV Mechanical Engineer, May 29, 2019.

DIAZ. Rey, NASA Level-IV Electrical Engineer, April 23, 2019.

DOLBY. Don, SEIS Level-IV Technician, July 11, 2019.

DUMOULIN. James "Jim", NASA Level-IV Electrical Engineer, December 19, 2019

DUREN. Riley, NASA Level-IV Electrical Engineer, June 20, 2020.

DURRANCE Samuel "Sam", Payload Specialist, May 16, 2019.

FERGUSON, Cynthia "Tia", NASA Level-IV Mechanical Engineer, March 15, 2019.

GILL. Tracy, NASA Level-IV Electrical Engineer, February 20, 2019.

GRUNSFELD. John, NASA Mission Specialist, July 22, 2019.

HILL, Michael, NASA Level-IV Mechanical Engineer, April 23, 2019.

HUET. Sharolee, NASA Level-IV Mechanical Engineer, [Questionnaire] December 9, 1999

JACOBSON. Craig, NASA Level-IV Electrical Engineer, November 4, 2019.

JASNOCHA. Richard "Rich", NASA Level-IV Electrical Engineer, July 18, 2019.

JEWEL. William "Bill", NASA, "Father of Level-IV", December 14, 2017.

KIENLEN. E. Michael, NASA Level-IV Operations, June 14, 2019.

KOVALCHIK. Dan, McDonnell Douglas Computer Design, [Questionnaire] April 21, 2018

KURSCHEID. Hermann, Spacelab-1 ESA-FSLP Head of Ground Operations, October 14, 2019

KRUG. Eugene "Gene", NASA Level-IV Mechanical Engineering Support, August 21, 2019.

LAVOIE. Maurice "Mo", NASA Level-IV Electrical Engineer, July 16, 2019.

LICHTENBERG. Byron, Payload Specialist, July 30, 2019.

MARTIN-BRENNAN. Cynthia, NASA Level-IV Computer Science, August 16, 2019.

MATHIS. Johnny, NASA Level-IV Mechanical Engineer, May 3, 2019.

McPHILLIPS. Cheryl, NASA Level-IV Electrical Engineer, February 23, 2019.

MOCTEZUMA. Luis, NASA Level-IV Civil Engineer, [Questionnaire] July 19, 2019.

MORRIS. Kenneth "Bruce", NASA Level-IV Mechanical Engineer, June 4, 2019.

NAFFAH. Eli, NASA Level-IV Mechanical Engineer, February 19, 2019.

NELSON. Damon, NASA Level-IV Mechanical Engineer, May 13, 2019.

ORNELAS. Antonio, "Tony", NASA Level-IV Mechanical Engineer, March 5, 2019.

OTERO. Angel, NASA Level-IV Mechanical Engineer, March 6, 2019.

POWERS. Gary, NASA Level-IV Test Director, May 28, 2019.

RAYMOND. Robert, "Bob" NASA Level-IV Quality, April 16, 2019.
RICE. Herb, NASA Level-IV Lead Electrical Engineering, July 29, 2019.
RICHARDS. Joni, NASA Level-IV Electrical Engineer, August 28, 2019.
RIVERA. Geraldo "Gerry" NASA Level-IV Electrical Engineer, May 22, 2019.
RIVERA. Juan, NASA Level-IV Software, May 10, 2019.
RUETHER. Mark, NASA Level-IV Mechanical Engineer, March 1, 2019.
RUIZ. Jeannie, NASA Level-IV Mechanical Engineer, August 12, 2019
RUIZ. Robert, "Bob", NASA Level-IV Mechanical Engineer, February 21, 2019
SCHLIERF. Roland, NASA Level-IV Electrical Engineer, August 15, 2019
SMITH. Jay, SEIS Level-IV Technician, April 21, 2019.
SMITH. Maynette, NASA Level-IV Electrical Engineer, April 11, 2019.
STELZER. Michael, NASA Level-IV Mechanical Engineer, [Questionnaire] December 14, 2019
SUDERMANN. James "Jim", NASA Level-IV Electrical Engineer, July 23, 2019.
VALDEZ. Frank, NASA Level-IV Mechanical Engineer, March 8, 2019
VANGEN. Scott, NASA Level-IV Electrical Engineer, March 5, 2019.
VEAUDRY. George, NASA Level-IV Mechanical Engineer, June 11, 2020
VICTOR. Elias, NASA Level-IV Electrical Engineer, August 5, 2019.
WILSON. Lori, NASA Level-IV Operations, [Questionnaire] December 11, 2019.
WRIGHT. Michael, NASA Level-IV Electrical Engineer, [Questionnaire] July 26, 2019
ZARI. Kevin, NASA Electrical Engineer, May 31, 2019.

Published Works
For further information on the Space Shuttle, Spacelab and Space Station facilities at the Kennedy Space Center, the reader is directed to the following publications.

NASA Publications
1984   *STS Facilities and Operations,* Kennedy Space Center, Florida, NASA KSC, K-STSM-01 Appendix A, April 1984, Revision A.
1987   *Spacelab: An International Success Story*, Douglas R. Lord, NASA SP-487.
1988   *Science in Orbit: The Shuttle & Spacelab Experience: 1981–1986*, NASA NP-119 *National Space Transportation System Overview*, September 1988.
1993   *Orbiter Processing Facility Payload Processing and Support Capabilities*, NASA KSC, K-STS-M-14.1.13REVD-OPF, October 1993.
1999   *NASA Historical Data Book* Volume V, NASA Launch Systems, Space Transportation, Human Spaceflight and Space Science 1979–1988, Judy A. Rumerman, NASA SP-4012.

ESA Publications

1979 *Spacelab User's Manual*, DP/ST(79)3.

Reports

1984 *Space Transportation System*, Press Information, January 1984, Rockwell International.

2007 *John F. Kennedy Space Center, Brevard County, Florida. Survey and Evaluation of NASA-owned Historical Facilities and Properties in the Context of the U.S. Space Shuttle Program.* Prepared by Archaeological Consultants Inc., Sarasota, Florida, October 2007.

2010 *Historical Survey and Evaluation of the Space Station Processing Facility, John F. Kennedy Space Center, Brevard County, Florida*, Prepared by Archaeological Consultants Inc., Sarasota, Florida, September 2010.

Conference Papers

1983 *Spacelab Program Preparations for the First Flight and Projected Utilization.* James C. Harrington, Space Congress Proceedings.

Other Books

1984 *Spacelab Research in Orbit*, David Shapland and Michael Rycroft, Cambridge University Press.

2007 *Praxis Manned Spaceflight Log 1961–2006,* Tim Furniss and David J. Shayler, with Michael D. Shayler, Springer/Praxis.

About the Authors

Michael E. Haddad

Mikey Haddad, standing in front of the Space Shuttle *Atlantis* at the Kennedy Space Center Visitors Complex.

Mike (Mikey) Haddad was born in Garfield Heights, Ohio in 1958, and from an early age was very interested in the Space Program. Watching the launch of humans into space on TV during the Mercury and Gemini programs made him more determined that somehow he had to be part of the Space Program. In May

© Springer Nature Switzerland AG 2022

M. E. Haddad, D. J. Shayler, *Spacelab Payloads*, Springer Praxis Books,
https://doi.org/10.1007/978-3-030-86775-1

1968, his family relocated to the small town of Titusville in Florida, located directly across the Indian River from Kennedy Space Center (KSC). His dream had come true and basically set the course for his future. Now he could see the large Vertical Assembly Building (later changed to Vehicle Assembly Building) and launch towers in person, and watch those Apollo and unmanned launches live from just a few miles away.

Starting almost from day one in Florida, Mike would design and build his own model rockets made from cardboard tubes, plastic cups and balsa wood. Designing and building his own launch pad, he would launch his rockets on the large field at the Junior High School, just down the road from their home.

Mikey, with his brother Fred on the left, connecting the launch cable to the engine igniter. This rocket was built to look like a Saturn V and was made from gift wrap tube, plastic cups and paper towel roll. c. 1970.

In grade school he took math and science courses in an effort to get himself ready for college. Graduating from Titusville High School in June 1976, he began college at Florida Technological University (later changed to the University of Central Florida) in Orlando, Florida, in the fall of 1976.

The university had a Co-operative Education (Co-op) Program, where students would get hired by a company while still attending school. Alternating semesters between working and attending school, this would give college students some real-life experience before graduating, as well as earning money. This meant it took longer to graduate, but was well worth the extra time. For Mike, Co-op at KSC and nowhere else was the goal, and that was achieved in the fall of 1979, landing a job in the Launch Control Center (LCC). He would Co-op there through till he graduated in April 1982 with a degree in Mechanical and Aerospace Engineering. His Co-op role provided him the opportunity to be in the LCC for STS-1, the first launch of the Space Shuttle, on April 12, 1981. Upon graduation, he would work in the LCC or somewhere else at KSC. His cousin, Eli Naffah, was

Co-op at a different location at KSC where they prepared the payloads for the Space Shuttle, and he told Mike they urgently needed new engineers. That location was called Level-IV, and sounded interesting because NASA would do the work there with little contractor involvement.

Mike began working with NASA at KSC in May 1982 as a mechanical and fluids engineer, assembling, modifying, servicing and testing flight hardware for domestic and international payloads, as part of the Spacelab program.

From 1988 and through the early 1990s, he specialized in hypergolic propellants, and was responsible for the ground processing flow for Shuttle planetary, observatory and satellite class payloads such as Galileo, Ulysses, Compton Gamma Ray Observatory, TDRS communication satellites and the Hubble Space Telescope.

Mike obtained a Master of Science Degree in Industrial Engineering and Management Systems from the University of Central Florida in August 1994.

In 1995, he performed a six-month detail assignment to Johnson Space Center (JSC), with responsibilities that included flight console training and operations for STS-70 and -71 from the Mission Control Center.

Beginning in 1996, he served as a senior systems engineer working International Space Station (ISS), with primary responsibilities that included Extra-Vehicular Activities (EVA). As KSC Point-of-Contact for EVA and Crew Systems, he was responsible for test, verification and on-orbit support of Space Station EVA systems. His EVA worked earned him a Silver Snoopy Award in 2001. Mike was also an integration engineer for the ISS Multi-Element Integrated Testing (MEIT), performing requirements generation, test and verification activities.

Mike was a finalist for the Astronaut Class of 1998. He obtained an interview at JSC in January 1998 for the week of mental and physical tests required for astronaut selection. He was a member of the final group of 121 candidates, from which 25 astronauts were selected.

Early in 2002 Mike worked as an Independent Assessment Senior Systems Engineer in the Safety, Health and Independent Assessment (SHIA) Directorate, performing technical engineering (including S&MA) assessments of KSC and NASA-wide processes for Flight Hardware and Ground Support Equipment (GSE).

Moving to the KSC Constellation Project Office/Systems Engineering and Integration Office in 2007, he continued his previous EVA and Crew Systems responsibilities with additional work that included participation in new EVA suit design, EVA Lunar Operations, KSC Emergency Egress studies and Lunar Surface Systems design, build and operations. Mike had responsibilities that included generating Constellation requirements, participation in technical trade studies (Orion and SLS), determining future ground operations for Orion and SLS, and continuing to serve as KSC Point-of-Contact for EVA Systems. He was a member of the KSC team working Lunar Surface Systems to include Lunar Base design, Assembly, Integration, Testing and Operations.

Mike was also a member of NASA's Desert Research and Technology Studies (D-RaTS) 2011 Team and co-writer for the Habitat Demonstration Unit (HDU)/ Deep Space Habitat (DSH) Operations Manual to be used for HDU/DSH operations in Building 220 at Johnson Space Center and at Black Point Lava Flow, Arizona. He also supported development of the HDU/DSH Integrations Systems Test Plan and participated in the D-RaTS 2011 Integrated Systems Test at Johnson Space Center and field operations at Black Point Lava Flow, Arizona. Tasks performed during field operations included final integration of D-RaTS elements following post shipment, checkout testing, practice testing, and full test days with flight and ground crews, including two overnight stays by flight crews in the DSH complex.

Mikey Haddad retired from NASA KSC in December 2011. He is currently a Space Industry Subject Matter Expert and Vice President at Haddad Management Advisors LLC, specializing in: hands-on spaceflight hardware assembly, integration, servicing and testing; input to spacecraft/payload design and associated Ground Support Equipment (GSE); developing and performing operational scenarios for ground and flight activities; providing systems engineering and integration to flight and ground systems; working directly with astronauts for ground processing tasks and support during flight missions; and providing real-time answers to operational problems, flight and ground systems.

He lives in Cocoa Beach Florida with his wife Kathy and enjoys spending time visiting family, traveling and volunteering as a Docent at the *Atlantis* Exhibit in the KSC Visitors Center.

David J. Shayler

Dave Shayler

Space historian David J. Shayler, FBIS (Fellow of the British Interplanetary Society or – as Dave likes to call it – Future Briton In Space!) was born in England in 1955. His lifelong interest in space exploration began at the age of five by drawing rockets, but it was not until the launch of Apollo 8 to the Moon in December 1968 that this interest in human space exploration became a passion. He recalls staying up all night with his grandfather to watch the Apollo 11 moonwalk. Dave joined the British Interplanetary Society as a Member in January 1976, became an Associate Fellow in 1983, and a Fellow in 1984. He served on the Council of the BIS between 2013 and 2019. From 2020, Dave became the third Editor of *Space Chronicle*, the BIS history magazine. His first articles were published by the Society in the late 1970s, and then in 1982 he set up Astro Info Service in order to focus his research efforts (www.astroinfoservice.co.uk).

Dave's first book was published in 1987, and has been followed by over 30 other titles on the American and Russian space programs, spacewalking, women in space, and the human exploration of Mars. His authorized biography of Skylab 4 astronaut Jerry Carr was published in 2008.

In 1989, Dave applied as a cosmonaut candidate for the UK's Project Juno program in cooperation with the Soviet Union (now Russia). The mission was to spend seven days in space aboard the Mir space station. Dave did not reach the final selection, but progressed further than he expected. The mission was flown in May 1991 by Helen Sharman.

Dave, with KSC PAO guide Manny Virata, near the parked crawler transporter at LC-39, November 1990.

In support of his research, Dave has visited NASA field centers in Houston and Florida in the USA, and the Yuri Gagarin Cosmonaut Training Center in Russia. It was during these trips that he was able to conduct in-depth research, interview many space explorers and workers, tour training facilities, and handle real space hardware. He also gained a valuable insight into the activities of a space explorer, as well as the realities of not only flying and living in space but also what goes into preparing for a mission and planning future programs.

Dave is a friend of many former astronauts and cosmonauts, some of whom have accompanied him on visits to schools all across the UK. For over 30 years, he has delivered space-themed presentations and workshops to members of the public in an effort to increase popular awareness of the history and development of human space exploration. Dave has a particular desire to help the younger generation to develop an interest in science and technology and the world around them.

Dave lives in the West Midlands region of the UK and enjoys spending time with his wife Bel, a youthful and enormous white German Shepherd that answers to the name of Shado, and indulging in his loves of cooking, fine wines, and classical music. His other interests are in reading, especially about military history and in particular the Napoleonic Wars, visiting historical sites and landmarks, and following Formula 1 motor racing.

Index

© Springer Nature Switzerland AG 2022
M. E. Haddad, D. J. Shayler, *Spacelab Payloads*, Springer Praxis Books,
https://doi.org/10.1007/978-3-030-86775-1

Printed in the United States
by Baker & Taylor Publisher Services